D0919273

The Electrical Engineering Handbook
Third Edition

Computers,
Software Engineering,
and
Digital Devices

The Electrical Engineering Handbook Series

Series Editor
Richard C. Dorf
University of California, Davis

Titles Included in the Series

The Electrical Engineering Handbook
Third Edition

Edited by
Richard C. Dorf

Circuits, Signals, and Speech and Image Processing

*Electronics, Power Electronics, Optoelectronics,
Microwaves, Electromagnetics, and Radar*

*Sensors, Nanoscience, Biomedical Engineering,
and Instruments*

Broadcasting and Optical Communication Technology

Computers, Software Engineering, and Digital Devices

*Systems, Controls, Embedded Systems, Energy,
and Machines*

The Electrical Engineering Handbook
Third Edition

Computers, Software Engineering, and Digital Devices

Edited by

Richard C. Dorf

University of California
Davis, California, U.S.A.

Taylor & Francis
Taylor & Francis Group
Boca Raton   London   New York

A CRC title, part of the Taylor & Francis imprint, a member of the
Taylor & Francis Group, the academic division of T&F Informa plc.

Published in 2006 by
CRC Press
Taylor & Francis Group
6000 Broken Sound Parkway NW, Suite 300
Boca Raton, FL 33487-2742

International Standard Book Number-10: 0-8493-7340-9 (Hardcover)
International Standard Book Number-13: 978-0-8493-7340-4 (Hardcover)
Library of Congress Card Number 2005054349

Library of Congress Cataloging-in-Publication Data

Computers, software engineering, and digital devices / edited by Richard C. Dorf.
    p. cm.
  Includes bibliographical references and index.
  ISBN 0-8493-7340-9 (alk. paper)
   1. Computer engineering. 2. Digital electronics. 3. Software engineering. I. Dorf, Richard C. II. Title.

TK7885.C657 2005
621.39--dc22   2005054349

Taylor & Francis Group
is the Academic Division of Informa plc.

Visit the Taylor & Francis Web site at
http://www.taylorandfrancis.com

and the CRC Press Web site at
http://www.crcpress.com

Preface

Purpose

The purpose of *The Electrical Engineering Handbook, 3rd Edition* is to provide a ready reference for the practicing engineer in industry, government, and academia, as well as aid students of engineering. The third edition has a new look and comprises six volumes including:

Circuits, Signals, and Speech and Image Processing
Electronics, Power Electronics, Optoelectronics, Microwaves, Electromagnetics, and Radar
Sensors, Nanoscience, Biomedical Engineering, and Instruments
Broadcasting and Optical Communication Technology
Computers, Software Engineering, and Digital Devices
Systems, Controls, Embedded Systems, Energy, and Machines

Each volume is edited by Richard C. Dorf, and is a comprehensive format that encompasses the many aspects of electrical engineering with articles from internationally recognized contributors. The goal is to provide the most up-to-date information in the classical fields of circuits, signal processing, electronics, electromagnetic fields, energy devices, systems, and electrical effects and devices, while covering the emerging fields of communications, nanotechnology, biometrics, digital devices, computer engineering, systems, and biomedical engineering. In addition, a complete compendium of information regarding physical, chemical, and materials data, as well as widely inclusive information on mathematics is included in each volume. Many articles from this volume and the other five volumes have been completely revised or updated to fit the needs of today and many new chapters have been added.

The purpose of this volume, *Computers, Software Engineering, and Digital Devices*, is to provide a ready reference to subjects in the fields of digital and logical devices, displays, testing, software, and computers. Here we provide the basic information for understanding these fields. We also provide information about the emerging fields of programmable logic, hardware description languages, and parallel computing.

Organization

The information is organized into three sections. The first two sections encompass 20 chapters and the last section summarizes the applicable mathematics, symbols, and physical constants.

Most articles include three important and useful categories: defining terms, references, and further information. *Defining terms* are key definitions and the first occurrence of each term defined is indicated in boldface in the text. The definitions of these terms are summarized as a list at the end of each chapter or article. The *references* provide a list of useful books and articles for follow-up reading. Finally, *further information* provides some general and useful sources of additional information on the topic.

Locating Your Topic

Numerous avenues of access to information are provided. A complete table of contents is presented at the front of the book. In addition, an individual table of contents precedes each section. Finally, each chapter begins with its own table of contents. The reader should look over these tables of contents to become familiar with the structure, organization, and content of the book. For example, see Section II: Computer Engineering, then Chapter 17: Parallel Processors, and then Chapter 17.2: Parallel Computing. This tree-and-branch table of contents enables the reader to move up the tree to locate information on the topic of interest.

Two indexes have been compiled to provide multiple means of accessing information: subject index and index of contributing authors. The subject index can also be used to locate key definitions. The page on which the definition appears for each key (defining) term is clearly identified in the subject index.

The Electrical Engineering Handbook, 3rd Edition is designed to provide answers to most inquiries and direct the inquirer to further sources and references. We hope that this handbook will be referred to often and that informational requirements will be satisfied effectively.

Acknowledgments

This handbook is testimony to the dedication of the Board of Advisors, the publishers, and my editorial associates. I particularly wish to acknowledge at Taylor & Francis Nora Konopka, Publisher; Helena Redshaw, Editorial Project Development Manager; and Richard Tressider, Project Editor. Finally, I am indebted to the support of Elizabeth Spangenberger, Editorial Assistant.

Richard C. Dorf
Editor-in-Chief

Editor-in-Chief

Richard C. Dorf, Professor of Electrical and Computer Engineering at the University of California, Davis, teaches graduate and undergraduate courses in electrical engineering in the fields of circuits and control systems. He earned a Ph.D. in electrical engineering from the U.S. Naval Postgraduate School, an M.S. from the University of Colorado, and a B.S. from Clarkson University. Highly concerned with the discipline of electrical engineering and its wide value to social and economic needs, he has written and lectured internationally on the contributions and advances in electrical engineering.

Professor Dorf has extensive experience with education and industry and is professionally active in the fields of robotics, automation, electric circuits, and communications. He has served as a visiting professor at the University of Edinburgh, Scotland; the Massachusetts Institute of Technology; Stanford University; and the University of California, Berkeley.

Professor Dorf is Fellow of The Institute of Electrical and Electronics Engineers and a Fellow of the American Society for Engineering Education. Dr. Dorf is widely known to the profession for his *Modern Control Systems, 10th Edition* (Addison-Wesley, 2004) and *The International Encyclopedia of Robotics* (Wiley, 1988). Dr. Dorf is also the co-author of *Circuits, Devices and Systems* (with Ralph Smith), *5th Edition* (Wiley, 1992), and *Electric Circuits, 7th Edition* (Wiley, 2006). He also is author of *Technology Ventures* (McGraw-Hill, 2005) and *The Engineering Handbook, 2nd Edition* (CRC Press, 2005).

Advisory Board

Contributors

M. Abdelguerfi
University of New Orleans
New Orleans, Louisiana

Cajetan M. Akujuobi
Prairie View A&M University
Prairie View, Texas

Carl A. Argila
Software Engineering Consultant
Pico Rivera, California

B.R. Bannister
University of Hull (retired)
Hull, U.K.

Bill D. Carroll
University of Texas
Arlington, Texas

Michael D. Ciletti
University of Colorado
Colorado Springs, Colorado

George A. Constantinides
Imperial College of Science
London, U.K.

J. Arlin Cooper
Sandia National Laboratories
Albuquerque, New Mexico

Edward W. Czeck
Chrysatis Symbolic Design
North Billerica, Massachusetts

Bulent I. Dervisoglu
Silicon Graphics, Inc.
Mountain View, California

R. Eskicioglu
University of Alberta
Edmonton, Canada

James M. Feldman
Northeastern University
Boston, Massachusetts

Tse-yun Feng
Pennsylvania State University
University Park, Pennsylvania

James F. Frenzel
University of Idaho
Moscow, Idaho

Raphael Finkel
University of Kentucky
Lexington, Kentucky

James M. Gilbert
University of Hull
Hull, U.K.

Peter Graham
University of Minnesota
Saint Paul, Minnesota

Chris G. Guy
University of Reading
Reading, U.K.

Carl Hamacher
Queen's University
Kingston, Canada

H.S. Hinton
Utah State University
Logan, Utah

Barry W. Johnson
University of Virginia
Charlottesville, Virginia

Anna M. Johnston
Sandia National Laboratories
Albuquerque, New Mexico

Paul C. Jorgensen
Grand Valley State University
Rockford, Michigan

Miro Kraetzl
Defence Science and Technology
    Organisation
Salisbury, Australia

**Dhammika
Kurumbalapitiya**
Harvey Mudd College
Claremont, California

Peter A. Lee
East of England Development Agency
Cambridge, U.K.

Young Choon Lee
University of Sydney
Sydney, Australia

Ted G. Lewis
Naval Postgraduate School
Monterey, California

Albert A. Liddicoat
California Polytechnic State University
San Luis Obispo, California

Jay Liebowitz
Johns Hopkins University
Rockville, Maryland

M. Mansuripur
University of Arizona
Tucson, Arizona

Johannes J. Martin
University of New Orleans
New Orleans, Louisiana

James E. Morris
Portland State University
Lake Oswego, Oregon

Gregory L. Moss
Purdue University
West Lafayette, Indiana

Franco P. Preparata
Brown University
Providence, Rhode Island

W. David Pricer
Pricer Business Services
Charlotte, Vermont

Jacques Raymond
University of Ottawa
Ottawa, Canada

Evelyn P. Rozanski
Rochester Institute of Technology
Rochester, New York

Matthew N.O. Sadiku
Prairie View A&M University
Prairie View, Texas

Richard S. Sandige
California Polytechnic State University
San Luis Obispo, California

Nan C. Schaller
Rochester Institute of Technology
Rochester Center, New York

Michaela Serra
University of Victoria
Victoria, Canada

Mostafa Hashem Sherif
AT&T
Tinton Falls, New Jersey

Solomon Sherr
Westland Electronics
Old Chatham, New York

Lynne A. Slivovsky
California Polytechnic State University
San Luis Obispo, California

John Staudhammer
University of Florida
Gainesville, Florida

Ronald J. Tallarida
Temple University
Philadelphia, Pennsylvania

Charles W. Therrien
Naval Postgraduate School
Monterey, California

Richard F. Tinder
Washington State University
Pullman, Washington

Zvonko Vranesic
University of Toronto
Toronto, Canada

Larry F. Weber
The Society for Information
Highland, New York

D.G. Whitehead
University of Hull
Hull, U.K.

Phillip J. Windley
Brigham Young University
Provo, Utah

S.N. Yanushkevich
University of Calgary
Calgary, Canada

Safwat Zaky
University of Toronto
Toronto, Canada

Albert Y. Zomaya
University of Sydney
Sydney, Australia

Contents

SECTION II   Computer Engineering

SECTION III   Mathematics, Symbols, and Physical Constants

Indexes

I

Digital Devices

Logic Elements

Gregory L. Moss
Purdue University

Peter Graham
University of Minnesota

Richard S. Sandige
California Polytechnic State University

Lynne A. Slivovsky
California Polytechnic State University

H.S. Hinton
Utah State University

1.1 IC Logic Family Operation and Characteristics

Gregory L. Moss

Digital logic circuits can be classified as belonging to one of two categories, either combinational (also called combinatorial) or sequential logic circuits. The output logic level of a combinational circuit depends only on the current logic levels at the circuit's inputs. Conversely, sequential logic circuits have a memory characteristic, making the sequential circuit's output dependent not only on current input conditions but also on the current output state of the circuit. The primary building block of combinational circuits is the logic gate. The three simplest logic gate functions are the inverter (or NOT), AND and OR. Other basic logic functions are derived from these three. See Table 1.1 for truth table definitions of the various types of logic gates. The memory elements used to construct sequential logic circuits are called latches and flip-flops.

The integrated circuit switching logic used in modern digital systems generally comes from one of three families: transistor-transistor logic (TTL), complementary metal oxide semiconductor logic (CMOS) or emitter coupled logic (ECL). Each logic family has its advantages and disadvantages. The three major families are divided into various subfamilies derived from performance improvements in IC design technology. Bipolar transistors provide switching action in both the TTL and ECL families, while enhancement-mode MOS transistors form the basis for the CMOS family. Recent improvements in switching-circuit performance are also attained using BiCMOS technology, the merging of bipolar and CMOS technologies on a single chip. A particular logic family is usually selected by digital designers based on criteria such as:

1. Switching speed
2. Power dissipation
3. PC board-area requirements (level of integration)
4. Output drive capability (fan-out)
5. Noise immunity characteristics
6. Product breadth
7. Sourcing of components

TABLE 1.1   Defining Truth Tables for Logic Gates

1-Input Function		2-Input Functions							
Input	Output	Inputs		Output Functions					
A	NOT	A	B	AND	OR	NAND	NOR	XOR	XNOR
0	1	0	0	0	0	1	1	0	1
1	0	0	1	0	1	1	0	1	0
		1	0	0	1	1	0	1	0
		1	1	1	1	0	0	0	1

IC Logic Families and Subfamilies

Integrated circuit logic families actually consist of several subfamilies of ICs that differ in performance characteristics. The TTL logic family has been the most widely used family type for applications employing small scale integration (SSI) or medium scale integration (MSI) integrated circuits. Lower power consumption and higher levels of integration are the principal advantages of the CMOS family. The ECL family is generally used in applications requiring high-speed switching logic. Today, the most common device-numbering system used in the TTL and CMOS families has a prefix of 54 (generally used in military applications and having an operating temperature range from −55 to 125°C) and 74 (generally used in industrial/commercial applications and having an operating temperature range from 0 to 70°C). Table 1.2 identifies various logic families and subfamilies.

TTL Logic Family

The TTL family has been the most widely used logic family for many years in applications employing SSI and MSI. It is moderately fast and offers a great variety of standard chips, but it is a mature technology that is generally no longer used in new circuit designs.

TABLE 1.2   Logic Families and Subfamilies

Family (Subfamily)	Description
TTL	Transistor-Transistor Logic
74xx	Standard TTL
74Lxx	Low power TTL
74Hxx	High speed TTL
74Sxx	Schottky TTL
74LSxx	Low power Schottky TTL
74Asxx	Advanced Schottky TTL
74ALSxx	Advanced low power Schottky TTL
74Fxx	Fast TTL
CMOS	Complementary Metal Oxide Semiconductor
4xxx	Standard CMOS
74Cxx	Standard CMOS using TTL numbering system
74HCxx	High speed CMOS
74HCTxx	High speed CMOS – TTL compatible
74FCTxx	Fast CMOS – TTL compatible
74Acxx	Advanced CMOS
74ACTxx	Advanced CMOS – TTL compatible
74AHCxx	Advanced high speed CMOS
74AHCTxx	Advanced high speed CMOS – TTL compatible
ECL (or CML)	Emitter Coupled (Current Mode) Logic
10xxx	Standard ECL
10Hxxx	High speed ECL

The active switching element used in all TTL family circuits is the NPN (not pointing in) bipolar junction transistor (BJT). The transistor is turned on when the base is approximately 0.7 volts more positive than the emitter and there is a sufficient flow of base current. The turned-on transistor (in non-Schottky subfamilies) is said to be in saturation and, ideally, acts like a closed switch between collector and emitter terminals. The transistor is turned off when the base is not biased with a high enough voltage with respect to the emitter. In this condition, the transistor acts like an open switch between the collector and emitter terminals.

Figure 1.1 illustrates the transistor circuit blocks used in a standard TTL inverter. Four transistors are used to achieve the inverter function. The gate input connects to the emitter of transistor Q1, the input-coupling transistor. A clamping diode on the input prevents negative input-voltage spikes from damaging Q1. The collector voltage (and current) of Q1 controls Q2, the phase-splitter transistor. Q2 in turn controls the Q3 and Q4 transistors, forming the output circuit called a totem-pole arrangement. Q4 serves as a pull-up transistor, pulling the output high when it is turned on. Q3 does the opposite to the output, acting as a pull-down transistor. Q3 pulls the output low when it is turned on. Only one of the two transistors in the totem pole can be turned on at a time. This is the function of the phase-splitter transistor.

When a high-logic level is applied to the inverter's input, Q1's base-emitter junction will be reverse-biased and the base-collector junction will be forward-biased. This circuit condition will allow Q1 collector current to flow into the base of Q2, saturating Q2 and providing base current into Q3, and turning on Q3 as well.

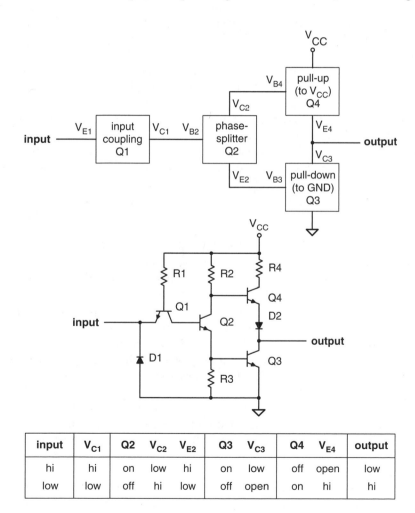

input	V_{C1}	Q2	V_{C2}	V_{E2}	Q3	V_{C3}	Q4	V_{E4}	output
hi	hi	on	low	hi	on	low	off	open	low
low	low	off	hi	low	off	open	on	hi	hi

FIGURE 1.1   TTL inverter circuit and operation.

The collector voltage of Q2 is too low to turn on Q4, so that it appears as an opening in the top part of the totem pole. A diode between the two totem-pole transistors provides an extra voltage drop, in series with the base-emitter junction of Q4, to ensure that Q4 will be turned off when Q2 is turned on. The saturated Q3 transistor brings the output near ground potential, producing a low-output result for a high input into the inverter.

When a low logic level is applied to the inverter's input, Q1's base-emitter junction will be forward-biased and the base-collector junction will be reverse-biased. This circuit condition will turn on Q1, shorting the collector terminal to the emitter and, therefore, to ground (low-level). This low voltage also acts on the base of Q2 and turns Q2 off. With Q2 off, insufficient base current flows into Q3, turning it off also. The Q2 leakage current is shunted to ground with a resistor to prevent the partial turning on of Q3. The collector voltage of Q2 is pulled to a high potential with another resistor and then turns on Q4, making it appear as a short in the top of the totem pole. The saturated Q4 transistor provides a low-resistance path from VCC to the output, producing high output for a low input into the inverter.

A TTL NAND gate is similar to the inverter circuit. The only exception is that the input-coupling transistor Q1 is constructed with multiple emitter-base junctions and each input to the NAND is connected to a separate emitter terminal. Any of the transistor's multiple emitters can turn on Q1. The TTL NAND gate thus functions in the same manner as the inverter, in that if any of the NAND gate inputs are low, the same circuit action will take place as with a low input to the inverter. Therefore, any time a low input is applied to the NAND gate, it will produce high output. Only if all the NAND gate inputs are simultaneously high, will it produce the same circuit action as the inverter, with its single input high and the resultant output low. Input coupling transistors with up to eight emitter-base junctions and, therefore, eight-input NAND gates are constructed.

Storage time (the time it takes for the transistor to come out of saturation) is a major factor of propagation delay for saturated BJT transistors. A long storage time limits switching speed of a standard TTL circuit. Propagation delay can be decreased and the switching speed increased by placing a Schottky diode between the base and collector of each transistor that might saturate. The resulting Schottky-clamped transistors then will not go into saturation, effectively eliminating storage time, since the diode shunts current from the base into the collector before the transistor can achieve saturation. Digital circuit designs implemented with TTL logic almost exclusively use one of the Schottky subfamilies to take advantage of a significant improvement in switching speed.

CMOS Logic Family

The vast majority of new circuit designs today utilize CMOS family devices. The active switching element in all CMOS family circuits is the metal-oxide semiconductor field-effect transistor (MOSFET). CMOS stands for complementary MOS transistors and refers to both types of MOSFET transistors, n-channel and p-channel, used to design this type of switching circuit. While the physical construction and internal physics of a MOSFET differ from the BJT, the circuit switching action of the two transistor types is quite similar. The MOSFET switch is essentially turned off and has a very high channel resistance by applying the same potential to the gate terminal as to the source. An n-channel MOSFET is turned on and has a very low channel resistance when a high voltage with respect to the source is applied to the gate. A p-channel MOSFET operates in the same fashion but with opposite polarities; the gate must be more negative than the source to turn on the transistor.

A block diagram and schematic for a CMOS inverter circuit is shown in Figure 1.2. Note that the circuit has a simpler and more compact design than that for the TTL inverter. That is a major reason why MOSFET integrated circuits have a higher circuit density than BJT integrated circuits and is one advantage that MOSFET ICs have over BJT ICs. As a result, CMOS is used in all levels of integration, from SSI through Very Large Scale Integration (VLSI).

When a high logic level is applied to the inverter's input, the p-channel MOSFET Q1 will be turned off and the n-channel MOSFET Q2 will be turned on. This causes the output to be shorted to ground through the low-resistance path of Q2's channel. The turned-off Q1 has a very high channel resistance and acts almost like an open channel.

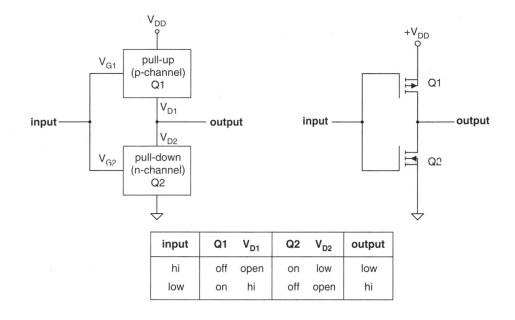

input	Q1	V_{D1}	Q2	V_{D2}	output
hi	off	open	on	low	low
low	on	hi	off	open	hi

FIGURE 1.2 CMOS inverter circuit and operation.

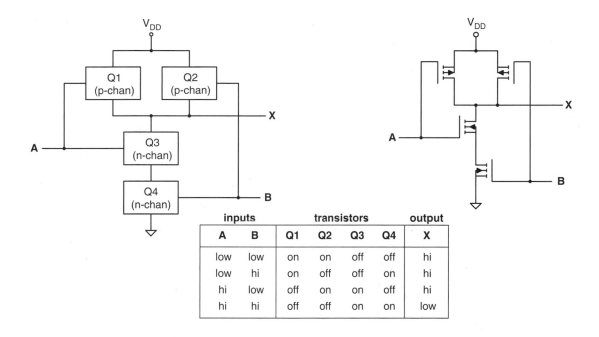

inputs		transistors				output
A	B	Q1	Q2	Q3	Q4	X
low	low	on	on	off	off	hi
low	hi	on	off	off	on	hi
hi	low	off	on	on	off	hi
hi	hi	off	off	on	on	low

FIGURE 1.3 CMOS two-input NAND circuit and operation.

When a low logic level is applied to the inverter's input, the p-channel MOSFET Q1 will be turned on and the n-channel MOSFET Q2 will be turned off. This causes the output to be shorted to VDD through the low-resistance path of Q1's channel. The turned-off Q2 has a very high channel resistance and acts almost like an open channel.

CMOS NAND gates are constructed by paralleling p-channel MOSFETs, one for each input, and putting in series an n-channel MOSFET for each input, as shown in the block diagram and schematic of Figure 1.3.

The NAND gate will produce a low output only when both Q3 and Q4 are turned on, creating a low-resistance path from the output to ground through the two series channels. This can be achieved by having a high input on both A and B. This input condition will also turn off Q1 and Q2. If either input A or input B or both are low, the respective parallel MOSFET will be turned on, providing a low resistance path for the output to VDD. This will also turn off at least one of the series MOSFETs, resulting in a high resistance path for the output to ground.

ECL Logic Family

ECL is the highest-speed logic family available. While it does not offer as large a variety of IC chips as are available in the TTL or CMOS families, it has been popular for logic applications requiring high-speed switching, although its power consumption is also relatively high. ECL power consumption, however, does not increase as the switching frequency increases. At frequencies above 20 MHz, the dynamic power consumption of CMOS gates will continue to increase and exceed the per-gate consumption of ECL devices. Newer ECL family devices are available that can be switched at a rate faster than 3GHz.

The active switching element used in ECL family circuits is also the NPN BJT. But unlike the TTL family, which switches the transistors into saturation while turning them on, ECL switching is designed to prevent driving the transistors into saturation. Whenever bipolar transistors are driven into saturation, their switching speed will be limited by the charge-carrier storage delay, a transistor operational characteristic. Thus, the switching speed of ECL circuits will be significantly higher than that for TTL circuits. ECL operation is based on switching a fixed amount of bias current, which is less than the saturation amount between two different transistors. The basic circuit found in the ECL family is the differential amplifier. A bias circuit controls one side of the differential amplifier, while the other is controlled by the logic inputs to the gate. This logic family is also referred to as current-mode logic (CML), due to its current switching operation.

Logic Family Circuit Parameters

Digital circuits and systems operate in only two states, logic 1 and 0, usually represented by two different voltage levels, a HIGH and a LOW. The two logic levels consist of a range of values with numerical quantities dependent upon the specific family used. Minimum high-logic levels and maximum low-logic levels are established by specifications for each family. Minimum device output levels for a logic high are called $V_{OH(min)}$, and minimum input levels are called $V_{IH(min)}$. The abbreviations for maximum output and input

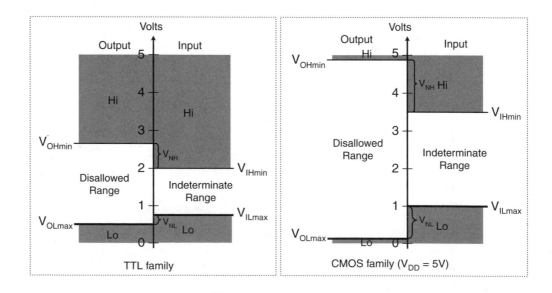

FIGURE 1.4   TTL and CMOS family logic levels.

TABLE 1.3   Logic Signal Voltage Parameters for Selected Logic Subfamilies (in Volts)

Subfamily	$V_{OH(min)}$	$V_{OL(max)}$	$V_{IH(min)}$	$V_{IL(max)}$
74xx	2.4	0.4	2.0	0.8
74LSxx	2.7	0.5	2.0	0.8
74ASxx	2.5	0.5	2.0	0.8
74ALSxx	2.5	0.4	2.0	0.8
74Fxx	2.5	0.5	2.0	0.8
74HCxx	4.9	0.1	3.15	0.9
74HCTxx	4.9	0.1	2.0	0.8
74ACxx	3.8	0.4	3.15	1.35
74ACTxx	3.8	0.4	2.0	0.8
74AHCxx	4.5	0.1	3.85	1.65
74AHCTxx	3.65	0.1	2.0	0.8
10xxx	−0.96	−1.65	−1.105	−1.475
10Hxxx	−0.98	−1.63	−1.13	−1.48

low-logic levels are $V_{OL(max)}$ and $V_{IL(max)}$. Figure 1.4 shows the relationships between these parameters. Logic voltage-level parameters for selected prominent logic subfamilies are illustrated in Table 1.3. As seen in this illustration, there are many operational incompatibilities between major logic family types.

Noise margin is a quantitative measure of a device's **noise immunity**. High-level noise margin (V_{NH}) and low-level noise margin (V_{NL}) are defined in Equation (1.1) and Equation (1.2).

$$V_{NH} = V_{OH(min)} - V_{IH(min)} \tag{1.1}$$

$$V_{NL} = V_{IL(max)} - V_{OL(max)} \tag{1.2}$$

Using the logic voltage values in Table 1.3 for the selected subfamilies reveals that the highest noise immunity is obtained with logic devices in the CMOS family while the lowest noise immunity is endemic to the ECL family.

Switching circuit outputs are loaded by the inputs of the devices they are driving, as illustrated in Figure 1.5. Worst-case input loading current levels and output driving current capabilities are listed in Table 1.4 for various logic subfamilies. The **fan-out** of a driving device is the ratio between its output current capabilities at each logic level and the corresponding gate-input current loading value.

Switching circuits based on bipolar transistors have fan-out that is limited primarily by the current-sinking and current-sourcing capabilities of the driving device.

CMOS switching circuits are limited by the charging and discharging times associated with the output resistance of the driving gate and the input capacitance of the load gates. Thus, CMOS fan-out depends on switching frequency. With fewer capacitive loading inputs to drive, the maximum switching frequency of CMOS devices will increase.

The switching speed of logic devices depends on the device's **propagation delay time**. The propagation delay of a logic device limits the frequency at which it can be operated. There are two propagation delay times specified for logic gates: t_{PHL}, delay time for the output to change from high to low, and t_{PLH}, delay time for the output to change from low to high. Average typical propagation delay times for a single gate are listed in Table 1.5 for several logic subfamilies. The ECL family has the fastest switching speed.

The amount of power required by an IC is normally specified in terms of the amount of current I_{CC} (TTL family), I_{DD} (CMOS family) or I_{EE} (ECL family) drawn from the power supply. For complex IC devices, the required supply current is given under specified test conditions. For TTL chips containing simple gates, the average power dissipation $P_{D(ave)}$ is normally calculated from two measurements, I_{CCH} (when all gate outputs are high) and I_{CCL} (when all gate outputs are low). Table 1.5 compares the static power dissipation of several logic subfamilies. The ECL family has the highest power dissipation for switching frequencies below about

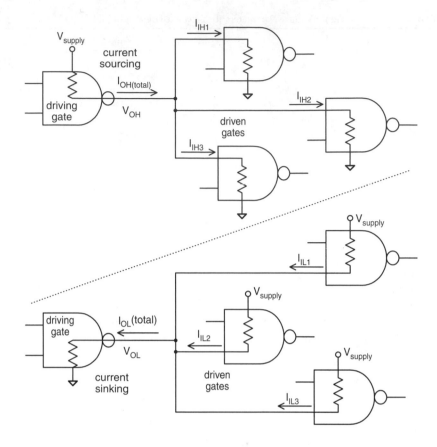

FIGURE 1.5   Current loading of driving gates.

TABLE 1.4   Worst Case Current Parameters for Selected Logic Subfamilies

Subfamily	$I_{OH(max)}$	$I_{OL(max)}$	$I_{IH(max)}$	$I_{IL(max)}$
74xx	−400 μA	16 mA	40 μA	−1.6 mA
74LSxx	−400 μA	8 mA	20 μA	−400 μA
74ASxx	−2 mA	20 mA	20 μA	−0.5 mA
74ALSxx	−400 μA	8 mA	20 μA	−100 μA
74Fxx	−1 mA	20 mA	20 μA	−0.6 mA
74HCxx	−4 mA	4 mA	1 μA	−1 μA
74HCTxx	−4 mA	4 mA	1 μA	−1 μA
74ACxx	−24 mA	24 mA	1 μA	−1 μA
74ACTxx	−24 mA	24 mA	1 μA	−1 μA
74AHCxx	−8 mA	8 mA	1 μA	−1 μA
74AHCTxx	−8 mA	8 mA	1 μA	−1 μA
10xxx	50 mA	−50 mA	−265 μA	0.5 μA
10Hxxx	50 mA	−50 mA	−265 μA	0.5 μA

20 MHz, while the lowest dissipation is found in the CMOS family. Power dissipation for the CMOS family is directly proportional to gate-input signal frequency. For example, typically, the power dissipation for a CMOS logic circuit will increase by a factor of 100 if input signal frequency is increased from 1 kHz to 100 kHz.

It is desirable to implement high speed (and, therefore, low propagation delay time) switching devices that consume low amounts of power. Because of the nature of transistor switching circuits, it is difficult to attain

TABLE 1.5   Speed-Power Comparison for a Single Gate in
Selected Logic Subfamilies

Subfamily	Propagation Delay Time, ns (ave.)	Static Power Dissipation, mW (per gate)
74xx	10	10
74LSxx	9.5	2
74Asxx	1.5	8.5
74ALSxx	4	1.2
74Fxx	3	6
74HCxx	8	0.003
74HCTxx	14	0.003
74Acxx	5	0.010
74ACTxx	5	0.010
74AHCxx	5.5	0.003
74AHCTxx	5	0.003
10xxx	2	25
10Hxxx	1	25

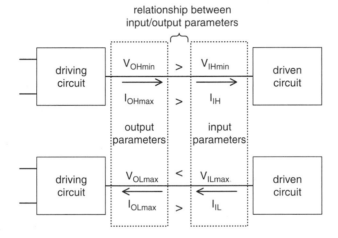

FIGURE 1.6   Circuit interfacing requirements.

high-speed switching with low power dissipation. The continued development of new IC logic families and subfamilies is due largely to the trade-offs between these two device-switching parameters.

Interfacing between Logic Families

The interconnection of logic chips requires that input and output specifications be satisfied. Figure 1.6 illustrates voltage and current requirements. The driving chip's V_{OHmin} must be greater than the driven circuit's V_{IHmin}, and the driver's V_{OLmax} must be less than V_{ILmax} for the loading circuit. Voltage level shifters must be used to interface the circuits if these voltage requirements are not met. Of course, a driving circuit's output must not exceed the maximum- and minimum-allowable input voltages for the driven circuit. The current sinking and sourcing ability of the driver circuit's output must be greater than the total current requirements for the loading circuit. Buffer gates or stages must be used if current requirements are not satisfied. All chips within a single logic family are designed to be compatible with other chips in that family. Mixing chips from multiple subfamilies together within a single digital circuit can have adverse effects on the overall circuit's switching speed and noise immunity.

Defining Terms

Fan-out: The specification used to identify the limit to the number of loading inputs that can be reliably driven by a driving device's output.

Logic Level: The high or low value of a voltage variable, assigned as a 1 or a 0 state.

Noise Immunity: A logic device's ability to tolerate input voltage fluctuation caused by noise without changing its output state.

Propagation Delay Time: The time delay from when the input logic level to a device is changed until that device produces the resultant output change.

Truth Table: A listing of the relationship of a circuit's output produced for various combinations of logic levels at the inputs.

References

N.P. Cook, *Practical Digital Electronics*, Upper Saddle River, NJ: Pearson Prentice-Hall, 2004.

R.K. Dueck, *Digital Design with CPLD Applications and VHDL*, 2nd ed., Albany, NY: Delmar Thomson Learning, 2005.

T.L. Floyd, *Digital Fundamentals*, 8th ed., Upper Saddle River, NJ: Pearson Prentice-Hall, 2003.

D.D. Givone, *Digital Principles and Design*, New York, NY: McGraw-Hill, 2003.

W. Kleitz, *Digital Electronics: A Practical Approach*, 7th ed., Upper Saddle River, NJ: Pearson Prentice-Hall, 2005.

M.M. Mano, *Digital Design*, 3rd ed., Upper Saddle River, NJ: Pearson Prentice-Hall, 2002.

R.J. Tocci, N.S. Widmer, and G.L. Moss, *Digital Systems: Principles and Applications*, 9th ed., Upper Saddle River, NJ: Pearson Prentice-Hall, 2004.

J.F. Wakerly, *Digital Design: Principles and Practices*, 3rd ed., Upper Saddle River, NJ: Pearson Prentice-Hall, 2001.

Further Information

Journals & Trade Magazines:

EDN, Highlands Ranch, CO: Reed Business Information.

Electronic Design, Cleveland, OH: Penton Media.

Electronic Engineering Times, Manhasset, NY: CMP Publications.

Internet Addresses for Digital Device Data Sheets:

Texas Instruments, Inc.: <http://focus.ti.com/general/docs/scproducts.jsp>.

ON Semiconductor: <http://www.onsemi.com/site/products/taxonomy/>.

1.2   Logic Gates (IC)[1]

Peter Graham

This section introduces and analyzes the electronic circuit realizations of the basic gates of the three technologies: transistor-transistor logic (TTL), emitter-coupled logic (ECL), and complementary metal-oxide semiconductor (CMOS) logic. These circuits are commercially available on small-scale integration chips and are also the building blocks for more elaborate logic systems. The three technologies are compared with regard to speed, power consumption, and noise immunity, and parameters are defined which facilitate these comparisons. Also included are recommendations which are useful in choosing and using these technologies.

---

[1]Based on P. Graham, "Gates," in *Handbook of Modern Electronics and Electrical Engineering*, C. Belove, Ed., New York: Wiley-Interscience, 1986, pp. 864–876. With permission.

Gate Specification Parameters

Theoretically almost any logic device or system could be constructed by wiring together the appropriate configuration of the basic gates of the selected technology. In practice, however, the gates are interconnected during the fabrication process to produce a desired system on a single chip. The circuit complexity of a given chip is described by one of the following four rather broad classifications:

- **Small-Scale Integration (SSI).** The inputs and outputs of every gate are available for external connection at the chip pins (with the exception that exclusive OR and AND-OR gates are considered SSI).
- **Medium-Scale Integration (MSI).** Several gates are interconnected to perform somewhat more elaborate logic functions such as flip-flops, counters, multiplexers, etc.
- **Large-Scale Integration (LSI).** Several of the more elaborate circuits associated with MSI are interconnected within the integrated circuit to form a logic system on a single chip. Chips such as calculators, digital clocks, and small microprocessors are examples of LSI.
- **Very-Large-Scale Integration (VLSI).** This designation is usually reserved for chips having a very high density, 1000 or more gates per chip. These include the large single-chip memories, gate arrays, and microcomputers.

Specifications of logic speed require definitions of switching times. These definitions can be found in the introductory pages of most data manuals. Four of them pertain directly to gate circuits. These are (see also Figure 1.7):

- **LOW-to-HIGH Propagation Delay Time (t_{PLH}).** The time between specified reference points on the input and output voltage waveforms when the output is changing from low to high.
- **HIGH-to-LOW Propagation Delay Tune (t_{PHL}).** The time between specified reference points on the input and output voltage waveforms when the output is changing from high to low.
- **Propagation Delay Time (t_{PD}).** The average of the two propagation delay times: $t_{PD} = (t_{PD} + t_{PHL})/2$.
- **LOW-to-HIGH Transition Time (t_{TLH}).** The rise time between specified reference points on the LOW-to-HIGH shift of the output waveform.

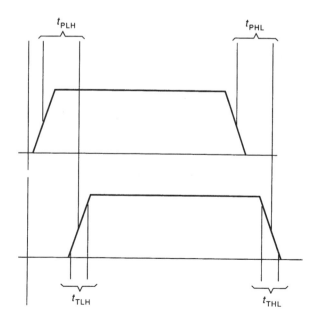

FIGURE 1.7 Definitions of switching times. (*Source:* P. Graham, "Gates," in *Handbook of Modern Electronics and Electrical Engineering*, C. Belove, Ed., New York: Wiley-Interscience, 1986, p. 865. With permission.)

- **HIGH-to-LOW Transition Time (t_{THL}).** The fall time between specified reference points on the HIGH-to-LOW shift of the output waveform. The reference points usually are 10 and 90% of the voltage level difference in each case.

Power consumption, driving capability, and effective loading of gates are defined in terms of currents.

- **Supply Current, Outputs High (I_{xxH}).** The current delivered to the chip by the power supply when all outputs are open and at the logical 1 level. The xx subscript depends on the technology.
- **Supply Current, Outputs Low (I_{xxL}).** The current delivered to the chip by the supply when all outputs are open and at the logical 0 level.
- **Supply Current, Worst Case (I_{xx}).** When the output level is unspecified, the input conditions are assumed to correspond to maximum supply current.
- **Input HIGH Current (I_{IH}).** The current flowing into an input when the specified HIGH voltage is applied.
- **Input LOW Current (I_{IL}).** The current flowing into an input when the specified LOW voltage is applied.
- **Output HIGH Current (I_{OH}).** The current flowing into the output when it is in the HIGH state. I_{OHmax} is the largest I_{OH} for which $V_{\text{OH}} \geq V_{\text{OHmin}}$ is guaranteed.
- **Output LOW Current (I_{OL}).** The current flowing into the output when it is in the LOW state. I_{OLmax} is the largest I_{OL} for which $V_{\text{OL}} \geq V_{\text{OLmax}}$ is guaranteed.

The most important voltage definitions are concerned with establishing ranges on the logical 1 (HIGH) and logical 0 (LOW) voltage levels.

- **Minimum High-Level Input Voltage (V_{IHmin}).** The least positive value of input voltage guaranteed to result in the output voltage level specified for a logical 1 input.
- **Maximum Low-Level Input Voltage (V_{ILmax}).** The most positive value of input voltage guaranteed to result in the output voltage level specified for a logical 0 input.
- **Minimum High-Level Output Voltage (V_{OHmin}).** The guaranteed least positive output voltage when the input is properly driven to produce a logical 1 at the output.
- **Maximum Low-Level Output Voltage (V_{OLmax}).** The guaranteed most positive output voltage when the input is properly driven to produce a logical 0 at the output.
- **Noise Margins.** $\text{NM}_{\text{H}} = V_{\text{OHmin}} - V_{\text{IHmin}}$ is how much larger the guaranteed least positive output logical 1 level is than the least positive input level that will be interpreted as a logical 1. It represents how large a negative-going glitch on an input 1 can be before it affects the output of the driven device. Similarly, $\text{NM}_{\text{L}} = V_{\text{ILmax}} - V_{\text{OLmax}}$ is the amplitude of the largest positive-going glitch on an input 0 that will not affect the output of the driven device.

Finally, three important definitions are associated with specifying the load that can be driven by a gate. Since in most cases the load on a gate output will be the sum of inputs of other gates, the first definition characterizes the relative current requirements of gate inputs.

- **Load Factor (LF).** Each logic family has a reference gate, each of whose inputs is defined to be a unit load in both the HIGH and the LOW conditions. The respective ratios of the input currents I_{IH} and I_{IL} of a given input to the corresponding I_{IH} and I_{IL} of the reference gate define the HIGH and LOW load factors of that input.
- **Drive Factor (DF).** A device output has drive factors for both the HIGH and the LOW output conditions. These factors are defined as the respective ratios of I_{OHmax} and I_{OLmax} of the gate to I_{OHmax} and I_{OLmax} of the reference gate.
- **Fan-Out.** For a given gate the fan-out is defined as the maximum number of inputs of the same type of gate that can be properly driven by that gate output. When gates of different load and drive factors are interconnected, fan-out must be adjusted accordingly.

Bipolar Transistor Gates

A logic circuit using bipolar junction transistors (BJTs) can be classified either as saturated or as nonsaturated logic. A saturated logic circuit contains at least one BJT that is saturated in one of the stable modes of the circuit.

In nonsaturated logic circuits none of the transistors is allowed to saturate. Since bringing a BJT out of saturation requires a few additional nanoseconds (called the storage time), nonsaturated logic is faster. The fastest circuits available at this time are emitter-coupled logic (ECL), with transistor-transistor logic (TTL) having Schottky diodes connected to prevent the transistors from saturating (Schottky TTL) being a fairly close second. Both of these families are nonsaturated logic. All TTL families other than Schottky are saturated logic.

Transistor-Transistor Logic

TTL evolved from resistor-transistor logic (RTL) through the intermediate step of diode-transistor logic (DTL). All three families are catalogued in data books published in 1968, but of the three only TTL is still available.

The basic circuit of the standard TTL family is typified by the two-input NAND gate shown in Figure 1.8(a). To estimate the operating levels of voltage and current in this circuit, assume that any transistor in saturation has $V_{CE} = 0.2$ and $V_{BE} = 0.75$ V. Let drops across conducting diodes also be 0.75 V and transistor current gains (when nonsaturated) be about 50. As a starting point, let the voltage levels at both inputs A and B be high enough that T_1 operates in the reversed mode. In this case the emitter currents of T_1 are negligible, and

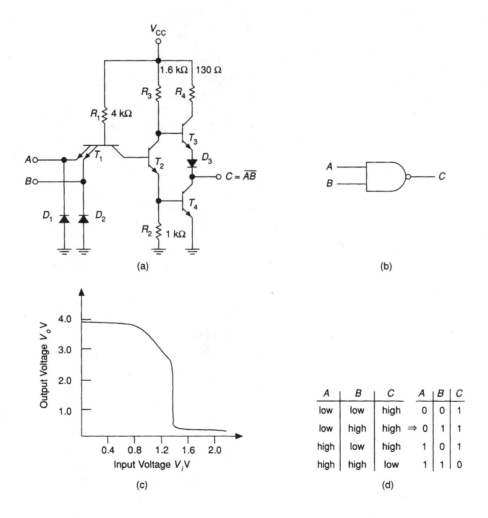

A	B	C		A	B	C
low	low	high		0	0	1
low	high	high	$\Rightarrow$	0	1	1
high	low	high		1	0	1
high	high	low		1	1	0

FIGURE 1.8 Two-input transistor-transistor logic (TTL) NAND gate type 7400: (a) circuit, (b) symbol, (c) voltage transfer characteristic (V_i to both inputs), (d) truth table. (*Source:* P. Graham, "Gates," in *Handbook of Modern Electronics and Electrical Engineering*, C. Belove, Ed., New York: Wiley-Interscience, 1986, p. 867. With permission.)

the current into the base of T_1 goes out the collector to become the base current of T_2. This current is readily calculated by observing that the base of T_1 is at $3 \times 0.75 = 2.25$ V so there is a 2.75-V drop across the 4-kΩ resistor. Thus $I_{BI} = I_{B2} = 0.7$ mA, and it follows that T_2 is saturated. With T_2 saturated, the base of T_3 is at $V_C + V_{BE4} = 0.95$ V. If T_4 is also saturated, the emitter of T_3 will be at $V_{D3} + V_{CE4} = 0.95$ V, and T_3 will be cut off. The voltage across the 1.6-kΩ resistor is $5 - 0.95 = 4.05$ V, so the collector current of T_2 is about 2.5 mA. This means the emitter current of T_2 is 3.2 mA. Of this, 0.75 mA goes through the 1-kΩ resistor, leaving 2.45 mA as the base current of T_4. Since the current gain of T_4 is about 50, it will be well into saturation for any collector current less than 100 mA, and the output at C is a logic 0. The corresponding minimum voltage levels required at the inputs are estimated from $V_{BE4} + V_{ECI}$, or about 1.7 V.

Now let either or both of the inputs be dropped to 0.2 V. T_1 is then biased to saturation in the normal mode, so the collector current of T_1 extracts the charge from the base region of T_2. With T_2 cut off, the base of T_4 is at 0 V and T_4 is cut off. T_3 will be biased by the current through the 1.6-kΩ resistor (R_3) to a degree regulated by the current demand at the output C. The drop across R_3 is quite small for light loads, so the output level at C will be $V_{CC} - V_{BE3} - V_{D3}$, which will be about 3.5 V corresponding to the logical 1.

The operation is summarized in the truth table in Figure 1.8(d), identifying the circuit as a two-input NAND gate. The derivation of the input-output voltage transfer characteristic [Figure 1.8(c)], where V_i is applied to inputs A and B simultaneously, can be found in most digital circuit textbooks. The sloping portion of the characteristic between $V_i = 0.55$ and 1.2 V corresponds to T_2 passing through the active region in going from cutoff to saturation.

Diodes D_1 and D_2 are present to damp out "ringing" that can occur, for example, when fast voltage level shifts are propagated down an appreciable length (20 cm or more) of microstripline formed by printed circuit board interconnections. Negative overshoots are clamped to the 0.7 V across the diode.

The series combination of the 130-Ω resistor, T_3, D_3, and T_4 in the circuit of Figure 1.8(a), forming what is called the totem-pole output circuit, provides a low impedance drive in both the source (output $C = 1$) and sink (output $C = 0$) modes and contributes significantly to the relatively high speed of TTL. The available source and sink currents, which are well above the normal requirements for steady state, come into play during the charging and discharging of capacitive loads. Ideally T_3 should have a very large current gain and the 130-Ω resistor should be reduced to 0. The latter, however, would cause a short-circuit load current which would overheat T_3, since T_3 would be unable to saturate. All TTL families other than the standard shown in Figure 1.8(a) use some form of Darlington connection for T_3, providing increased current gain and eliminating the need for diode D_3. The drop across D_3 is replaced by the base emitter voltage of the added transistor T_5. This connection appears in Figure 1.9(a), an example of the 74Hxx series of TTL gates that increases speed at the expense of increased power consumption, and in Figure 1.9(b), a gate from the 74Lxx series that sacrifices speed to lower power dissipation.

A number of TTL logic function implementations are available with open collector outputs. For example, the 7403 two-input NAND gate shown in Figure 1.10 is the open collector version of Figure 1.8(a). The open collector output has some useful applications. The current in an external load connected between the open collector and V_{CC} can be switched on and off in response to the input combinations. This load, for example, might be a relay, an indicator light, or an LED display. Also, two or more open collector gates can share a common load, resulting in the anding together of the individual gate functions. This is called a "wired-AND connection." In any application, there must be some form of load or the device will not function. There is a lower limit to the resistance of this load which is determined by the current rating of the open collector transistor. For wired-AND applications the resistance range depends on how many outputs are being wired and on the load being driven by the wired outputs. Formulas are given in the data books. Since the open collector configuration does not have the speed enhancement associated with an active pull-up, the low to high propagation delay (t_{PLH}) is about double that of the totem-pole output. It should be observed that totem-pole outputs should not be wired, since excessive currents in the active pull-up circuit could result.

Nonsaturated TTL.   Two TTL families, the Schottky (74Sxx) and the low-power Schottky (74LSxx), can be classified as nonsaturating logic. The transistors in these circuits are kept out of saturation by the connection of Schottky diodes, with the anode to the base and the cathode to the collector.

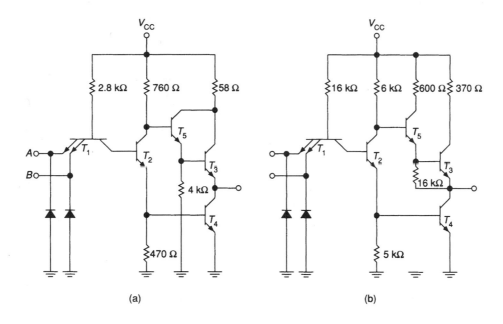

FIGURE 1.9 Modified transistor-transistor logic (TTL) two-input NAND states: (a) type 74Hxx, (b) type 74L00. (*Source:* P. Graham, "Gates," in *Handbook of Modern Electronics and Electrical Engineering,* C. Belove, Ed., New York: Wiley-Interscience, 1986, p. 868. With permission.)

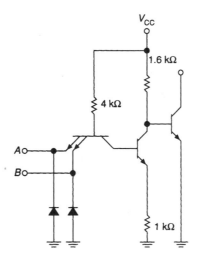

FIGURE 1.10 Open collector two-input NAND gate. (*Source:* P. Graham, "Gates," in *Handbook of Modern Electronics and Electrical Engineering,* C. Belove, Ed., New York: Wiley-Interscience, 1986, p. 868. With permission.)

Schottky diodes are formed from junctions of metal and an n-type semiconductor, the metal fulfilling the role of the p-region. Since there are thus no minority carriers in the region of the forward-biased junction, the storage time required to bring a pn junction out of saturation is eliminated. The forward-biased drop across a Schottky diode is around 0.3 V. This clamps the collector at 0.3 V less than the base, thus maintaining V_{CE} above the 0.3-V saturation threshold. Circuits for the two-input NAND gates 74LS00 and 74S00 are given in Figure 1.11(a) and (b). The special transistor symbol is a short-form notation indicating the presence of the Schottky diode, as illustrated in Figure 1.11(c).

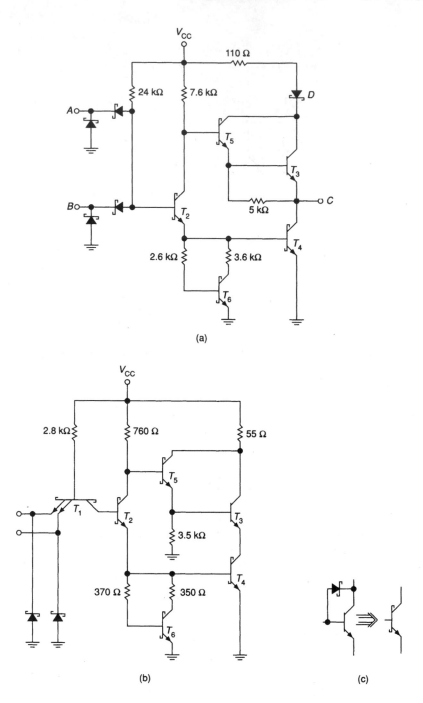

FIGURE 1.11 Transistor-transistor logic (TTL) nonsaturated logic. (a) Type 74LS00 two-input NAND gate, (b) type 74S00 two-input NAND gate, (c) significance of the Schottky transistor symbol. (*Source*: P. Graham, "Gates," in *Handbook of Modern Electronics and Electrical Engineering*, C. Belove, Ed., New York: Wiley-Interscience, 1986, p. 870. With permission.)

Note that both of these circuits have an active pull-down transistor T_6 replacing the pull-down resistance connected to the emitter of T_2 in Figure 1.9. The addition of T_6 decreases the turn-on and turn-off times of T_4. In addition, the transfer characteristic for these devices is improved by the squaring off of the sloping region between $V_i = 0.55$ and 1.2 V [see Figure 1.8(c)]. This happens because T_2 cannot become active until T_6 turns on, which requires at least 1.2 V at the input.

TABLE 1.6   Comparison of TTL Two-Input NAND Gates

TTL Type	I_{CCH}[a] (mA)	I_{CCL} (mA)	Propagation Delay Time		Noise Margins		Load Factor, H/L	Drive Factor, H/L	Fan-Out
			t_{PLH} (ns)	t_{PHL} (ns)	NM_H (V)	NM_L (V)			
74F00	2.8	10.2	2.9	2.6	0.7	0.3	0.5/0.375	25/12.5	33
74S00	10	20	3	3	0.7	0.3	1.25/1.25	25/12.5	10
74H00	10	26	5.9	6.2	0.4	0.4	1.25/1.25	12.5/12.5	10
74LS00	0.8	2.4	9	10	0.7	0.3	0.5/0.25	10/5	20
7400	4	12	11	7	0.4	0.4	1/1	20/10	10
74L00	0.44	1.16	31	31	0.4	0.5	0.24/0.1125	5/2.25	20

[a]See text for explanation of abbreviations.

Source: P. Graham, "Gates," in *Handbook of Modern Electronics and Electrical Engineering*, C. Belove, Ed., New York: Wiley-Interscience, 1986, p. 871. With permission.

The diode AND circuit of the 74LS00 in place of the multi-emitter transistor will permit maximum input levels substantially higher than the 5.5-V limit set for all other TTL families. Input leakage currents for 74LSxx are specified at $V_i = 10$ V, and input voltage levels up to 15 V are allowed. The 74LSxx has the additional feature of the Schottky diode D_1 in series with the 100-Ω output resistor. This allows the output to be pulled up to 10 V without causing a reverse breakdown of T_5. The relative characteristics of the several versions of the TTL two-input NAND gate are compared in Table 1.6. The 74F00 represents one of the new technologies that have introduced improved Schottky TTL in recent years.

TTL Design Considerations.   Before undertaking construction of a logic system, the wise designer consults the information and recommendations provided in the data books of most manufacturers. Some of the more significant tips are provided here for easy reference.

1. **Power supply, decoupling, and grounding.** The power supply voltage should be 5 V with less than 5% ripple factor and better than 5% regulation. When packages on the same printed circuit board are supplied by a bus there should be a 0.05-μF decoupling capacitor between the bus and the ground for every five to ten packages. If a ground bus is used, it should be as wide as possible, and should surround all the packages on the board. Whenever possible, use a ground plane. If a long ground bus is used, both ends must be tied to the common system ground point.

2. **Unused gates and inputs.** If a gate on a package is not used, its inputs should be tied either high or low, whichever results in the least supply current. For example, the 7400 draws three times the current with the output low as with the output high, so the inputs of an unused 7400 gate should be grounded. An unused input of a gate, however, must be connected so as not to affect the function of the active inputs. For a 7400 NAND gate, such an input must either be tied high or paralleled with a used input. It must be recognized that paralleled inputs count as two when determining the fan-out. Inputs that are tied high can be connected either to V_{CC} through a 1-kΩ or more resistance (for protection from supply voltage surges) or to the output of an unused gate whose input will establish a permanent output high. Several inputs can share a common protective resistance. Unused inputs of low-power Schottky TTL can be tied directly to V_{CC}, since 74LSxx inputs tolerate up to 15 V without breakdown. If inputs of low-power Schottky are connected in parallel and driven as a single input, the switching speed is decreased, in contrast to the situation with other TTL families.

3. **Interconnection.** Use of line lengths of up to 10 in. (5 in. for 74S) requires no particular precautions, except that in some critical situations lines cannot run side by side for an appreciable distance without causing cross talk due to capacitive coupling between them. For transmission line connections, a gate should drive only one line, and a line should be terminated in only one gate input. If overshoots are a problem, a 25- to 50-Ω resistor should be used in series with the driving gate input and the receiving gate input should be pulled up to 5 V through a 1-kΩ resistor. Driving and receiving gates should

have their own decoupling capacitors between the V_{CC} and ground pins. Parallel lines should have a grounded line separating them to avoid cross talk.

4. **Mixing TTL subfamilies.** Even synchronous sequential systems often have asynchronous features such as reset, preset, load, and so on. Mixing high-speed 74S TTL with lower speed TTL (74LS for example) in some applications can cause timing problems resulting in anomalous behavior. Such mixing is to be avoided, with rare exceptions which must be carefully analyzed.

Emitter-Coupled Logic

ECL is a nonsaturated logic family where saturation is avoided by operating the transistors in the common collector configuration. This feature, in combination with a smaller difference between the HIGH and LOW voltage levels (less than 1 V) than other logic families, makes ECL the fastest logic available at this time. The circuit diagram of a widely used version of the basic two-input ECL gate is given in Figure 1.12. The power supply terminals V_{CC1}, V_{CC2}, V_{EE}, and V_{TT} are available for flexibility in biasing. In normal operation, V_{CC1} and V_{CC2} are connected to a common ground, V_{EE} is biased to -5.2 V, and V_{TT} is biased to -2 V. With these values the nominal voltage for the logical 0 and 1 are, respectively, -1.75 and -0.9 V. Operation with the V_{CC} terminals grounded maximizes the immunity from noise interference.

A brief description of the operation of the circuit will verify that none of the transistors saturates. For the following discussion, V_{CC1} and V_{CC2} are grounded, V_{EE} is -5.2 V, and V_{TT} is -2 V. Diode drops and base-emitter voltages of active transistors are 0.8 V.

First, observe that the resistor-diode (D_1 and D_2) voltage divider establishes a reference voltage of -0.55 V at the base of T_3, which translates to -1.35 V at the base of T_2. When either or both of the inputs A and B are at the logical 1 level of -0.9 V, the emitters of T_{1A}, T_{1B}, and T_2 will be 0.8 V lower, at -1.7 V. This establishes the base-emitter voltage of T_2 at $-1.35-(-1.7) = 0.35$ V, so T_2 is cut off. With T_2 off, T_4 is biased into the active region, and its emitter will be at about -0.9 V, corresponding to a logical 1 at the $(A + B)$ output. Most of the current through the 365-Ω emitter resistor, which is $[-1.7-(-5.2)]/0.365 = 9.6$ mA, flows through the 100-Ω collector resistor, dropping the base voltage of T_5 to -0.96 V. Thus the voltage level at the output terminal designated $(A + B)$ is -1.76 V, corresponding to a logical 0.

When both A and B inputs are at the LOW level of -1.75 V, T_2 will be active, with its emitter voltage at $-1.35-0.8 = -2.15$ V. The current through the 365-Ω resistor becomes $[-2.15-(-5.2)]/0.365 = 8.2$ mA.

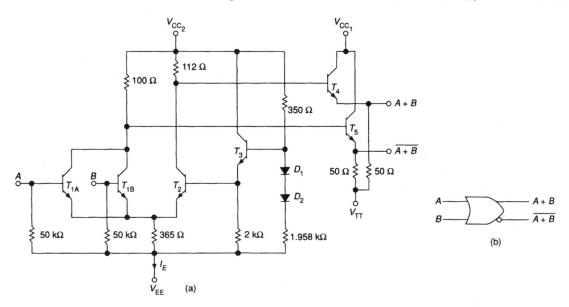

FIGURE 1.12 Emitter-coupled logic basic gate (ECL 10102): (a) circuit, (b) symbol. (*Source:* P. Graham, "Gates," in *Handbook of Modern Electronics and Electrical Engineering*, C. Belove, Ed., New York: Wiley-Interscience, 1986, p. 872. With permission.)

This current flows through the 112-Ω resistor pulling the base of T_4 down to -0.94 V, so that the $(A + B)$ output will be at the LOW level of -1.75 V. With T_{1A} and T_{1B} cut off, the base of T_5 is close to 0.0 V, and the $(A + B)$ output will therefore be at the nominal HIGH level of -0.9 V.

Observe that the output transistors T_4 and T_5 are always active and function as emitter followers, providing the low-output impedances required for driving capacitive loads. As T_{1A} and/or T_{1B} turn on, and T_2 turns off as a consequence, the transition is accomplished with very little current change in the 365-Ω emitter resistor. It follows that the supply current from V_{EE} does not undergo the sudden increases and decreases prevalent in TTL, thus eliminating the need for decoupling capacitors. This is a major reason why ECL can be operated successfully with the low noise margins which are inherent in logic having a relatively small voltage difference between the HIGH and LOW voltage levels (see Table 1.7). The small level shifts between LOW and HIGH also permit low propagation times without excessively fast rise and fall times. This reduces the effects of residual capacitive coupling between gates, thereby lessening the required noise margin. For this reason the faster ECL (100xxx) should not be used where the speed of the 10xxx series is sufficient. A comparison of three ECL series is given in Table 1.7. The propagation times t_{PLH} and t_{PHL} and transition times t_{TLH} and t_{THL} are defined in Figure 1.7. Transitions are between the 20 and 80% levels.

The 50-Ω pull-down resistors shown in Figure 1.12 are connected externally. The outputs of several gates can therefore share a common pull-down resistor to form a wired-OR connection. The open emitter outputs also provide flexibility for driving transmission lines, the use of which in most cases is mandatory for interconnecting this high-speed logic. A twisted pair interconnection can be driven using the complementary outputs $(A + B)$ and $(A + B)$ as a differential output. Such a line should be terminated in an ECL line receiver (10114).

Since ECL is used in high-speed applications, special techniques must be applied in the layout and interconnection of chips on circuit boards. Users should consult design handbooks published by the suppliers before undertaking the construction of an ECL logic system.

While ECL is not compatible with any other logic family, interfacing buffers, called translators, are available. In particular, the 10124 converts TTL output levels to ECL complementary levels, and the 10125 converts either single-ended or differential ECL outputs to TTL levels. Among other applications of these translators, they allow the use of ECL for the highest speed requirements of a system while the rest of the system uses the more rugged TTL. Another translator is the 10177, which converts the ECL output levels to n-channel metal-oxide semiconductor (NMOS) levels. This is designed for interfacing ECL with n-channel memory systems.

TABLE 1.7   Comparison of ECL Quad Two-Input NOR Gates ($V_{TT} = V_{EE} = 5.2$ V, $V_{CC1} = 0$ V)

ECL Type	Power Supply Terminal V_{EE} (V)	Power Supply Current I_E (mA)	Propagation Delay Time t_{PLH}[a] (ns)	Propagation Delay Time t_{PHL} (ns)	Transition Time t_{TLH}[b] (ns)	Transition Time t_{THL}[b] (ns)	Noise Margins NM_H (V)	Noise Margins NM_L (V)	Test Load
ECL II									
1012	-5.2	18[c]	5	4.5	4	6	0.175	0.175	Fan-out of 3
95102	-5.2	11	2	2	2	2	0.14	0.145	50 Ω
10102	-5.2	20	2	2	2.2	2.2	0.135	0.175	50 Ω
ECL III									
1662	-5.2	56[c]	1	1.1	1.4	1.2	0.125	0.125	50 Ω
100102[d]	-4.5	55	0.75	0.75	0.7	0.7	0.14	0.145	50 Ω
11001[e]	-5.2	24	0.7	0.7	0.7	0.7	0.145	0.175	50 Ω

[a]See text for explanation of abbreviations.
[b]20 to 80% levels.
[c]Maximum value (all other typical).
[d]Quint 2-input NOR/OR gate.
[e]Dual 5/4-input NOR/OR gate.

Source: P. Graham, "Gates," in *Handbook of Modern Electronics and Electrical Engineering*, C. Belove, Ed., New York: Wiley-Interscience, 1986, p. 873. With permission.

Complementary Metal-Oxide Semiconductor (CMOS) Logic

Metal-oxide semiconductor (MOS) technology is prevalent in LSI systems due to the high circuit densities possible with these devices. *p*-Channel MOS was used in the first LSI systems, and it still is the cheapest to produce because of the higher yields achieved due to the longer experience with PMOS technology. PMOS, however, is largely being replaced by NMOS (*n*-channel MOS), which has the advantages of being faster (since electrons have greater mobility than holes) and having TTL compatibility. In addition, NMOS has a higher function/chip area density than PMOS, the highest density in fact of any of the current technologies. Use of NMOS and PMOS, however, is limited to LSI and VLSI fabrications. The only MOS logic available as SSI and MSI is CMOS (complementary MOS).

CMOS is faster than NMOS and PMOS, and it uses less power per function than any other logic. While it is suitable for LSI, it is more expensive and requires somewhat more chip area than NMOS or PMOS. In many respects it is unsurpassed for SSI and MSI applications. Standard CMOS (the 4000 series) is as fast as low-power TTL (74Lxx) and has the largest noise margin of any logic type.

A unique advantage of CMOS is that for all input combinations the steady-state current from V_{DD} to V_{SS} is almost zero because at least one of the series FETs is open. Since CMOS circuits of any complexity are interconnections of the basic gates, the quiescent currents for these circuits are extremely small, an obvious advantage which becomes a necessity for the practicality of digital watches, for example, and one which alleviates heat dissipation problems in high-density chips. Also a noteworthy feature of CMOS digital circuits is the absence of components other than FETs. This attribute, which is shared by PMOS and NMOS, accounts for the much higher function/chip area density than is possible with TTL or ECL. During the time the output of a CMOS gate is switching there will be current flow from V_{DD} to V_{SS}, partly due to the charging of junction capacitances and partly because the path between V_{DD} and V_{SS} closes momentarily as the FETs turn on and off. This causes the dc supply current to increase in proportion to the switching frequency in a CMOS circuit. Manufacturers specify that the supply voltage for standard CMOS can range over $3 \text{ V} \leq V_{DD} - V_{SS} \leq 18 \text{ V}$, but switching speeds are slower at the lower voltages, mainly due to the increased resistances of the "on" transistors. The output switches between low and high when the input is midway between V_{DD} and V_{SS}, and the output logical 1 level will be V_{DD} and the logical 0 level V_{SS} [Figure 1.13(c)]. If CMOS is operated with $V_{DD} = 5 \text{ V}$ and $V_{SS} = 0 \text{ V}$, the V_{DD} and V_{SS} levels will be almost compatible with TTL except that the TTL totem-pole output high of 3.4 V is marginal as a logical 1 for CMOS. To alleviate this, when CMOS is driven with TTL a 3.3-kΩ

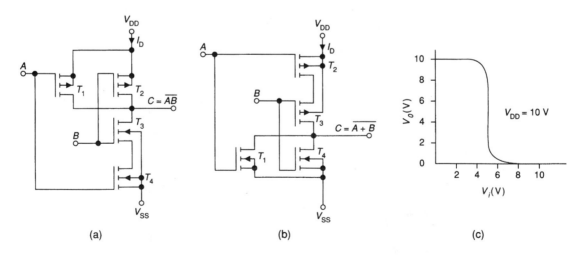

(a)   (b)   (c)

FIGURE 1.13   (a) Complementary metal-oxide semiconductor (CMOS) NAND gate, (b) NOR gate, and (c) inverter transfer characteristic. (*Source*: P. Graham, "Gates," in *Handbook of Modern Electronics and Electrical Engineering*, C. Belove, Ed., New York: Wiley-Interscience, 1986, p. 874. With permission.)

pull-up resistor between the TTL output and the common V_{CC}, V_{DD} supply terminal should be used. This raises V_{OH} of the TTL output to 5 V.

All CMOS inputs are diode protected to prevent static charge from accumulating on the FET gates and causing punch-through of the oxide insulating layer. A typical configuration is illustrated in Figure 1.14. Diodes D_1 and D_2 clamp the transistor gates between V_{DD} and V_{SS}. Care must be taken to avoid input voltages that would cause excessive diode currents. For this reason manufacturers specify an input voltage constraint from $V_{SS} - 0.5$ V to $V_{DD} + 0.5$ V. The resistance R_s helps protect the diodes from excessive currents but is introduced at the expense of switching speed, which is deteriorated by the time constant of this resistance and the junction capacitances.

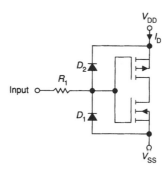

FIGURE 1.14 Diode protecion of input transistor gates. $200 \, \Omega < R_s < k \, \Omega$. (*Source*: P. Graham, "Gates," in *Handbook of Modern Electronics and Electrical Engineering*, C. Belove, Ed., New York: Wiley-Interscience, 1986, p. 875. With permission.)

Advanced versions of CMOS have been developed which are faster than standard CMOS. The first of these to appear were designated 74HCxx and 74HCTxx. The supply voltage range for this series is limited to $2 \, \text{V} \leq V_{DD} - V_{SS} \leq 6 \, \text{V}$. The pin numbering of a given chip is the same as its correspondingly numbered TTL device. Furthermore, gates with the HCT code have skewed transfer characteristics which match those of its TTL cousin, so that these chips can be directly interchanged with low-power Schottky TTL.

More recently, a much faster CMOS has appeared and carries the designations 74ACxx and 74ACTxx. These operate in the same supply voltage range and bear the same relationship with TTL as the HCMOS. The driving capabilities (characterized by I_{OH} and I_{OL}) of this series are much greater, such that they can be fanned out to 10 low-power Schottky inputs.

The three types of CMOS are compared in Table 1.8. The relative speeds of these technologies are best illustrated by including in the table the maximum clock frequencies for D flip-flops. In each case, the frequency given is the maximum for which the device is guaranteed to work. It is worth noting that a typical maximum clocking of 160 MHz is claimed for the 74ACT374 D flip-flop.

TABLE 1.8 Comparison of Standard, High-Speed, and Advanced High-Speed CMOS

Parameter	Symbol	Unit	Standard CMOS NORGates		High-Speed CMOS Inverter		Advanced CMOS Inverter	
			4001B	4011UB	74HC04	74HCT04	74AC04	74ACT04
Supply voltage	V_{DD}-V_{SS}	V	15	15	6	5.5	5.5	5.5
Input voltage thresholds	V_{IHmin}	V	11	12.5	4.2	2	3.85	2
	V_{ILmax}	V	4	2.5	1.8	0.8	1.65	0.8
Guaranteed output	V_{OHmin}	V	13.5	13.5	5.9	4.5	4.86	4.76
levels at maximum IO	V_{OLmax}	V	1.5	1.5	0.1	0.26	0.32	0.37
Maximum output currents	I_{OH}	mA	−8.8	−3.5	−4	−4	−24	−24
	I_{OL}	mA	8.8	8.8	4	4	24	24
Noise margins	NM_L	V	2.5	2.5	1.7	0.54	1.33	0.43
	NM_H	V	2.5	2.5	1.7	2.5	1.01	1.24
Propagation times	t_{PLH}	ns	40	40	16	15	4	4.3
	t_{PHL}	ns	40	40	16	17	3.5	3.9
Max input current leakage	I_{INmax}	µA	0.1	0.1	0.1	0.1	0.1	0.1
D-flip-flop max frequency			4013 B		74HC3 74	74HCT374 A	74AC37 4	74ACT37 4
(guaranteed minimum)	f_{max}	MHz	7.0	N.A.	35	30	100	100

CMOS Design Considerations

Design and handling recommendations for CMOS, which are included in several of the data books, should be consulted by the designer using this technology. A few selected recommendations are included here to illustrate the importance of such information.

1. All unused CMOS inputs should be tied either to V_{DD} or V_{SS}, whichever is appropriate for proper operation of the gate. This rule applies even to inputs of unused gates, not only to protect the inputs from possible static charge buildup, but to avoid unnecessary supply current drain. Floating gate inputs will cause all the FETs to be conducting, wasting power and heating the chip unnecessarily.
2. CMOS inputs should never be driven when the supply voltage V_{DD} is off, since damage to the input-protecting diodes could result. Inputs wired to edge connectors should be shunted by resistors to V_{DD} or V_{SS} to guard against this possibility.
3. Slowly changing inputs should be conditioned using Schmitt trigger buffers to avoid oscillations that can arise when a gate input voltage is in the transition region.
4. Wired-AND configurations cannot be used with CMOS gates, since wiring an output HIGH to an output LOW would place two series FETs in the "on" condition directly across the chip supply.
5. Capacitive loads greater than 5000 pF across CMOS gate outputs act as short circuits and can overheat the output FETs at higher frequencies.
6. Designs should be used that avoid the possibility of having low impedances (such as generator outputs) connected to CMOS inputs prior to power-up of the CMOS chip. The resulting current surge when V_{DD} is turned on can damage the input diodes.

While this list of recommendations is incomplete, it should alert the CMOS designer to the value of the information supplied by the manufacturers.

Choosing a Logic Family

A logic designer planning a system using SSI and MSI chips will find that an extensive variety of circuits is available in all three technologies: TTL, ECL, and CMOS. The choice of which technology will dominate the system is governed by what are often conflicting needs, namely, speed, power consumption, noise immunity, cost, availability, and the ease of interfacing. Sometimes the decision is easy. If the need for a low static power drain is paramount, CMOS is the only choice. It used to be the case that speed would dictate the selection; ECL was high speed, TTL was moderate, and CMOS low. With the advent of advanced TTL and, especially, advanced CMOS the choice is no longer clear-cut. All three will work at 100 MHz or more. ECL might be used since it generates the least noise because the transitions are small, yet for that same reason it is more susceptible to externally generated noise. Perhaps TTL might be the best compromise between noise generation and susceptibility. Advanced CMOS is the noisiest because of its rapid rise and fall times, but the designer might opt to cope with the noise problems to take advantage of the low standby power requirements.

A good rule is to use devices which are no faster than the application requires and which consume the least power consistent with the needed driving capability. The information published in the manufacturers' data books and designer handbooks is very helpful when choice is in doubt.

Defining Term

Logic gate: Basic building block for logic systems that controls the flow of pulses.

References

Advanced CMOS Logic Designers Handbook, Dallas: Texas Instruments, Inc., 1987.
C. Belove and D. Schilling, *Electronic Circuits, Discrete and Integrated,* 2nd ed., New York: McGraw-Hill, 1979.
FACT Data, Phoenix: Motorola Semiconductor Products, Inc., 1989.
Fairchild Advanced Schottky TTL, California: Fairchild Camera and Instrument Corporation, 1980.

W.I. Fletcher, *An Engineering Approach to Digital Design*, Englewood Cliffs, N.J.: Prentice-Hall, 1980.
High Speed CMOS Logic Data, Phoenix: Motorola Semiconductor Products, Inc., 1989.
P. Horowitz and W. Hill, *The Art of Electronics*, 2nd ed., New York: Cambridge University Press, 1990.
MECL System Design Handbook, Phoenix: Motorola Semiconductor Products, Inc., 1988.
H. Taub and D. Schilling, *Digital Integrated Electronics*, New York: McGraw-Hill, 1977.
The TTL Data Book for Design Engineers, Dallas: Texas Instruments, Inc., 1990.

Further Information

An excellent presentation of the practical design of logic systems using SSI and MSI devices is developed in the referenced book *An Engineering Approach to Digital Design* by William I. Fletcher. The author pays particular attention to the importance of device speed and timing.

The Art of Electronics by Horowitz and Hill is particularly helpful for its practical approach to interfacing digital with analog.

Everything one needs to know about digital devices and their interconnection can be found somewhere in the data manuals, design handbooks, and application notes published by the device manufacturers. Unfortunately, no single publication has it all, so the serious user should acquire as large a collection of these sources as possible.

1.3 Bistable Devices

Richard S. Sandige and Lynne A. Slivovsky

This section explores bistable devices, also commonly referred to as bistables, latches or flip-flops. Bistable devices are memory elements and can store one bit of information, such as a logic 1 or a logic 0 state. Latches and flip-flops are used to implement finite-state machines, counters and registers and are part of the configurable logic in complex programmable-logic devices and field programmable-gate arrays. Distinguishing behavior between a latch and a flip-flop is when the output changes due to a change in one or more inputs. A latch is considered **transparent** when changes in inputs, and hence stored data, immediately appear at the output. Edge-triggered flip-flops that change state with respect to a clock signal are not transparent. Output changes are triggered by a clock event.

The simplest bistable device consists of a pair of cross-coupled inverters where the output from one inverter feeds the input of the other, as depicted in Figure 1.15. After this circuit is powered up the value stored in the device, or the Q state, becomes indeterminate and will randomly fall into one of three states shown in Figure 1.15(b). Logic 0 or logic 1, corresponding to a low- or high-output voltage at Q, are stable states. Once the circuit moves to one of these states, it will never leave it. There is one metastable point in the center of the graph that would also satisfy the device's physical properties. But the likelihood of the circuit spending significant time in the metastable state is low, since any noise applied to the circuit would cause it to change to one of the stable states.

Latches

Replacing the inverters in the bistable element in Figure 1.15 with NOR gates provides the inputs to the bistable that can cause a change in state. Figure 1.16 shows an example of a basic set-reset (S-R) NOR latch implementation using two cross-coupled NOR gates. The logic symbol recommended for the S-R NOR latch by the Institute of Electrical and Electronics Engineers (IEEE) is shown to the right of the logic circuit implementation.

The S-R latch consists of a set (S) input, a reset (R) input and two outputs (Q and QN) that are normally complements of each other. Table 1.9 shows the operation of the S-R circuit. For S R = 00, Q = last Q, illustrating that the output for the next state Q is the same as for the present state output. For S R = 01, Q = 0, specifying that the output for the next state is reset. For S R = 10, Q = 1, indicating that the output

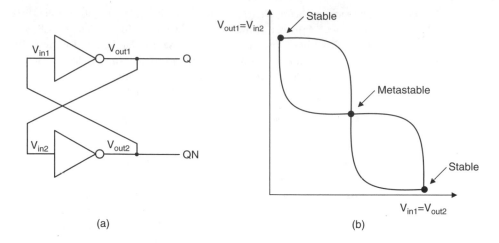

(a)   (b)

FIGURE 1.15   (a) The bistable element is composed of two inverter gates. (b) This shows the relationship between the input and output inverter voltages of the bistable element. The circuit has two stable operating points and one metastable operating point, all satisfying the circuit's transfer functions.

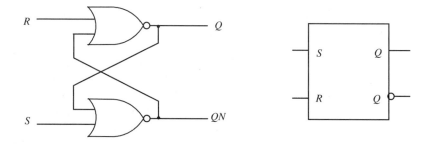

FIGURE 1.16   Set-Reset (S-R). This latch is constructed using NOR gates, in a configuration similar to the bistable element and its corresponding circuit symbol.

for the next state is set. In most cases, the input conditions S R = 11 are not allowed for two reasons. If S R = 11, then the QN output for the bistable element is not logically correct, as it is for all other input combinations. The second reason is more subtle since the next state of the bistable can be set or reset due to a critical race condition when the inputs are changed from 11 to 00. Such unpredictability is not desirable and therefore, the S R = 11 condition is generally not allowed. Latches and flip-flops that contain both a Q and a QN output (complementary outputs) provide double-rail outputs.

The S-R NAND latch in Figure 1.17 uses two cross-coupled NAND gates. In most cases, the input conditions $\bar{S}\bar{R}$ = 00 (S R = 11) are not allowed, for the same reasons provided above for the S-R NOR latch. For $\bar{S}\bar{R}$ = 01 (S R = 10), Q = 1 indicating that the output for the next state is set. For $\bar{S}\bar{R}$ = 10 (S R = 01), Q = 0 specifying that the output for the next state is reset. For $\bar{S}\bar{R}$ = 11 (S R = 00), Q = last illustrating that the output for the next state Q is the same as for the present stateoutput.

A gated, S-R latch is generated by AND-ing the inputs S and R with input C, as depicted in Figure 1.18. The C input acts to enable the latch. When C is asserted, the S-R latch behaves as described. When C is negated, both data inputs are logic 0, and the latch maintains its current state. Whatever value the output has when C goes to 0 is latched, captured or stored (memory mode).

The D latch in Figure 1.19 avoids the SR = 11 input conditions by guaranteeing that the data inputs are complements of each other. The S and R inputs are reduced to a single input, named D for data. The schematic symbol and characteristic table for the gated D latch circuit are shown in Figure 1.20. When input C is

TABLE 1.9 Operation of the S-R Latch

S	R	Q	QN
0	0	last Q	last QN
0	1	0	1
1	0	1	0
1	1	0	0

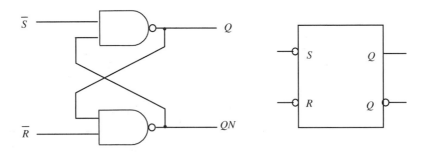

FIGURE 1.17 Basic S-R NAND latch and corresponding circuit symbol.

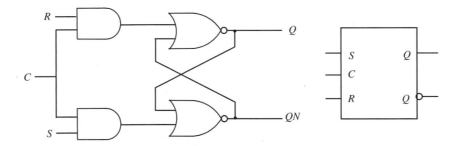

FIGURE 1.18 Gated S-R latch, where input C behaves like an enable signal.

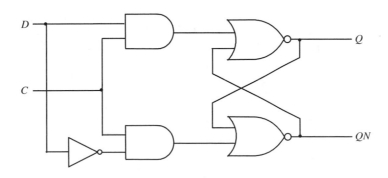

FIGURE 1.19 Gated D latch circuit based on the S-R NOR latch.

asserted, the D latch is transparent and the value of data input D appears at output Q. When input C is negated, the last value of data input D is stored in the latch. The D latch is level-sensitive with respect to C. The next section discusses devices that are edge-sensitive.

Flip-Flops

Early types of flip-flops were master-slave, pulse-triggered devices that had no data-lockout circuitry and caused a storage error if improperly used due to 1s and 0s catching. To prevent 1s and 0s catching, data-lockout (also called variable-skew) circuitry was added to some master-slave flip-flop types. Due to the improved design features and popularity of edge-triggered flip-flops, master-slave flip-flops are not recommended for newer designs and, in some cases, have been made obsolete by manufacturers, making them difficult to obtain even for repair parts. For this reason, only edge-triggered flip-flops will be discussed.

Four types of edge-triggered flip-flops are presented here. These are the D, J-K, T and S-R flip-flops. The D type is the most commonly used because its circuitry generally takes up less space on an IC chip and because most engineers consider it an easier device to use as the excitation equation to drive the D input is identical to the next state equation. An example of a positive, edge-triggered D flip-flop circuit is shown in Figure 1.21.

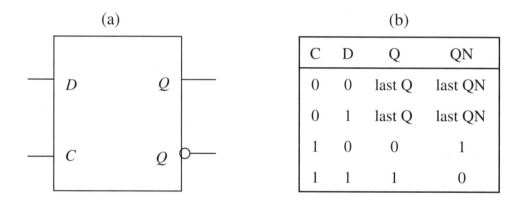

C	D	Q	QN
0	0	last Q	last QN
0	1	last Q	last QN
1	0	0	1
1	1	1	0

FIGURE 1.20   Gated D latch: (a) schematic symbol and (b) characteristic table.

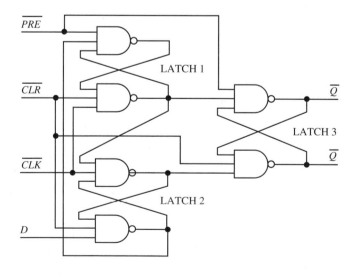

FIGURE 1.21   Positive, edge-triggered, D flip-flop circuit. (*Source*: Modified from R.S. Sandige, *Modern Digital Design*, New York: McGraw-Hill, 1990, p. 490.)

TABLE 1.10   Characteristic Table of a Positive Edge-Triggered, D Flip-Flop

D	CLK	Q	QN
0	↑	0	1
1	↑	1	0
x	0	last Q	last QN
x	1	last Q	last QN

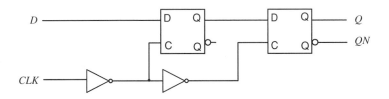

FIGURE 1.22   Positive edge-triggered D flip-flop circuit constructed from two D latches.

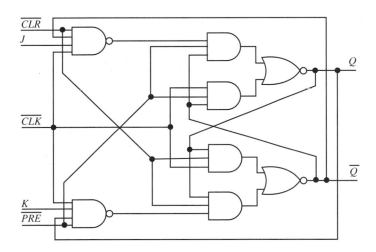

FIGURE 1.23   Negative, edge-triggered, J-K flip-flop circuit. (*Source*: Modified from R.S. Sandige, *Modern Digital Design*, New York: McGraw-Hill, 1990, p. 493.)

The characteristic table illustrating the operation of this flip-flop is shown in Table 1.10. At the rising edge of the clock input, the value at D is stored in the flip-flop. The D flip-flop can also be constructed by using two D latches, as shown in Figure 1.22.

The main difference between a latch and an edge-triggered flip-flop is their transparency. The gated D latch is transparent (the Q output follows the D input when the control input C = 1) and it latches, captures or stores the value at the D input when the control input C shifts to 0. The positive edge-triggered D flip-flop is never transparent from the time of its data input D to that of its output Q. When the clock is 0, the output Q does not follow the D input and remains unchanged; however, the value at the D input is latched, captured or stored when the clock makes a transition from 0 to 1. The flip-flop changes state only on the rising edge of the clock. Edge-triggered flip-flops are desirable for feedback applications due to their lack of transparency. Their outputs can be fed back as inputs to the device without causing oscillation. This is true for all types of edge-triggered flip-flops. A negative, edge-triggered, J-K flip-flop circuit is shown in the circuit diagram in Figure 1.23 with its corresponding IEEE symbol. Notice that the J-K flip-flop requires eight logic gates,

TABLE 1.11   Characteristic Table of a Positive, Edge-
Triggered, J-K Flip-Flop

J	K	Q
0	0	Iast Q (hold)
0	1	0 (clear)
1	0	1 (set)
1	1	$\frac{1}{Q}$ (toggle)

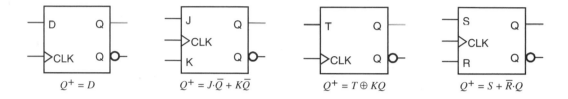

$$Q^+ = D \qquad Q^+ = J\cdot\overline{Q} + K\overline{Q} \qquad Q^+ = T \oplus KQ \qquad Q^+ = S + \overline{R}\cdot Q$$

FIGURE 1.24   Common edge-triggered flip-flops and their characteristic equations.

compared to only six logic gates for the D flip-flop in Figure 1.21. The characteristic table for this negative, edge-triggered flip-flop is shown in Table 1.11. When the J and K inputs are both 1 and the clock makes a 1 to 0 transition, the flip-flop toggles, and the next state output Q changes to the complement of the present state. By connecting J and K together and renaming it T for toggle, one can obtain a negative, edge-triggered, T flip-flop.

The behavior of each flip-flop in the characteristic tables can be captured in a characteristic equation. This equation describes the behavior of a flip-flop at a clock edge. Figure 1.24 shows the D and J-K flip-flops with their characteristic equations along with the T (or toggle) flip-flop and the S-R flip flop.

Since bistable devices are asynchronous, fundamental-mode, sequential logic circuits, only one input is allowed to change at a time. This means that for proper operation for a basic latch, only one of the data inputs S or R for an S-R NOR latch (and the NAND implementation) may be changed at one time. For proper operation of a gated latch, the data inputs S and R or data input D must meet minimum setup and hold-time requirements; i.e., the data input(s) must be stable for a minimum period before the control input C changes the latch from the transparent mode to the memory mode. For proper operation of an edge-triggered flip-flop, data inputs must meet minimum setup and hold time requirements relative to the clock changing from 0 to 1 (positive edge-triggered) or from 1 to 0 (negative edge-triggered).

An interesting exercise is to design a circuit for a D flip-flop using a J-K flip-flop and some additional gates. In general, a circuit can be designed that implements the characteristic equation of any flip-flop by using any other flip-flop and some added logic.

Defining Terms

Bistable, latch and flip-flop: Substitutions for the term bistable device.
Critical race: A change in two input variables resulting in an unpredictable output value for a bistable device.
Edge-triggered: Term describing the edge of a positive or negative pulse applied to the control input of a nontransparent bistable device to latch, capture or store the value indicated by the data input(s).
Fundamental mode: Operating mode of a circuit allowing only one input to change at a time.
Memory element: A bistable device or element providing data storage for a logic 1 or a logic 0 state.
Characteristic table: A tabular representation that illustrates the operation of various bistable devices.
Setup and hold time: The time required for the data input(s) to be held stable before or after the control input C changes to latch, capture or store the value indicated by the data input(s).

Toggle: Change in state from logic 0 to logic 1 or from logic 1 to logic 0 in a bistable device.
Transparent mode: Mode of a bistable device where the output responds to data-input signal changes.
Volatile device: A memory or storage device that loses its storage capability when power is removed.

References

ANSI/IEEE Std 91-1984, *IEEE Standard Graphic Symbols for Logic Functions*, New York, NY: Institute of Electrical and Electronics Engineers.

ANSI/IEEE Std 991-1986, *IEEE Standard for Logic Circuit Diagrams*, New York, NY: Institute of Electrical and Electronics Engineers.

R.S. Sandige, *Digital Design Essentials*, Upper Saddle River, NJ: Prentice Hall, 2002.

Texas Instruments, *The TTL Data Book, Advanced Low-Power Schottky, Advanced Schottky*, vol. 3, Dallas, TX: Texas Instruments, 1984.

J.F. Wakerly, *Digital Design Principles and Practices*, 3rd ed., Upper Saddle River, NJ: Prentice Hall, 2001 (Updated).

Further Information

Journals published by the IEEE contain the latest information on a variety of topics related to computer design and realization, including digital devices, logic and circuit design. Look in *IEEE Transactions on Computers*, *IEEE Transactions on Computer-Aided Design of Integrated Circuits* and *IEEE Transactions on Very Large-Scale Integration Systems*.

1.4 Optical Devices

H.S. Hinton

Since the first demonstration of optical logic devices in the late 1970s, there have been many different experimental devices reported. Figure 1.25 categorizes optical logic devices into four main classes. The first division is between all-optical and optoelectronic devices. All-optical devices are devices that do not use electrical currents to create the nonlinearity required by digital devices. These devices can be either single-pass devices (light passes through the nonlinear material once) or they can use a resonant cavity to further enhance the optical nonlinearity (multiple passes through the same nonlinear material). Optoelectronic devices, on the other hand, use electrical currents and electronic devices to process a signal that has gone through an optical-to-electrical conversion process. The output of these devices is either provided by electrically driving an optical source such as a laser or LED (detect/emit) or by modulating some external light source (detect/modulate). Below each of these categories are listed some of the devices that have been experimentally demonstrated.

All-Optical Devices

To create an all-optical logic device requires a medium that will allow one beam of light to affect another. This phenomenon can arise from the cubic response to the applied field. These third-order processes can lead to purely dielectric phenomena, such as irradiance-dependent refractive indices. By exploiting purely dielectric third-order nonlinearities, such as the optical Kerr effect, changes can be induced in the optical constants of the medium which can be read out directly at the same wavelength as that inducing them. This then opens up the possibilities for digital optical circuitry based on cascadable all-optical logic gates. Although there have been many different all-optical gates demonstrated, this section will only briefly review the soliton gate (single-pass) and one example of the **nonlinear Fabry–Perot** structures (cavity-based).

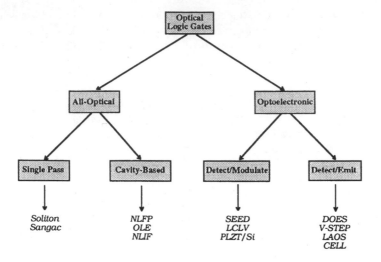

FIGURE 1.25   Classification of optical logic devices.

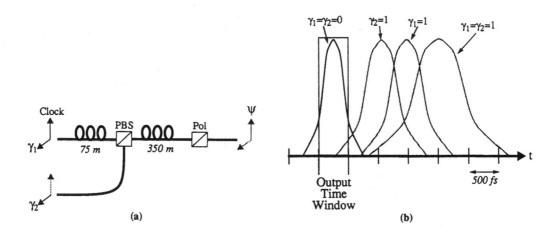

FIGURE 1.26   Soliton NOR gate: (a) physical implementation, (b) timing diagram.

Single-Pass Devices

An example of an all-optical single-pass optical logic gate is the soliton NOR gate. It is an all-fiber logic gate based on time shifts resulting from soliton dragging. A NOR gate consists of two birefringent fibers connected through a polarizing beamsplitter with the output filtered by a polarizer as shown in Figure 1.26. The clock pulse, which provides both gain and logic level restoration, propagates along one principal axis in both fibers. For the NOR gate the fiber length is trimmed so that in the absence of any signal the entering clock pulse will arrive within the output time window corresponding to a "1." When either or both of the input signals are incident, they interact with the clock pulse through soliton dragging and shift the clock pulse out of the allowed output time window creating a "0" output. In soliton dragging two temporally coincident, ortho-gonally polarized pulses interact in the fiber through cross-phase modulation and shift each other's velocities. This velocity shift converts into a time shift after propagating some distance in the fiber. To implement the device, the two input signal pulses γ_1 and γ_2 are polarized orthogonal to the clock. The signals are timed so that γ_1 and the clock pulse coincide at the input to the first fiber and γ_2 and the clock pulse coincide (in the absence of γ_1) at the input to the second fiber. At the output the two input signals are blocked by the polarizer, allowing only the temporally modified clock pulse to pass. In a prototyped demonstration this all-optical NOR gate required 5.8 pJ of signal energy and provided an effective gain of 6.

Cavity-Based Devices

Cavity-based optical logic devices are composed of two highly reflective mirrors that are separated by a distance d [Figure 1.27(a)]. The volume between the mirrors, referred to as the cavity of the etalon, is filled with a nonlinear material possessing an index of refraction that varies with intensity according to $n_c = n_0 + n_2 \gamma_c$ where n_0 is the linear index of refraction, n_2 is the nonlinear index of refraction, and γ_c is the intensity of light within the cavity. In the ideal case, the characteristic response of the reflectivity of a Fabry–Perot cavity, R_{fp}, is shown in Figure 1.27(b). At low intensities, the cavity resonance peak is not coincident with the wavelength of the incident light; thus the reflectivity is high, which allows little of the incident light to be transmitted [solid curves in Figure 1.27(b)]. As the intensity of the incident light γ increases, so does the intercavity light intensity which shifts the resonance peak [dotted curve in Figure 1.27(b)]. This shift in the resonant peak increases the transmission which in turn reduces the reflectivity. This reduction in ψ will continue with increasing γ until a minimum value is reached. It should be noted that in practice all systems of interest have both intensity-dependent absorption and n_2.

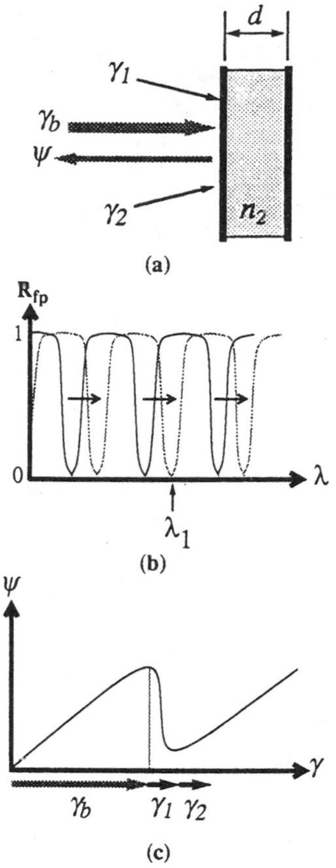

To implement a two-input NOR gate using the characteristic curve shown in Figure 1.27(c) requires a third input which is referred to as the *bias beam,* γ_b. This energy source biases the etalon at a point on its operating curve such that any other input will

FIGURE 1.27 (a) Nonlinear Fabry–Perot etalon, (b) reflection peaks of NLFP, and (c) NLFP in reflection (NOR).

exceed the nonlinear portion of the curve moving the etalon from the high reflection state. This is illustrated in Figure 1.27(c) where the γ_b combines with the inputs γ_1 and γ_2 to exceed the threshold of the nonlinear characteristic curve.

The first etalon-based optical logic device was in the form of a non-linear interference filter (NLIF). A simple interference filter has a general form similar to a Fabry–Perot etalon, being constructed by depositing a series of thin layers of transparent material of various refractive indices on a transparent substrate. The first several layers deposited form a stack of alternating high and low refractive indices, all of optical thickness equal to one quarter of the operating wavelength. The next layer is a low integer (1–20) number of half wavelengths thick and finally a further stack is deposited to form the filter. The two outer stacks have the property of high reflectivity at one wavelength, thus playing the role of mirrors forming a cavity. A high finesse cavity is usually formed when both mirrors are identical, i.e., of equal reflectivity. However, unlike a Fabry–Perot etalon with a nonabsorptive material in the cavity, matched (equal) stack reflectivities do not give the optimum cavity design to minimize switch power because of the absorption in the spacer (which may be necessary to induce nonlinearity). A balanced design which takes into account the effective decrease in back mirror reflectivity due to the double pass through the absorbing cavity is preferable and also results in greater contrast between bistable states. The balanced design is easily achieved

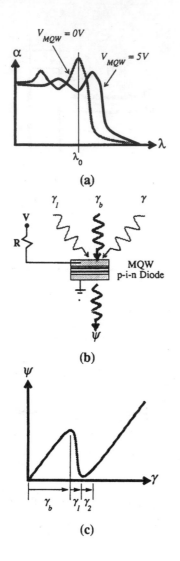

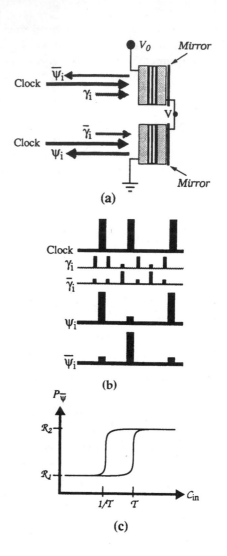

FIGURE 1.28 (a) Absorption spectra of MQW material for both 0 and 5 V, (b) schematic of MQW *pin* diode, (c) input/output characteristics of MQW *pin* diode.

FIGURE 1.29 Symmetric self-electro-optic effect device (S-SEED). (a) S-SEED with inputs and outputs, (b) power transfer characteristics, and (c) optically enabled S-SEED.

by varying one or all of the available parameters: number of periods, thickness and refractive index of each layer within either stack.

Optoelectronic Devices

Optoelectronic devices take advantage of both the digital processing capabilities of electronics and communications capabilities of the optical domain. This section will review both the SEED-based optical logic gates and the pnpn structures that have demonstrated optical logic.

Detect/Modulate Devices

In the most general terms the self-electro-optic effect device (**SEED**) **technology** corresponds to any device based on **multiple quantum well** (MQW) modulators. The basic physical mechanism used by this technology is the quantum confined Stark effect. This mechanism creates a shift in the bandedge of a semiconductor with an applied voltage. This is illustrated in Figure 1.28(a). This shift in the bandedge is then used to vary the absorption of incident light on the MQW material. When this MQW material is placed in the intrinsic region of a *pin* diode and electrically connected to a resistor as shown in Figure 1.28(b) the characteristic curve shown in Figure 1.28(c) results. When the incident intensity, γ_i, is low there is no current flowing through the *pin* diode or resistor; thus the majority of the voltage is across the *pin* diode. If the device is operating at the wavelength λ_0, the device will be in a low absorptive state. As the incident intensity increases so does the current flowing in the *pin* diode; this in turn reduces the voltage across the diode which increases the absorption and current flow. This state of increasing absorption creates the nonlinearity in the output signal, ψ, shown in Figure 1.28(c). Optical logic gates can be formed by optically biasing the R-SEED close to the nonlinearity, γ_b, and then applying lower level data signals γ_1 and γ_2 to the device.

The S-SEED, which behaves like an optical inverting S–R latch, is composed of two electrically connected MQW *pin* diodes as illustrated in Figure 1.29(a). In this figure, the device inputs include the signal, γ_i (Set), and its complement, $\bar{\gamma}$ (Reset), and a clock signal. To operate the S-SEED the γ_i and $\bar{\gamma}_i$ inputs are also separated in time from the clock inputs as shown in Figure 1.28(b). The γ_i and $\bar{\gamma}_i$ inputs, which represent the incoming data and its complement, are used to set the state of the device. When $\bar{\gamma}_i > \gamma_i$, the S-SEED will enter a state where the upper MQW *pin* diode will be reflective, forcing the lower diode to be absorptive. When $\gamma_i > \bar{\gamma}_i$ the opposite condition will occur. Low switching intensities are able to change the device's state when the clock signals are not present. After the device has been put into its proper state, the clock beams are applied to both inputs. The ratio of the power between the two clock beams should be approximately one, which will prevent the device from changing states. These higher energy clock pulses, on reflection, will transmit the state of the device to the next stage of the system. Since the inputs γ_i and $\bar{\gamma}_i$ are low-intensity pulses and the clock signals are high-intensity pulses, a large differential gain may be achieved. This type of gain is referred to as time-sequential gain.

The operation of an S SEED is determined by the power transfer characteristic shown in Figure 1.29(c). The optical power reflected by the ψ_i window, when the clock signal is applied, is plotted against the ratio of the total optical signal power impinging on the γ_i and $\bar{\gamma}_i$ windows (when the clock signal is not applied). Assuming the clock power incident on both signal windows, γ_i and $\bar{\gamma}_i$, the output power is proportional to the reflectivity, R_i. The ratio of the input signal powers is defined as the input contrast ratio $C_{in} = P_\gamma/P_{\bar{\gamma}_i}$. As C_{in} is increased from zero, the reflectivity of the ψ_i window switches from a low value, R_1, to a high value, R_2, at a C_{in} value approximately equal to the ratio of the absorbances of the two optical windows: $T = (1-R_1)/(1-R_2)$. Simultaneously, the reflectivity of the other window (ψ_i) switches from R_2 to R_1. The return transition point (ideally) occurs when $C_{in} = (1-R_2)/(1-R_1) = l/T$. The ratio of the two reflectivities, R_2/R_1, is the output contrast, C_{out}. Typical measured values of the preceding parameters include $C_{out} = 3.2$, $T = 1.4$, $R_2 = 50\%$ and $R_1 = 15\%$. The switching energy for these devices has been measured at ~ 7 fJ/μm^2.

The S-SEED is also capable of performing optical logic functions such as NOR, OR, NAND, and AND. The inputs will also be differential, thus still avoiding any critical biasing of the device. A method of achieving logic gate operation is shown in Figure 1.30. The logic level of the inputs will be defined as the ratio of the optical power on the two optical windows. When the power of the signal incident on the γ_i input is greater than the power of the signal on the $\bar{\gamma}_i$ input, a logic "1" will be present on the input. On the other hand, when the power of the signal incident on the γ_i input is less than the power of the signal on the $\bar{\gamma}_i$ input, a logic "0" will be incident on the input.

For the noninverting gates, OR and AND, we can represent the output logic level by the power of the signal coming from the ψ output relative to the power of the signal coming from the $\bar{\psi}$ output. As before, when the power of the signal leaving the ψ output is greater than the power of the signal leaving the $\bar{\psi}$ output, a logic "1" will be represented on the output. To achieve AND operation, the device is initially set to its "off" or logic "0" state (i.e., ψ low and $\bar{\psi}$ high) with preset pulse, $Preset_\psi$ incident on only one *pin* diode as shown in Figure 1.30. If both input signals have logic levels of "1" (i.e., set = 1, reset = 0), then the S-SEED AND gate is set to its "on" state. For any other input combination, there is no change of state, resulting in AND operation. After the signal beams determine the state of the device, the clock beams are then set high to read out the state of the AND gate. For NAND operation, the logic level is represented by the power of the $\bar{\psi}$ output signal relative to the power of the ψ output signal. That is, when the power of the signal leaving the $\bar{\psi}$ output is

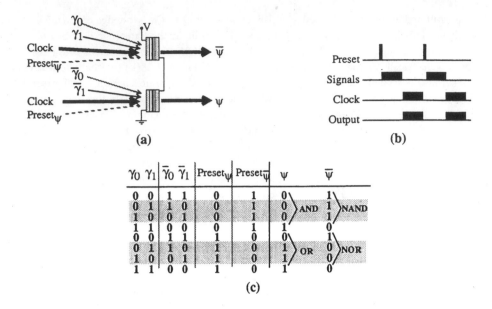

FIGURE 1.30 Logic using S-SEED devices.

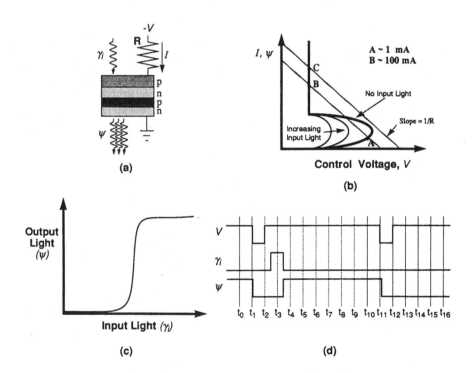

FIGURE 1.31 pnpn devices: (a) basic structure, (b) voltage/output characteristics, (c) input/output characteristics, and (d) timing diagram of device operation.

greater than the power of the signal leaving the ψ output, a logic "1" is present on the output. The operation of the OR and NOR gates is identical to the AND and NAND gates, except that preset pulse $Preset_\psi$ is used instead of the preset pulse $Preset_{\bar\psi}$. Thus, a single array of devices can perform any or all of the four logic functions and memory functions with the proper optical interconnections and preset pulse routing.

Detect/Emit Devices

Detect/emit devices are optoelectronic structures that detect the incoming signal, process the information, and then transfer the information off the device through the use of active light emitters such as LEDs or lasers. An example of a detect/emit device is the "thyristor-like" pnpn device as illustrated in Figure 1.31(a). It is a digital active optical logic device with "high" and "low" light-emitting optical output states corresponding to electrical states of high impedance (low optical output) or low impedance (high optical output). The device can be driven from one state to the other either electrically or optically. The optical output can be either a lasing output or light-emitting diode output. There are several devices that are based on this general structure. The double heterostructure optoelectronic switch (DOES) is actually an *npnp* structure that is designed as an integrated bipolar inversion channel heterojunction field-effect transistor (BICFET) phototransistor controlling and driving either an LED or microlaser output. The second device is a *pnpn* structure referred to as a vertical-to-surface transmission electrophotonic device (VSTEP).

The operation of these pnpn structures can be illustrated through the use of load lines. For the simplest device, the load consists of a resistor and a power supply. In Figure 1.31(b), we see that for small amounts of light, the device will be at point A. Point A is in a region of high electrical impedance with little or no optical output. As the input light intensity increases, there is no longer an intersection point near A and the device will switch to point B [Figure 1.31(c)]. At this point the electrical impedance is low and light is emitted. When the input light is removed, the operating point returns via the origin to point A by momentarily setting the electrical supply to zero [Figure 1.31(d)]. These devices can be used as either optical OR or AND gates using a bias beam and several other optical inputs.

The device can also be electrically switched. Assuming no input intensity, the initial operating point is at point A. By increasing the power supply voltage, the device will switch to point C. Point C like point B is in the region of light emission and low impedance. To turn off the device, the power supply must then be reduced to zero, after which it may be increased up to some voltage where switching occurs.

A differential pnpn device made by simply connecting two pnpn devices in parallel and connecting that combination in series with a resistive load is illustrated in Figure 1.32. The operation of the device can be described as follows. When the device is biased below threshold, that is, with the device unilluminated, both optical switches are "off." When the device is illuminated, the one with the highest power is switched "on." The increase in current leads to a voltage drop across the resistor which in turn leads to a lowering of the voltage across both optical switches. Therefore, the one with the lower input cannot be switched "on." Unless both inputs were illuminated with precisely the same power and both devices had identical characteristics (both of these are impossible), only one of the two optical switches will emit light.

The required input optical switching energy density can be quite low if the device without light is biased critically just below threshold. Since incoherent light from an LED cannot be effectively collected from small devices or focused onto small devices, a lasing pnpn is needed. Microlaser-based structures are also required to reduce the total power dissipation to acceptable levels. Surface-emitting microlasers provide an ideal laser because of their small size, single-mode operation, and low thresholds. The surface-emitting microlasers consist of two AlAs/AlGaAs dielectric mirrors with a thin active layer in between. This active layer typically consists of one or a few MQWs. The material can be etched vertically into small posts, typically 1–5 μm in diameter. Thresholds are typically on the order of milliwatts.

The switching speed of these devices is limited by the time it takes the photogenerated carriers to diffuse into the light-emitting region. Optical turn-off times are also limited by the RC time constant. For devices made so far, the RC time constants are in the range of 1–10 ns, and optical switch-on times were ~10 ns. Performance of the devices is expected to improve as the areas are reduced;

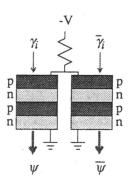

FIGURE 1.32 Differential *pnpn* device.

switching times comparable to the best electronic devices (~10 ps) are possible, although the optical turn-on times of at least the surface-emitting LED devices will continue to be slower since this time is determined by diffusion effects and not device capacitance and resistance. Lasing devices should offer improved optical turn-on times.

Another approach to active devices is to combine lasers/modulators with electronics and photodiodes as has been proposed for optical interconnections of electronic circuits. Since the logic function is implemented with electronic circuitry, any relatively complex functionality can be achieved. Several examples of logic gates have been made using GaAs circuitry and light-emitting diodes. Again surface-emitting microlasers provide an ideal emitter for this purpose, because of their small size and low threshold current. However, the integration of these lasers with the required electrical components has yet to be demonstrated.

Limitations

In the normal operating regions of most devices, a fixed amount of energy, the switching energy, is required to make them change states. This switching energy can be used to establish a relationship between both the switching speed and the power required to change the state of the device. Since the power required to switch the device is equal to the switching energy divided by the switching time, a shorter switching time will require more power. As an example, for a photonic device with an area of 100 μm^2 and a switching energy of 1 fJ/μm^2 to change states in 1 ps requires 100 mW of power instead of the 100 μW that would be required if the device were to switch at 1 ns. Thus, for high power signals the device will change states rapidly, while low power signals yield a slow switching response.

Some approximate limits on the possible switching times of a given device, whether optical or electrical, are illustrated in Figure 1.33. In this figure the time required to switch the state of a device is on the abscissa while the power/bit required to switch the state of a device is on the ordinate. The region of spontaneous switching is the result of the background thermal energy that is present in a device. If the switching energy for the device is too low, the background thermal energy will cause the device to change states spontaneously. To prevent these random transitions in the state of a device, the switching energy required by the device must be much larger than the background thermal energy. To be able to differentiate statistically between two states, this figure assumes that each bit should be composed of at least 1000 photons. Thus, the total energy of 1000 photons sets the approximate boundary for this region of spontaneous switching. For a wavelength of 850 nm, this implies a minimum switching energy on the order of 0.2 fJ. For the thermal transfer region, it is assumed that for continuous operation, the thermal energy present in the device cannot be removed any faster than 100 W/cm^2 (1 μW/μm^2). There has been some work done to indicate that this value could be as

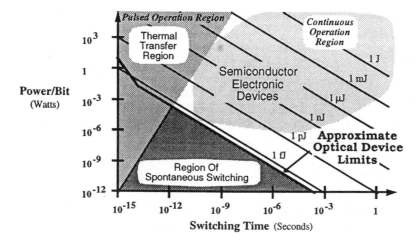

FIGURE 1.33 Fundamental limitations of optical logic devices.

large as 1000 W/cm^2. This region also assumes that there will be no more than an increase of 20°C in the temperature of the device. Devices can be operated in this region using a pulsed rather than continuous mode of operation. Thus, high energy pulses can be used if sufficient time is allowed between pulses to allow the absorbed energy to be removed from the devices. The cloud in Figure 1.33 represents the performance capabilities of current electronic devices. This figure illustrates that optical devices will not be able to switch states orders of magnitude faster than electronic devices when the system is in a continuous rather than a pulsed mode of operation. There are, however, other considerations in the use of photonic switching devices than how fast a single device can change states. Assume that several physically small devices need to be interconnected so that the state information of one device can be used to control the state of another device. To communicate this information, there needs to be some type of interconnection with a large bandwidth that will allow short pulses to travel between the separated devices. Fortunately, the optical domain can support the bandwidth necessary to allow bit rates in excess of 100 Gb/s, which will allow high-speed communication between these individual switching devices. In the electrical domain, the communications bandwidth between two or more devices is limited by the resistance, capacitance, and inductance of the path between the different devices. Therefore, even though photonic devices cannot switch orders of magnitude faster than their electronic counterparts, the communications capability or transmission bandwidth present in the optical domain should allow higher data rate systems than are possible in the electrical domain.

Defining Terms

Light-amplifying optical switch (LAOS): Vertically integrated heterojunction phototransistor and light-emitting diode which has latching thyristor-type current-voltage characteristics.

Liquid-crystal light valve (LCLV): Optical controlled spatial light modulator based on liquid crystals.

Multiple quantum well (MQW): Collection of alternating thin layers of semiconductors (e.g., GaAs and AlGaAs) that results in strong peaks in the absorption spectrum which can be shifted with an applied voltage.

Nonlinear Fabry–Perot (NLFP): Fabry–Perot etalon or interferometer that has an optically nonlinear medium in its cavity.

Optical logic etalon (OLE): Pulsed nonlinear Fabry–Perot etalon that requires two wavelengths (λ_1 = signal, λ_2 = clock).

PLZT/Si: Technology based on conventional silicon electronics using silicon detectors for the device inputs and PLZT modulators for the outputs.

Sagnac logic gate: An all-optical gate based on a Sagnac interferometer. A Sagnac interferometer is composed of two coils of optical fiber arranged so that light from a single source travels clockwise in one and counterclockwise in the other.

SEED technology: Any device based on multiple quantum well (MQW) modulators.

Soliton: Any isolated wave that propagates without dispersion of energy.

Surface-emitting laser logic (CELL): Device that integrates a phototransistor with a low threshold vertical-cavity surface-emitting laser.

References

H.S. Hinton, "Architectural consideration for photonic switching networks," *IEEE Journal on Selected Areas in Communications*, 6, 1988.

M.N. Islam et al., "Ultrafast all-optical fiber-soliton gates," in *Proceedings on Photonic Switching*, vol. 8, H.S. Hinton and J.W. Goodman, Eds., Washington, D.C.: Optical Society of America, 1991, pp. 98–104.

J.L. Jewell et al., "Use of a single nonlinear Fabry–Perot etalon as optical logic gates," *Applied Physics Letters*, 44, 1984.

K. Kasahara et al., "Double heterostructure optoelectronic switch as a dynamic memory with lowpower consumption," *Applied Physics Letters*, 52, 1988.

A.L. Lentine et al., "Symmetric self-electrooptic effect device: Optical set-reset latch, differential logic gate, and differential modulator/detector," *IEEE Journal of Quantum Electronics,* 25, 1989.

J.E. Midwinter, "Digital optics, optical logic or smart interconnect or optical logic," *Physics in Technology,* 19, 1988.

D.A.B. Miller, "Quantum well self-electro-optic effect devices," *Optical and Quantum Electronics,* 22, 1990.

P.W. Smith, "On the physical limits of digital optical switching and logic elements," *Bell System Technical Journal,* 61, 1982.

S.D. Smith, "Optical bistability, photonic logic, and optical computation," *Applied Optics,* 25, 1986.

G.W. Taylor et al., "A new double heterostructure optoelectronic device using molecular beam epitaxy," *Journal of Applied Physics,* 59, 1986.

Further Information

Books which cover this material in more detail include:

H.H. Arsenault, T. Szoplik, and B. Macukow, *Optical Processing and Computing,* New York: Academic Press, 1989.

H.M. Gibbs, *Optical Bistability: Controlling Light with Light,* New York: Academic Press, 1985.

M.N. Islam, *Ultrafast Fiber Switching Devices and Systems,* London: Cambridge University Press, 1992.

A.D. McAulay, *Optical Computer Architectures,* New York: John Wiley, 1991.

B.S. Wherrett and F.A.P. Tooley, *Optical Computing,* Scottish Universities Summer School in Physics, 1989.

2

Memory Devices

W. David Pricer
Pricer Business Services

Peter A. Lee
East of England Development Agency

M. Mansuripur
University of Arizona

2.1 Integrated Circuits (RAM, ROM)

W. David Pricer

The major forms of semiconductor memory, in descending order of present economic importance, are

1. Dynamic Random-Access Memories (DRAMs)
2. Nonvolatile Programmable Memories (PROMs, EEPROMs, Flash, EAROMs, EPROMs)
3. Static Random-Access Memories (SRAMs)
4. Read-Only Memories (ROMs)

Outwardly, DRAMs and SRAMs differ little, except in their relative density and performance. But internally, DRAMs are distinguished from SRAMs, in that no bi-stable electronic circuit maintains the information. Instead, DRAM information is stored "dynamically" as charge on a capacitor. Modern designs feature one field-effect transistor (FET) to access the information for both reading and writing and a thin film capacitor to store information. SRAMs maintain their bi-stability, as long as power is applied, through a cross-coupled pair of inverters within each storage cell. Almost always, two additional transistors access the internal nodes for reading and writing. Most modern cell designs are CMOS, with two P-channel and four N-channel FETs.

Programmable memories operate more like read-only memories in that they can be programmed at least once, and some can be reprogrammed a million times or more. Programming typically takes only milliseconds to complete. A floating-gate FET typically stores the information, which is not indefinitely nonvolatile, in its storage cells, The discharge time constant is of the order of 10 years. ROMs are generally programmed through a custom information mask within the fabrication sequence. As the read-only name implies, information can only be read. The stored information is truly nonvolatile, even when power is removed. This is the densest form of semiconductor storage, but also the least flexible. Other forms of semiconductor memories, such as associative memories and charge-coupled devices, are used rarely.

Dynamic RAMs (DRAMs)

The universally used storage-cell circuit of one transistor and one capacitor has remained unchanged for more than 30 years. Its implementation, however, has undergone much change and refinement. The innovations in physical implementation were driven primarily by the need to maintain a nearly constant value of stored charge while the surface area of the cell has decreased. A nearly fixed value of stored charge is needed to meet two important design goals. For one, the cell has no internal amplification. Once the information is accessed, the stored voltage is attenuated by the much larger bit line capacitance (see Figure 2.1). The resulting signal must be kept larger than the resolution limits of the sensing amplifier. DRAMs in particular are also sensitive to a problem called soft errors. These are typically initiated by atomic events, such as the incidence of a single alpha particle. That alpha particle can cause a spurious signal of 50,000 electrons or more. Modern DRAM designs resolve this problem by constructing the capacitor in space outside the plane of the transistors(see Figure 2.2 for examples). Placing the capacitor

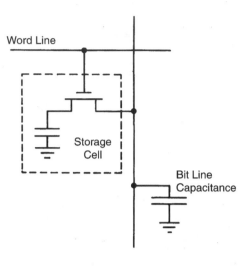

FIGURE 2.1   Cell and bit line capacitance.

in a space unusable for transistor fabrication has allowed great strides in DRAM density, but generally at the expense of fabrication complexity. DRAM chip capacity has historically increased by about a factor of four every three years.

DRAMs operate somewhat slower than SRAMs. This derives directly from the smaller signal available from DRAMs and from certain constraints put on the support circuitry by the DRAM array. DRAMs also need to periodically "refresh" the lost charge from the capacitor. This charge is lost primarily across the semiconductor junctions and must be replenished every few milliseconds. Manufacturers usually provide these "housekeeping" functions with on-chip circuitry.

Signal detection and amplification remain critical areas for good DRAM design. Figure 2.3 illustrates an arrangement called a "folded bit line." This design cancels many noise sources starting in the array and decreases circuit sensitivity to manufacturing process variations. It also yields a high ratio of storage cells per sense amplifier. The signal from an accessed cell is compared to a reference voltage halfway between a stored 1 and a stored 0. This reference voltage is typically supplied by an on-chip generator. Figure 2.3 shows an alternative reference-voltage scheme. Note the presence of dummy cells that store the reference voltage. The stored reference voltage in this case is created by shorting two driven bit lines after one of the storage cells has been written.

Large DRAM-integrated circuit chips frequently provide other useful features [Kalter et al., 1990]. Faster access is provided between certain adjacent addresses, usually along a common word line. Some designs feature on-chip buffer memories, low standby power modes or error-correction circuitry. A few DRAM chips are designed to mesh with the constraints of particular applications, such as providing image support for CRT displays. Some on-chip features are effectively hidden from the user. These may include redundant memory addresses that the manufacturer activates by laser to improve yield.

The largest single market for DRAMs is in microprocessors for personal computers. Modern microprocessors include an on-chip cache memory that is a small SRAM. The system is formulated so that almost all memory accesses are satisfied by information already stored in the on-chip cache memory. When the microprocessor's activity moves across a program task boundary, massive data transfer is needed between

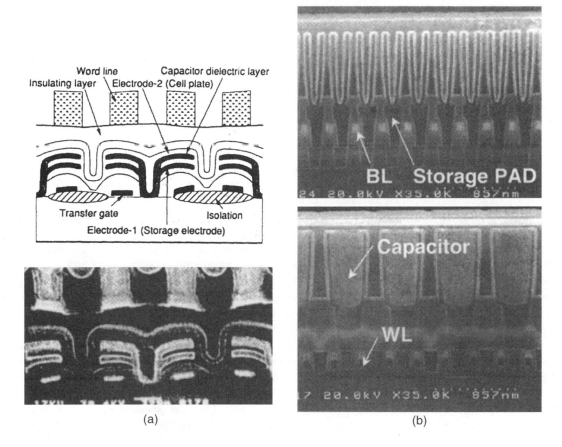

(a)   (b)

FIGURE 2.2 (a) Cross-section of stacked capacitors fabricated above the semiconductor surface of a DRAM integrated circuit. (*Source*: M. Taguchi et al., "A 40-ns 64-b parallel data bus architecture," *IEEE J. Solid State Circuits*, vol. 26, no. 11, p. 1495. Copyright 1991 IEEE. With permission.) (b) Cross-section of V-shaped stacked capacitors. (*Source*: H. Yoon et al., "A 2.5-V, 333-Mb/s/pin 1-Gbit Double-Data-Rate Synchronous DRAM," *IEEE J. Solid-State Circuits*, vol. 34, no. 11, p. 1596, Copyright 1999 IEEE. With permission.)

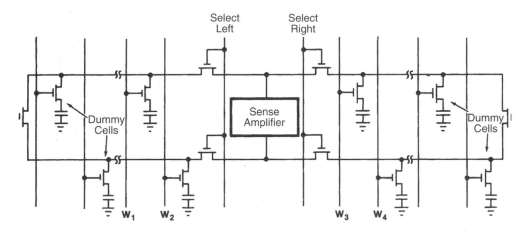

FIGURE 2.3 Folded bit-line array (with dummy cells).

the cache and the DRAM "backing store." Synchronous DRAM (SDRAM) allows the rapid, sequential transfer of large blocks of data between the microprocessor's cache and the much larger-capacity SDRAM without extensive signal "hand-shaking". While SDRAMs do not improve the access time to first data, they greatly improve the bandwidth between microprocessor and DRAM. Recent DRAM developments have focused on optimizing the SDRAM role in microprocessor applications. SDRAM features now include variable-size data block transfer, wrap-around modes that transfer the most urgently needed data in a block first and variable latency time to first data.

Static RAMs (SRAMs)

Compared to DRAMs, the primary advantages of SRAMs are high speed and ease of use. In addition, SRAMs fabricated with CMOS technology exhibit extremely low standby power. This latter feature is used effectively in portable equipment such as pocket calculators. Bipolar SRAMs are generally faster but less dense than FET versions. Figure 2.4 illustrates two cells. SRAM performance is dominated by the speed of the support circuits, leading some manufacturers to design bipolar support circuits to FET arrays.

Bipolar designs frequently incorporate circuit consolidation unavailable in FET technology, such as the multi-emitter cell shown in Figure 2.4(a). Here, one of the two lower emitters is normally forward-biased, turning one inverter on and the other off for bi-stability. The upper emitters can either extract a differential signal or discharge one collector to ground in order to write the cell. The word line is pulsed positive to both read and write the cell.

A few CMOS SRAMs use polysilicon load resistors of very high resistance value in place of the two P-channel transistors shown in Figure 2.4(b). Most SRAMs are full CMOS designs like the one shown. Sometimes, the P-channel transistors are constructed by thin-film techniques and are placed over the N-channel transistors to improve density. When both P- and N-channel transistors are produced in the same plane of the single-crystal semiconductor, the standby current can be extremely low. Typically, the current can be measured in microamps for megabit chips. The low standby current is possible becauseeach cell sources and sinks only that current needed to overcome the actual node leakage within the cell.

Selecting the proper transconductance of each transistor is an important design focus. The accessing transistors should be large enough to extract a large read signal but insufficiently large to disturb the stored information. During the write operation, these same transistors must be able to override the current drive of at least one of the internal CMOS inverters.

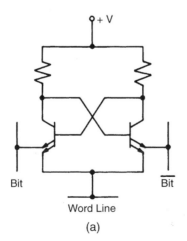

(a)

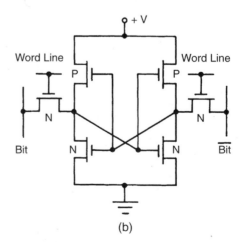

(b)

FIGURE 2.4   (a) Bipolar SRAM cell. (b) CMOS SRAM cell.

The superior performance of SRAMs comes from their larger signal and the lack of need to refresh the stored information as with a DRAM. As a result, SRAMs need fewer sense amplifiers. These amplifiers are not forced to match the array's cell pitch. SRAM design engineers have exploited this freedom to create higher-performance sense amplifiers.

Practical SRAM designs routinely achieve access times of a few nanoseconds [Kato, 1992]. The cycle time typically equals access time and, in at least one pipeline design, cycle time is actually less than access time.

SRAM integrated circuit chips have fewer special on-chip features than DRAM chips, primarily because no special performance enhancements are needed. Many ASICs (application-specific integrated circuits) use on-chip SRAMs because of their low power requirement and ease of use. Modern microprocessors include one or more on-chip, cache, SRAM memories, providing a high-speed link between processor and memory.

Nonvolatile Programmable Memories

Some nonvolatile memories can be programmed only once. These memories have arrays of diodes or transistors with fuses or antifuses in series with each semiconductor cross point. Aluminum, titanium, tungsten, platinum silicide and polysilicon have all been used successfully for fuse technology (see Figure 2.5).

Most nonvolatile cells rely on trapped charge stored on a floating gate in an FET [Atsumi et al., 2000]. These cells can be rewritten many times. The trapped charge is subject to very long-term leakage, of the order of 10 years. The number of times the cell may be rewritten is limited by programming stress-induced degradation of the dielectric. Charge reaches the floating gate either by tunneling or by avalanche injection from a region near the drain. Both phenomena are induced by overvoltage conditions and hence the degradation after repeated erase/write cycles. Commercially available chips typically promise 100 to 100,000 write cycles. Erasure of charge from the floating gate may be by tunneling or by exposure to ultraviolet light. Asperities on the polysilicon gate and silicon-rich oxide have both been shown to enhance charging and discharging of the gate. The cell's nomenclature is not entirely consistent throughout the industry. However, EPROM generally describes cells that are electronically written but UV-erased. EEPROM can describe cells that are electronically written and erased.

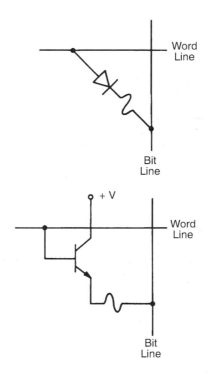

FIGURE 2.5 PROM cells.

Cells can be designed with either one or two transistors. Where two transistors are used, the second transistor is normally a conventional enhancement mode transistor (see Figure 2.6). The second transistor minimizes disturbance of unselected cells. It also removes some constraints on the writing limits of the programmable transistor, which in one state may be depletion mode. The two transistors in series then assume the threshold of the second, or enhancement, transistor, or a very high threshold as determined by the programmable transistor. Some designs are so well-integrated that the features of the two transistors are merged.

Flash EEPROMs describe a family of single-transistor cell EEPROMs. Cell sizes are about half that of two-transistor EEPROMs, an important economic consideration. Care must be taken to prevent these cells

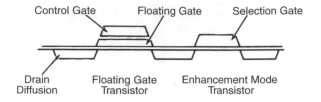

FIGURE 2.6　Cross-section of two-transistor EEPROM cell.

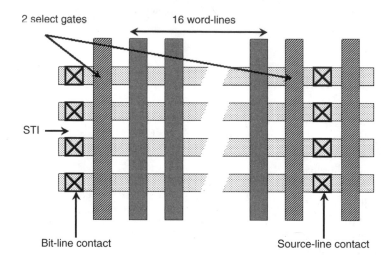

FIGURE 2.7　Top view of a NAND Flash memory segment. (*Source*: K. Imamiya et al., "A 130-mm2, 256-Mbit NAND Flash with Shallow Trench Isolation Technology," *IEEE J. Solid-State Circ* vol. 34, no. 11, p. 1537. Copyright 1999 IEEE. With permission.)

from being programmed into depletion mode. An array of many depletion-mode cells would confound the read operation by generating multiple signal paths. The cells can be programmed to enhancement-only thresholds through a sequence of partial-program and monitor subcycles, performed until the threshold complies with specification limits. Flash EEPROMs require the bulk erasure of large portions of the array.

A very dense form of Flash EEPROM, called NAND Flash, is achieved by wiring the cells in series. Figure 2.7 depicts a top view of this arrangement. A selected cell within a portion of this array is accessed by applying positive selection signals to all unselected word lines in a selected series. The selected cell's threshold cell can then be sensed. note the high order decoded addressing applied to the outside transistors. These are not floating-gate transistors.

The Flash EEPROM market has grown rapidly in recent years. Electronic cameras, camera phones, many hand-held consumer products and memory sticks have fueled this market.

NVRAM has become a generic term encompassing many forms of nonvolatility. These include SRAM or DRAM with nonvolatile circuit elements, the various programmable RAMs just described and some emerging technologies with combined semiconductor and magnetic effects.

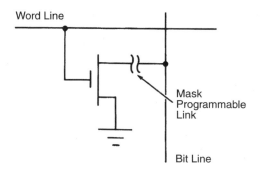

FIGURE 2.8　ROM cell.

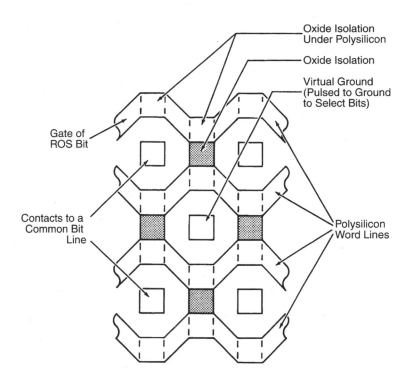

FIGURE 2.9 Layout of ROS X array.

Read-Only Memories (ROMs)

ROM is the only form of semiconductor storage that is permanently nonvolatile. Information is retained without applied power, and there is no information loss as in EEPROMs. It is also the densest form of semiconductor storage. However, ROMs are used less than RAMs or EEPROMs. ROMs must be personalized by a mask in the fabrication process. This method is cumbersome and expensive unless many identical parts are made. Also, much of the "permanent" information is not really permanent and must be occasionally updated.

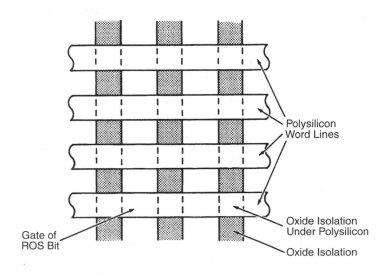

FIGURE 2.10 Layout of ROS AND array.

ROM cells can be formed as diodes or transistors at every intersection of the word and bit lines of a ROM array (see Figure 2.8). one of the masks in the chip-fabrication process programs which of these devices will be active. Clever layout and circuit techniques may provide further density. Two suchtechniques are illustrated in Figure 2.9 and Figure 2.10. The X array shares bit and virtual ground lines. The AND array places many ROM cells in series. This arrangement is similar to that of NAND Flash memory. Each of these series AND ROM cells is an enhancement- or depletion-channel FET. Sensing is performed by applying positive selection signals to the gates of all series cells except the gate to be interrogated. The current will flow through all series channels only if the interrogated channel is in depletion-mode [Kawagoe and Tsuji, 1976].

ROM applications include look-up tables, machine-level instruction code for computers and small arrays to perform logic. (See Sections 3.4, 3.5 and 3.7 of this volume for more on programmable logic and arrays).

Defining Terms

Antifuse: A fuse-like device that becomes low-impedance when activated.

Application-specific integrated circuits (ASICs): Integrated circuits specifically designed for a particular application.

Avalanche injection: A physical phenomenon where electrons highly energized in avalanche current at a semiconductor junction can penetrate into a dielectric.

Depletion mode: An FET that is on when zero-volts bias is applied from gate to source.

Enhancement mode: An FET that is off when zerovolts bias is applied from gate to source.

Polysilicon: Silicon in polycrystalline form.

Tunneling: A physical phenomenon where an electron can move instantly through a thin dielectric.

References

H. Kalter et al., "A 50 nsec 167 Mb DRAM with 10 nsec data rate and on-chip ECC," *IEEE J. Solid-State Circ.*, vol. SC 25, no. 5, 1990.

H. Yoon et al., "A 2.5-V, 333-Mb/s/pin, 1-Gbit, double-data-rate synchronous DRAM," *IEEE J. Solid-State Circ.*, vol. 34, no. 11, 1999.

S-B Lee et al., "A 1.6 Gb/s/pin double-data-rate SDRAM with wave-pipelined CAS latency control," *Dig. Tech. Pap. ISSCC*, 47, 2004.

H. Kato, "A 9 nsec 4 Mb BiCMOS SRAM with 3.3 V operation," *Dig. Tech. Pap. ISSCC*, 35, 1992.

H. Kawagoe and N. Tsuji, "Minimum size ROM structure compatible with silicon-gate E/D MOS LSI," *IEEE J. Solid State Circ.*, vol. SC 11, no. 2, 1976.

S. Atsumi et al., "A channel-erasing 1.8-V-only 32-Mb NOR flash EEPROM with a bitline direct sensing scheme," *IEEE J. Solid-State Circ.*, vol. 35, no. 11, 2000.

K. Imamiya et al., "A 130-mm 2 256-Mbit NAND flash with shallow trench isolation technology," *IEEE J. Solid-State Circ.*, vol. 34, no. 11, 1999.

Further Information

W. Donoghue et al., "A 256K H CMOS ROM using a four state cell approach," IEEE J. Solid-State Circ., vol. SC20, no. 2, 1985.

D. Frohmann-Bentchkowsky, "A fully decoded 2048 bit electronically programmable MOS-ROM," Digest of Technical Papers ISSCC, vol. 14, 1971.

F. Masuoka, "Are you ready of the next generation dynamic RAM chips," *IEEE Spectrum Magazine*, vol. 27, no. 11, 1990.

R.D. Pashley and S.K. Lai, "Flash memories: The best of two worlds," *IEEE Spectrum Magazine*, vol. 26, no. 12, 1989.

B. Keeth, and J. Baker, "DRAM Circuit Design, A Tutorial" IEEE Press, ISBN 0–7803–6014–1, 2001.

2.2 Magnetic Tape[1]

Peter A. Lee

Computers depend on memory to execute programs and to store program code and data. They also need access to stored program code and data in a **nonvolatile memory** (i.e., a form in which the information is not lost when the power is removed from the computer system). Different types of memory have been developed for different tasks. This memory can be categorized according to its price per bit, **access time**, and other parameters. Table 2.1 shows a typical hierarchy for memory which places the smallest and fastest memory at the top in level 0 and in general the largest, slowest, and cheapest at the bottom in level 4 [Ciminiera and Valenzano, 1987]. Auxiliary (secondary or mass) memory of level 4 forms the large storage capacity for program and code that are not currently required by the CPU. This is usually nonvolatile and is at a low cost per bit. Computer **magnetic tape** falls within this category and is the subject of this section.

A Brief Historical Review

Probably the first recorded storage device, developed by Schickard in 1623, used mechanical positions of cogs and gears to work a semi-automatic calculator. Then came Pascal's calculating machine based on 10 digits per wheel. In 1812 punched cards were used in weaving looms to store patterns for woven material. Since that time there have been many mechanical and, latterly, electromechanical devices developed for memory and storage.

In 1948 at Manchester University in England the cathode ray tube (Williams) and the magnetic drum were developed. These consisted of 1024 bits and 1280 bits and a magnetic drum capacity of 120K bits. Cambridge University developed the mercury delay line in 1949, which represented the first fully operational delay line memory, consisting of 576 bits per tube with a total capacity of 18K bits and a circulation time of 1.1 ms.

The first commercial computer with a magnetic tape system was introduced in 1951. The UNIVAC I had a magnetic tape system of 1.44M bits on 150 feet of tape and was capable of storing 128 characters per inch. The tape could be read at a rate of 100 ips. Optical memories are now available as very fast storage devices and will replace magnetic storage in the next few years. At present these devices are expensive although it is envisaged that optical disks with large silicon caches will be the storage arrangement of the future where computer systems utilizing CAD software and image processing can take advantage of the large storage capacities with fast access times. In the future, semiconductor memories are likely to continue their advancing trend.

Introduction

Today's microprocessors are capable of addressing up to 16 Mbytes of main memory. To take advantage of this large capacity, it is usual to have several programs residing in memory at the same time. With intelligent memory management units (MMUs), the programs can be swopped in and out of the main memory to the auxiliary memory when required. For the system to keep pace with this program swopping, it must have a fast auxiliary memory to write to. In the past, most auxiliary systems like magnetic tape and disks have had slow access times, and this has meant that expensive systems have evolved to cater for this requirement. Now that auxiliary memory has improved, and access times are fast and the memory cheap, computer systems have been developed that provide memory swopping with large nonvolatile storage systems. Although the basic technology has not changed over the last 20 years, new materials and different approaches have meant that a new form of auxiliary memory has been brought to the market at a very cheap cost.

---

[1]Based on P.A. Lee, "Memory subsystems," in *Digital Systems Reference Book*, B. Holdworth and G.R. Martins, Eds., Oxford: Butterworth-Heinemann, 1991, chap. 2.6. With permission.

TABLE 2.1   Memory Hierarchy

	Data	Code	MMU
Level 0	CPU register	Instruction registers	MMU registers
Level 1	Data cache	Instruction cache	MMU memory
Level 2		On-board cache	
Level 3		Main memory	
Level 4		Auxiliary memory	

Source: P.A. Lee, "Memory subsystems," in *Digital Systems Reference Book*, B. Holdsworth and G.R. Martin, Eds., Oxford: Butterworth-Heinemann, 1991, p. 2.6/3. With permission.

Magnetic Tape

Magnetic tape currently provides the cheapest form of storage for large quantities of computer data in a nonvolatile form. The tape is arranged on a reel and has several different packaging styles. It is made from a polyester transportation layer with a deposited layer of oxide having a property similar to a ferrite material with a large hysteresis. Magnetic tape is packaged either in a cartridge, on a reel, or in a cassette. The magnetic cartridge is manufactured in several tape lengths and cartridge sizes capable of storing up to 2 G (giga) bytes of data. These can be purchased in many popular preformatted styles.

The magnetic tape reel is usually 1/2 inch or 1 inch wide and has lengths of 600, 1200, and 2400 feet. Most reels can store data at rates from 800 bits per inch (bpi) up to 6250 bpi. The reel-to-reel magnetic tape reader is generally bulkier and more expensive than the cartridge readers due to the complicated pneumatic drive mechanisms, but it provides a large data storage capacity with high access speeds [Wiehler, 1974]. An example of a typical magnetic tape drive with the reel-to-reel arrangement is shown in Figure 2.11.

A cheap storage medium is the magnetic cassette. Based on the audio cassette, this uses the normal audio cassette recorder for reading and writing data via the standard Kansas City interface through a serial computer

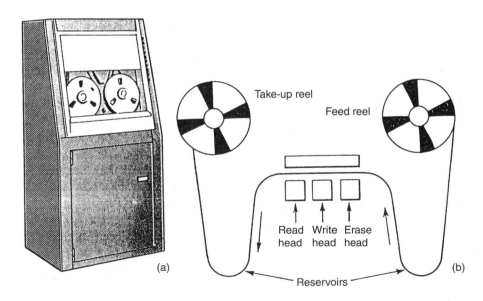

FIGURE 2.11   (a) Magnetic tape drive. (b) Magnetic tape reel arrangement. (*Source:* K. London, *Introduction to Computers*, London: Faber and Faber, 1986, p. 141. With permission.)

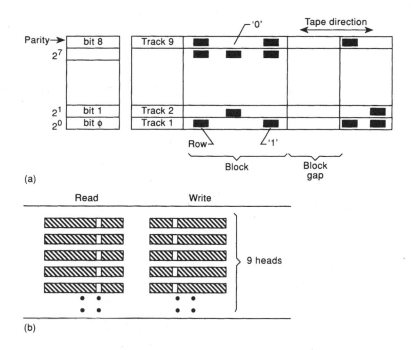

FIGURE 2.12 Magnetic tape format. (*Source:* P.A. Lee, "Memory subsystems," in *Digital Systems Reference Book,* B. Holdsworth and G.R. Martin, Eds., Oxford: Butterworth-Heinemann, 1991, p. 2.6/11. With permission.)

I/O line. A logic data "1" is recorded by a high frequency and a logic data "0" by a lower frequency. High-density cassettes can store up to 60 Mbytes of data on each tape and are popular with the computer games market as a cheap storage medium for program distribution.

Both reel-to-reel and cartridge tapes are generally organised by using nine separate tracks across the tape as shown in Figure 2.12(a).

Each track has its own read and write head operated independently from other tracks [see Figure 2.12(b)]. Tracks 1 to 8 are used for data and track nine for the parity bit. Data is written on the tape in rows of magnetized islands, using for example EBCDIC (Extended Binary Coded Decimal Interchange Code).

Each read/write head is shaped from a **ferromagnetic material** with an air gap 1 μm wide as seen in Figure 2.13. The writing head is concerned with converting an electrical pulse into a magnetic state and can be magnetized in one of two directions. This is done by passing a current through the magnetic coil which sets up

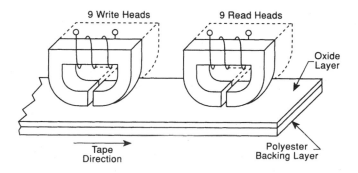

FIGURE 2.13 Read/write head layout. (*Source:* P.A. Lee, "Memory subsystems," in *Digital Systems Reference Book,* B. Holdsworth and G.R. Martin, Eds., Oxford: Butterworth-Heinemann, 1991, p. 2.6/12. With permission.)

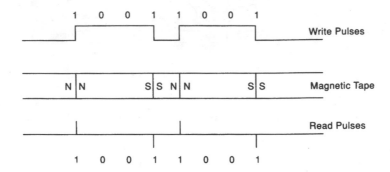

FIGURE 2.14 Write and read pulses on magnetic tape. (*Source:* P.A. Lee, "Memory subsystems," in *Digital Systems Reference Book,* B. Holdsworth and G.R. Martin, Eds., Oxford: Butterworth-Heinemann, 1991, p. 2.6/12. With permission.)

a leakage field across the 1-μm gap. When the current is reversed the field across the gap is changed, reversing the polarity of the magnetic field on the tape. The head magnetizes the passing magnetic tape recording the state of the magnetic field in the air gap. A logic 1 is recorded as a change in polarity on the tape, and a logic 0 is recorded as no change in polarity, as seen in Figure 2.14. Reading the magnetic tape states from the tape and converting them to electrical signals is done by the read head. The bit sequences in Figure 2.14 show the change in magnetic states on the tape. When the tape is passed over the read head, it induces a voltage into the magnetic coil which is converted to digital levels to retrieve the original data.

Tape Format

Information is stored on magnetic tape in the form of a coherent sequence of rows forming a block. This usually corresponds to a page of computer memory and is the minimum amount of data written to or read from magnetic tape with each program statement. Each block of data is separated by a block gap which is approximately 15 mm long and has no data stored in it. This is shown in Figure 2.15.

Block gaps are used to allow the tape to accelerate to its operational speed and for the tape to decelerate when stopping at the end of a block. Block gaps use up to 50% of the tape space available for recording, although this may be reduced by making the block sizes larger but has the disadvantage of requiring larger memory buffers to accommodate the data.

A number of blocks make up a file identified by a tape file marker which is written to the tape by the tape controller. The entire length of tape is enclosed between the beginning and end of tape markers.

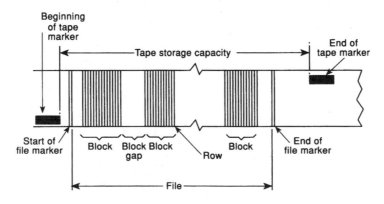

FIGURE 2.15 Magnetic tape format. (*Source:* P.A. Lee, "Memory subsystems," in *Digital Systems Reference Book,* B. Holdsworth and G.R. Martin, Eds., Oxford: Butterworth-Heinemann, 1991, p. 2.6/12. With permission.)

These normally consist of a photosensitive material that triggers sensors on the read/write heads. When a new tape is loaded, it normally advances to the beginning of a tape marker and then it is ready for access by the CPU. The end of tape marker is used to prevent the tape from running off the end of the tape spool and indicates the limit of the storage length.

Recording Modes

Several recording modes are used with the express objective of storing data at the highest density and with the greatest reliability of noncorruption of retrieved data. Two popular but contrasting modes are the *non-return-to-zero* (NRZ) and *phase encoding* (PE) modes. These are incompatible although some magnetic tape drives have detectors to sense the mode and operate in a bimodal way. The NRZ technique is shown in Figure 2.14, where only the 1 bit is displayed by a reversal of magnetization on the tape. The magnetic polarity remains unchanged for logic 0. An external clock track is also required for this mode because a pulse is not always generated for each row of data on the tape.

The PE technique allows both the 0 and 1 states to be displayed by changes of magnetization. A 1 bit is given by a north-to-north pole on the tape, and a 0 bit is given by a south-to-south pole on the tape. PE provides approximately double the recording density and processor speed of NRZ. PE tapes carry an identification mark called a *burst*, which consists of successive magnetization changes at the beginning of track 4. This allows the tape drive to recognize the tape mode and configure itself accordingly.

Defining Terms

Access time: The cycle time for the computer store to present information to the CPU. Access times vary from less than 40 ns for level 0 register storage up to tens of seconds for magnetic tape storage.

Auxiliary (secondary, mass, or backing) storage: Computer stores which have a capacity to store enormous amounts of information in a *nonvolatile* form. This type of memory has an access time usually greater than main memory and consists of magnetic tape drives, magnetic disk stores, and optical disk stores.

Ferromagnetic material: Materials that exhibit high magnetic properties. These include metals such as cobalt, iron, and some alloys.

Magnetic tape: A polyester film sheet coated with a *ferromagnetic* powder, which is used extensively in auxiliary memory. It is produced on a reel, in a cassette, or in a cartridge transportation medium.

Nonvolatile memory: The class of computer memory that retains its stored information when the power supply is cut off. It includes magnetic tape, magnetic disks, flash memory, and most types of ROM.

References

L. Ciminiera and A. Valenzano, *Advanced Microprocessor Architectures*, Reading, Mass.: Addison-Wesley, 1987.

B. Holdsworth and G. Martin, Eds, *Digital Systems Reference Book*, Oxford: Butterworth-Heinemann, 1991, pp 2.6/1–2.6/11.

R. Hyde, "Overview of memory management," *Byte*, pp. 219–225, April 1988.

J. Isailovíc, *Video Disc and Optical Memory Systems*, Englewood Cliffs, N.J.: Prentice-Hall, 1985.

K. London, *Introduction to Computers*, London: Faber and Faber Press, 1986, p. 141.

M. Mano, *Computer Systems Architecture*, Englewood Cliffs, N.J.: Prentice-Hall, 1982.

R. Matick, *Computer Storage Systems & Technology*, New York: John Wiley, 1977.

A. Tanenbaum, *Structured Computer Organisation*, Englewood Cliffs, N.J.: Prentice-Hall, 1990.

G. Wiehler, *Magnetic Peripheral Data Storage*, Heydon & Son, 1974.

Further Information

The *IEEE Transactions on Magnetics* is available from the IEEE Service Center, Customer Service Department, 445 Hoes Lane, Piscataway, NJ 08855–1331; 800–678-IEEE (outside the USA: 908–981–0060). An IEEE-sponsored Conference on Magnetism and Magnetic Materials was held in December 1992. The British Tape Industry Association (BTIA) has a computer media committee, and further information on standards, etc. can be obtained from British Tape Industry Association, Carolyn House, 22–26 Dingwall Road, Croydon CR0 9XF, England. The equivalent American Association also provides information on computer tape and can be contacted at International Tape Manufacturers' Association, 505 Eighth Avenue, New York, NY 10018.

2.3   Magneto-Optical Disk Data Storage

M. Mansuripur

Since the early 1940s, magnetic recording has been the mainstay of electronic information storage worldwide. Audio tapes provided the first major application for the storage of information on magnetic media. Magnetic tape has been used extensively in consumer products such as audio tapes and video cassette recorders (VCRs); it has also found application in backup/archival storage of computer files, satellite images, medical records, etc. Large volumetric capacity and low cost are the hallmarks of tape data storage, although sequential access to the recorded information is perhaps the main drawback of this technology. Magnetic hard disk drives have been used as mass storage devices in the computer industry ever since their inception in 1957. With an areal density that has doubled roughly every other year, hard disks have been and remain the medium of choice for secondary storage in computers.[1] Another magnetic data storage device, the floppy disk, has been successful in areas where compactness, removability, and fairly rapid access to the recorded information have been of prime concern. In addition to providing backup and safe storage, inexpensive floppies with their moderate capacities (2 Mbyte on a 3.5-in. diameter platter is typical nowadays) and reasonable transfer rates have provided the crucial function of file/data transfer between isolated machines. All in all, it has been a great half-century of progress and market dominance for magnetic recording devices, which are only now beginning to face a potentially serious challenge from the technology of optical recording.

Like magnetic recording, a major application area for optical data storage systems is the secondary storage of information for computers and computerized systems. Like the high-end magnetic media, optical disks can provide recording densities in the range of 10^7 bits/cm^2 and beyond. The added advantage of optical recording is that, like floppies, these disks can be removed from the drive and stored on the shelf. Thus the functions of the hard disk (i.e., high capacity, high data transfer rate, rapid access) may be combined with those of the floppy (i.e., backup storage, removable media) in a single optical disk drive. Applications of optical recording are not confined to computer data storage. The enormously successful audio **compact disk (CD)**, which was introduced in 1983 and has since become the de facto standard of the music industry, is but one example of the tremendous potentials of the optical technology.

A strength of optical recording is that, unlike its magnetic counterpart, it can support read-only, write-once, and erasable/rewritable modes of data storage. Consider, for example, the technology of optical audio/video disks. Here the information is recorded on a master disk which is then used as a stamper to transfer the embossed patterns to a plastic substrate for rapid, accurate, and inexpensive reproduction. The same process is employed in the mass production of read-only files (CD-ROM, O-ROM) which are now being used to distribute software, catalogues, and other large databases. Or consider the write-once read-many (WORM)

---

[1] At the time of this writing, achievable densities on hard disks are in the range of 10^7 bits/cm^2. Random access to arbitrary blocks of data in these devices can take on the order of 10 ms, and individual read/write heads can transfer data at the rate of several megabits per second.

technology, where one can permanently store massive amounts of information on a given medium and have rapid, random access to them afterwards. The optical drive can be designed to handle read-only, WORM, and erasable media all in one unit, thus combining their useful features without sacrificing performance and ease of use or occupying too much space. What is more, the media can contain regions with prerecorded information as well as regions for read/write/erase operations, both on the same platter. These possibilities open new vistas and offer opportunities for applications that have heretofore been unthinkable; the interactive video disk is perhaps a good example of such applications.

In this article we will lay out the conceptual basis for optical data storage systems; the emphasis will be on disk technology in general and magneto optical disk in particular. The first section is devoted to a discussion of some elementary aspects of disk data storage including the concept of track and definition of the access time. The second section describes the basic elements of the optical path and its functions; included are the properties of the semiconductor laser diode, characteristics of the beamshaping optics, and certain features of the focusing objective lens. Because of the limited depth of focus of the objective and the eccentricity of tracks, optical disk systems must have a closed-loop feedback mechanism for maintaining the focused spot on the right track. These mechanisms are described in the third and fourth sections for automatic focusing and automatic track following, respectively. The physical process of thermomagnetic recording in magneto-optic (MO) media is described next, followed by a discussion of the MO readout process in the sixth section. The final section describes the properties of the MO media.

Preliminaries and Basic Definitions

A disk, whether magnetic or optical, consists of a number of **tracks** along which the information is recorded. These tracks may be concentric rings of a certain width, W_t, as shown in Figure 2.16. Neighboring tracks may be separated from each other by a guard band whose width we shall denote by W_g. In the least sophisticated recording scheme imaginable, marks of length Δ_0 are recorded along these tracks. Now, if each mark can be in either one of two states, present or absent, it may be associated with a binary digit, 0 or 1. When the entire disk surface of radius R is covered with such marks, its capacity C_0 will be

$$C_0 = \frac{\pi R^2}{(W_t + W_g)\Delta_0} \qquad \text{bits per surface} \qquad (2.1)$$

Consider the parameter values typical of current optical disk technology: $R = 67$ mm corresponding to 5.25-in. diameter platters, $\Delta_0 = 0.5$ μm which is roughly determined by the wavelength of the read/write laser diodes,

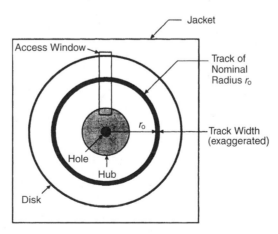

FIGURE 2.16 Physical appearance and general features of an optical disk. The read/write head gains access to the disk through a window in the jacket; the jacket itself is for protection purposes only. The hub is the mechanical interface with the drive for mounting and centering the disk on the spindle. The track shown at radius r_0 is of the concentric-ring type.

and $W_t + W_g = 1$ μm for the track pitch. The disk capacity will then be around 28×10^9 bits, or 3.5 gigabytes. This is a reasonable estimate and one that is fairly close to reality, despite the many simplifying assumptions made in its derivation. In the following paragraphs we examine some of these assumptions in more detail.

The disk was assumed to be fully covered with information-carrying marks. This is generally not the case in practice. Consider a disk rotating at $\mathcal{N}$ revolutions per second (rps). For reasons to be clarified later, this rotational speed should remain constant during the disk operation. Let the electronic circuitry have a fixed clock duration T_c. Then only pulses of length T_c (or an integer multiple thereof) may be used for writing. Now, a mark written along a track of radius r, with a pulse-width equal to T_c, will have length ℓ where

$$\ell = 2\pi \mathcal{N} r T_c \tag{2.2}$$

Thus for a given rotational speed $\mathcal{N}$ and a fixed clock cycle T_c, the minimum mark length ℓ, is a linear function of track radius r, and ℓ decreases toward zero as r approaches zero. One must, therefore, pick a minimum usable track radius, r_{min}, where the spatial extent of the recorded marks is always greater than the minimum allowed mark length, Δ_0. Equation (2.3) yields

$$r_{min} = \frac{\Delta_0}{2\pi \mathcal{N} T_c} \tag{2.3}$$

One may also define a maximum usable track radius r_{max}, although for present purposes $r_{max} = R$ is a perfectly good choice. The region of the disk used for data storage is thus confined to the area between r_{min} and r_{max}. The total number N of tracks in this region is given by

$$N = \frac{r_{max} - r_{min}}{W_t + W_g} \tag{2.4}$$

The number of marks on any given track in this scheme is independent of the track radius; in fact, the number is the same for all tracks, since the period of revolution of the disk and the clock cycle uniquely determine the total number of marks on any individual track. Multiplying the number of usable tracks N with the capacity per track, we obtain for the usable disk capacity

$$C = \frac{N}{\mathcal{N} T_c} \tag{2.5}$$

Replacing for N from Equation (2.4) and for $\mathcal{N} T_c$ from Equation (2.3), we find,

$$C = \frac{2\pi r_{min}(r_{max} - r_{min})}{(W_t + W_g)\Delta_0} \tag{2.6}$$

If the capacity C in Equation (2.6) is considered a function of r_{min} with the remaining parameters held constant, it is not difficult to show that maximum capacity is achieved when

$$r_{min} = 1/2 \, r_{max} \tag{2.7}$$

With this optimum r_{min}, the value of C in Equation (2.6) is only half that of C_0 in Equation (2.1). In other words, the estimate of 3.5 gigabyte per side for 5.25-in. disks seems to have been optimistic by a factor of two.

One scheme often proposed to enhance the capacity entails the use of multiple zones, where either the rotation speed $\mathcal{N}$ or the clock period T_c is allowed to vary from one zone to the next. In general, zoning schemes can reduce the minimum usable track radius below that given by Equation (2.7). More importantly, however, they allow tracks with larger radii to store more data than tracks with smaller radii. The capacity of the zoned disk is somewhere between C of Equation (2.6) and C_0 of Equation (2.1), the exact value depending on the number of zones implemented.

A fraction of the disk surface area is usually reserved for **preformat** information and cannot be used for data storage. Also, prior to recording, additional bits are generally added to the data for **error correction coding** and other housekeeping chores. These constitute a certain amount of overhead on the user data and must be allowed for in determining the capacity. A good rule of thumb is that overhead consumes approximately 20% of the raw capacity of an optical disk, although the exact number may vary among the systems in use. Substrate defects and film contaminants during the deposition process can create bad **sectors** on the disk. These are typically identified during the certification process and are marked for elimination from the sector directory. Needless to say, bad sectors must be discounted when evaluating the capacity.

Modulation codes may be used to enhance the capacity beyond what has been described so far. Modulation coding does not modify the minimum mark length of Δ_0, but frees the longer marks from the constraint of being integer multiples of Δ_0. The use of this type of code results in more efficient data storage and an effective number of bits per Δ_0 that is greater than unity. For example, the popular (2, 7) modulation code has an effective bit density of 1.5 bits per Δ_0. This or any other modulation code can increase the disk capacity beyond the estimate of Equation (2.6).

The Concept of Track

The information on magnetic and optical disks is recorded along tracks. Typically, a track is a narrow annulus at some distance r from the disk center. The width of the annulus is denoted by W_t, while the width of the guard band, if any, between adjacent tracks is denoted by W_g. The track pitch is the center-to-center distance between neighboring tracks and is therefore equal to $W_t + W_g$. A major difference between the magnetic floppy disk, the magnetic hard disk, and the optical disk is that their respective track pitches are presently of the order of 100, 10, and 1 μm. Tracks may be fictitious entities, in the sense that no independent existence outside the pattern of recorded marks may be ascribed to them. This is the case, for example, with the audio compact disk format where prerecorded marks simply define their own tracks and help guide the laser beam during readout. In the other extreme are tracks that are physically engraved on the disk surface before any data is ever recorded. Examples of this type of track are provided by pregrooved WORM and magneto-optical disks.

It is generally desired to keep the read/write head stationary while the disk spins and a given track is being read from or written onto. Thus, in an ideal situation, not only should the track be perfectly circular, but also the disk must be precisely centered on the spindle axis. In practical systems, however, tracks are neither precisely circular, nor are they concentric with the spindle axis. These eccentricity problems are solved in low-performance floppy drives by making tracks wide enough to provide tolerance for misregistrations and misalignments. Thus the head moves blindly to a radius where the track center is nominally expected to be and stays put until the reading or writing is over. By making the head narrower than the track pitch, the track center is allowed to wobble around its nominal position without significantly degrading the performance during the read/write operation. This kind of wobble, however, is unacceptable in optical disk systems, which have a very narrow track, about the same size as the focused beam spot. In a typical situation arising in practice, the eccentricity of a given track may be as much as ± 50 μm while the track pitch is only about 1 μm, thus requiring active track-following procedures.

One method of defining tracks on an optical disk is by means of pregrooves that are either etched, stamped, or molded onto the substrate. In **grooved media of optical storage**, the space between neighboring grooves is the so-called land [see Figure 2.17(a)]. Data may be written in the grooves with the land acting as a guard band. Alternatively, the land regions may be used for recording while the grooves separate adjacent tracks. The groove depth is optimized for generating an optical signal sensitive to the radial position of the read/write laser beam. For the push-pull method of track-error detection the groove depth is in the neighborhood of $\lambda/8$, where λ is the wavelength of the laser beam.

In digital data storage applications, each track is divided into small segments or sectors, intended for the storage of a single block of data (typically either 512 or 1024 bytes). The physical length of a sector is thus a few millimeters. Each sector is preceded by header information such as the identity of the sector, identity of the corresponding track, synchronization marks, etc. The header information may be preformatted onto the substrate, or it may be written on the storage layer prior to shipping the disk. Pregrooved tracks may be "carved" on the optical disk either as concentric rings or as a single continuous spiral. There are certain advantages to each format. A spiral track can contain a succession of sectors without interruption, whereas concentric rings may each end up with some empty space that is smaller than the required length for a sector. Also, large files may be written onto (and read from) spiral tracks without jumping to the next track, which occurs when concentric tracks are used. On the other hand, multiple-path operations such as write-and-verify or erase-and-write, which require two paths each for a given sector, or still-frame video are more conveniently handled on concentric-ring tracks.

Another track format used in practice is based on the sampled-servo concept. Here the tracks are identified by occasional marks placed permanently on the substrate at regular intervals, as shown in Figure 2.17. Details of track following by the sampled-servo scheme will follow shortly; suffice it to say at this point that servo marks help the system identify the position of the focused spot relative to the track center. Once the position is determined it is fairly simple to steer the beam and adjust its position.

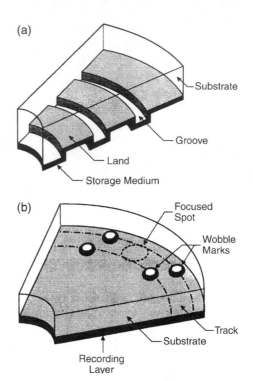

FIGURE 2.17   (a) Lands and grooves in an optical disk. The substrate is transparent, and the laser beam must pass through it before reaching the storage medium. (b) Sampled-servo marks in an optical disk. These marks which are offset from the track-center provide information regarding the position of focused spot.

Disk Rotation Speed

When a disk rotates at a constant angular velocity ω, a track of radius r moves with the constant linear velocity $V = r\omega$. Ideally, one would like to have the same linear velocity for all the tracks, but this is impractical except in a limited number of situations. For instance, when the desired mode of access to the various tracks is sequential, such as in audio and video disk applications, it is possible to place the head in the beginning at the inner radius and move outward from the center thereafter while continuously decreasing the angular velocity. By keeping the product of r and ω constant, one can thus achieve constant linear velocity for all the tracks.[1] Sequential access mode, however, is the exception rather than the norm in data storage systems. In most applications, the tracks are accessed randomly with such rapidity that it becomes impossible to adjust the rotation speed for constant linear velocity. Under these circumstances, the angular velocity is best kept constant during the normal operation of the disk. Typical rotation speeds are 1200 and 1800 rpm for slower drives and 3600 rpm for the high data rate systems. Higher rotation rates (5000 rpm and beyond) are certainly feasible and will likely appear in future storage devices.

---

[1] In compact disk players the linear velocity is kept constant at 1.2 m/s. The starting position of the head is at the inner radius $r_{min} = 25$ mm, where the disk spins at 460 rpm. The spiral track ends at the outer radius $r_{max} = 58$ mm, where the disk's angular velocity is 200 rpm.

Access Time

The direct-access storage device or DASD, used in computer systems for the mass storage of digital information, is a disk drive capable of storing large quantities of data and accessing blocks of this data rapidly and in arbitrary order. In read/write operations it is often necessary to move the head to new locations in search of sectors containing specific data items. Such relocations are usually time-consuming and can become the factor that limits performance in certain applications. The access time τ_a is defined as the average time spent in going from one randomly selected spot on the disk to another. τ_a can be considered the sum of a seek time, τ_s, which is the average time needed to acquire the target track, and a latency, τ_l, which is the average time spent on the target track waiting for the desired sector. Thus,

$$\tau_a = \tau_s + \tau_l \tag{2.8}$$

The latency is half the revolution period of the disk, since a randomly selected sector is, on the average, halfway along the track from the point where the head initially lands. Thus for a disk rotating at 1200 rpm $\tau_l = 25$ ms, while at 3600 rpm $\tau_l \simeq 8.3$ ms. The seek time, on the other hand, is independent of the rotation speed, but is determined by the traveling distance of the head during an average seek, as well as by the mechanism of head actuation. It can be shown that the average length of travel in a random seek is one third of the full stroke. (In our notation the full stroke is $r_{max} - r_{min}$.) In magnetic disk drives where the head/actuator assembly is relatively light-weight (a typical Winchester head weighs about 5 grams) the acceleration and deceleration periods are short, and seek times are typically around 10 ms in small drives (i.e., 5.25 and 3.5 in.). In optical disk systems, on the other hand, the head, being an assembly of discrete elements, is fairly large and heavy (typical weight $\simeq$ 100grams), resulting in values of τ_s that are several times greater than those obtained in magnetic recording systems. The seek times reported for commercially available optical drives presently range from 20 ms in high-performance 3.5-in. drives to about 80 ms in larger drives. We emphasize, however, that the optical disk technology is still in its infancy; with the passage of time, the integration and miniaturization of the elements within the optical head will surely produce lightweight devices capable of achieving seek times of the order of a few milliseconds.

The Optical Path

The **optical path** begins at the light source which, in practically all laser disk systems in use today, is a semiconductor GaAs diode laser. Several unique features have made the laser diode indispensable in optical recording technology, not only for the readout of stored information but also for writing and erasure. The small size of this laser has made possible the construction of compact head assemblies, its coherence properties have enabled diffraction-limited focusing to extremely small spots, and its direct modulation capability has eliminated the need for external modulators. The laser beam is modulated by controlling the injection current; one applies pulses of variable duration to turn the laser on and off during the recording process. The pulse duration can be as short as a few nanoseconds, with rise and fall times typically less than 1 ns. Although readout can be accomplished at constant power level, i.e., in CW mode, it is customary for noise reduction purposes to modulate the laser at a high frequency (e.g., several hundred megahertz during readout).

Collimation and Beam Shaping

Since the cross-sectional area of the active region in a laser diode is only about one micrometer, diffraction effects cause the emerging beam to diverge rapidly. This phenomenon is depicted schematically in Figure 2.18(a). In practical applications of the laser diode, the expansion of the emerging beam is arrested by a collimating lens, such as that shown in Figure 2.18(b). If the beam happens to have aberrations (astigmatism is particularly severe in diode lasers), then the collimating lens must be designed to correct this defect as well.

In optical recording it is most desirable to have a beam with circular cross section. The need for shaping the beam arises from the special geometry of the laser cavity with its rectangular cross section. Since the emerging beam has different dimensions in the directions parallel and perpendicular to the junction, its cross section at the collimator becomes elliptical, with the initially narrow dimension expanding more rapidly to become the major axis of the ellipse. The collimating lens thus produces a beam with elliptical cross section. Circularization may be achieved by bending various rays of the beam at a prism, as shown in Figure 2.18(c). The bending changes the beam's diameter in the plane of incidence but leaves the diameter in the perpendicular direction intact.

Focusing by the Objective Lens

The collimated and circularized beam of the diode laser is focused on the surface of the disk using an **objective lens**. The objective is designed to be aberration-free, so that its focused spot size is limited only by the effects of diffraction. Figure 2.19(a) shows the design of a typical objective made from spherical optics. According to the classical theory of diffraction, the diameter of the beam, d, at the objective's focal plane is given by

$$d \simeq \frac{\lambda}{\mathrm{NA}} \qquad (2.10)$$

where λ is the wavelength of light, and NA is the numerical aperture of the objective.[1]

In optical recording it is desired to achieve the smallest possible spot, since the size of the spot is directly related to the size of marks recorded on the medium. Also, in readout, the spot size determines the resolution of the system. According to Equation (2.10) there are two ways to achieve a small spot: first by reducing the wavelength and, second, by increasing the numerical aperture of the objective. The wavelengths currently available from GaAs lasers are in the range of 670–840 nm. It is possible to use a nonlinear optical device to double the frequency of these diode lasers, thus achieving blue light. Good efficiencies have been demonstrated by frequency doubling. Also recent developments in II–VI materials have improved the prospects for obtaining green and blue light

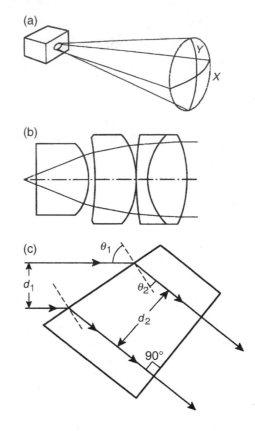

FIGURE 2.18 (a) Away from the facet, the output beam of a diode laser diverges rapidly. In general, the beam diameter along X is different from that along Y, which makes the cross section of the beam elliptical. Also, the radii of curvature R_x and R_y are not the same, thus creating a certain amount of astigmatism in the beam. (b) Multi-element collimator lens for laser diode applications. Aside from collimating, this lens also corrects astigmatic aberrations of the beam. (c) Beam shaping by deflection at a prism surface. θ_1 and θ_2 are related by snell's law, and the ratio d_2/d_1 is the same as $\cos \theta_2 / \cos \theta_1$. Passage through the prism circularizes the elliptical cross section of the beam.

directly from semiconductor lasers. Consequently, there is hope that in the near future optical storage systems will operate in the wavelength range of 400–500 nm. As for the numerical aperture, current practice is to use a lens with NA $\simeq 0.5 - 0.6$. Although this value might increase slightly in the coming years, much higher numerical apertures are unlikely, since they put strict constraints on the other characteristics of the system and limit the tolerances. For instance, the working distance at high numerical aperture is relatively short, making access to the recording layer through the substrate more difficult. The smaller depth of focus of a high numerical aperture lens

---

[1] Numerical aperture is defined as NA $= n \sin \theta$, where n is the refractive index of the image space, and θ is the half-angle subtended by the exit pupil at the focal point. In optical recording systems the image space is air whose index is very nearly unity; thus for all practical purposes NA $= \sin \theta$.

will make attaining/maintaining proper focus more of a problem, while the limited field of view might restrict automatic track-following procedures. A small field of view also places constraints on the possibility of read/write/erase operations involving multiple beams.

The depth of focus of a lens, δ, is the distance away from the focal plane over which tight focus can be maintained [see Figure 2.19(b)]. According to the classical diffraction theory

$$\delta \simeq \frac{\lambda}{\text{NA}^2} \qquad (2.11)$$

Thus for a wavelength of $\lambda = 700$ nm and NA = 0.6, the depth of focus is about ± 1 μm. As the disk spins under the optical head at the rate of several thousand rpm, the objective lens must stay within a distance of $f \pm \delta$ from the active layer if proper focus is to be maintained. Given the conditions under which drives usually operate, it is impossible to make rigid enough mechanical systems to yield the required positioning tolerances. On the other hand, it is fairly simple to mount the objective lens in an actuator capable of adjusting its position with the aid of closed-loop feedback control. We shall discuss the technique of **automatic focusing** in the next section. For now, let us emphasize that by going to shorter wavelengths and/or larger numerical apertures (as is required for attaining higher data densities) one will have to face a much stricter regime as far as automatic focusing is concerned. Increasing the numerical aperture is particularly worrisome, since δ drops with the square of NA.

A source of spherical aberrations in optical disk systems is the substrate through which the light must travel to reach the active layer of the disk. Figure 2.19(c) shows the bending of the rays at the disk surface that causes the aberration. This problem can be solved by taking into account the effects of the substrate in the design of the objective, so that the lens is corrected for all aberrations including those

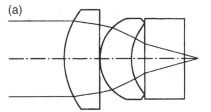

(a)

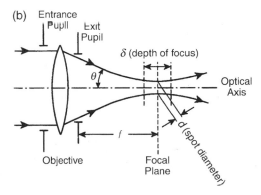

(b)

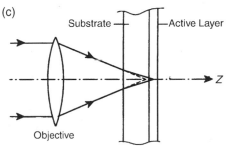

(c)

FIGURE 2.19 (a) Multi-element lens design for a high numerical aperture video disk objective. (*Source:* D. Kuntz, "specifying laser diode optics," *Laser Focus,* March 1984. With permission.) (b)Various parmeters of the objective lens. The numerical aperture is NA = $\sin \theta$. The spot diameter d and the depth of focus δ are given by Equation (2.10) and Equation (2.11), respectively. (c) Focusing through the substrate can cause spherical aberration at the active layer. The problem can be corrected if the substrate is taken into account while designing the objective.

arising at the substrate. Recent developments in molding of aspheric glass lenses have gone a long way in simplifying the lens design problem. Figure 2.20 shows a pair of molded glass aspherics designed for optical disk system applications; both the collimator and the objective are single-element lenses and are corrected for aberrations.

Automatic Focusing

We mentioned in the preceding section that since the objective has a large numerical aperture (NA $\geq$ 0.5), its depth of focus δ is rather shallow ($\delta \simeq \pm 1$ μm at $\lambda = 780$ nm). During all read/write/erase operations, therefore, the disk must remain within a fraction of a micrometer from the focal plane of the objective.

In practice, however, the disks are not flat and they are not always mounted rigidly parallel to the focal plane, so that movements away from focus occur a few times during each revolution. The peak-to-peak movement in and out of focus may be as much as 100 μm. Without automatic focusing of the objective along the optical axis, this runout (or disk flutter) will be detrimental to the operation of the system. In practice, the objective is mounted on a small motor (usually a voice coil) and allowed to move back and forth in order to keep its distance within an acceptable range from the disk. The spindle turns at a few thousand rpm, which is a hundred or so revolutions per second. If the disk moves in and out of focus a few times during each revolution, then the voice coil must be fast enough to follow these movements in real time; in other words, its frequency response must extend to several kilohertz.

The signal that controls the voice coil is obtained from the light reflected from the disk. There are several techniques for deriving the focus error signal, one of which is depicted in Figure 2.21(a). In this so-called obscuration method a secondary lens is placed in the path of the reflected light, one-half of its aperture is covered, and a split detector is placed at its focal plane. When the disk is in focus, the returning beam is collimated and the secondary lens will focus the beam at the center of the split detector, giving a difference signal ΔS equal to zero. If the disk now moves away from the objective, the returning beam will become converging, as in Figure 2.21(b), sending all the light to detector #1. In this case ΔS will be positive and the voice coil will push the lens towards the disk. On the other hand, when the disk moves close to the objective, the returning beam becomes diverging and detector #2 receives the light [see Figure 2.21(c)]. This results in a negative ΔS that forces the voice coil to pull back in order to return ΔS to zero. A given focus error detection scheme is generally characterized by the shape of its focus error signal ΔS versus the amount of defocus Δz; one such curve is shown in Figure 2.21(d). The slope of the focus error signal (FES) curve near the origin is of particular importance, since it determines the overall performance and stability of the servo loop.

Automatic Tracking

Consider a track at a certain radial location, say r_0, and imagine viewing this track through the access window shown in Figure 2.16. It is through this window that the head gains access to arbitrarily selected tracks. To a viewer looking through the window, a perfectly circular track centered on the spindle axis will look stationary, irrespective of

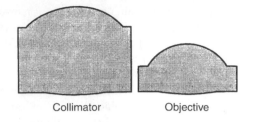

FIGURE 2.20   Molded glass aspheric lens pair for optical disk applications. These singlets can replace multi-element spherical lenses.

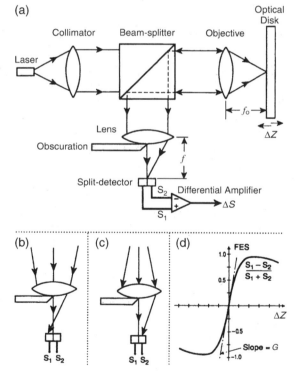

FIGURE 2.21   Focus error detection by the obscuration method. In (a) the disk is in focus, and the two halves of the split detector receive equal amounts of light. When the disk is too far from the objective (b) or too close to it (c) the balance of detector signals shifts to one side or the other. A plot of the focus error signal (FES) versus defocus is shown in (d) and its slope near the origin is identified as the FES gain, G.

the rotation rate. However, any eccentricity will cause an apparent radial motion of the track. The peak-to-peak distance traveled by a track (as seen through the window) depends on a number of factors including centering accuracy of the hub, deformability of the substrate, mechanical vibrations, manufacturing tolerances, etc. For a typical 3.5-in. disk, for example, this peak-to-peak motion can be as much as 100 μm during one revolution. Assuming a revolution rate of 3600 rpm, the apparent velocity of the track in the radial

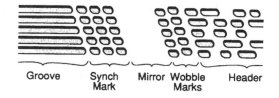

FIGURE 2.22 Servo fields in continuous/composite format contain a mirror area and offset marks for tracking (*Source:* A.B. Marchant, *Optical Recording*, Reading, Mass.: Addison-Wesley, 1990, p. 264. With permission.)

direction will be several millimeters per second. Now, if the focused spot remains stationary while trying to read from or write to this track, it is clear that the beam will miss the track for a good fraction of every revolution cycle.

Practical solutions to the above problem are provided by **automatic tracking** techniques. Here the objective is placed in a fine actuator, typically a voice coil, which is capable of moving the necessary radial distances and maintaining a lock on the desired track. The signal that controls the movement of this actuator is derived from the reflected light itself, which carries information about the position of the focused spot. There exist several mechanisms for extracting the track error signal (TES); all these methods require some sort of structure on the disk surface in order to identify the track. In the case of read-only disks (CD, CD-ROM, and video disk), the embossed pattern of data provides ample information for tracking purposes. In the case of write-once and erasable disks, tracking guides are "carved" on the substrate in the manufacturing process. As mentioned earlier, the two major formats for these tracking guides are pregrooves (for continuous tracking) and sampled-servo marks (for discrete tracking). A combination of the two schemes, known as continuous/composite format, is often used in practice. This scheme is depicted in Figure 2.22 which shows a small section containing five tracks, each consisting of the tail end of a groove, synchronization marks, a mirror area used for adjusting focus/track offsets, a pair of wobble marks for sampled tracking, and header information for sector identification.

Tracking on Grooved Regions

As shown in Figure 2.17(a), grooves are continuous depressions that are either embossed or etched or molded onto the substrate prior to deposition of the storage medium. If the data is recorded on the grooves, then the lands are not used except for providing a guard band between neighboring grooves. Conversely, the land regions may be used to record the information, in which case grooves provide the guard band. Typical track widths are about one wavelength. The guard bands are somewhat narrower than the tracks, their exact shape and dimensions depending on the beam size, required track-servo accuracy, and the acceptable levels of cross-talk between adjacent tracks. The groove depth is usually around one-eighth of one wavelength ($\lambda/8$), since this depth can be shown to give the largest TES in the push-pull method. Cross sections of the grooves may be rectangular, trapezoidal, triangular, etc.

When the focused spot is centered on track, it is diffracted symmetrically from the two edges of the track, resulting in a balanced far field pattern. As soon as the spot moves away from the center, the symmetry breaks down and the light distribution in the far field tends to shift to one side or the other. A split photodetector placed in the path of the reflected light can therefore sense the relative position of the spot and provide the appropriate feedback signal. This strategy is depicted schematically in Figure 2.23.

Sampled Tracking

Since dynamic track runout is usually a slow and gradual process, there is actually no need for continuous tracking as done on grooved media. A pair of embedded marks, offset from the track center as in Figure 2.17(b), can provide the necessary information for correcting the relative position of the focused spot. The reflected intensity will indicate the positions of the two servo marks as two successive short pulses. If the beam happens to be on track, the two pulses will have equal magnitudes and there will be no need for correction. If, on the other hand, the beam is off-track, one of the pulses will be stronger than the other. Depending on which pulse is the stronger, the system will recognize the direction in which it has to move and will correct the error accordingly. The servo marks must appear frequently enough along the track to ensure proper track following. In a typical application, the track might be divided into groups of 18 bytes, 2 bytes dedicated as servo offset areas and 16 bytes filled with other format information or left blank for user data.

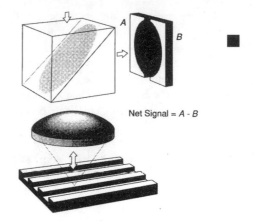

Net Signal = A - B

Thermomagnetic Recording Process

Recording and erasure of information on a magneto-optical disk are both achieved by the **thermomagnetic process**. The essence of thermomagnetic recording is shown in Figure 2.24. At the ambient temperature the film has a high magnetic coercivity[1] and therefore does not respond to the externally

FIGURE 2.23 Push-pull sensor for tracking on grooves. (*Source:* A.B. Marchant, *Optical Recording*, Reading, Mass.: Addison-Wesley, 1990, p. 175. With permission.)

applied field. When a focused beam raises the local temperature of the film, the hot spot becomes magnetically soft (i.e., its coercivity drops). As the temperature rises, coercivity drops continuously until such time as the field of the electromagnet finally overcomes the material's resistance to reversal and switches its magnetization. Turning the laser off brings the temperatures back to normal, but the reverse-magnetized domain remains frozen in the film. In a typical situation in practice, the film thickness may be around 300 Å, laser power at the disk $\simeq$10 mW, diameter of the focused spot $\simeq$1 μm, laser pulse duration $\simeq$50 ns, linear velocity of the track $\simeq$10 m/s, and the magnetic field strength $\simeq$200 gauss. The temperature may reach a peak of 500 K at the center of the spot, which is sufficient for magnetization reversal, but is not nearly high enough to melt or crystalize or in any other way modify the material's structure.

The materials of magneto-optical recording have strong perpendicular magnetic anisotropy. This type of anisotropy favors the "up" and "down" directions of magnetization over all other orientations. The disk is initialized in one of these two directions, say up, and the recording takes place when small regions are selectively reverse-magnetized by the thermomagnetic process. The resulting magnetization distribution then represents the pattern of recorded information. For instance, binary sequences may be represented by a mapping of zeros to up-magnetized regions and ones to down-magnetized regions (non-return to zero or NRZ). Alternatively, the NRZI scheme might be used, whereby transitions (up-to-down and down-to-up) are used to represent the ones in the bit-sequence.

Recording by Laser Power Modulation (LPM)

In this traditional approach to thermomagnetic recording, the electromagnet produces a constant field, while the information signal is used to modulate the power of the laser beam. As the disk rotates under the focused spot, the on/off laser pulses create a sequence of up/down domains along the track. The domains are highly stable and may be read over and over again without significant degradation. If, however, the user decides to discard a recorded block and to use the space for new data, the LPM scheme does not allow direct overwrite; the system must erase the old data during one disk revolution cycle and record the new data in a subsequent revolution cycle.

---

[1] Coercivity of a magnetic medium is a measure of its resistance to magnetization reversal. For example, consider a thin film with perpendicular magnetic moment saturated in the $+Z$ direction. A magnetic field applied along $-Z$ will succeed in reversing the direction of magnetization only if the field is stronger than the coercivity of the film.

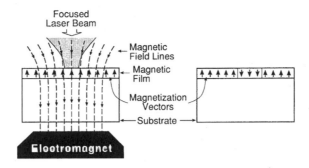

FIGURE 2.24 Thermomagnetic recording process. The field of the electromagnet helps reverse the direction of magnetization in the area heated by the focused laser beam. (*Source:* F. Greidanus et al., Paper 26B-5, presented at the International Symposium on Optical Memory, Kobe, Japan, September 1989. With permission.)

During erasure, the direction of the external field is reversed, so that up-magnetized domains in Figure 2.24 now become the favored ones. Whereas writing is achieved with a modulated laser beam, in erasure the laser stays on for a relatively long period of time, erasing an entire sector. Selective erasure of individual domains is not practical, nor is it desired, since mass data storage systems generally deal with data at the level of blocks, which are recorded onto and read from individual sectors. Note that at least one revolution period elapses between the erasure of an old block and its replacement by a new block. The electromagnet therefore need not be capable of rapid switchings. (When the disk rotates at 3600 rpm, for example, there is a period of 16 ms or so between successive switchings.) This kind of slow reversal allows the magnet to be large enough to cover all the tracks simultaneously, thereby eliminating the need for a moving magnet and an actuator. It also affords a relatively large gap between the disk and the magnet, which enables the use of double-sided disks and relaxes the mechanical tolerances of the system without overburdening the magnet's driver.

The obvious disadvantage of LPM is its lack of direct overwrite capability. A more subtle concern is that it is perhaps unsuitable for the PWM (pulse width modulation) scheme of representing binary waveforms. Due to fluctuations in the laser power, spatial variations of material properties, lack of perfect focusing and track following, etc., the length of a recorded domain along the track may fluctuate in small but unpredictable ways. If the information is to be encoded in the distance between adjacent domain walls (i.e., PWM), then the LPM scheme of thermomagnetic writing may suffer from excessive domain-wall jitter. Laser power modulation works well, however, when the information is encoded in the position of domain centers (i.e., pulse position modulation or PPM). In general, PWM is superior to PPM in terms of the recording density, and, therefore, recording techniques that allow PWM are preferred.

Recording by Magnetic Field Modulation

Another method of thermomagnetic recording is based on magnetic field modulation (MFM) and is depicted schematically in Figure 2.25(a). Here the laser power may be kept constant while the information signal is used to modulate the magnetic field. Crescent-shaped domains are the hallmark of the field modulation technique. If one assumes (using a much simplified model) that the magnetization aligns itself with the applied field within a region whose temperature has passed a certain critical value, T_{crit}, then one can explain the crescent shape of these domains in the following way: With the laser operating in the CW mode and the disk moving at constant velocity, temperature distribution in the magnetic medium assumes a steady-state profile, such as that shown in Figure 2.25(b). Of course, relative to the laser beam, the temperature profile is stationary, but in the frame of reference of the disk the profile moves along the track with the linear track velocity. The isotherm corresponding to T_{crit} is identified as such in the figure; within this isotherm the magnetization aligns itself with the applied field. Figure 2.25(c) shows a succession of critical isotherms along the track, each obtained at the particular instant of time when the magnetic field switches direction. From this picture it is easy to infer

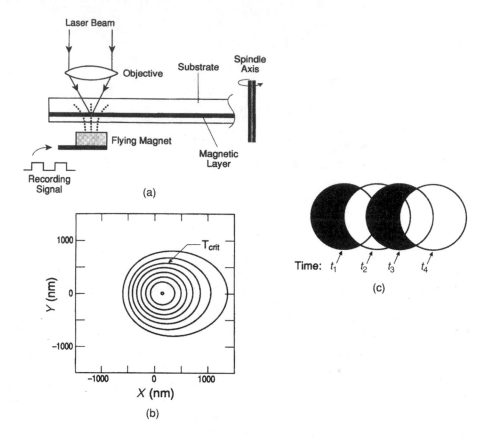

FIGURE 2.25   (a) Thermomagnetic recording by magnetic field modulation. The power of the beam is kept constant, while the magnetic field direction is switched by the data signal. (b) Computed isotherms produced by a CW laser beam, focused on the magnetic layer of a disk. The disk moves with constant velocity under the beam. The region inside the isotherm marked as T_{crit} is above the critical temperature for writing, that is, its magnetization aligns with the direction of the applied field. (c) Magnetization within the heated region (above T_{crit}) follows the direction of the applied field, whose switchings occur at times t_n. The resulting domains are crescent-shaped.

how the crescent-shaped domains form and also understand the relation between the waveform that controls the magnet and the resulting domain pattern.

The advantages of magnetic field modulation recording are that (1) direct overwriting is possible and (2) domain-wall positions along the track, being rather insensitive to defocus and laser power fluctuations, are fairly accurately controlled by the timing of the magnetic field switchings. On the negative side, the magnet must now be small and fly close to the disk surface, if it is to produce rapidly switched fields with a magnitude of a hundred gauss or so. Systems that utilize magnetic field modulation often fly a small electromagnet on the opposite side of the disk from the optical stylus. Since mechanical tolerances are tight, this might compromise the removability of the disk. Moreover, the requirement of close proximity between the magnet and the storage medium dictates the use of single-sided disks in practice.

Magneto-Optical Readout

The information recorded on a perpendicularly magnetized medium may be read with the aid of the polar **magneto-optical Kerr effect**. When linearly polarized light is normally incident on a perpendicular magnetic medium, its plane of polarization undergoes a slight rotation upon reflection. This rotation of the plane of

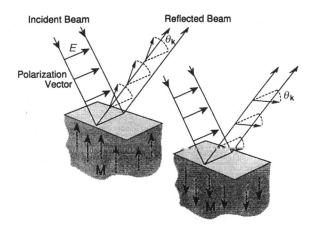

FIGURE 2.26   Schematic diagram describing the polar magneto-optical Kerr effect. Upon reflection from the surface of a perpendicularly magnetized medium, the polarization vector undergoes a rotation. The sense of rotation depends on the direction of magnetization, **M**, and switches sign when **M** is reversed.

polarization, whose sense depends on the direction of magnetization in the medium, is known as the polar Kerr effect. The schematic representation of this phenomenon in Figure 2.26 shows that if the polarization vector suffers a counterclockwise rotation upon reflection from an up-magnetized region, then the same vector will rotate clockwise when the magnetization is down. A magneto-optical medium is characterized in terms of its reflectivity R and its Kerr rotation angle θ_k. R is a real number (between 0 and 1) that indicates the fraction of the incident power reflected back from the medium at normal incidence. θ_k is generally quoted as a positive number, but is understood to be positive or negative depending on the direction of magnetization; in MO readout, it is the sign of θ_k that carries the information about the state of magnetization, i.e., the recorded bit pattern.

The laser used for readout is usually the same as that used for recording, but its output power level is substantially reduced in order to avoid erasing (or otherwise obliterating) the previously recorded information. For instance, if the power of the write/erase beam is 20 mW, then for the read operation the beam is attenuated to about 3 or 4 mW. The same objective lens that focuses the write beam is now used to focus the read beam, creating a diffraction-limited spot for resolving the recorded marks. Whereas in writing the laser was pulsed to selectively reverse-magnetize small regions along the track, in readout it operates with constant power, i.e., in CW mode. Both up- and down-magnetized regions are read as the track passes under the focused spot. The reflected beam, which is now polarization-modulated, goes back through the objective and becomes collimated once again; its information content is subsequently decoded by polarization-sensitive optics, and the scanned pattern of magnetization is reproduced as an electronic signal.

Differential Detection

Figure 2.27 shows the differential detection system that is the basis of magneto-optical readout in practically all erasable optical storage systems in use today. The beam splitter (BS) diverts half of the reflected beam away from the laser and into the detection module.[1] The polarizing beam splitter (PBS) splits the beam into two parts, each carrying the projection of the incident polarization along one axis of the PBS, as shown in Figure 2.27(b). The component of polarization along one of the axes goes straight through, while the component along the other axis splits off and branches to the side. The PBS is oriented such that in the absence of the Kerr effect its two branches will receive equal amounts of light. In other words, if the polarization, upon reflection

---

[1] The use of an ordinary beam splitter is an inefficient way of separating the incoming and outgoing beams, since half the light is lost in each pass through the splitter. One can do much better by using a so-called "leaky" polarizing beam splitter.

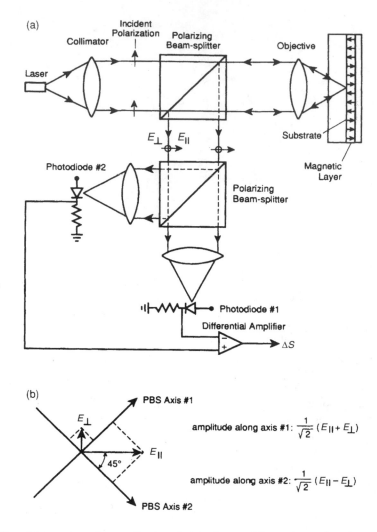

FIGURE 2.27   Differential detection scheme utilizes a polarizing beam splitter and two photodetectors in order to convert the rotation of polarization to an electronic signal. $E_{\parallel}$ and $E_{\perp}$ are the reflected components of polarization; they are, respectively, parallel and perpendicular to the direction of incident polarization. The diagram in (b) shows the orientation of the PBS axes relative to the polarization vectors.

from the disk, did not undergo any rotations whatsoever, then the beam entering the PBS would be polarized at 45° to the PBS axes, in which case it would split equally between the two branches. Under this condition, the two detectors generate identical signals and the differential signal ΔS will be zero. Now, if the beam returns from the disk with its polarization rotated clockwise (rotation angle $= \theta_k$), then detector #1 will receive more light than detector #2, and the differential signal will be positive. Similarly, a counterclockwise rotation will generate a negative ΔS. Thus, as the disk rotates under the focused spot, the electronic signal ΔS reproduces the pattern of magnetization along the scanned track.

Materials of Magneto-Optical Data Storage

Amorphous rare earth transition metal alloys are presently the media of choice for erasable optical data storage applications. The general formula for the composition of the alloy may be written $(\text{Tb}_y\text{Gd}_{1-y})_x(\text{Fe}_z\text{Co}_{1-z})_{1-x}$ where terbium and gadolinium are the rare earth (RE) elements, while iron and cobalt are the transition metals (TM). In practice, the transition metals constitute roughly 80 atomic percent of the alloy (i.e., $x \simeq 0.2$).

In the transition metal subnetwork, the fraction of cobalt is usually small, typically around 10%, and iron is the dominant element ($z \simeq 0.9$). Similarly, in the rare earth subnetwork Tb is the main element ($y \simeq 0.9$) while the gadolinium content is small or it may even be absent in some cases. Since the rare earth elements are highly reactive to oxygen, RE-TM films tend to have poor corrosion resistance and, therefore, require protective coatings. In multilayer disk structures, the dielectric layers that enable optimization of the medium for the best optical/thermal behavior also perform the crucial function of protecting the MO layer from the environment.

The amorphous nature of the material allows its composition to be continuously varied until a number of desirable properties are achieved. In other words, the fractions x, y, z of the various elements are not constrained by the rules of stoichiometry. Disks with very large areas can be coated uniformly with thin films of these media, and, in contrast to polycrystalline films whose grains and grain boundaries scatter the beam and cause noise, amorphous films are continuous, smooth, and substantially free from noise. The films are deposited either by sputtering from an alloy target or by co-sputtering from multiple elemental targets. In the latter case, the substrate moves under the various targets and the fraction of a given element in the alloy is determined by the time spent under each target as well as the power applied to that target. During film deposition the substrate is kept at a low temperature (usually by chilled water) in order to reduce the mobility of deposited atoms and thus inhibit crystal growth. The type of the sputtering gas (argon, krypton, xenon, etc.) and its pressure during sputtering, the bias voltage applied to the substrate, deposition rate, nature of the substarte and its pretreatment, and temperature of the substrate all can have dramatic effects on the composition and short-range order of the deposited film. A comprehensive discussion of the factors that influence film properties will take us beyond the intended scope here; the interested reader may consult the vast literature of this field for further information.

Defining Terms

Automatic focusing: The process in which the distance of the disk from the objective's focal plane is continuously monitored and fed back to the system in order to keep the disk in focus at all times.

Automatic tracking: The process in which the distance of the focused spot from the track center is continuously monitored and the information fed back to the system in order to maintain the read/write beam on track at all times.

Compact disk (CD):   A plastic substrate embossed with a pattern of pits that encode audio signals in digital format. The disk is coated with a metallic layer (to enhance its reflectivity) and read in a drive (CD player) that employs a focused laser beam and monitors fluctuations of the reflected intensity in order to detect the pits.

Error correction coding (ECC):   Systematic addition of redundant bits to a block of binary data, as insurance against possible read/write errors. A given error-correcting code can recover the original data from a contaminated block, provided that the number of erroneous bits is less than the maximum number allowed by that particular code.

Grooved media of optical storage: A disk embossed with grooves of either the concentric-ring type or the spiral type. If grooves are used as tracks, then the lands (i.e., regions between adjacent grooves) are the guard bands. Alternatively, lands may be used as tracks, in which case the grooves act as guard bands. In a typical grooved optical disk in use today the track width is 1.1 μm, the width of the guard band is 0.5 μm, and the groove depth is 70 nm.

Magneto-optical Kerr effect: The rotation of the plane of polarization of a linearly polarized beam of light upon reflection from the surface of a perpendicularly magnetized medium.

Objective lens: A well-corrected lens of high numerical aperture, similar to a microscope objective, used to focus the beam of light onto the surface of the storage medium. The objective also collects and recollimates the light reflected from the medium.

Optical path: Optical elements in the path of the laser beam in an optical drive. The path begins at the laser itself and contains a collimating lens, beam shaping optics, beam splitters, polarization-sensitive elements, photodetectors, and an objective lens.

Preformat: Information such as sector address, synchronization marks, servo marks, etc., embossed permanently on the optical disk substrate.

Sector: A small section of track with the capacity to store one block of user data (typical blocks are either 512 or 1024 bytes). The surface of the disk is covered with tracks, and tracks are divided into contiguous sectors.

Thermomagnetic process: The process of recording and erasure in magneto-optical media, involving local heating of the medium by a focused laser beam, followed by the formation or annihilation of a reverse-magnetized domain. The successful completion of the process usually requires an external magnetic field to assist the reversal of the magnetization.

Track: A narrow annulus or ring-like region on a disk surface, scanned by the read/write head during one revolution of the spindle; the data bits of magnetic and optical disks are stored sequentially along these tracks. The disk is covered either with concentric rings of densely packed circular tracks or with one continuous, fine-pitched spiral track.

References

A.B. Marchant, *Optical Recording*, Reading, Mass.: Addison-Wesley, 1990.

P. Hansen and H. Heitman, "Media for erasable magneto-optic recording," *IEEE Trans. Mag.*, vol. 25, pp. 4390–4404, 1989.

M.H. Kryder, "Data-storage technologies for advanced computing," *Scientific American*, vol. 257, pp. 116–125, 1987.

G. Bouwhuis, J. Braat, A. Huijser, J. Pasman, G. Van Rosmalen, and K.S. Immink, *Principles of Optical Disk Systems*, Bristol: Adam Hilger Ltd., 1985, chap. 2 and 3.

Special issue of *Applied Optics* on video disks, July 1, 1978.

E. Wolf, "Electromagnetic diffraction in optical systems. I. An integral representation of the image field," *Proc. R. Soc. Ser. A*, vol. 253, pp. 349–357, 1959.

M. Mansuripur, "Certain computational aspects of vector diffraction problems," *J. Opt. Soc. Am. A*, vol. 6, pp. 786–806, 1989.

D.O. Smith, "Magneto-optical scattering from multilayer magnetic and dielectric films," *Opt. Acta*, vol. 12, p. 13, 1965.

P.S. Pershan, "Magneto-optic effects," *J. Appl. Phys.*, vol. 38, pp. 1482–1490, 1967.

K. Egashira and R. Yamada, "Kerr effect enhancement and improvement of readout characteristics in MnBi film memory," *J. Appl. Phys.*, vol. 45, pp. 3643–3648, 1974.

H.S. Carslaw and J.C. Jaeger, *Conduction of Heat in Solids*, London: Oxford University Press, 1954.

P. Kivits, R. deBont, and P. Zalm, "Superheating of thin films for optical recording," *Appl. Phys.*, vol. 24, pp. 273–278, 1981.

M. Mansuripur, G.A.N. Connell, and J.W. Goodman, "Laser-induced local heating of multilayers," *Appl. Opt.*, vol. 21, p. 1106, 1982.

J. Heemskerk, "Noise in a video disk system: experiments with an (AlGa)As laser," *Appl. Opt.*, vol. 17, p. 2007, 1978.

A. Arimoto, M. Ojima, N. Chinone, A. Oishi, T. Gotoh, and N. Ohnuki, "Optimum conditions for the high frequency noise reduction method in optical video disk players," *Appl. Opt.*, vol. 25, p. 1398, 1986.

M. Mansuripur, G.A.N. Connell, and J.W. Goodman, "Signal and noise in magneto-optical readout," *J. Appl. Phys.*, vol. 53, p. 4485, 1982.

J.W. Beck, "Noise considerations of optical beam recording," *Appl. Opt.*, vol. 9, p. 2559, 1970.

S. Chikazumi and S.H. Charap, *Physics of Magnetism*, New York: John Wiley, 1964.

B.G. Huth, "Calculation of stable domain radii produced by thermomagnetic writing," *IBM J. Res. Dev.*, pp. 100–109, 1974.

A.P. Malozemoff and J.C. Slonczewski, *Magnetic Domain Walls in Bubble Materials*, New York: Academic Press, 1979.

A.M. Patel, "Signal and error-control coding," in *Magnetic Recording*, vol. II, C.D. Mee and E.D. Daniel, Eds., New York: McGraw-Hill, 1988.

K.A.S. Immink, "Coding methods for high-density optical recording," *Philips J. Res.*, vol. 41, pp. 410–430, 1986.

L.I. Maissel and R. Glang, Eds., *Handbook of Thin Film Technology*, New York: McGraw-Hill, 1970.

G.L. Weissler and R.W. Carlson, Eds., *Vacuum Physics and Technology*, vol. 14 of *Methods of Experimental Physics*, New York: Academic Press, 1979.

T. Suzuki, "Magneto-optic recording materials," *Mater. Res. Soc. Bull.*, pp. 42–47, Sept. 1996.

K.G. Ashar, *Magnetic Disk Drive Technology*, New York: IEEE Press, 1997.

Further Information

Proceedings of the *Optical Data Storage Conference* are published annually by SPIE, the International Society for Optical Engineering. These proceedings document the latest developments in the field of optical recording each year. Two other conferences in this field are the *International Symposium on Optical Memory* (ISOM), whose proceedings are published as a special issue of the *Japanese Journal of Applied Physics*, and the *Magneto-Optical Recording International Symposium* (MORIS), whose proceedings appear in a special issue of the *Journal of the Magnetics Society of Japan*.

3

Logical Devices

Franco P. Preparata
Brown University

Richard S. Sandige
California Polytechnic State University

Albert A. Liddicoat
California Polytechnic State University

B.R. Bannister
University of Hull (retired)

D.G. Whitehead
University of Hull

James M. Gilbert
University of Hull

George A. Constantinides
Imperial College of Science

Bill D. Carroll
University of Texas

Lynne A. Slivovsky
California Polytechnic State University

3.1 Combinational Networks and Switching Algebra[1]

Franco P. Preparata

Introduction to Binary Functions of Binary Variables

A digital system can be analyzed as an interconnection of functional components of basically two types:

1. Storage components, called memory or registers, as appropriate, which store information
2. *Combinational* components, which do not have any memory capability and whose function is the implementation of the required information processing activities (computing)

In this section we shall be concerned with the study of combinational components. Consider, for example, a circuit (the *adder*) designed to compute the sum of two integers represented in binary (Figure 3.1). The adder has

---

[1]All material in this section is adapted from F. Preparata, *Introduction to Computer Engineering*, New York: John Wiley, 1985, chap. 3. With permission.

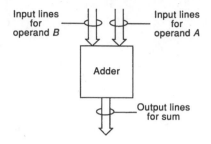

FIGURE 3.1   A binary adder.

two sets of input lines, one for each of the two operands A and B and one set of output lines for the sum. Each set of lines (input and output) consists of as many wires as there are bits in the number it is designed to carry.

Such an adder can be realized by means of a collection of simpler blocks, called *adder cells,* each of which is assigned to a fixed position of the operands (that is, the ith cell receives the ith bits of the addends). The adder is completed by connecting the carry-out output of the ith cell to the carry-in input of the $(i + 1)$th cell and setting c_0 permanently to 0 (see Figure 3.2).

Thus, all we need do to design the adder is to design the adder cell, whose behavior is specified in Figure 3.2(b). The adder cell receives three binary inputs A, B, and C, of which A and B are the operand bits

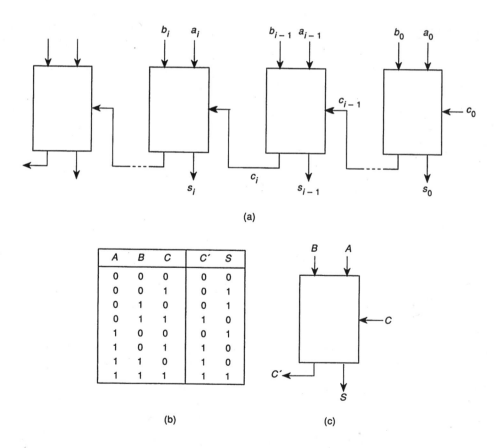

FIGURE 3.2   An adder (a); an adder cell (b); and its behavior (c).

and C is the carry-in bit, and produces two binary outputs C' and S, of which S is the sum bit and C' is the carry-out bit. A, B, and C (the input variables) are independent and therefore can appear in any one of the eight configurations shown in Figure 3.2(b); C' and S, instead, depend entirely upon the binary values of A, B, and C. Specifically, the ordered pair (C', S) corresponding to a given triple (A, B, C) will represent in binary the number of 1's appearing in the binary string ABC; for example, in $ABC = 110$ there are two 1's, whence (C', S) will represent in binary the number 2, that is, $C'S = 10$ [see Figure 3.2(b)].

We recognize a familiar notion; both C' and S are functions of (A, B, C), that is, for each of them we have a *domain* consisting of the eight possible triples of binary values for (A, B, C) [Figure 3.2(b), three left columns] and a range consisting of the two values 0,1. Thus C' and S are each binary functions of binary variables; now, we have the following:

> A binary function of binary variables is called a switching function. A combinational circuit or network is a digital subsystem which realizes a switching function.

The conventional way to display a switching function f is that shown in Figure 3.2(b). Specifically, the combinations of 0's and 1's are ordered so that, when viewed as binary numbers, they are in natural order. Next to each combination, the value of the function is given: this table of function values is called the *truth table* of f. The reason for this name is that a binary variable is said to be *true* when equal to 1 and *false* otherwise. So, the function table gives the *truth* values of the function.

Our objective is to develop a methodology for the design of a combinational network that realizes a given switching function.

Switching Functions of One and Two Variables

To gain insight into the nature of switching functions, we begin by considering binary functions of one binary variable x. This variable can assume only two values, 0 and 1, which form the domain (see Figure 3.3). All the functions of one variable are obtained by filling a two-place truth table in all possible ways, that is, in four ways, shown in Figure 3.3.

We say that a function is *degenerate* if it does not depend upon all of its arguments and *nondegenerate* otherwise. So we see that function f_0 is degenerate: in fact, it is constant and equal to 0, so we will call it the constant 0; similarly, f_3 will be called the constant 1.

Functions / Domain	f_0	f_1	f_2	f_3
0	0	0	1	1
1	0	1	0	1

FIGURE 3.3 Truth tables of all functions of one variable.

Instead, f_1 and f_2 are nondegenerate functions: notice that $f_1(0) = 0$ and $f_1(1) = 1$, thus $f_1(x) = x$, so f_1 will be called the identity; $f_2(0) = 1$ and $f_2(1) = 0$ (f_2 maps 0 to 1 and 1 to 0) and will be called the complement function and denoted by $f_2(x) = \bar{x}$.

We are now ready to consider the binary function of two binary variables x_2 and x_1. Here the pair (x_2, x_1) can assume four possible values (00, 01, 10, 11), which form the domain (see Figure 3.4). All functions are now obtained by filling a truth table with four entries in all possible ways (Figure 3.4). Obviously, this can be done in 16 ways; that is, we have 16 binary functions of 2 binary variables. Let us examine these functions g_0,

x_1	x_2	g_0	g_1	g_2	g_3	g_4	g_5	g_6	g_7	g_8	g_9	g_{10}	g_{11}	g_{12}	g_{13}	g_{14}	g_{15}
0	0	0	0	0	0	0	0	0	0	1	1	1	1	1	1	1	1
0	1	0	0	0	0	1	1	1	0	0	0	0	0	1	1	1	1
1	0	0	0	1	1	0	0	1	1	0	0	1	1	0	0	1	1
1	1	0	1	0	1	0	1	0	1	0	1	0	1	0	1	0	1

FIGURE 3.4 Truth tables of all functions of two variables.

x_2	x_1	g_1	g_1
0	0	0	0
0	1	0	1
1	0	0	1
1	1	1	1

FIGURE 3.5 Study of the AND and OR functions.

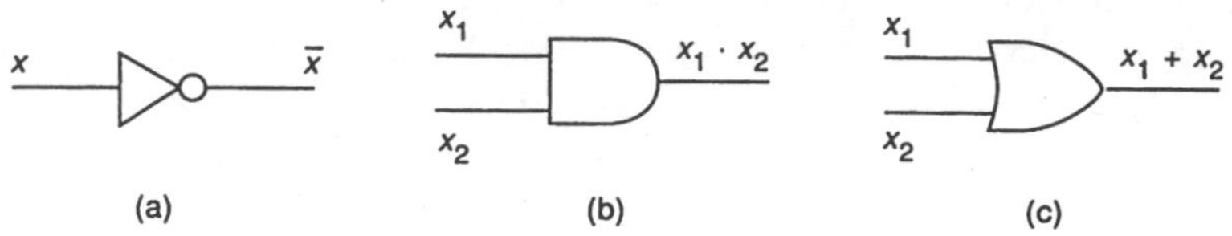

FIGURE 3.6 Symbols for inverter (a), AND gate (b), and OR gate (c).

$g_1, \ldots, g_{15}$. We realize that g_0 and g_{15} are, respectively, the constants 0 and 1; moreover, we notice that $g_3 = x_2$, $g_5 = x_1$, $g_{10} = \bar{x}_1$, and $g_{12} = \bar{x}_2$, that is, the latter are actually nondegenerate functions of *only one* variable. The remaining 10 functions $g_1, g_2, g_4, g_6, g_7, g_8, g_9, g_{11}, g_{13}, g_{14}$ are nondegenerate functions of two variables (i.e., *each* of them depends upon *both* variables). We could analyze all of them, but temporarily we content ourselves with the study of g_1 and g_7.

For convenience, we redraw the truth tables of functions g_1 and g_7 in Figure 3.5. Function g_1 is equal to 1 only when both x_2 and x_1 are equal to 1; for this reason it is called the AND function. Function g_7 is equal to 1 when either x_2 or x_1, or both, are equal to 1; for this reason it is called the OR function.

We can now imagine that special devices are available for the realization of some of the functions we have considered. Specifically, we have a one-input–one-output device, called an inverter [in Figure 3.6(a) we give the conventional symbol for this device], which realizes the function COMPLEMENT; a two-input–one-output device [Figure 3.6(b)], called *AND gate*, which realizes the function AND, and an analogous device [Figure 3.6(c)], called *OR gate,* which realizes the function OR.

The output of an AND gate with inputs x_1 and x_2 will be denoted by $x_1 \cdot x_2$ or simply $x_1 x_2$; analogously, the output of an OR gate with inputs x_1 and x_2 will be denoted by $x_1 + x_2$. (The context will avoid confusion with the symbols "·" and "+" when used in ordinary arithmetic.)

Networks and Expressions

Consider an interconnection of the basic building blocks, AND gates, OR gates, and inverters, such as that shown in Figure 3.7(a). Such an interconnection we call a network. Notice that each gate output, except the single-network output, feeds exactly one gate input and that there are no loops; that is, when tracing a path in the obvious way, in no case will this path traverse the same gate twice. The input terminals are all the unconnected gate inputs. We may think of constructing this network by connecting to the two input terminals of gate G_1 [see Figure 3.7(b)] the output terminals of two smaller networks. In turn, the latter networks could be decomposed into even smaller networks, and so on until we reach the simplest networks of all: terminals. This analysis actually enables us to give an (inductive) definition of a network.

Definition of Combinational Networks

1. Input terminals are networks (elementary networks).
2. If N_1 and N_2 are networks (represented as black boxes), so are the following:

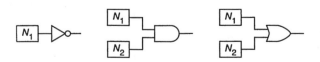

In analogy with the preceding definition, consider now the following definition of **Boolean expressions**.

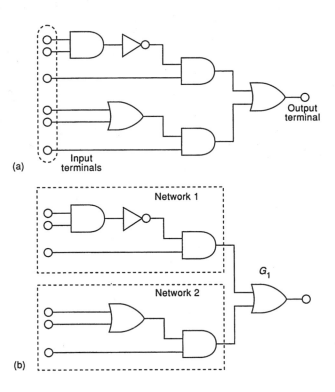

FIGURE 3.7 A combinatorial network.

Definition of Boolean Expressions

1. Variables (both complemented and uncomplemented) and constants are expressions (elementary expressions).
2. If E_1 and E_2 are expressions, so are $\bar{E}_1$, $E_1 \cdot E_2$, and $E_1 + E_2$.[1]

Example 3.1. x_1, 0, x_2 are elementary expressions; is an expression; specifically, it is the AND of expressions $E_1 = x_1 + (\bar{x}_2 x_3)$ and $E_2 = x_4$.

Suppose that we now assign either a variable (complemented or otherwise) or a constant to each input terminal of a nonelementary network. Then, with the output of each gate whose inputs are connected to the input terminals, we can associate an expression, and so on downstream until we associate an expression with the output of the network [for example see Figure 3.8(a)]. Therefore, we see that there is a one-to-one correspondence between expressions and networks whose inputs have been assigned (input-assigned network). Specifically we say that

$$\text{An expression} \leftrightarrow \text{an input} - \text{assigned network}$$

Normally we will drop the qualifier *input-assigned* whenever the context makes it obvious.

Consider the network of Figure 3.8(a). We may now assign to each binary input variable, that is, to x_1, x_2, and x_3, one of the two possible values 0 or 1. Once this assignment has been made we can easily trace the

---

[1]To avoid any ambiguity, the new expressions $E_1 \cdot E_2$ and $E_1 + E_2$ should be parenthesized as $(E_1 \cdot E_2)$ and $(E_1 + E_2)$. However, we shall conform here to the familiar rules for parentheses adopted in ordinary algebra for " + " and " $\cdot$ ".

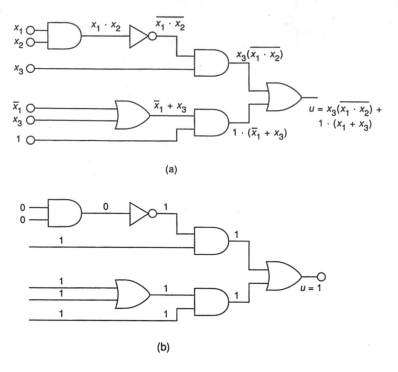

FIGURE 3.8   Determination of the output expression of a network.

network downstream and calculate a binary value on each internal wire of the network until we obtain a binary value at the output terminal [Figure 3.8(b)]. Consider what we have done: We have chosen a set of binary values for (x_1, x_2, x_3) (in our example 001) and have obtained a binary value of u, and for each different choice of input values the network embodies a well-defined rule for obtaining a value of u. This means that u is a function of (x_1, x_2, x_3), from which we conclude that *any given switching network computes a switching function of its inputs.*

A most remarkable fact—to be shown later—is that the converse of the above statement is also true, that is, for any given switching function we can design a network that realizes it. The corresponding design techniques will be presented in the next subsections.

The notion of combinational networks can be slightly generalized to encompass the class of networks in which a gate output may feed *more than one* gate.[1] A gate output feeding two or more gate inputs is said to have *multiple fan-out.* Notice, however, that a network with multiple fan-out gates does not correspond to a single Boolean expression: indeed, if we want to be able to reconstruct the network from its description, we must have a distinct expression for each gate having a multiple fan-out.

Switching Algebra

The techniques for designing combinational networks rest on the properties of a fundamental formal system called *switching algebra* (which is a special case of more general systems called Boolean algebras, although frequently switching algebra is referred to as *Boolean algebra*).

The objects that switching algebra deals with are the (Boolean) expressions defined in the preceding subsection. The basic axiom is as follows.

[1] The number of inputs driven by the output of a gate is called the *fan-out* of that gate. Also, the number of inputs of a gate is called the *fan-in* of the gate.

x_2	x_1	AND
0	0	0
0	1	0
1	0	0
1	1	1

(a)

AND	x_2 = 0	x_2 = 1
x_1 = 0	0	0
x_1 = 1	0	1

x_2	x_1	OR
0	0	0
0	1	1
1	0	1
1	1	1

(b)

OR	x_2 = 0	x_2 = 1
x_1 = 0	0	1
x_1 = 1	1	1

FIGURE 3.9 Operational tables for (a) AND (b) OR.

Axiom. Each expression assumes either the value 0 or the value 1 for all assignments of values (0 or 1) to its variables.

We begin by regarding the function of one variable, COMPLEMENT, as a *unary* operation, that is, as an operation with *one* operand x, the function's argument, which is itself to be regarded as an expression and can only assume the two values 0 and 1. The table of the operation is repeated below.

Operand	Complement
0	1
1	0

Notice that the complement of 0 is 1, $\bar{0} = 1$; similarly $\bar{1} = 0$. It follows that

$$\overline{(\bar{0})} = \bar{1} = 0, \quad \overline{(\bar{1})} = \bar{0} = 1$$

This is summarized by the identity

$$\bar{\bar{x}} = x \quad \text{Involution} \tag{3.1}$$

which describes a fundamental property of COMPLEMENT. Notice also that the constants 0 and 1 are mutually complementary.

We now regard the functions AND and OR of two variables x_1 and x_2 as *binary* operations, that is, as operations with two operands. Here again x_1 and x_2 are to be regarded as expressions, and by the axiom, each can only assume either the value 0 or the value 1. The transformation of each of the function tables to the corresponding operation table, shown in Figure 3.9, should be self-explanatory. From the inspection of these operation tables, we can now deduce their characteristic properties. First of all, both tables are *symmetric* with respect to the main diagonal, that is, we can exchange the role of x_1 and x_2; we shall summarize this as follows:

$$x_1 \cdot x_2 = x_2 \cdot x_1 \quad x_1 + x_2 = x_2 + x_1 \quad \text{Commutativity} \tag{3.2}$$

Next, we notice that $0 \cdot 0 = 0 + 0 = 0$ and $1 \cdot 1 = 1 + 1 = 1$, which leads to the property, for any expression x,

$$xx = x \quad x + x = x \quad \text{Idempotency} \tag{3.3}$$

Since $0 \cdot 1 = 0$ and $1 \cdot 1 = 1$ we extract the rule $x \cdot 1 = x$; similarly $0 + 0 = 0$ and $1 + 0 = 1$ gives $x + 0 = x$, and we have the properties

$$x \cdot 1 = x \quad x + 0 = x \tag{3.4}$$

Also, $0 \cdot 0 = 0$ and $1 \cdot 0 = 0$ yields $x \cdot 0 = 0$; similarly $0 + 1 = 1$ and $1 + 1 = 1$ yields $x + 1 = 1$, thus the properties

$$x \cdot 0 = 0 \quad x + 1 = 1 \tag{3.5}$$

Finally, considering the off-diagonal elements in both operation tables, we see that $0 \cdot 1 = 1 \cdot 0 = 0$ and $0 + 1 = 1 + 0 = 1$; therefore

$$x \cdot \bar{x} = 0 \quad x + \bar{x} = 1 \quad \text{Complementarity} \tag{3.6}$$

There is now a collection of additional properties that can be easily derived by means of a useful proof mechanism called *perfect induction,* which is stated as follows.

Perfect Induction.   *Two* expressions E_1 and E_2 on the same set of variables are equivalent (denoted by $E_1 = E_2$) if, for all possible assignments of values to the variables, the values of E_1 and E_2 coincide.

Perfect induction will now be used to prove the following identities:

$$x \cdot (y + z) = xy + xz \quad x + yz = (x + y)(x + z) \quad \text{Distributivity} \tag{3.7}$$

Indeed, the claim is proved by the tables in Figure 3.10.

By perfect induction we can also prove the identities

$$x(x + y) = x \quad x + xy = x \quad \text{Absorption} \tag{3.8}$$

$$(xy)z = x(yz) \quad (x + y) + z = x + (y + z) \quad \text{Associativity} \tag{3.9}$$

$$\overline{xy} = \bar{x} + \bar{y} \quad \overline{x + y} = \bar{x} \cdot \bar{y} \tag{3.10}$$

(Notice that because associativity holds, we will omit parentheses when writing the AND or the OR of more than two variables.)

Consider now the identities (3.2)–(3.10) which we have established. They are offered in pairs such that one term of the pair is obtained from the other by interchanging AND and OR and by interchanging the constants 0 and 1. This fact is summarized as follows.

Principle of Duality.   Given a valid identity, we obtain another valid identity by:

1. Interchanging the operators AND and OR
2. Interchanging the constants 0 to 1

We can now concisely summarize the properties of switching algebra which we have just established. *Switching algebra* is a set $\mathfrak{B}$ of elements (Boolean expressions) containing the constants 0 and 1, with the following operations:

1. Two binary operations, AND and OR, which are commutative (3.2), associative (3.9), idempotent (3.3), absorptive (3.8), and mutually distributive (3.7).
2. A unary operation, COMPLEMENT (or NEGATION), with the properties of involution (3.1), complementarity (3.6), De Morgan's law (3.10).

x	y	z	y + z	x(y + z)	xy	xz	xy + xz	yz	x + yz	x + y	x + z	(x + y)(x + z)
0	0	0	0	0	0	0	0	0	0	0	0	0
0	0	1	1	0	0	0	0	0	0	0	1	0
0	1	0	1	0	0	0	0	0	0	1	0	0
0	1	1	1	0	0	0	0	1	1	1	1	1
1	0	0	0	0	0	0	0	0	1	1	1	1
1	0	1	1	1	0	1	1	0	1	1	1	1
1	1	0	1	1	1	0	1	0	1	1	1	1
1	1	1	1	1	1	1	1	1	1	1	1	1

FIGURE 3.10   Perfect induction proofs of identities (3.7).

The constants 0 and 1 have the following properties (3.4) and (3.5):

$$\bar{1} = 0$$

$$x \cdot 1 = x \quad x + 0 = x$$

$$x \cdot 0 = 0 \quad x + 1 = 1$$

Identities (3.1)–(3.10) given above represent a set of rules—given in dual pairs—that can be applied to transform an expression into an equivalent expression. It can be shown that rules (3.1)–(3.10) are not independent and that we can select five of them and derive the others from these; this, however, is outside our present scope.

Example 3.2. Prove the following identities, without using perfect induction and by transforming the left side to the right side.

$$a + \bar{a}b = a + b \tag{3.11}$$

$$a + \bar{a}b = a \cdot 1 + \bar{a}b \qquad \text{[by (3.4)]}$$

$$= a(b + \bar{b}) + \bar{a}b \qquad \text{[by (3.6)]}$$

$$= ab + a\bar{b} + \bar{a}b \qquad \text{[by (3.7)]}$$

$$= a\bar{b} + ab + \bar{a}b \qquad \text{[by (3.2)]}$$

$$= a\bar{b} + ab + ab + \bar{a}b \qquad \text{[by (3.3)]}$$

$$= a(\bar{b} + b) + (a + \bar{a})b \qquad \text{[by (3.7)]}$$

$$= a \cdot 1 + 1 \cdot b \qquad \text{[by (3.6)]}$$

$$= a + b \qquad \text{[by (3.4)]}$$

An alternative and simpler proof of (3.11) runs as follows:

$$a + \bar{a}b = (a + \bar{a})(a + b) \qquad \text{[by (3.7)]}$$

$$= 1 \cdot (a + b) \qquad \text{[by (3.6)]}$$

$$= a + b \qquad \text{[by (3.4)]}$$

$$ab + bc + \bar{a}c = ab + \bar{a}c \tag{3.12}$$

$$ab + bc + \bar{a}c = ab + 1 \cdot bc + \bar{a}c \qquad \text{[by (3.4)]}$$

$$= ab + (a + \bar{a})bc + \bar{a}c \qquad \text{[by (3.6)]}$$

$$= ab + abc + \bar{a}bc + \bar{a}c \qquad \text{[by (3.7)]}$$

$$= ab \cdot 1 + abc + \bar{a}bc + \bar{a}c \cdot 1 \qquad \text{[by (3.4)]}$$

TABLE 3.1 Switching Algebra Summary

(P1) $XY = YX$	(S1) $X + Y = Y + X$	Commutativity
(P2) $X(YZ) = (XY)Z$	(S2) $X + (Y + Z) = (X + Y) + Z$	Associativity
(P3) $XX = X$	(S3) $X + X = X$	Idempotency
(P4) $X(X + Y) = X$	(S4) $X + XY = X$	Absorption
(P5) $X(Y + Z) = XY + XZ$	(S5) $X + YZ = (X + Y)(X + Z)$	Distributivity
(P6) $X\bar{X} = 0$	(S6) $X + \bar{X} = 1$	Complementarity
(C1) $\bar{\bar{X}} = X$		Involution
(P7) $\overline{XY} = \bar{X} + \bar{Y}$	(S7) $\overline{X + Y} = \bar{X}\bar{Y}$	De Morgan's
(P8) $X(\bar{X} + Y) = XY$	(S8) $X + \bar{X}Y = X + Y$	
(B1) $\bar{1} = 0$		
(P10) $X \cdot 0 = 0$	(S10) $X + 1 = 1$	
(P11) $X \cdot 1 = X$	(S11) $X + 0 = X$	
(P13) $(X + Y)(Y + Z)(\bar{X} + Z) = (X + Y)(\bar{X} + Z)$	(S13) $XY + YZ + \bar{X}Z = XY + \bar{X}Z$	Consensus

$$= ab(1 + c) + \bar{a}c(b + 1) \qquad \text{[by (3.7)]}$$

$$= ab \cdot 1 + \bar{a}c \cdot 1 \qquad \text{[by (3.5)]}$$

$$= ab + \bar{a}c \qquad \text{[by (3.4)]}$$

Identities (3.11) and (3.12) are actually theorems that have been proved by using the valid identities (3.1)–(3.10); (3.11) is sometimes, but improperly, called *absorption* because of its similarity with (3.8), and (3.12) is known as the *consensus* identity. These two identities are quite convenient because they are relatively easy to memorize and can themselves be applied to accomplish transformations of Boolean expressions. We could continue deriving identities of this kind to be included in our bag of valid rules; however, the burden of memorization will rapidly reach the point of diminishing return.

Table 3.1 is a summary of the manipulative rules of switching algebra.

Boolean Expressions: Normal and Canonical Forms

We saw earlier that every Boolean expression involving n distinct variables describes a switching function of those n variables. We shall now show a very important fact, namely, the converse of the above statement: *given any binary function of n binary variables we can construct a Boolean expression describing that function.* Since every Boolean expression corresponds to a combinational circuit consisting of single fan-out gates, we obtain the far-reaching result that every switching function is realizable by means of a combinational network.

An expression involves variables in *uncomplemented or complemented* forms: we call *literal* any occurrence of a variable in either form. For example, the expression $[x_1 = x_2(x_3 + x_4\bar{x}_1)]\bar{x}_3 + \bar{x}_2 x_4$ has four variables and six literals.

An expression is in *normal sum-of-products* (SOP) *form* when it is the OR (sum) of ANDs (products) of literals. We shall now describe how an arbitrary Boolean expression E can be transformed into an equivalent expression in normal SOP form.

Reduction to Normal SOP Form

Let expression $E = [x_1 + x_2(\overline{x_3 + x_4\bar{x}_1})]x_3 + \overline{\bar{x}_2 x_4}$ be given.

1. Place all complements directly on variables (by using De Morgan's laws). In our example,

$$E = (x_1 + x_2 \cdot \bar{x}_3 \cdot \overline{x_4\bar{x}_1})x_3 + \bar{\bar{x}}_2 + \bar{x}_4$$
$$= (x_1 + x_2 \cdot \bar{x}_3(\bar{x}_4 + \bar{\bar{x}}_1))x_3 + x_2 + \bar{x}_4$$
$$= [x_1 + x_2 \bar{x}_3(\bar{x}_4 + x_1)]x_3 + x_2 + \bar{x}_4$$

2. Apply the distributive law. In our example,

$$E = (x_1 + x_2\,\bar{x}_3\,\bar{x}_4 + x_1\,x_2\,\bar{x}_3)x_3 + x_2 + \bar{x}_4$$
$$= x_1\,x_3 + x_2\,\bar{x}_3\,x_3\,\bar{x}_4 + x_1 x_2 \bar{x}_3 x_3 + x_2 + \bar{x}_4$$

3. Eliminate redundant terms (using idempotency and complementarity). In our example notice that, by (3.6), $x_3\bar{x}_3 = 0$ and that, by (3.5), all product terms containing a factor 0 are 0 themselves, whereby

$$E = x\,x_1\,x_3 + 0 + 0 + x_2 + \bar{x}_4$$
$$= x\,x_1\,x_3 + x_2 + \bar{x}_4$$

The latter expression is in normal SOP form and is equivalent to the given expression.

With reference to expressions on n variables, a special type of product term (AND term) is one that contains as a factor each variable, either uncomplemented or complemented: these terms are called *fundamental products* or *minterms*. For example, for $n = 4$, $\bar{x}_1\bar{x}_2\bar{x}_3 x_4$ and $\bar{x}_1 x_2\bar{x}_3 x_4$ are minterms but $\bar{x}_1\bar{x}_2 x_4$ is not. A normal SOP expression is said to be in *canonical* (SOP) *form* if its product terms are all minterms. We shall now describe how to transform a normal form expression into a canonical form expression.

Transformation from Normal Form to Canonical Form

Let the normal form expression $x_1x_3 + x_2 + \bar{x}_3$ be given.

1. If a product term contains neither x_i nor $\bar{x}_i$ "multiply" it by $(x_i + \bar{x}_i)$. [Notice that this transforms the product term into an equivalent expression since $x_i + \bar{x}_i = 1$ by (3.6).] In our example x_1x_3 does not contain a literal with index 2: x_2 does not contain literals with indices 1 and 3; $\bar{x}_3$ does not contain literals with indices 1 and 2. Thus,

$$x_1x_3 + x_2 + \bar{x}_3 = x_1x_3(x_2 + \bar{x}_2) + x_2(x_1 + \bar{x}_1)(x_3 + \bar{x}_3) + x_3(x_1 + \bar{x}_1)(x_2 + \bar{x}_2)$$

2. Apply the distributive law. In our example,

$$E = x_1x_3\bar{x}_2 + x_1x_3x_2 + x_2\bar{x}_1\bar{x}_3 + x_2\bar{x}_1x_3 + x_2x_1\bar{x}_3 + x_2x_1x_3 + \bar{x}_3\bar{x}_1\bar{x}_2$$
$$+ \bar{x}_3\bar{x}_1x_2 + \bar{x}_3x_1\bar{x}_2 + \bar{x}_3x_1x_2$$

3. Eliminate repeated product term using idempotency. In our example, the following sets of terms are sets of identical terms: (2nd, 6th) (3rd, 8th) (5th, 10th). Thus, after eliminating the repeated terms and rearranging the order of the indices as (3,2,1), we obtain the canonical expression for $x_1x_3 + x_2 + \bar{x}_3$:

$$E = x_3\bar{x}_2x_1 + x_3x_2x_1 + \bar{x}_3x_2\bar{x}_1 + x_3x_2\bar{x}_1 + \bar{x}_3x_2x_1 + \bar{x}_3\bar{x}_2\bar{x}_1 + \bar{x}_3\bar{x}_2x_1$$

We begin by introducing a useful notation for minterms. Consider, for example, the minterm $\bar{x}_4x_3x_2\bar{x}_1$; we associate with this minterm an ordered binary 4-tuple $(b_4b_3b_2b_1)$, where b_i corresponds either to x_i or $\bar{x}_i$, and $b_i = 1$ if x_i is uncomplemented and is 0 otherwise. In our example, with $\bar{x}_4x_3x_2\bar{x}_1$ we associate 0110: this string is the binary equivalent of the integer 6, so that we shall denote $\bar{x}_4x_3x_2\bar{x}_1$ by m_6. Referring to the previous example, the expression E in this new notation becomes $m_5 + m_7 + m_2 + m_6 + m_3 + m_0 + m_1$, or equivalently, OR $(m_0, m_1, m_2, m_3, m_5, m_6, m_7)$.

Suppose now that we have combinational networks F and G, which respectively compute switching functions $f(x_1, \ldots, x_n)$ and $g(x_1, \ldots, x_n)$, and that we connect the outputs of these networks to the inputs of gates or inverters. Clearly, if f and g are fed, say, to an AND gate, then the function u at the output of this gate will be 1 only when both f and g are 1, and similarly for the other cases. Obviously u is a function of the same set of variables $\{x_1, \ldots, x_n\}$ as f and g, so we obtain the following simple rules for its truth table:

x_3	x_2	x_1	f	f_1	f_2	f_3
0	0	0	0	0	0	0
0	0	1	0	0	0	0
0	1	0	0	0	0	0
0	1	1	1	1	0	0
1	0	0	1	0	1	0
1	0	1	0	0	0	0
1	1	0	1	0	0	1
1	1	1	0	0	0	0

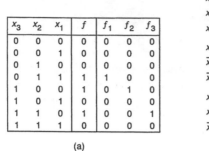

(a)

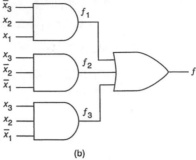

(b)

FIGURE 3.11   A switching function and its corresponding AND-to-OR network.

The truth table of $\bar{f}$ is the entry-by-entry (component-wise) complement of the truth table of f; the truth table of $f \cdot g$[or$(f + g)$] is the component-wise AND (or OR) of the truth tables of f and g.

Given a minterm m of $x_1, x_2, \ldots x_n$ (the minterm itself obviously describes a function of these variables), we recognize that $m = 1$ exactly when the following conditions hold: if x_i appears in m in uncomplemented form, then $x_i = 1$; if x_i appears in m in complemented form, then $\bar{x}_i = 1$, that is, $x_i = 0$. Thus $m = 1$ only when each variable attains a specific value, that is, $m = 1$ for a unique combination of the variables or, equivalently, *the truth table of a minterm has exactly one "1."* (Incidentally, this explains the denomination *minterm*: a nondegenerate function with the minimum number of 1's in its truth table.)

Example 3.3.   For $n = 4$, $\bar{x}_4\bar{x}_3\bar{x}_2x_1 = m_1$ is 1 when and only when $\bar{x}_4 = 1$, $\bar{x}_3 = 1$, $\bar{x}_2 = 1$, $x_1 = 1$, that is, when $(x_4x_3x_2x_1) = (0001)$.

Now, let f be a switching function of arguments $x_1, \ldots, x_n$, given by means of its truth table (see Figure 3.11 where $n = 3$). We may view the truth table of f as the OR of as many distinct truth tables as it has 1's, each of the latter truth tables having exactly a single 1 [Figure 3.11(a)]. However, each such table is the table of a minterm! Moreover, each minterm is a product of literals, which for $i = 1, 2, \ldots, n$ contains either x_i or $\bar{x}_i$, depending upon whether in the combination corresponding to the single 1 in the table the x_i entry is 1 or 0. In Figure 3.11, $f_1 = 1$ in correspondence to $(x_3x_2x_1) = (011)$, whence $f_1 = \bar{x}_3x_2x_1$. Similarly, we obtain $f_2 = x_3\bar{x}_2\bar{x}_1$ and $f_3 = x_3x_2\bar{x}_1$.

In conclusion, since $f = f_1 + f_2 + f_3$, we have

$$f = \bar{x}_3x_2x_1 + x_3\bar{x}_2\bar{x}_1 + x_3x_2\bar{x}_1$$

Notice that this is a most remarkable finding: given a function f by means of its truth table (i.e., as a binary function of binary variables), we have obtained an expression (actually, a canonical expression) describing that function!

Once we have an expression for the given function, we shall design the corresponding combinational network. Before proceeding, however, we recall that in the subsection "Switching Functions of One and Two Variables" we have introduced AND gates and OR gates as two-input–one-output devices; since we have proved [identity (3.9)] that the AND and OR operations are associative, we may think of using in our networks devices that realize the AND (or OR) *of more than two* inputs; this is indeed technically possible, although, for physical reasons, the number of inputs may not be too large. Therefore, we see that by using these newly introduced gates, we can construct the network of Figure 3.11(b), which computes the given function. (This network is a collection of AND gates feeding a single OR gate and is therefore called an AND-to-OR network.) Notice that we have achieved the objective set forth at the end of the subsection "Networks and Expressions" and summarized below:

Given a switching function f by means of its truth table (i.e., as a binary function of binary variables), we can construct a switching network that computes it.

TABLE 3.2   Duality of Canonical SOP and POS Expressions

Canonical SOP Expressions	Canonical POS Expressions
Minterm—a *product* of literals that has as a "factor" each of the *n* variables either true or complemented	*Maxterm*—a *sum* of literals that has as an "addend" each of the *n* variables either true or complemented
Canonical SOP expressions—*OR of minterms*	Canonical *POS* expression—AND of maxterms
A *minterm* is a canonical *SOP* expression that is 1 for exactly one combination of the variables	A *maxterm* is a canonical *POS* expression that is 0 for exactly one combination of the variables
A *minterm* corresponds to one switching function whose truth table has exactly one "1"	A *maxterm* corresponds to one switching function whose truth table has exactly one "0"
There exists a one-to-one correspondence between switching functions and canonical *SOP* expressions	There exists a one-to-one correspondence between switching functions and canonical *POS* expressions

In Table 3.2, left side, we summarize the important notions concerning SOP canonical expressions. A discussion analogous to the one just completed can be carried out with reference to *normal product-of-sums* (POS) expressions, that is, an AND of ORs of literals. Indeed, all we need in the preceding discussion is a set of substitutions as dictated by the principle of duality

$$AND \leftrightarrow OR$$

$$Product \leftrightarrow Sum$$

$$Minterm \leftrightarrow Maxterm \text{ (see below)}$$

$$0 \leftrightarrow 1$$

The conclusions are summarized in Table 3.2, right side. Notice the perfect duality of corresponding statements in this table. The only novel term in this table is *maxterm,* the dual of minterm: the reason for the denomination is that a maxterm describes a nondegenerate function with the maximum number of 1's in its truth table. A maxterm is usually denoted by the symbol M_j, specifically $M_j = \bar{m}_j$, that is, $M_j = 0$ if and only if $m_j = 1$, and vice versa. For example, for variables x_3, x_2, x_1, $M_5 = \bar{m}_5 = \overline{x_3\bar{x}_2x_1} = (x_3 + x_2 + \bar{x}_1)$, that is, M_5 is the maxterm which is 0 exactly for $(x_3x_2x_1) = (101)$ and is 1 otherwise.

In conclusion, a **Boolean function** can be specified either as an *OR of minterms* (corresponding to the 1's in the truth table) or as an *AND of maxterms* (corresponding to the 0's in the truth table).

Example 3.4.

Decimal equivalent	x_3	x_2	x_1	f
0	0	0	0	0
1	0	0	1	0
2	0	1	0	0
3	0	1	1	1
4	1	0	0	1
5	1	0	1	0
6	1	1	0	1
7	1	1	1	0

$$f = OR\,(m_3, m_4, m_6),\ SOP$$

$$f = AND\,(M_0, M_1, M_2, M_5, M_7),\ POS$$

Consider now a Boolean function $f(x_1, x_2, \ldots, x_n)$ expressed in canonical SOP form. Each minterm of f contains either $\bar{x}_n$ or x_n; therefore, we associate into two separate expressions, F_0 and F_1, the minterms of f depending upon whether they contain $\bar{x}_n$ or x_n, respectively, that is,

$$f = F_0 + F_1$$

Now from all terms of F_0 we can factor out $\bar{x}_n$, i.e., F_0 can be written as the AND of $\bar{x}_n$ and an expression f_0, which consists exactly of minterms over the variables $x_1, x_2, \ldots, x_{n-1}$, that is, $F_0 = \bar{x}_n f_0(x_1, x_2, \ldots, x_{n-1})$. Similarly, we can express F_1 as $F_1 = x_n f_1(x_1, x_2, \ldots, x_{n-1})$. It follows that f can be expressed as

$$f(x_1, \ldots, x_{n-1}, x_n) = \bar{x}_n f_0(x_1, \ldots, x_{n-1}) + x_n f_1(x_1, \ldots, x_{n-1}) \qquad (3.13)$$

In the above relation, we now set $x_n = 0$ and obtain

$$f(x_1, \ldots, x_{n-1}, 0) = 1 \cdot f_0(x_1, \ldots, x_{n-1}) + 0 \cdot f_1(x_1, \ldots, x_{n-1})$$

that is, $f_0(x_1, \ldots, x_{n-1}) = f(x_1, \ldots, x_{n-1}, 0)$. Similarly, if we set $x_n = 1$ in (3.13) we have

$$f_1(x_1, \ldots, x_{n-1}, 1) = 0 \cdot f_0(x_1, \ldots, x_{n-1}) + 1 \cdot f_1(x_1, \ldots, x_{n-1})$$

that is, $f_1(x_1, \ldots, x_{n-1}) = f_1(x_1, \ldots, x_{n-1}, 1)$. This result is called the fundamental theorem of Boolean algebra and can be stated as follows.

Fundamental Theorem of Boolean Algebra. Every function $f(x_1, \ldots, x_n)$ of $x_1, \ldots, x_n$, for any x_i can be expressed as

$$f = \bar{x}_i f_0 + x_i f_1$$

where $f_0 = f(x_1, \ldots, x_{i-1}, 0, x_{i+1}, \ldots, x_n)$ and $f_1 = f(x_1, \ldots, x_{i-1}, 1, x_{i+1}, \ldots, x_n)$ are both functions of the $(n-1)$ variables $x_1, \ldots, x_{i-1}, x_{i+1}, \ldots, x_n$.

Other Important Boolean Connectives

Although the operators AND, OR, and NOT are perfectly adequate for the realization of any combinational network, there are other connectives that are quite important and are now introduced.

The NAND and NOR Connectives

The first of these connectives, called NAND (AND followed by NOT), realizes the function $\overline{x \cdot y}$ of two variables x and y; its circuit symbol is given, for two inputs, in Figure 3.12(a). (Note that, as already has been done for the AND and OR connectives, the function NAND can be generalized to any number of variables, as $\overline{x \cdot y \ldots w}$.) The connective NAND is interesting because alone it can be used to realize any combinational network. (We refer to this property by saying that NAND is *logically complete*.) Indeed, since we know that AND, OR, and NOT are adequate for realizing combinational networks, all we need to show is that each of these three connectives can, in turn, be realized by an expression involving only NAND. This is readily shown below: (rules C1 and P3, Table 3.1)

$$a \cdot b = \overline{\overline{a \cdot b}} = \overline{\overline{a \cdot b} \cdot \overline{a \cdot b}} = \text{NAND}[\text{NAND}(a, b), \text{NAND}(a, b)]$$

(rules P7 and P3, Table 3.1)

$$a + b = \overline{\overline{a} \cdot \overline{b}} = \overline{\overline{aa} \cdot \overline{bb}} = \text{NAND}[\text{NAND}(a, a), \text{NAND}(b, b)]$$

(rule P3, Table 3.1)

$$\bar{a} = \overline{aa} = \text{NAND}(a, a)$$

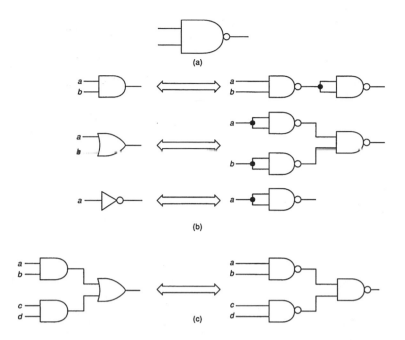

FIGURE 3.12 (a) Circuit symbol of NAND gate. (b) and (c) NAND gate realization of AND, OR, and NOT gates and SOP expression.

These transformations are illustrated in Figure 3.12(b). Thus, given a network consisting of AND, OR, and NOT gates, by using the above rules, one can transform it into an equivalent one consisting of NAND gates alone. Besides this rather cumbersome transformation, there is a more direct and useful correspondence between NAND expressions and SOP expressions, as shown below:

$$ ab + cd = \overline{\overline{ab + cd}} = \overline{\overline{ab} \cdot \overline{cd}} = \text{NAND}[\text{NAND}(a, b),\ \text{NAND}(c, d)] $$

So we see that any SOP expression can be realized by a network consisting of NAND gates alone [see Figure 3.12(c)] by simply replacing with NAND gates both the AND gates and the OR gate in the standard AND-to-OR realization of the given expression.

As we may expect from duality, there is another connective, NOR, which enjoys analogous properties. This connective NOR (OR followed by NOT) realizes the function $\overline{x + y}$ of x and y, and its symbol is given, for two inputs, in Figure 3.13(a). Transformations analogous to those obtained above can be easily derived and the results are shown in Figure 3.13(b) and 3.13(c). Notice that a two-level NAND network corresponds to an SOP expression, and a two-level NOR network corresponds to a POS expression. These properties make NOR and NAND gates very attractive and popular in digital design, since entire systems can be realized by using just one type of component.

The XOR Connective

Finally, we introduce the connective exclusive-OR (frequently abbreviated as XOR), which realizes the function $(x\bar{y} + \bar{x}y)$ of two variables x and y. The symbol used for this connective is $\oplus$, while the circuit symbol is given in Figure 3.14. The reason for the name exclusive-OR is that $(x \oplus y)$ is equal to 1 if and only if either x or y, but not both, is equal to 1. (By contrast, the ordinary OR is correctly called *inclusive* OR, although the adjective inclusive is normally omitted.)

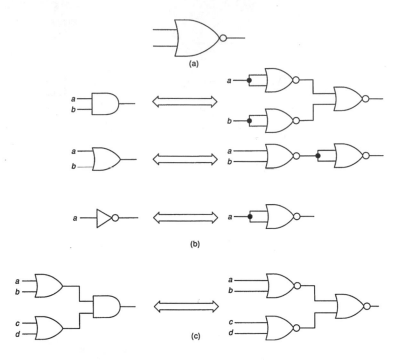

FIGURE 3.13 (a) Circuit symbol of NOR gate. (b) and (c) NOR gate realizations of AND, OR, and NOT gates and POS expression.

The exclusive-OR has several interesting properties, whose proof is left as an exercise. First, exclusive-OR is associative

$$(x \oplus y) \oplus z = x \oplus (y \oplus z)$$

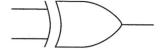

FIGURE 3.14 Symbol for the exclusive OR gate.

so that we may omit parentheses and write $x \oplus y \oplus z$. Therefore, the exclusive-OR is generalized to an arbitrary number of variables, and we shall have exclusive-OR gates with correspondingly many input lines. Other important properties, also left as an exercise, are

$$x(y \oplus z) = xy \oplus xz$$

$$x \oplus 1 = \bar{x}$$

It is appropriate to introduce at this point an alternative set of standard symbols for the logic gates discussed in this chapter. They are displayed in Table 3.3 vis-à-vis their by now familiar counterparts (symbol set 1) and deserve no additional comments.

Notes and References

Boolean algebra, which—as we saw—provides the formalism for the description of (binary) digital networks, was developed in the last century, originating with the English mathematician George Boole, who in 1854 published his fundamental work, *An Investigation of the Laws of Thought.* Boole's goal was essentially the development of a formalism to compute the truth or falsehood (i.e., the *truth value*) of complex compound statements from the truth values of their component statements. The discipline developed later into a more complex body of knowledge, known as symbolic logic.

Apparently, early in this century, more than one scientist perceived the applicability of Boolean algebra to the design of telephone circuits [Ehrenfest, 1910]. It was only in the thirties that the potential was fully

TABLE 3.3

Name	Symbol Set 1	Symbol Set 2
AND		
OR		
NOT		
NAND		
NOR		
EXCLUSIVE OR		

realized, when C.E. Shannon [1938] published his paper "A Symbolic Analysis of Relay and Switching Circuits," which became the foundation of switching theory and logical design. Because of the context in which it was originally used (telephone networks, also called switching networks), the name *switching algebra* has become standard for the algebra of functions of two-valued variables. Although initially the interests of researchers focused on relay networks (also called contact networks), as mechanical devices were gradually replaced by electronic devices the techniques were tailored to gate networks of the type described above. The term *gate* was already in use in the forties to denote the logical elements discussed earlier.

There are very many good references on Boolean algebra and we may quote only a selected few of them. Suffice it to mention the texts by Hill and Peterson [1974], Kohavi [1978], and Hohn [1966]. These books give a sufficiently rigorous formulation of the subject, tailored to the analysis and the design of combinational networks. In addition, like most of the earlier books, Hohn's and Kohavi's texts also contain a discussion of the Boolean techniques used in connection with relay circuits. (Some of the more recent works completely omit this topic, which has been but totally overshadowed by the impressive development of electronic networks.) The reader interested in studying the relation of switching algebra to Boolean algebras in general is referred to Preparata and Yeh [1973] for an elementary introduction.

Defining Terms

Boolean algebra:　The algebra of logical values enabling the logical designer to obtain expressions for digital circuits.

Boolean expressions:　Expressions of logical variables constructed using the connectives *and, or,* and *not.*

Boolean functions:　Common designations of binary functions of binary variables.

Combinational logic:　Interconnections of memory-free digital elements.

Switching theory:　The theory of digital circuits viewed as interconnections of elements whose output can switch between the logical values of 0 and 1.

References

G. Boole, *An Investigation of the Laws of Thought,* New York: Dover Publication, 1954.
F.J. Hill and G.R. Peterson, *Introduction to Switching Theory and Logical Design,* New York: Wiley, 1974.
F.E. Hohn, *Applied Boolean Algebra,* New York: Macmillan, 1966.
Z. Kohavi, *Switching and Finite Automata Theory,* New York: McGraw-Hill, 1978.
F.P. Preparata and R.T. Yeh, *Introduction to Discrete Structures,* Reading, Mass.: Addison-Wesley, 1973.
C.E. Shannon, "A symbolic analysis of relay and switching circuits," *Trans. AIEE,* vol. 57, pp. 713–723, 1938.

3.2 Logic Circuits

Richard S. Sandige and Albert A. Liddicoat

This section discusses two-state (high or low, 1 or 0, or true or false) digital logic circuits. There are two classifications of two-state logic circuits: *combinational logic circuits* and *sequential logic circuits.* By definition, the external output signals of combinational logic circuits are only dependent on the external input signals applied to the circuit. Combinational circuits are memory-less because the output states are only dependent on the current input values. An example of a combinational circuit is an adder. An adder circuit evaluates to the sum of the current values on the inputs regardless of the previous inputs. In contrast, the output signals of sequential logic circuits are dependent on both the present state of the circuit and any external input signals that may exist. Since sequential circuits depend on their present state value, they have feedback and exhibit memory. An up/down counter is an example of a sequential logic circuit that maintains the current count and has an external input that indicates if the counter should count up or count down. For an up/down counter, the next output state (or next count) is a function of both the current count and the external input that indicates if the counter is counting up or down.

Sequential logic circuits can be subdivided into *synchronous* circuits and *asynchronous* circuits. Synchronous circuits change state when a clock input is activated, while asynchronous circuits change state without a clock. Asynchronous circuits can be further divided into fundamental-mode circuits and pulse-mode circuits. Figure 3.15 shows the classification or taxonomy of the logic circuits that have been introduced.

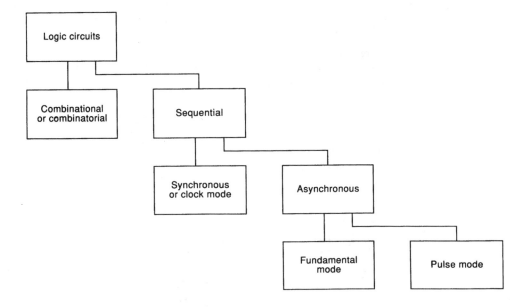

FIGURE 3.15 Graphic classification of logic circuits. (*Source*: R.S. Sandige, *Modern Digital Design,* New York: McGraw-Hill, 1990, p. 440. With permission.)

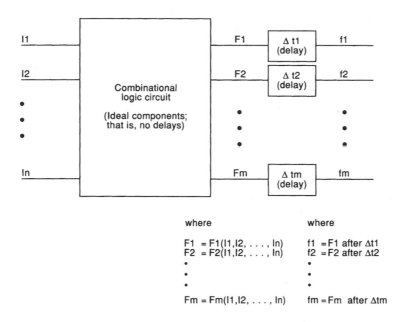

FIGURE 3.16 Block diagram model for combinational logic circuits. (*Source*: R.S. Sandige, *Modern Digital Design*, New York: McGraw-Hill, 1990, p. 440. With permission.)

Combinational Logic Circuits

The block diagram in Figure 3.16 illustrates the model for combinational logic circuits. Logic elements exist inside the block entitled *combinational logic circuit*. The combinational logic circuit may consist of any configuration of two-state logic elements where the output signals are totally dependent on the input signals to the circuit, as indicated by the functional relationships in the figure. Combinational logic circuits do not have feedback loops because these feedback loops introduce memory or state information.

The logic elements can be anything from relays with slow on and off switching action to modern integrated circuit (IC) transistor switches with extremely fast switching action. Modern ICs exist in various technologies and circuit configurations, for example transistor-transistor logic (TTL), complementary metal-oxide semiconductor (CMOS), emitter-coupled logic (ECL), and integrated injection logic (I^2L).

The combinational logic circuits do not evaluate instantaneously. There is a delay from the time the inputs are set to the time that the output has evaluated. Figure 3.16 models this delay using the delay output blocks that represent a lumped delay. In the figure, the outputs F1 through Fm are assumed to evaluate instantaneously. Then, at a time $\Delta t1$ through Δtm later, the outputs f1 through fm are known. The worst-case delays through the longest delay path from the inputs to each respective output of the combinational logic circuit are used for the lumped delay Δt values. The lumped delays provide an approximate measure of circuit speed or settling time (the time it takes an output signal to become stable after the input signals have become stable).

Figure 3.17 illustrates the gate-level method (random logic method) of implementing a combinational logic circuit. Each shape in the gate-level logic circuit represents a different logic function. The external inputs (D through A) are all on the left side of each logic diagram, while the external outputs (OA through OG) are all shown on the right side of each logic diagram. Notice that there are no feedback loops in the logic diagrams. This gate-level logic circuit implements a binary to seven-segment hexadecimal character generator suitable for driving a seven-segment common cathode LED display like the one in Figure 3.18.

The propagation delays of logic circuits are seldom shown on logic circuit diagrams because these diagrams are used to indicate the functionality of the circuit. However, delays are inherent in each logic element and must be considered while designing logic systems. This gate-level combinational logic circuit converts the

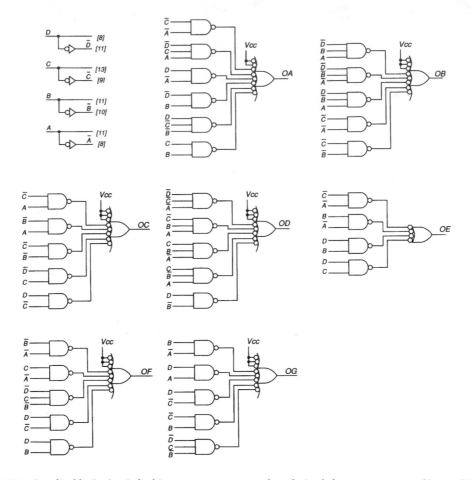

FIGURE 3.17 Gate-level logic circuit for binary to seven-segment hexadecimal character generator. (*Source*: R.S. Sandige, *Modern Digital Design*, New York: McGraw-Hill, 1990, pp. 258–259. With permission.)

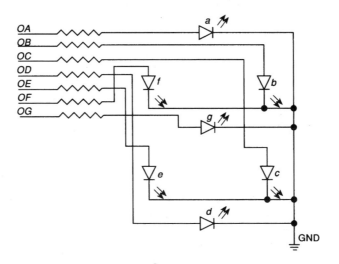

FIGURE 3.18 Seven-segment common cathode LED display. (*Source*: R.S. Sandige, *Modern Digital Design*, New York: McGraw-Hill, 1990, p. 255. With permission.)

TABLE 3.4   Truth Table for Binary to Seven-Segment Hexadecimal Character Generator

Binary Inputs				Seven-Segment Outputs							Displayed Characters
D	C	B	A	OA	OB	OC	OD	OE	OF	OG	
0	0	0	0	1	1	1	1	1	1	0	0
0	0	0	1	0	1	1	0	0	0	0	1
0	0	1	0	1	1	0	1	1	0	1	2
0	0	1	1	1	1	1	1	0	0	1	3
0	1	0	0	0	1	1	0	0	1	1	4
0	1	0	1	1	0	1	1	0	1	1	5
0	1	1	0	1	0	1	1	1	1	1	6
0	1	1	1	1	1	1	0	0	0	0	7
1	0	0	0	1	1	1	1	1	1	1	8
1	0	0	1	1	1	1	1	0	1	1	9
1	0	1	0	1	1	1	0	1	1	1	A
1	0	1	1	0	0	1	1	1	1	1	b
1	1	0	0	1	0	0	1	1	1	0	C
1	1	0	1	0	1	1	1	1	0	1	d
1	1	1	0	1	0	0	1	1	1	1	E
1	1	1	1	1	0	0	0	1	1	1	F

Source: R.S. Sandige, *Modern Digital Design*, New York: McGraw-Hill, 1990, p. 252. With permission.

binary input code into the proper binary output code to drive a seven-segment display. The binary codes 0000 though 1111 represented on the signal inputs D(MSB) C B A(LSB) are converted to the binary output code on the outputs OA through OG. These outputs generate the hexadecimal characters 0 through F when applied to a seven-segment common cathode LED display. Each of the signal lines D through A (and D bar through A bar) must be capable of driving the number of gate inputs shown in the brackets (*fan-out requirement*) to both the proper high-level and low-level voltages. The gate-level circuits in Figure 3.17 are designed using the minimum *sum of products* (SOP) implementation derived from the 1s of the functions OA though OG, respectively, represented by the truth table in Table 3.4.

Implementing the circuit shown in Figure 3.17 would require 18 IC chips if the circuit was built out of discrete logic gates. A more efficient way (in terms of IC package or chip count) to implement the same combinational logic function would be to use a *medium-scale integration* (MSI) IC component. MSI components are single IC chips that contain more complicated logic functions than individual logic gates. For example, a 4- to 16-line decoder is used with some discrete logic gates in Figure 3.19 to implement the same seven-segment decoder as is shown in Figure 3.17. The tildes in Figure 3.19 are used to indicate the logical complements of D0 through D15.

The decoder circuit in Figure 3.19 requires only eight IC packages, as compared to the gate-level circuit in Figure 3.17 that requires 18 IC packages. Functionally, both circuits perform the same. The output equations for the circuit in Figure 3.19 are the canonical or standard SOP equations for the 0s of the functions OA though OG, respectively, represented by the truth table (Table 3.4). The gates that are shown in Figure 3.19 with more than four inputs are eight-input NAND gates. Each unused input is tied to VCC via a pull-up resistor (not shown on the logic diagram). Tying the unused input of a NAND gate effectively disables that particular input.

An even more efficient way to implement the same combinational logic function would be to utilize part of a simple programmable read-only memory (PROM) circuit such as the 27S19 fuse programmable PROM in Figure 3.20. An equivalent architectural gate structure for a portion of the PROM is shown in Figure 3.21. A full 4- to 16-line decoder exists on the left side of Figure 3.21 which decodes each line in the truth table (Table 3.4) so that every row line on the right side of the figure corresponds to the equivalent row in the truth table for each output function. The right half of Figure 3.21 is programmed using fuses. Intact fuses are represented by Xs. For every row in the truth table in which a 1 exists for a respective function output,

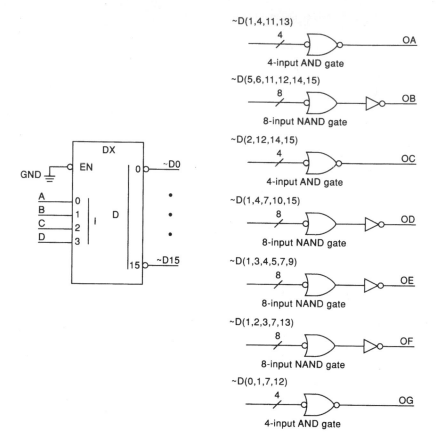

~D(1,4,11,13)
4-input AND gate — OA

~D(5,6,11,12,14,15)
8-input NAND gate — OB

~D(2,12,14,15)
4-input AND gate — OC

~D(1,4,7,10,15)
8-input NAND gate — OD

~D(1,3,4,5,7,9)
8-input NAND gate — OE

~D(1,2,3,7,13)
8-input NAND gate — OF

~D(0,1,7,12)
4-input AND gate — OG

FIGURE 3.19 Decoder logic circuit for binary to seven-segment hexadecimal character generator. (*Source*: R.S. Sandige, *Modern Digital Design*, New York: McGraw-Hill, 1990, p. 359. With permission.)

an X should appear in that output's column. The code for programming the PROM or generating the truth table of the function can be read either from the truth table (Table 3.4) or directly from each row of the circuit diagram in Figure 3.21 (expressed in hexadecimal: 7E, 30, 6D, 79, 33, 5B, 5F, 70, 7F, 7B, 77, 1F, 4E, 3D, 4F, and 47) beginning with the first line, which represents the output when the binary input is set to 0000, down to the last line, which represents the output when the binary input is set to 1111. The PROM solution for the combinational logic circuit requires only one IC package.

Programmable logic devices (PLDs), such as PROM, programmable array logic (PAL), programmable logic array (PLA), and field programmable logic array (FPGA) devices, are fast becoming the preferred devices for implementing combinational and sequential logic circuits when *application specific integrated circuits* (ASICs) are not used. These devices (a) use less real estate on a PC board, (b) shorten design time, (c) allow design changes to be made more easily, and (d) improve reliability because of fewer connections as compared to discrete logic gates.

Figure 3.22 shows a PAL16L8 implementation for the binary to seven-segment hexadecimal character generator that also requires a single IC package. The fuse map for this design was obtained

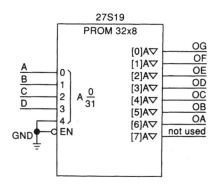

FIGURE 3.20 PROM implementation for binary to seven-segment hexadecimal character generator. (*Source*: R.S. Sandige, *Modern Digital Design*, New York: McGraw-Hill, 1990, p. 382. With permission).

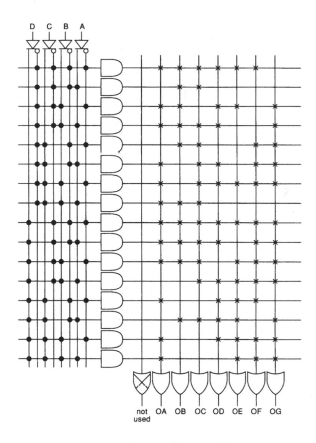

FIGURE 3.21 PROM logic circuit. (*Source*: R.S. Sandige, *Modern Digital Design*, New York: McGraw-Hill, 1990, p. 381. With permission.)

using the software program PLDesigner-XL. Karnaugh maps are handy tools that allow a designer to easily obtain minimum SOP equations for either the 1s or 0s of Boolean functions with up to four or five input variables. However, there is a host of commercially available software programs that provide not only Boolean reduction but also equation simulation and fuse map generation for PLDs and FPGAs. Xilinx Integrated Software Environment (ISE) is an example of a premier commercial software package available for logic synthesis for PLDs and FPGAs, for both combinational logic and sequential logic circuits.

Sequential Logic Circuits

A sequential logic circuit is a circuit that has feedback in which the output signals of the circuit are functions of all or part of the present state output signals of the circuit, in addition to any external input signals to the circuit. The vast majority of sequential logic circuits designed for industrial applications are synchronous or clock-mode circuits.

Synchronous Sequential Logic Circuits

Synchronous sequential logic circuits change states only at the rising or falling edge of the synchronous clock signal. To allow proper circuit operation, any external input signals to the synchronous sequential logic circuit must generate excitation inputs that occur with the proper setup time (*tsu*) and hold time (*th*) requirements, relative to the designated clock edge for the memory elements being used. Synchronous or clock-mode

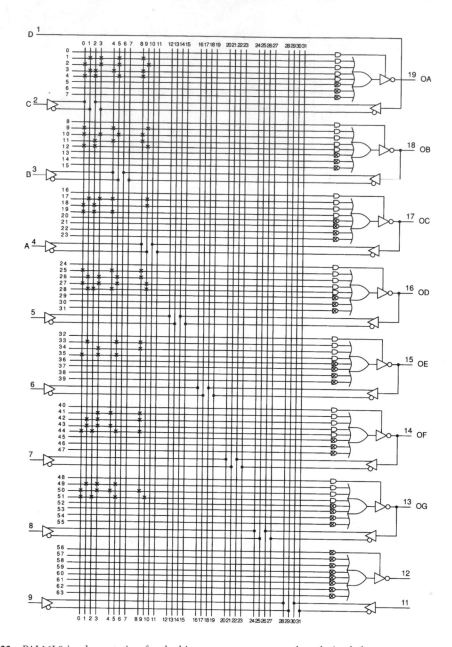

FIGURE 3.22   PAL16L8 implementation for the binary to seven-segment hexadecimal character generator. (*Source: PAL Device Data Book*, Advanced Micro Devices, Sunnyvale, Calif., 1988, pp. 5–46.)

sequential logic circuits depend on the present state of memory devices, called bistable devices or flip-flops (asynchronous sequential logic circuits), that are driven by a system clock as illustrated by the synchronous sequential logic circuit in Figure 3.23.

With the availability of edge-triggered D flip-flops and edge-triggered J-K flip-flops in IC packages, a designer can choose which flip-flop type to use for the memory devices in the memory section of a synchronous sequential logic circuit. Many designers prefer to design with edge-triggered D flip-flops rather than edge-triggered J-K flip-flops because D flip-flops are (a) more cost efficient, (b) easier to design with, and (c) more convenient, since many of the available PAL devices incorporate edge-triggered D flip-flops in the

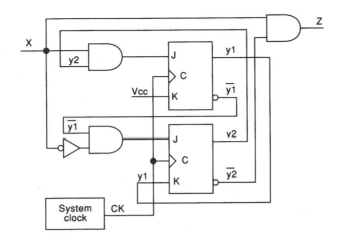

FIGURE 3.23 Synchronous sequential logic circuit using positive edge-triggered J-K flip-flops.

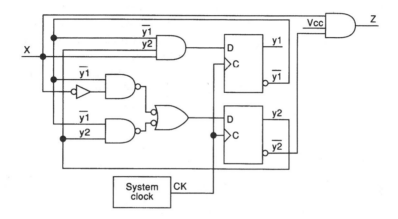

FIGURE 3.24 Synchronous sequential logic circuit using positive edge-triggered D flip-flops.

output section of their architectures. PAL devices that contain flip-flops in their output section are referred to as registered PALs (or, in general, registered PLDs). The synchronous sequential logic circuit shown in Figure 3.24 using edge-triggered D flip-flops functionally performs the same as the circuit in Figure 3.23.

Notice that, in general, more combinational logic gates will be required to implement synchronous sequential logic circuits using D flip-flops than to implement circuits using J-K flip-flops as memory devices. Using a registered PAL such as a PAL16RP4A would only require one IC package to implement the circuit in Figure 3.24. The PAL16RP4A has four edge-triggered D flip-flops in its output section, of which only two are required for this design.

Generally speaking, synchronous sequential logic circuits can be designed much more quickly than fundamental-mode asynchronous sequential logic circuits. Synchronous sequential logic circuits are the most common digital circuits today. With a system clock and edge-triggered flip-flops, a designer does not have to worry about *hazards* or *glitches* (momentary error conditions that occur at the outputs of combinational logic circuits), since outputs are allowed to become stable before the next clock edge occurs. Thus, sequential logic circuit designs allow the use of combinational hazardous circuits as well as the use of arbitrary state assignments, provided the resulting combinational logic gate count or package count is acceptable.

Asynchronous Sequential Logic Circuits

Asynchronous sequential logic circuits may change states immediately when any input signal changes state (either a level change for a fundamental mode circuit or a pulse for a pulse mode circuit). No other input signal change (either level change or pulse) is allowed until the circuit reaches a stable internal state. Latches and edge-triggered flip-flops are asynchronous sequential logic circuits and must be designed with care by utilizing hazard-free combinational logic circuits and race-free or critical *race-free state assignments*. Both hazards and race conditions interfere with the proper operation of asynchronous logic circuits. The gated D latch circuit illustrated in Figure 3.25 is an example of a fundamental-mode asynchronous sequential logic circuit that is used extensively in microprocessor systems for the temporary storage of data.

Quad, octal, 9-bit, and 10-bit transparent latches are readily available as off-the-shelf IC devices for these types of applications. For proper asynchronous circuit operation, the signal applied to the data input D of the fundamental-mode circuit in Figure 3.25 must meet a minimum setup time and hold time requirement relative to the control input C, changing the latch to the memory mode when C goes to 0. This is a basic requirement for asynchronous circuits with level inputs, i.e., only one input signal is allowed to change at one time. Another restriction is letting the circuit reach a stable state before allowing the next input signal to change.

An example of a reliable pulse-mode asynchronous sequential logic circuit is shown in Figure 3.26. While the inputs to asynchronous fundamental-mode circuits are logic levels, the inputs to asynchronous pulse-mode circuits are pulses. Pulse-mode circuits have the restriction that the maximum pulse width of any input pulse must be sufficiently narrow that an input pulse is no longer present when the new present state output signal becomes available. The purpose of the double-rank circuit in Figure 3.26 is to ensure that the maximum pulse width requirement is easily met, since the output is not fed back until the input pulse is removed, (i.e., goes low or goes to logic 0). The input signal to pulse-mode circuits must also meet the following restrictions: (a) only one input pulse may be applied at one time, (b) the circuit must be allowed to reach a new stable state before applying the next input pulse, and (c) the minimum pulse width of an input pulse is determined by the time it takes to change the slowest flip-flop used in the circuit to a new stable state.

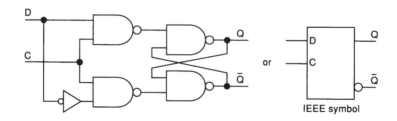

FIGURE 3.25   Fundamental-mode asynchronous sequential logic circuit. (*Source*: R.S. Sandige, *Modern Digital Design*, New York: McGraw-Hill, 1990, p. 470. With permission.)

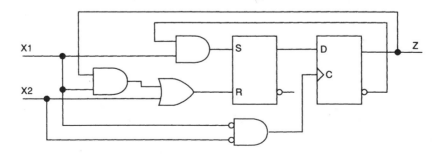

FIGURE 3.26   Double-rank pulse-mode asynchronous sequential logic circuit. (*Source*: R.S. Sandige, *Modern Digital Design*, New York: McGraw-Hill, 1990, p. 615. With permission.)

Defining Terms

Application specific integrated circuits (ASICs): Integrated circuits designed using standard cell libraries for logic gates.

Asynchronous circuit: A sequential logic circuit without a system clock.

Combinational logic circuit: A circuit with external output signals that are totally dependent on the external input signals applied to the circuit.

Fan-out requirement: The maximum number of loads (gate inputs) a device output can drive and still provide dependable 1 and 0 logic levels.

Hazard or glitch: A momentary output error that occurs in a logic circuit because of input signal propagation along different delay paths in the circuit.

Hexadecimal: The name of the number system with a base or radix of 16 with the usual symbols of 0 . . . 9,A,B,C,D,E,F.

Medium-scale integration: A single packaged IC device with 12 to 99 gate-equivalent circuits.

Race-free state assignment: A state assignment made for asynchronous sequential logic circuits such that no more than a 1-bit change occurs between each stable state transition, thus preventing possible critical races.

Sequential logic circuit: A circuit with output signals that are dependent on all or part of the present state output signals fed back as input signals, as well as any external input signals if they should exist.

Sum of products (SOP): A standard form for writing a Boolean equation that contains product terms (complemented or uncomplemented input variables connected with AND) that are logically summed (connected with OR).

Synchronous or clock-mode circuit: A sequential logic circuit that is synchronized with a system clock.

References

Advanced Micro Devices, *PAL Device Data Book*, Sunnyvale, CA: Advanced Micro Devices, Inc., 1988.

ANSI/IEEE Std 91–1984, *IEEE Standard Graphic Symbols for Logic Functions*, New York, NY: The Institute of Electrical and Electronics Engineers, 1984.

ANSI/IEEE Std 991–1986, *IEEE Standard for Logic Circuit Diagrams*, New York, NY: The Institute of Electrical and Electronics Engineers, 1986.

K.J. Breeding, *Digital Design Fundamentals*, 2nd ed., Englewood Cliffs, NJ: Prentice-Hall, 1992.

F.J. Hill and G.R. Peterson, *Introduction to Switching Theory & Logical Design*, 3rd ed., New York, NY: John Wiley, 1981.

M.M. Mano, *Digital Design*, 3rd ed., Englewood Cliffs, NJ: Prentice-Hall, 2001.

E.J. McCluskey, *Logic Design Principles*, Englewood Cliffs, NJ: Prentice-Hall, 1986.

Minc, *PLDesigner-XL, The Next Generation in Programmable Logic Synthesis*, Version 3.5, User's Guide, Colorado Springs, CO: Minc, Incorporated, 1996.

R.S. Sandige, *Digital Design Essentials*, Upper Saddle River, NJ: Prentice-Hall, 2002.

J.F. Wakerly, *Digital Design: Principles and Practices*, 3rd ed., Upper Saddle River, NJ: Prentice-Hall, 2001.

Further Information

The monthly magazine *IEEE Transactions on Computers* presents papers discussing logic circuits, for example, "On the Number of Tests to Detect All Path Delay Faults in Combinational Logic Circuits," in its January 1996 issue, Volume 45, Number 1, pp. 50–62.

The monthly magazine *IEEE Journal on Solid-State Circuits* presents papers discussing logic circuits, for example, "Automating the Design of Asynchronous Sequential Logic Circuits," in its March 1991 issue, pp. 364–370.

Also, the monthly magazine *Electronics and Wireless World* presents articles discussing logic circuits, for example, "DIY PLD," in its June 1989 issue, pp. 578–581.

3.3   Registers and Their Applications

B.R. Bannister, D.G. Whitehead and J.M. Gilbert

The basic building block of any register is the flip-flop. But just as there are several types of flip-flops, there are many different register arrangements. An idea of the vast range and their interrelationships is shown in Figure 3.27.

The simplest type of flip-flop is the set-reset flip-flop, constructed simply by cross-connecting two NAND/NOR gates. This forms an asynchronous flip-flop, in which the set or reset signal determines *what* the flip-flop is to do and *when* it is to operate. If a state change is required, the flip-flop begins to change states when the input change is detected. This flip-flop is useful as a latch to detect when an event has occurred. It is often referred to as a flag since it indicates to other circuitry that the event has occurred and remains set until the controlling circuitry responds by resetting it.

Flags are widely used in digital systems to indicate a state change. All microprocessors have a set of flags which are used in deciding whether a program branch should be made. The flags available in microprocessors typically include those indicating the results of the most recent operation: zero, carry, sign and overflow [Intel, 1989; Renesas, 2001]. Specific microprocessors may also include flags associated with a particular operation: debug, register bank select [Renesas, 2001] or parity [Intel, 1989]. For convenience, although they all act independently, these flags are grouped together into what is known as the flag register or program status-word register.

Gated Registers

The conventional meaning of register applies to a collection of identical flip-flops activated simultaneously as a set rather than individually. Controlling when flip-flops set or reset makes use of synchronous flip-flops, leaving the D or J-K inputs to determine *what* the flip-flop is to do logically. That can include setting or resetting while some other signal, typically the clock, determines *when* it does it.

The signal controlling the register's input is applied to all flip-flops simultaneously, and its action depends on the type of flip-flop used. Edge-triggered flip-flops set or reset according to the value on the data inputs when the control signal changes. After the flip-flops settle into their new values, the register content is available

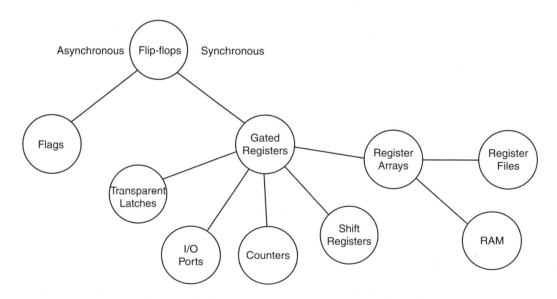

FIGURE 3.27   The register family.

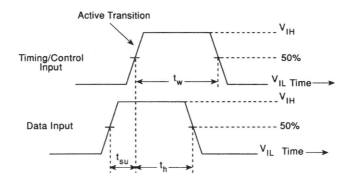

FIGURE 3.28   Control timing parameters.

at the output gating. The correct operation of the circuitry depends upon timing criteria being satisfied and minimum values quoted by the manufacturers. Each is the smallest time that *w* the device is guaranteed to operate correctly, but in practice, the device probably functions satisfactorily with smaller time intervals on at least some of the parameters. The main timing constraints occur at the flip-flop's inputs and are illustrated in Figure 3.28. The interval preceding the active transition of the control input is the setup time, t_{su}, during which the data signal must be held steady; t_h is the hold time and is the interval during which the data signal must be retained following the active transition of the control input; t_w is a minimum pulsewidth indication that applies to the control inputs such as the clock, reset and clear. The clock pulse width is usually quoted both for the high state ($t_{w(H)}$) and for the low ($t_{w(L)}$) and is related to the maximum clocking frequency of the flip-flops in the register.

In general, these registers are available as 4-bit, 8t-bit or higher powers of two and are used in multiples of 8 bits in most cases. The number of flip-flops in each register determines the width of the data bus in a microprocessor or other bus-based system and describes the microprocessor. Early

microprocessors, such as the Z80, as well as many modern microcontrollers, are said to be 8-bit processors, indicating that the working registers are 8 bits wide while the registers in modern processors are typically 32 bits wide. The bit-values in the register normally represent a numerical value in standard fixcd-point binary, floating point or some other coded form. Alternatively, they may indicate a logical pattern, such as the settings of switches in an industrial controller.

Such collections of flip-flops form the core of any synchronous circuits, as shown in Figure 3.29. The data stored in the registers is processed though a combinational logic circuit and returned to one or more register(s) which may or may not include the register from where the data was taken.

A bus-organized system allows flexibility to control the register that provides the data source and the register(s) that are the recipients of the data. In such systems it is necessary to control when the data held in a register is fed on to the output bus. This usually is achieved by means of three-state (3S) gates at the register outputs which are disabled, or set to their

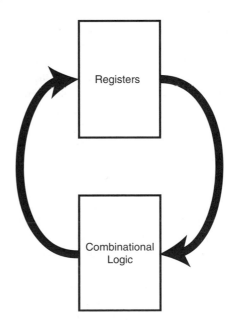

FIGURE 3.29   General synchronous circuit.

high-impedance state, until the data is required. This allows other devices connected to the common bus to assert their outputs.

A multi-register, bus-structured digital system will have the *n*th bit of each register connected to bit *n* of the data bus at both the input and output of the register. To transfer data from one register to another, or to transfer a copy of the contents of one register to another register, the output gates of the source register must be enabled so that the data is fed on to the bus. This data becomes available at the register inputs and is latched in under the control of the appropriate signals. Under normal circumstances, only one register can assert the value on to a common bus, while several registers may read data from it. The possibility of a bus contention, when more than one register attempts to control the bus, should be avoided in the design.

Sets of registers are often grouped together as register arrays. The advantage of using arrays of registers instead of individual ones is that the array may share some common connections. This reduces the number of connections required. Registers in an array may be arranged as files where multiple read/write operations take place simultaneously or, for Random Access Memory (RAM), where only a single read/write operation is permitted at one time. Array files permit greater operational flexibility than RAM but at the price of additional data, address and control lines.

An alternative flip-flop is the transparent latch, developed from the simple latch. When enabled by a control signal, *C*, by setting *C* high or at 1, the latch becomes a transparent section of the data path. The data value at the input reappears at the output. When the control signal disables the latch, *C* is low or "0". However, the last value applied to the latch is "frozen" and held until the control signal is taken high again. The 74LS373, Figure 3.30(a), consists of eight transparent latches with a common control input labeled ENABLE. The 74LS374, Figure 3.30(b), is a typical 8-bit register using positive edge-triggering for all flip-flops. It includes 3S output gates designed for driving highly capacitive loads, such as those found in bus-organized systems, and that respond to an output control signal operating quite independently of the flip-flops. Typical minimum timing figures for the 74LS373 and 74LS374 are shown in Figure 3.30(c), and the waveforms occurring for the two different types of register are illustrated in Figure 3.30(d).

Shift Registers

There are two modes of operation for a register, either serial or parallel. The registers considered so far have operated in parallel mode. In parallel operation, the group of bits held in the register may be altered independently during a single clock pulse. In serial operation, data bits are input (or output) sequentially to (or from) the register, one bit for every clock pulse. A register that has the facility to move the stored bits left or right and one place at a time under the control of the clock pulse is called a shift register (Figure 3.31). It is possible to combine serial and parallel operations in a single register. Registers may be designed that allow any combination of parallel and/or serial loads, along with parallel and/or serial reads as required.

Shift registers are normally implemented by means of D, S-R or J-K flip-flops. As an example, the 74LS165A (Figure 3.32) consists of eight S-R-type flip-flops with clock, clock inhibit and shift/load control inputs. Data presented to the eight separate inputs is loaded into the register in parallel when the shift/load input is taken low. Shifting occurs when the shift/load input is high and the clock pulse is applied, the action taking place on the low-to-high transition of the clock pulse. Registers are available that switch on the other clock edge. For example, the 74LS295A, which is a 4-bit shift register with serial and parallel operating modes, carries out all data transfers and shifting operations on the high-to-low clock transition. This device also provides 3S operation. Selection of the mode of operation is carried out by suitable combinations of the MODE SELECT inputs.

One of the most common applications of shift registers is in serial communication between devices. Such serial communication is particularly important when devices are physically separated and it is not convenient to have a large number of electrical connections between the two devices. By utilizing a serial connection, one may use significantly fewer connections than would be required for parallel communication, in which all bits are transferred simultaneously along individual connections. There are many types of serial communication but they generally possess the features illustrated in Figure 3.33. Data for transmission is loaded in parallel into the parallel-in, serial-out shift register. It is then shifted out and transmitted one bit

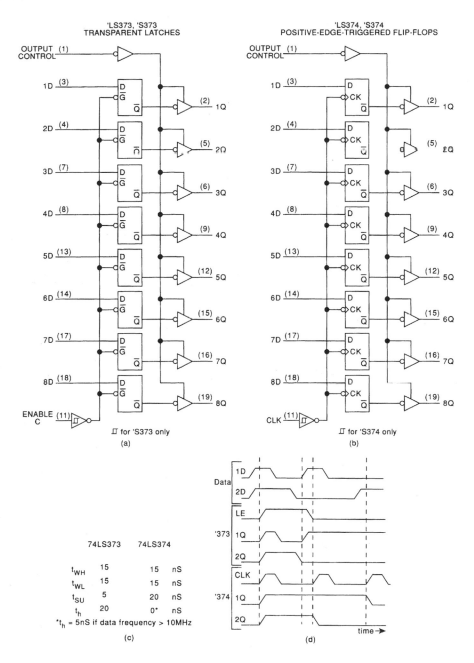

FIGURE 3.30 (a) 74LS373 and (b) 74LS374. (c) Typical minimum timing values and (d) *I/O* waveforms for the 74LS373/374. (*Source*: *TTL Data Book*, vol. 1, Texas Instruments, Inc. With permission.)

at a time. The receiver loads the received bits into the serial-in, parallel-out shift register and, once a complete set of data bits has been received, it will allow the receiver to read the contents in parallel. If the transmitter and receiver share a common clock signal then the transmission is said to be synchronous, while if each generates its own clock then the transmission is asynchronous. In the latter case the device is called a universal asynchronous receiver transmitter (UART). Many devices can act in both synchronous and asynchronous modes and are referred to as universal synchronous/asynchronous receiver transmitters (USARTs).

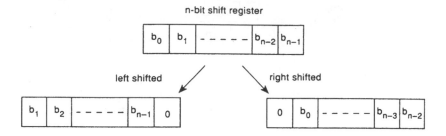

FIGURE 3.31 Shift register operation.

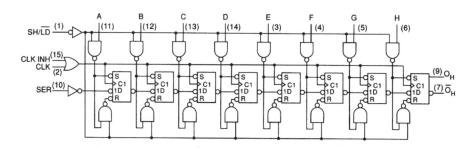

FIGURE 3.32 The 74LS165A shift register. (*Source*: *TTL Data Book*, Vol. 1, Texas Instruments, Inc. With permission.)

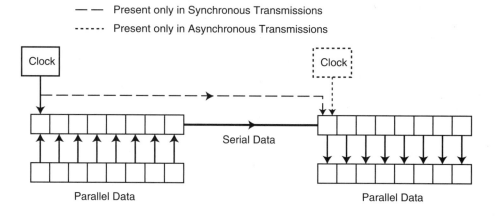

FIGURE 3.33 Shift registers used in serial communications.

Irrespective of whether synchronous or asynchronous transmission is used, it is often necessary to allow two-way communications. A connection which allows transfer in only one direction is referred to as a simplex line and if bi-directional communication is required then two simplex lines are required. A connection which allows bi-directional communication but only in one direction at a time is called a half duplex line, while one which allows simultaneous transmission in both directions is a full duplex.

In synchronous mode transmission the common clock may be generated by either the receiver or the transmitter circuit. In the case of two-way communications, one device (the master) may generate a common clock for both transmission to and reception from slave devices, as in the Serial Peripheral Interface (SPI), or each may generate a clock when it wishes to transmit, as in the Inter IC (I^2C) bus. In the case of asynchronous transmission, such as RS-232, the transmitter and receiver have independent clocks which are designed to operate at approximately the same frequency. Once the receiver detects the beginning of a transmission, it uses its own clock circuit to determine when each subsequent bit should arrive. In several of these serial protocols additional lines may be used to indicate, for instance, that a device is ready to send or when a particular device is being addressed.

As in parallel registers, data may be modified as it is transferred from one register to another. For example, contents of one register may be added to the contents of another, the resulting sum then being returned to one of the two registers. The example shown in Figure 3.34 is that of a serial adder. The registers could each be made up of 74LS165A devices as previously described. The data to be added is transferred to the two shifting registers, A and B, using parallel loading of data into the registers, carried out prior to the application of the shift clock pulses shown.

On the rising edge of each clock pulse the data is right-shifted one place. The resultant sum of the two bits plus any carry bit is, on the same clock edge, entered back into register A. The D-type flip-flop is used to delay the carry bit until the next add time: data entered at D does not appear at Q until the falling edge of the clock pulse has occurred, and at such time it is, therefore, too late to modify the previous addition. At the end of the addition process, when all the data bits have been shifted through the registers, register A contains the sum, and register B is unchanged.

A single shift register can be arranged to provide its own input by means of feedback circuits, and its action then becomes autonomous, since the only external signal required is the clock signal. There are only a finite number of states of the feedback shift register (FSR), and the output sequence from a register with a single feedback path will, therefore, repeat with a cycle length not greater than $2n$ bits, where n is the number of flip-flops in the register. This property can be used to create a counter known as a Johnson counter in which the shift register has the J and K inputs of the first stage fed directly from the $\bar{Q}$ and Q outputs, respectively, of the last stage. This simple form of feedback leads to the name twisted-ring counter, and the result is the generation of a creeping or stepping code with $2n$ different states. This form of counter is convenient only when the count is small, as the number of flip-flops quickly becomes excessive, but is ideal for a simple decade counter. Unlike standard binary decade counters, the Johnson decade counter requires five flip-flops but no additional feedback circuitry. Gating needed to detect specific settings of the counter is also very simple [Bannister and Whitehead, 1987].

Another set of sequences is obtained if the feedback uses exclusive-OR, that is, modulo-2 functions. By correct choice of function, the linear feedback shift register (LFSR) so formed generates a maximal length sequence, or m-sequence. A maximal length sequence has a length of $2n - 1$ bits (the all-zeros state is not included, since the mod-2 feedback would not allow any escape from that state, so the sequence has a 0 missing) with useful properties of repeatable randomness and is, therefore, described as a pseudorandom binary sequence (PRBS). The number of maximal length sequences for a register of length n, and the feedback

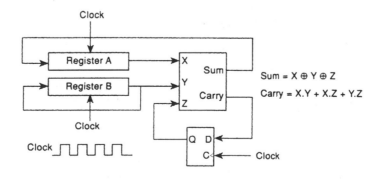

FIGURE 3.34   The serial adder using shifting registers.

arrangements to achieve them, are not at all obvious, but have been worked out for a large number of cases [Messina, 1972]. A 4-bit LFSR will produce only one maximal length sequence, but a 10-bit register can produce 30 distinct m-sequences, and a 30-bit register produces no less than 8,910,000 distinct sequences!

Shift registers can be used in parallel to form a first-in, first-out (FIFO) memory. These are typically 128×8-bit register memories with independent input and output buses. At the input port, data is controlled by a shift-in clock operating in conjunction with an input ready signal which indicates whether the memory is able to accept further words or is now full. The data entered is automatically shifted in parallel to the adjacent memory location if it is empty and as this continues the data words stack up at the output end of the memory. At the output port, data transfers are controlled by a shift-out clock and its associated output ready signal. The output ready signal indicates either that a data word is ready to be shifted out or that the memory is now empty. FIFOs can easily be cascaded to any desired depth and operated in parallel to give any required word length. This type of memory is widely used in controlling transfers of data between digital subsystems which operate at different clock rates and is often known as an elastic buffer.

Register Transfer Level (RTL) Notation

The transfer of data between registers may be described using a simple notation termed the register transfer level (or register transfer language) (RTL) notation. For data transferred from register A to register B we write: $B \leftarrow A$. The symbol $\leftarrow$ is called the transfer operator. Note that this statement does not indicate how many bits are to be transferred. To define the size of the register we declare the size thus: A[8], B[16], here defining an 8-bit register and a 16-bit register. If the action to be taken is the transfer of the most significant bit (7th bit) of register A to the least significant bit (bit 0) of register B, then we write: $B[0] \leftarrow A[7]$. Usually data is transferred by the control signal or a clock pulse. If such a signal is designated "C" then we would describe the action by C: $B \leftarrow A$.

Returning to the serial adder circuit shown earlier, we could describe the register transfers thus:

$$A[8], B[8], D[1]$$

$$C : A[7] \leftarrow A[0] \oplus B[0] \oplus D[0], B[7] \leftarrow B[0], D[0] \leftarrow \text{Carry}$$

Here in the declaration statement we refer to the D-type flip-flop as a single-bit register. Simultaneous processes are separated by a comma; sequential processes would be separated by a semicolon. The symbol "$\oplus$" is the exclusive-or (XOR) operator. Other logical operations include NOT, AND, and OR. The AND operation is also called the masking operation because it can be used to remove (or select) specific sections of data from a register. Thus the operation $A[8] \leftarrow A[8]$ and 3CH will result in the most significant two bits and the least significant two bits of the eight-bit register A being set to zero. Note that 3CH refers to the hexadecimal number 3C; i.e., 00111100 in binary. Some other terms commonly used are as follows:

$D \leftarrow \overline{A}$	transfer the complement of A to D
$A \leftarrow A + 1$	increment A
$A[8:15] \leftarrow B[8:15]$	transfer bits 8 through 15 from B to A

In order to differentiate between arithmetic and logical operations it is usual to represent logical OR and AND by $\wedge$ and $\vee$. Table 3.5 lists some typical RTL examples that include arithmetic, bit-by-bit logic, shift, rotate, scale and conditional operations. It is assumed that the three registers are set initially to A = 10110, B = 11000 and C = 00001.

The *rotate* operations included in Table 3.5 differ from the standard shift in that the data which is shifted out of one end of the register is inserted into the opposite end of the register. Thus rotating left 110000 by 1 bit results in 100001.

RTL provides a precise means of specifying circuit behavior and so is often used as part of hardware description languages (HDL). See section 9.4 for further details.

TABLE 3.5 Typical RTL Examples

Type of Operation	Meaning	Register Bits after Operation
General		
$A_3 \leftarrow A_2$	Bit 2 of A to bit 3 of A	A = 11110
$A_3 \leftarrow B_4$	Bit 4 of B to bit 3 of A	A = 11110
$A_{1-3} \leftarrow B_{1-3}$	Bits 1 through 3 of B to bits 1 through 3 of A	A = 11000
$A_{1,4} \leftarrow B_{1,4}$	Bits 1 and 4 of B to bits 1 and 4 of A	A = 10100
$A_{1-3} \leftarrow B_z$	Groups of bit Z of B to bits 1 through 3 of A	A = 11000
Arithmetic		
$B \leftarrow 0$	Clear B	B = 00000
$A \leftarrow B + C$	Sum of B and C to A	A = 11001
$A \leftarrow B - C$	Difference B – C to A	A = 10111
$C \leftarrow C + 1$	Increment C by 1	C = 00010
Logic		
$A \leftarrow B \wedge C$	Bit-by-bit AND result of B and C to A	A = 00000
$A \leftarrow B \vee C_4$	OR operation result of B with bit 4 of C to A	A = 11000
$C \leftarrow \bar{C}$	Complement C	C = 11110
$B \leftarrow \bar{B} + 1$	2's complement of B	B = 01000
$B \leftarrow A \oplus C$	XOR operation result of A and C to B	B = 10111
Serial		
$B \leftarrow$ sr B	Shift right B one bit	B = 01100
$B \leftarrow$ sl B	Shift left B one bit	B = 10000
$B \leftarrow$ sr2 B	Shift right B two bits	B = 00110
$B \leftarrow$ rr B	Rotate right B one bit	B = 01100
$B \leftarrow$ rl2 B	Rotate left two bits	B = 00011
$B \leftarrow$ scr B	Scale B one bit (shift right with sign bit unchanged)	B = 11100
$B \leftarrow$ scl B	Scale B one bit (shift left with sign bit unchanged)	B = 10000
B, C $\leftarrow$ sr2 B, C	Shift right concatenated B and C two bits	B, C = 0011000000
Conditional		
If $(B_4 = 1)$ then $(C \leftarrow 0)$	If bit 4 of B is a 1, then C is cleared	C = 00000
If $(B \geq C)$ then $(B \leftarrow 0, C_1 \leftarrow 1)$	If B is greater than or equal to C, then B is cleared and C is set to 1	B = 00000

Initial values: A = 10110, B = $\underset{z}{\underline{11000}}$ and C = 00001.

Source: E.L. Johnson and M.A. Karim, *Digital Design*, Boston: PWS Publishers, 1987. With permission.

Input/Output Ports

The working registers provided in microprocessors may be thought of as high-speed extensions to the memories used for storing programs and data. The random access memories (RAMs) themselves are also arrays of registers, though the form of circuit used differs considerably from the more conventional register. The need to transfer data in and out of the system has led manufacturers to produce special registers which are further extensions to the internal memory and are known as input and output ports. These ports may be integrated within the processor and connected to the internal bus or may be implemented as a separate device attached to the external data bus and controlled by external control lines and selected using the external address lines.

A range of external input/output devices was developed for early microprocessors. One of the simplest input/output ports is the Intel 8212 (Figure 3.35). This has two modes of operation selected by the mode input, MD. With MD at 0 the device acts as an input port and a peripheral unit can enter data on the DI lines by sending a high strobe signal, STB. When the central processor is ready for the data it selects the port by setting the correct address bits on the device select inputs. This enables the 3S output buffers and data is routed to the processor data bus via the DO lines. This device also includes a service request flip-flop to generate an

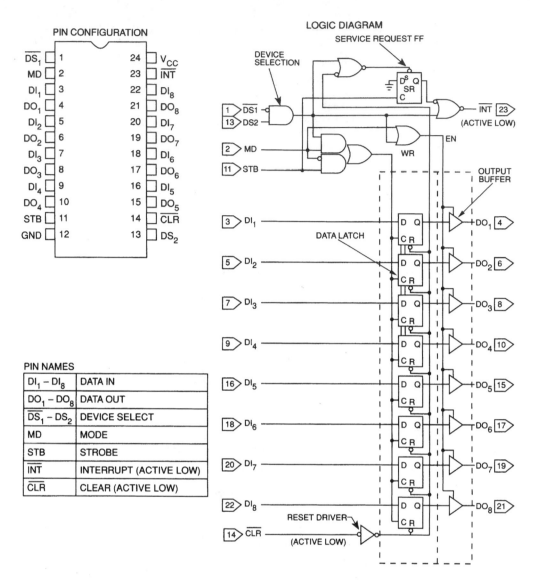

FIGURE 3.35 The Intel 8212 parallel inout/output port. (*Source*: *Microprocessor Component Handbook*, Intel Corp. With permission.)

interrupt signal to the processor when the data is ready. In the alternative mode of use, with the mode input at 1, the device select logic routes data from the processor, now connected to the DI inputs, so the 8212 acts as an output port. The data is immediately available to the peripheral unit on the DO lines, as the 3S output buffers are permanently enabled. A more sophisticated range of input and output facilities is provided by most microprocessor manufacturers in the form of programmable input/output ports or peripheral interfaces. These are special registers with appropriate buffers and additional built-in control and status registers to facilitate proper system operation. Typical features include not just parallel ports but USARTs based around shift registers and timer/counter circuits. Such external peripherals have become less popular as microprocessors have been developed with more sophisticated on-board interfaces.

In a typical modern microprocessor each pin may have several different functions and within each function may operate in a number of different modes. Figure 3.36 shows the structure of a parallel input/output port of the M16C microprocessor [Renesas, 2001]. The operation of each bit of this port is controlled by three

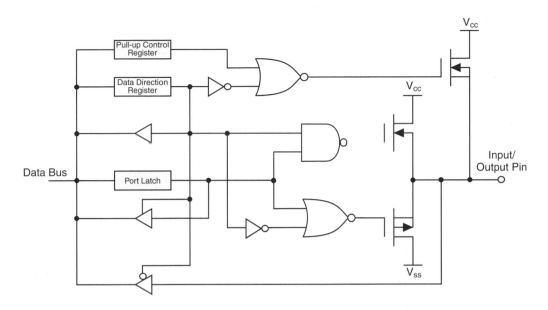

FIGURE 3.36   Parallel input/output port structure of M16C microprocessor.

registers. The data direction register (DDR) determines whether each bit of the port acts as an input or an output while the port latch stores the data to be output by the port. When DDR is low the data applied to the pin by the external circuit may be read by the microprocessor. When DDR is high the port acts as an output and the data stored in the port latch is asserted on the port pin (this data may also be read back by the microprocessor if desired). The third register is the pull-up control register which is only effective when the port is in input mode. When the corresponding bit of this register is high the pull-up FET acts as a pull-up resistor while if the bit is low, the pull-up FET is turned off and there is no pull-up resistance on the input.

Counters

A register can be loaded with any combination by applying the correct bit pattern to the input data lines and activating the control line. As with the feedback shift registers, it is then only a small step to arrange that the register itself provides the input data by use of feedback connections and, if other circuitry is included to increment the value each time, we have a synchronous counter. The 74LS191 (Figure 3.37) is a programmable counter which retains the facility for parallel loading of external data.

   Each output may be preset to either level by entering the data at the inputs while the LOAD signal is low. The outputs change to the new values independently of the count pulses, and counting continues when pulses are applied to the clock input. The master-slave flip-flops are triggered by a low-to-high transition of the clock. The "terminal count" and "ripple clock" outputs facilitate cascading of several counters. The ripple clock carry/borrow output signal, RCO, is a pulse equal in length to the clock pulse when the counter overflows or underflows, that is, when it is incremented from 1111 or decremented from 0000. By using this signal to reload the value at the data inputs we create a counter of modulus less than 16. Figure 3.38, for example, shows the arrangement to give a modulo-5 count, by reloading 1011 each time the ripple clock pulse occurs.

Registered ASICs

Developments in application specific integrated circuits (ASICs), programmable logic devices (PLDs) and field-programmable gate arrays (FPGAs) over recent years have provided digital system designers with a wide range of flexible devices which can be programmed for the specific job in hand. These devices typically have a

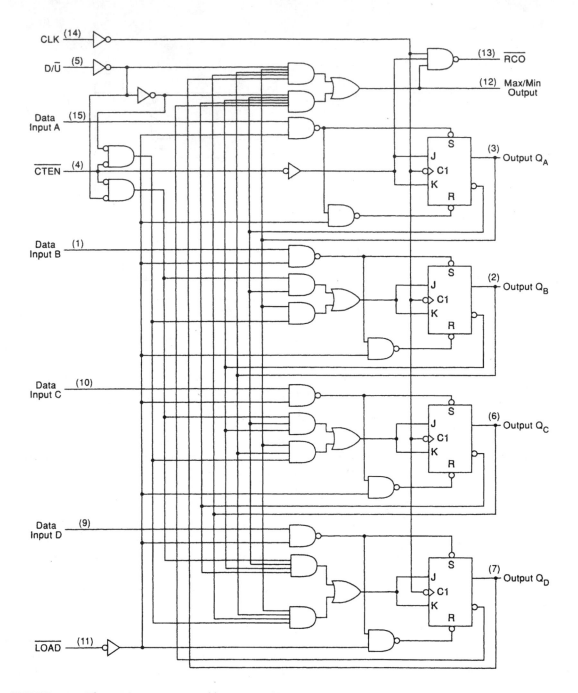

FIGURE 3.37 The 74LS191 programmable counter. (*Source*: *TTL Data Book*, Vol. 1, Texas Instruments, Inc. With permission.)

set of input/output blocks which may be programmed to link to a set of configurable logic blocks which contain both combinational elements and flip-flops. Devices also often include clock, initialization, self-test and program/read-back circuits. Over time, the structure of these devices has not changed radically but they have increased significantly in terms of the number of elements contained. The Xilinx Spartan XC3S5000, for example, has approximately 5 million logic gates and over 8000 configurable logic blocks (each of which

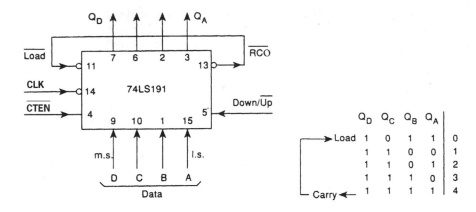

FIGURE 3.38 Programmable counter giving modulo-5.

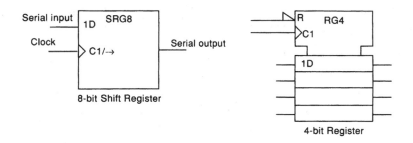

8-bit Shift Register

4-bit Register

FIGURE 3.39 Standard graphic symbols. (*Source*: IEEE, *Standard Graphic Symbols for Logic Functions*, ANSI/IEEE Std. 91–1984, New York, 1984. With permission.)

contains 8 flip-flops) which may be linked to 784 user input/output pins. Clearly to design circuits involving this level of complexity requires sophisticated CAD tools which are supplied by the device vendors.

Standard Graphic Symbols

The use of standardized graphical symbols is becoming widespread and the family of registers have their own coherent set of symbols. Two representative examples are given in Figure 3.39. As shown, the 8-bit shift register is designated SGR8. The direction of shift is given by the arrow. The "1D" is part of a notation called dependency notation. In the 4-bit parallel register, designated RG4, clock input C1 controls the inputs labeled 1D, of which only one of four is shown. The reset "R" and the clock are common to all units and are shown as inputs to the common block. The external reset line carries a polarity symbol which indicates that a low signal must be applied to reset the 4-bit register. For further details see the References at the end of this section.

Defining Terms

Autonomous operation: Operation of a sequential circuit in which no external signals, other than clock signals, are applied. The necessary logic inputs are derived internally using feedback circuits.

Duplex: Mode of serial communication which allows bidirectional transmission across a connection. Full duplex allows simultaneous communications in both directions while half duplex permits transmission in only one direction at a time.

Input/output port: A form of register designed specifically for data input-output purposes in a bus-oriented system.

Linear feedback shift register (LFSR):   An autonomous feedback shift register in which the feedback function involves only exclusive-OR operations.

Parallel operation: Data bits on separate lines (often in multiples of eight) are transferred simultaneously under control of signals common to all lines.

Register: A circuit formed from several identical gated flip-flops or latches and capable of storing several bits of data.

Serial operation: Data bits on a single line are transferred sequentially under the control of a single signal.

Simplex: Mode of serial communication which allows unidirectional transmission across a connection.

References

B.R. Bannister and D.G. Whitehead, *Fundamentals of Modern Digital Systems*, London: Macmillan, 1987.

M.D. Ercegovac, T. Lang, and J. Moreno, *Introduction to Digital Systems*, New York, NY: Wiley, 1999.

IEEE, Standard Graphic Symbols for Logic Functions, ANSI/IEEE Std. 91–1984, New York, 1984.

Intel Corporation, *Microprocessor and Peripheral Handbook*, Santa Clara, CA, 1983.

E.L. Johnson and M.A. Karim, *Digital Design*, Boston, MA: PWS Engineering, 1987.

M.M. Mano, *Digital Design*, 3rd ed., Englewood Cliffs NJ: Prentice Hall, 2002.

A. Messina, "Considerations for Non-binary Counter Applications," *Computer Design*, vol. 11, no. 11, 1972.

Renesas M16C/62 group datasheet, 2001.

Texas Instruments, Inc., *TTL Data Book*.

Xilinx Spartan-3 FPGA Data Sheet, 2004.

Further Information

The monthly journal *IEEE Transactions on Computers* regularly has articles involving the design and application of registers and associated systems. Further information can be obtained from IEEE Service Center, 445 Hoes Lane, P.O. Box 1331, Piscataway, NJ 08855–1331.

The *IEE Proceedings, Computers and Digital Techniques*, published bimonthly by the Institution of Electrical Engineers (Michael Faraday House, Six Hills Way, Stevenage, Herts. SG1 2AY, UK), is also a useful source of information on the application of register devices.

3.4   Programmable Arrays

George A. Constantinides

Introduction

This article presents an overview of historical and modern programmable arrays of logic, their applications, and the tools required to make use of them. Programmable logic devices have long been used for prototyping application specific integrated circuit designs, and are fast becoming the medium of choice for small to medium end-product production runs.

The first section of this article surveys the evolution of the programmable logic device from the programmable logic array of the 1970s to the platform field-programmable gate array of 2005. With the ever-increasing density and complexity of these devices, it is important to consider methods to automate or partially automate design for them; this is the subject of the second section of the article. Finally, we shall examine some of the areas of application of this impressive technology.

Architectures

Architectures of programmable logic devices have evolved over the years. The first programmable arrays specifically targeting logic implementation were called programmable logic arrays (PLAs), introduced in the 1970s. PLAs allow the direct implementation of Boolean sum-of-product expressions, through their programmable AND plane and programmable OR plane. In time the PLA evolved into the programmable array logic (PAL) device, where the programmability lies only in the AND plane while the OR plane is fixed, providing better speed and silicon density, as illustrated in Figure 3.40. PALs are very successful at performing simple logical functions, but the complexity of the programmable AND plane grows rapidly with the size of the circuit. The further evolution of the PAL into the complex programmable logic device (CPLD), integrating several PAL-like structures onto a single device, is an attempt to overcome this limitation. The CPLD has become widespread, and is still commonly used today.

A different type of field-programmable device, and one that is attracting significant attention from researchers and industrial developers alike, is the field-programmable gate array (FPGA). A traditional FPGA consists of a sea of small identical logic elements, surrounded by configurable routing, as shown in Figure 3.41. By wiring these logic elements using the configurable routing, the user can construct any logic circuit desired. The logic elements themselves may also be configurable, adding further flexibility to the device. The main logic element used in commercial devices is based on a small read-only memory (ROM) with four address lines and one data line, known as a 4-input lookup table, or 4-LUT. The programmability of the ROM allows each logic element to perform any logical function of no more than four inputs, and the configurability of the routing structure allows these ROMs to be connected in order to form far more complex circuits. In addition to the 4-LUT, it is usual to include a D-type register in a logic element in order that sequential logic circuits can be constructed [1], leading to the common logic element structure shown in Figure 3.42.

In recent years the size of such devices has grown very rapidly. For example, the latest FPGA from Xilinx contains up to 180,000 4-LUTs compared to just 7,500 available only 5 years ago [2]. This has led to a take up of FPGA technology for computational purposes, as described in the applications section below. A result of the increasing use of FPGAs for particular computations such as those in digital signal processing (DSP), is that manufacturers have tried to tune their architectures towards these applications by altering the logic element structure, or having a variety of different logic elements within the same FPGA.

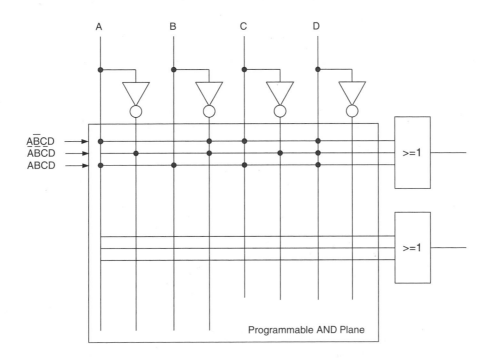

FIGURE 3.40   PAL structure.

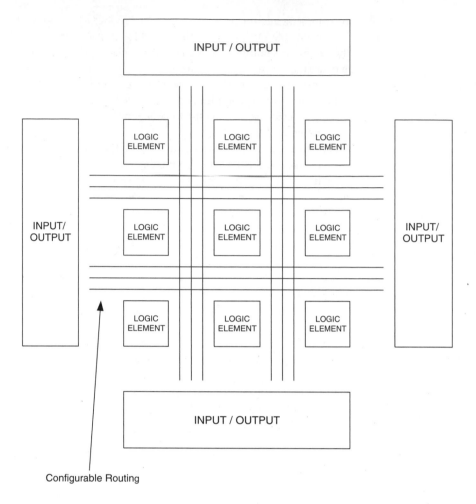

Configurable Routing

FIGURE 3.41   FPGA structure.

The first modifications that were seen to the basic structure of Figure 3.42 were to enable the efficient implementation of ripple-carry adders. The ripple-carry adder is built through the replication of a single-bit adder, or full adder circuit capable of summing any three binary digits to form a sum and a carry output. Since each bit slice has two outputs, whereas the circuit in Figure 3.42 has only one output, $2n$ such logic elements are required for an n-bit adder. In addition, such an adder would be quite slow as a result of the critical computational path, the carry chain, requiring the use of configurable routing structures. The solution adopted by Xilinx to this problem is shown in Figure 3.43. Some extra fixed-functionality circuitry has been introduced into the logic cell and some fixed; i.e., non-configurable routing has been introduced specifically to support the carry chain. In this logic cell, if the 4-LUT is configured to perform a two-bit XOR function between the lower two of its four inputs, then it can be seen that the logic-cell acts as a single-bit adder. In addition, the performance of the adder is much enhanced by the simple logic through which the carry must pass (a single

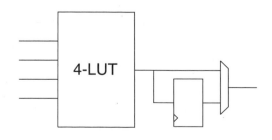

FIGURE 3.42   A common logic element.

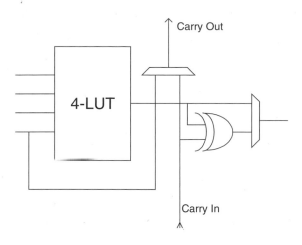

FIGURE 3.43   A modified logic element capable of a single-bit binary addition.

multiplexer per bit), and the fixed nature of the carry routing. The penalty for this achievement is the introduction of an extra multiplexer and XOR gate in fixed logic; if the logic cell is not used for addition, these will go to waste.

More recently, there has been a move to integrate word-level arithmetic operators as special purpose logic elements on the device. Modern reconfigurable arrays may thus be classified according to their granularity: fine-grain, coarse-grain or mixed-grain. This classification refers to the type of data operated on by each logic element: a fine-grain element operates at the bit level, such as a 4-LUT, while a coarse-grain element operates at the word level, such as a multiplier. Modern high-end reconfigurable devices from Xilinx and Altera, the two leading FPGA manufacturers, contain a mixture of these elements, and may therefore be classified as mixed-grain devices [2,3], while coarse-grain devices are an active area of research [4].

In recent years, with the emergence of deep sub-micron (DSM) effects, it has become particularly important to move away from a purely fine-grain architecture, as the routing structure has started to consume a significant portion of the energy required by the device, and has also become the major component of delay, outweighing the logic delay itself. In the effort to achieve power-efficient designs, it becomes necessary to jettison some of the reconfigurability afforded by early-generation PLDs. However, in so doing, the PLD becomes more specialised to a particular domain of application. The recent announcement of the Virtex 4 device from Xilinx [2], which comes in several families dedicated to logic, signal processing, and embedded systems, is a move in this direction.

Design Tools

Field programmable logic is often contrasted to general-purpose microprocessors as a way of obtaining fast, low power consumption and flexible implementations of particular algorithms. However, the automation tools that exist for hardware design are currently not at the comparatively advanced stage of optimising software compilers. The industrial state-of-the-art in hardware design for PLDs is the use of Verilog [5] or Very High Speed Integrated Circuit Hardware Description Language (VHDL) [6] at the register-transfer level. This is a form of description in which the clock-cycle by clock-cycle behaviour of the circuit must be explicitly specified. Synthesis of circuits from this type of specification is now well understood; mapping to LUT-based FPGA technology in particular was first considered in detail by Murgai [7].

However, there has been an increasing amount of attention paid to the so-called design productivity gap; the rate of increase in the design complexity manageable by a team of engineers is lagging behind the rate of increase in the design complexity which can be manufactured. Much research has therefore focussed on closing this gap.

Recently developed design tools for reconfigurable logic can be classified as domain-specific or domain-general tools. A domain-specific design tool is one that targets a particular application domain such as image processing, while a domain-general tool is one applicable across a wide range of application domains. The advantage of domain-specific compilation is that design methods and resulting architectures can be tailored to prior knowledge about the application domain, for example the concept of sample period in signal processing applications. This means that, often, more efficient and higher-level designs can be produced than with a domain-general approach. This perspective has become even more important recently with the advent of more domain-specific PLDs such as the Virtex IV series [2].

For domain-general design tools, often the focus has been on trying to migrate software designs into FPGA-based implementations. In this case, it makes sense to look at the most common software languages as potential input specifications for hardware design automation tools. To this end, C-based hardware design for PLDs has become an active area of research [8], and has led to several C-based products such as Handel-C [9] and Catapult [10]. Many of these approaches are either based on modifying the language to allow hardware-specific constructs such as true parallelism and configurable data-path width, such as in Handel-C, or through the insertion of pragmas and other external steering mechanisms to guide the synthesis tool.

Most work on domain-specific compilation has focussed on DSP applications, as they are often characterised by high levels of potential parallelism combined with simple control flow structures. As such, attention has been drawn to those languages and specification formats most widely used in the DSP design community. MATLAB is an extremely popular language in this community [11], and has been the focus of efforts to compile from MATLAB to FPGAs started at Northwestern University, and now appearing as a product of Accelchip [12]. Within MATLAB is embedded a graphical programming environment called Simulink. Simulink has also been the target of FPGA-based design automation, starting with the Synoptix tool [13]. Simulink is now an accepted format for design entry for FPGAs, with both Xilinx [14] and Altera [15] offering Simulink-based design flows. In addition to MATLAB-based design, other DSP-specific approaches have been postulated, including a modified version of C with image-processing extensions called SA-C [16]. Outside the realm of DSP, networking is also considered an application domain that can make use of the fine granularity parallelism offered by programmable logic devices; Kulkarni [17] and Lee [18] have proposed domain-specific compilation routes for networking.

Applications

In addition to their traditional role as glue logic, programmable logic devices have been demonstrated to be appropriate platforms for many computationally intensive tasks. Compared to the alternative of a microprocessor-based algorithm implementation, very high performance is often achieved from a PLD implementation when the algorithm can be highly parallelized. For example, cryptographic applications have shown speedups of 18 times [19], number theory applications 28 times [20], and similar speedups have been obtained for automatic target recognition [21], pattern matching [8], and Boolean satisfiability [22]. In addition, even when the degree of parallelism is not very large, there may be a case for a PLD-based implementation of algorithms that perform many non-standard operations such as bit-level shifting and masking, which are more efficiently implemented in a bit-level device [4].

Indeed, the only alternative implementation strategy that equals or surpasses the PLD in these measures is the application specific integrated circuit (ASIC). However, the so-called "nonrecurring engineering" (NRE) cost of manufacturing an ASIC in the latest technology is now beyond the reach of all but the truly mass-market products such as games consoles or some graphics cards. Using PLDs allows this NRE cost to be spread over the entire consumer-base of the PLD, making PLDs the affordable way to achieve high performance. The application areas of cryptography, video processing and networking have been shown many times to benefit from the parallelism afforded by modern FPGA devices [4]. Each of these achievements is described in a little more detail below.

Cryptography has long been recognised as a good match for the computational power of PLDs. Parallelism comes from the block-based processing inherent in most cryptographic standards. In addition, the finite-field arithmetic operations performed are typically a poor match for standard microprocessors, but can be efficiently implemented in bit-level PLDs [4].

Video processing remains one of the main uses of high-end FPGAs. Video processing tasks, such as edge detection, are often characterised by a combination of very large data sets, real-time constraints, a large number of operations on relatively small word-length data, and massive opportunities for parallelism. All these characteristics fit well with the computational capabilities of modern programmable hardware. PLDs have been embedded into system-level architectures specifically targeting video processing, such as the Sonic [23] and UltraSonic [24] boards.

In recent years, one of the growing areas of application for PLDs has been in network security. The high data rate of network traffic, combined with the large number of possible attack strategies against an insecure network, result in a large computational burden for firewalls and network intrusion detections systems. The reconfigurability of PLDs has also been taken advantage of in [18], where rapid customisation can be made to conform to firewall policies. Network intrusion detection systems typically take advantage of the hardware-efficiency of detecting the regular expressions corresponding to known attacks [25].

Conclusion

This article has provided an overview of the technologies behind historical and modern programmable logic devices. While PLDs and FPGAs are now industry-standard components, there remain many questions on how to obtain the best benefit from heterogeneous PLDs, and indeed what form of PLD provides the best speed, power, cost, and flexibility trade-off for a given application domain.

References

1. V. Betz, J. Rose, and A. Marquardt, *Architecture and CAD for Deep-Submicron FPGAs*, Dordrecht, Holland: Kluwer Academic Publishers, 1999.
2. Xilinx, Virtex-4 Family Overview, http://www.xilinx.com, accessed November 2004.
3. Altera, Stratix II Device Handbook, http://www.altera.com, accessed November 2004.
4. T.J. Todman, G.A. Constantinides, S.J.E. Wilton, O. Mencer, W. Luk, and P.Y.K. Cheung, "Reconfigurable computing: architectures, design methods, and applications," *IEE Proc. Comput. Digit. Tech.*, 152, 2, 193, 2005.
5. D.E. Thomas, P.R. Moorby, and Z. Navabi, *The Verilog Hardware Description Language*, Dordrecht, Holland: Kluwer Academic Publishers, 2002.
6. *The IEEE Standard for VHDL Register Transfer Level Synthesis: IEEE1076.6*, New York, NY: IEEE Publications, 1999.
7. R. Murgai, R.K. Brayton, and A. Sangiovanni-Vincentelli, *Logic Synthesis for Field-Programmable Gate Arrays*, Dordrecht, Holland: Kluwer Academic Publishers, 1995.
8. M. Weinhardt and W. Luk, "Pipeline vectorization," *IEEE Trans. Comput. Aided Des.*, vol. 20, no. 2, p. 234, 2001.
9. I. Page and R. Dettmer, Software to silicon, *IEE Review*, vol. 46, no. 5, p. 15, 2000.
10. Mentor Graphics, Catapult C Synthesis, http://www.mentorg.com, accessed November 2004.
11. The Mathworks, Matlab and Simulink for Technical Computing, http://www.mathworks.com, accessed November 2004.
12. Accelchip, Accelchip DSP Synthesis, http://www.accelchip.com, accessed November 2004.
13. G.A. Constantinides, P.Y.K. Cheung, and W. Luk, "Multiple precision for resource minimization," in *Proc. IEEE Int. Symp. Field-Programmable Custom Comput. Mach.*, Napa, CA 2000.
14. J. Hwang, B. Milne, N. Shirazi, and J. Stroomer, "System level tools for DSP in FPGAs," in R. Woods, and G. Brebner, Eds., *Proc. Field-Programmable Logic Appl.*, Berlin, Germany: Springer, 2001.
15. Altera, DSP Builder, http://www.altera.com, accessed November 2004.
16. W. Bohm, J. Hammes, B. Draper, M. Chawathe, C. Ross, R. Rinker, and W. Najjar, "Mapping a single assignment programming language to reconfigurable systems," *J. Supercomput.*, 21, 117, 2002.

17. C. Kulkarni, G. Brebner, and G. Schelle, "Mapping a domain specific language to a platform FPGA," in *Proc. IEEE/ACM Design Autom. Conf.*, San Diego, CA, 2004.

18. T.K. Lee, S. Yusuf, W. Luk, M. Sloman, E. Lupu, and N. Dulay, "Compiling policy descriptions into reconfigurable firewall processors," in *Proc. IEEE Int. Symp. Field-Programmable Custom Comput. Mach.*, Napa, CA, 2003.

19. A.J. Elbirt, and C. Paar, "An FPGA implementation and performance evaluation of the serpent block cipher," in *Proc. ACM/SIGDA Int. Symp. Field-Programmable Gate Arrays*, Monterey, CA, 2000.

20. H.J. Kim and W.H. Mangione-Smith, "Factoring large numbers with programmable hardware," in *Proc. ACM/SIGDA Int. Symp. Field-Programmable Gate Arrays*, Monterey, CA, 2000.

21. M. Rencher and B.L. Hutchings, "Automated target recognition on SPLASH2," in *Proc. IEEE Int. Symp. Field-Programmable Custom Comput. Mach.*, Napa, CA, 1997.

22. I. Skliarova and A. de Brito Ferrari, "Reconfigurable hardware SAT solvers: a survey of systems," *IEEE Trans. Comput.*, vol. 53, no. 11, p. 1449, 2004.

23. S.D. Haynes, J. Stone, P.Y.K. Cheung, and W. Luk, "Video image processing with the SONIC architecture," *IEEE Comput.*, 33, 50–57, 2000.

24. S.D. Haynes, H.G. Epsom, R.J. Cooper, and P.L. McAlpine, "Ultra-SONIC: a reconfigurable architecture for video image processing," in M. Glesner, P. Zipf, and M. Renovell, Eds., *Field-Programmable Logic and Applications*, Berlin, Germany: Springer, 2002.

25. C.R. Clark and Schimmel, "A pattern-matching co-processor for network intrusion detection systems," in *Proc. IEEE Int. Conf. Field Programmable Technol.*, Tokyo, Japan, 2003.

3.5   Arithmetic Logic Units

Bill D. Carroll

Arithmetic logic units (ALUs) are combinational logic circuits that perform basic arithmetic (only addition and subtraction) or logical (AND, OR and NOT) operations on two n-bit operands. ALUs can be constructed from standard integrated circuits or programmable-logic devices and are available as single-chip, medium-scale, integrated circuits.

This section covers the design of arithmetic and logic circuits in sufficient detail for the reader to design and implement basic logic units and to understand the operation and use of commercial devices. The reader that requires more detail or an in-depth discussion of the subject should consult the reference list and sources provided at the end of this section.

In the following material, operands are assumed to be signed, n-bit binary numbers, with the left bit representing the sign (0 for positive and 1 for negative) in arithmetic operations or in circuits. Negative numbers are represented in two's-complement form. The two's complement of an n-bit number A is $A' + 1$, where A' represents the bit-wise complement of A. Unsigned n-bit binary numbers are assumed for logic operations and circuits.

Basic Adders and Subtracters

The basic building block for most arithmetic circuits is the *full adder*. A full adder is a logic circuit that produces a two-bit sum (S and C) of three one-bit binary numbers (X, Y and Z). Table 3.6 shows the truth table and logic equations for a full adder. A logic symbol and gate-level realization of a full adder are shown in Figure 3.44.

The addition of two n-bit binary numbers ($X = x_{n-1} \ldots x_1 x_0$ and $Y = y_{n-1} \ldots y_1 y_0$) can be calculated with n full adders cascaded, as shown in Figure 3.45. Such a circuit is called a *ripple-carry adder*, since carries produced by lower-order stages must propagate or ripple through the higher-order stages before completing the addition operation.

TABLE 3.6 Full Adder Truth Table and Logic Equations

X	Y	Z	S	C	
0	0	0	0	0	
0	0	1	1	0	$S = XYZ + XY' Z' + X' YZ' + X' Y' Z$
0	1	0	1	0	
0	1	1	0	1	
1	0	0	1	0	$C = XY + XZ + YZ$
1	0	1	0	1	
1	1	0	0	1	
1	1	1	1	1	

Ripple-carry adders are simple in operation and structure but are slow. In the worst case ($X = 1 \ldots 11$ and $Y = 0 \ldots 01$), a carry produced in the least-significant full adder must propagate through the more significant ones. The worst-case add time, t_{add}, is shown below, where t_{pd} is the propagation delay introduced at each stage.

$$t_{add} = nt_{pd}$$

This assumes that all addend bits are presented to the adder simultaneously. In the least significant full adder, t_{pd} represents the time to compute c_1 from x_0 and y_0. In the most significant full adder, the time to compute s_{n-1} after c_{n-1} is received. In the intermediate stages, t_{pd} is the time needed to compute c_{i+1} from c_i. The propagation delay is approximately equal to the delay of a three-level logic circuit, consistent with the realization of a full adder shown in Figure 3.44.

Subtraction easily can be performed by adding the minuend to the negative of the subtrahend. In a two's-complement number system, $X - Y$ can be obtained by computing $X + Y' + 1$. The ripple-carry adder

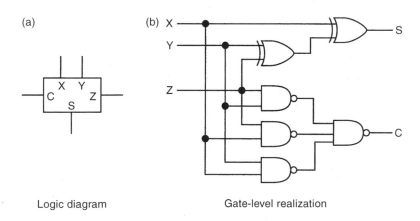

(a) Logic diagram

(b) Gate-level realization

FIGURE 3.44 Full adder.

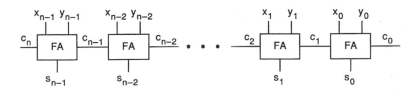

FIGURE 3.45 Ripple-carry adder for two *n*-bit binary numbers.

described above can be modified by placing inverters on the Y inputs of each full adder and by making the carry-in (c_0) equal to 1. The resulting two's-complement subtracter is shown in Figure 3.46.

A device capable of performing addition or subtraction can be built by replacing the inverters in the subtracter with exclusive-OR gates and using the carry-in (c_0) as a control signal. The resulting two's-complement adder/subtracter is shown in Figure 3.47. The device will function as a ripple-carry adder when $c_0 = 0$ and as a two's-complement subtracter when $c_0 = 1$.

High-Speed Adders

Several adder designs have been developed to perform high speed addition. These include *carry-lookahead adders* (CLAs), carry-completion adders, conditional-sum adders and carry-select adders. Carry-lookahead adders have gained wide acceptance in the design of ALUs due to their speed and because they can be conveniently implemented in integrated-circuit form.

This material covers only the carry-lookahead approach. Before discussing this approach, it is important to explain why fully parallel adders are not feasible. Addition is a combinational process, so it is theoretically possible to construct a $2n$-bit full adder that can be used with a three-level combinational logic circuit and perrform addition of two n-bit numbers in the time equal to the circuit's delay. However, such circuits, requiring gate fan-in, are too costly to be implemented for reasonable values of n. Carry-lookahead is a practical and effective compromise between fully parallel adders and ripple-carry adders. The block diagram of a four-bit CLA is shown in Figure 3.48(a).

CLAs are based on the observation that a carry-out (c_i) of the ith stage of a full adder is produced either by the propagation of the carry-in (c_{i-1}) through the ith stage or the generation of a carry in the ith stage. This can be seen in the following logic equations for c_i:

$$c_i = x_{i-1}y_{i-1} + x_{i-1}c_{i-1} + y_{i-1}c_{i-1}$$
$$= x_{i-1}y_{i-1} + (x_{i-1} + y_{i-1})c_{i-1}$$
$$= g_{i-1} + p_{i-1}c_{i-1}$$

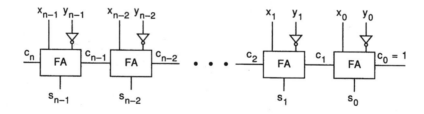

FIGURE 3.46   Two's-complement subtracter.

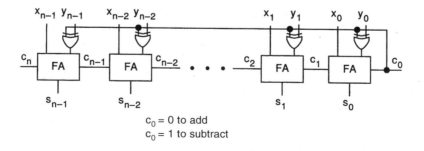

$c_0 = 0$ to add
$c_0 = 1$ to subtract

FIGURE 3.47   Two's-complement adder/subtracter.

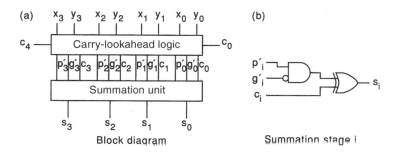

FIGURE 3.48 Carry-lookahead adder.

where $g_i = x_i y_i$ and $p_i = x_i + y_i$ are the generate and propagate terms, respectively, for stage i for $i = 0$ to $n - 1$.

The carry equations for an n-bit adder can be derived by repeatedly applying the above equation. The following set of equations results for the $n = 4$ case:

$$c_1 = g_0 + p_0 c_0$$
$$c_2 = g_1 + p_1 g_0 + p_1 p_0 c_0$$
$$c_3 = g_2 + p_2 g_1 + p_2 p_1 g_0 + p_2 p_1 p_0 c_0$$
$$c_4 = g_3 + p_3 g_2 + p_3 p_2 g_1 + p_3 p_2 p_1 g_0 + p_3 p_2 p_1 c_0$$

The carry equations can be calculated by three-level combinational logic circuits to form the carry-lookahead logic block shown in Figure 3.48(a). The sum (s_i) bits for the ith stage of an adder can be written in terms of g_i, p_i and c_i and generated by the logic circuit given in Figure 3.48(b).

The add-time, t_{add}, for a CLA will be examined next. Assume that both addends are applied to the CLA simultaneously and that $c_0 = 0$. Also, let t_{pd} represent the propagation delay of a three-level logic circuit. Two components contribute to the add time. First, the three-level carry-lookahead logic must produce the carries. This takes t_{pd}. Then, the summation unit must produce the final values for the sum bits. This step takes a time equal to the propagation delay of the exclusive-OR gate in the summation unit, which is t_{pd}, since an exclusive-OR gate can be calculated as a three-level, combinational logic circuit. Hence, the add time for a CLA is

$$t_{add} = 2 t_{pd}$$

The above result indicates that the add time of a CLA is not only faster than a ripple-carry adder but also is independent of the length (n) of the addends. It might be concluded that CLAs provide the final answer to the high-speed adder problem. However, a closer look at the carry equations above reveals that they become progressively more complex in the number of product terms and literals. Fan-in constraints eventually will limit the practicality of calculating the equations in three-level logic. The actual limit is technology-dependent. Standard single-chip, medium-scale ALUs typically handle four-bit operands, although longer lengths are certainly feasible with today's technology.

CLAs may be cascaded to produce an adder for longer operands. Figure 3.49 shows a cascade of four 4-bit CLAs that produce a 16-bit adder. Carries are produced using carry-lookahead logic within each CLA stage but must propagate between stages similar to a ripple-carry adder. Consequently, the add time of cascaded CLAs is dependent on the number of cascade stages. The four-stage adder in Figure 3.49 has a worst-case add time of $5 t_{pd}$. In general, the add time of an m-stage cascade is $(m + 1) t_{pd}$.

The carry-lookahead approach can be applied at a higher level to eliminate the propagation of carries between CLA stages or blocks. This approach uses *block carry-lookahead adders* (BCLAs) and block carry-lookahead (BCL) logic as shown in Figure 3.50. A BCLA is CLA-modified to produce block-carry

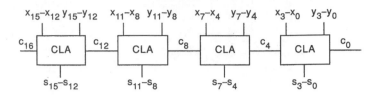

FIGURE 3.49   Cascaded carry-lookahead adders.

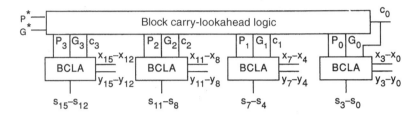

FIGURE 3.50   Block carry-lookahead adders.

propagate (P) and block-carry generate (G) outputs instead of a carry-out. BCL logic is a combinational logic circuit that generates block carries (Cj) for each BCLA from the P and G outputs of lower-order BCLAs and c_0. Logic equations for the block-carry logic can be derived by repeated application of the following equations for a typical block:

$$C_j = G_j + P_j C_{j-1}$$

where

$$G_j = [g_3 + p_3 g_2 + p_3 p_2 g_1 + p_3 p_2 p_1 g_0]_j$$

and

$$P_j = [p_3 p_2 p_1 p_0]_j$$

BCLAs and block-carry-logic units are available in standard, medium-scale integrated circuits. An extension of the carry-lookahead concept to k levels is possible in theory. However, applying more than three levels is usually not practical.

Multifunction Arithmetic Logic Units

Devices that can provide addition, subtraction and logical operations easily can be designed around the adders and subtracters in the previous sections. The logic diagram of the first two stages of an *n*-bit multifunction ALU is shown in Figure 3.51. Operand inputs for the device are $X = x_{n-1} \ldots x_1 x_0$ and $Y = y_{n-1} \ldots y_1 y_0$ and the output is $S = s_{n-1} \ldots s_1 s_0$. The function performed on the operands is determined by the values of the control inputs k_2, k_1, k_0 and c_{in}, as shown in Table 3.7. The given realization simply can be based on a ripple-carry adder. However, the same design approach can be used with other adders such as carry-lookahead.

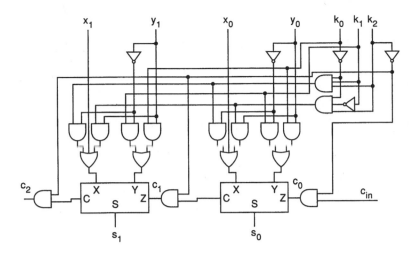

FIGURE 3.51 Multifunction ALU.

TABLE 3.7 Functions Performed by the Multifunction ALU

Control Inputs k_2	k_1	k_0	c_{in}	Result	Function
0	0	0	0	$S = X$	Transfer X
0	0	0	1	$S = X + 1$	Increment X
0	0	1	0	$S = X + Y$	Addition
0	0	1	1	$S = X + Y + 1$	Add with carry in
0	1	0	0	$S = X - Y - 1$	Subtract with borrow
0	1	0	1	$S = X - Y$	Subtraction
0	1	1	0	$S = X - 1$	Decrement X
0	1	1	1	$S = X$	Transfer X
1	0	0	...	$S = X$ OR Y	Logical OR
1	0	1	...	$S = X$ XOR Y	Exclusive-OR
1	1	0	...	$S = X$ AND Y	Logical AND
1	1	1	...	$S = $ NOT X	Bit-wise complement

TABLE 3.8 Typical Integrated Circuit Arithmetic and Logic Devices

Part Number	Function	Features
74F181	4-bit multifunction (16) ALU	BCL outputs
74F182	Carry-lookahead generator	Use with 74LS181 for BCL
74F183	Full adder	Two per package
74F283	4-bit binary adder	Internal CL
74F381	4-bit multifunction (8) ALU	BCL outputs
74F382	4-bit multifunction (8) ALU	Ripple-carry output

Standard Integrated Circuit ALUs

The devices described above are generic in nature but are similar in function and usefulness to many commercially available integrated-circuit products. Representative products are summarized in Table 3.8. Manufacturers are phasing out these types of products as newer technologies make them obsolete.

Defining Terms

Arithmetic logic unit (ALU):   A combinational logic circuit that can perform basic arithmetic and logical operations on n-bit binary operands.

Block carry-lookahead adder (BCLA):   An adder that uses two levels of carry-lookahead logic.

Carry-lookahead adder (CLA):   A high-speed adder that uses extra-combinational logic to generate all carries in an n-bit block in parallel.

Full adder (FA):   A combinational logic circuit that produces the two-bit sum of three one-bit binary numbers.

Ripple-carry adder (RCA):   A basic n-bit adder characterized by the need for carries to propagate from lower- to higher-order stages.

References

M.D. Erceqovac and T. Lang, *Digital Arithmetic*, San Mateo, CA: Morgan Kaufman, 2003.

H.A. Farhat, *Digital Design and Computer Organization*, Boca Raton, FL: CRC Press, 2004.

M.J. Flynn and S.F. Oberman, *Advanced Computer Arithmetic Design*, New York, NY: Wiley, 2001.

J.P. Hayes, *Computer Architecture and Organization*, 3rd ed., New York, NY: McGraw-Hill, 2003.

B. Holdsworth and C. Woods, *Digital Logic Design*, 4th ed., Oxford, England: Newnes, 2002.

R.H. Katz and G. Boriello, *Contemporary Logic Design*, 2nd ed., Upper Saddle River, NJ: Prentice-Hall, 2005.

V.P. Nelson, H.T. Nagle, B.D. Carroll and J.D. Irwin, *Digital Logic Circuit Analysis and Design*, Englewood Cliffs, NJ: Prentice-Hall, 1995.

E.E. Swartzlander, Jr., Ed., *Computer Arithmetic*, vol. I and vol. II, Los Alamitos, CA: IEEE Computer Society Press, 1980.

Further Information

Information on the theoretical aspects of computer arithmetic can be found in the *IEEE Transactions on Computers*, a monthly publication of the Institute for Electrical and Electronics Engineers, Inc., 445 Hoes Lane, P.O. Box 1331, Piscataway, NJ 08855–1331.

More information on the specifications and applications of integrated circuits is published in data books and application notes by electronics manufacturers such as Texas Instruments Inc., Fairchild Semiconductor International Inc. and National Semiconductor Corp.

Discussions of multiplication, division and floating-point arithmetic can be found in numerous textbooks on computer architecture.

3.6   Programmable Logic

Albert A. Liddicoat and Lynne A. Slivovsky

Introduction

Digital systems today are designed with drastically different implementation techniques. Each technique has unique advantages and disadvantages in cost, performance, power consumption, design time, manufacturing time and flexibility. Programmable logic often is used for hardware prototyping, for applications requiring concurrent hardware and software development and for applications where it is not economical to design a custom *integrated circuit* (IC). Systems using programmable logic often benefit from a performance advantage by using hardware implementation, compared to systems using a general-purpose processor or microcontroller. For these systems designers also avoid the large initial investment that custom integrated

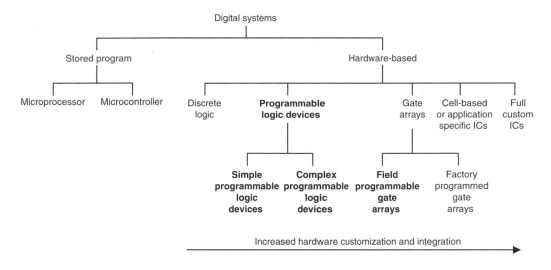

FIGURE 3.52 Taxonomy of digital systems.

circuits require. Figure 3.52 provides the taxonomy for common classifications of digital systems. *Programmable logic devices* (PLDs) and *field programmable gate arrays* (FPGAs) are shown in bold.

Digital systems can be implemented using stored programs that work with a microprocessor or microcontroller. For these digital systems, designers write custom software for a general-purpose microprocessor or microcontroller. This approach offers the advantage of using the same microprocessor or microcontroller for many different applications. This amortizes the development cost of a general-purpose device for a large market. However, this technique's disadvantage is with stored program execution. Execution is sequential in nature and adds overhead costs by fetching and decoding instructions, in contrast with a hardware-based system.

A primary advantage of hardware-based design is that independent computation can occur simultaneously. For parallel applications, hardware-based design often provides a significant performance advantage over stored programs executing on a processor. In Figure 3.52, five types of hardware-based designs are shown. Implementation techniques are arranged from left to right, with the least amount of customization over the integrated circuits shown on the left side and the greatest amount of customization and integration on the right. Each hardware-based implementation technique has a range of well-suited applications.

Discrete logic generally is preferred when less than five logic gates are needed for the application. Discrete-logic ICs often provide one to four logic gates on a single IC. If the application only requires one to four logic gates of a single type, a discrete-logic IC would be the most cost-effective solution. However, if the application requires from 5 to 1,000 logic gates, then *programmable logic devices* (PLDs) often are preferred. PLDs are integrated circuit devices that can be programmed to implement arbitrary logic functions. *Simple PLDs* (SPLDs) are devices implementing two-level logic functions. Simple PLDs come in many configurations and use different programming methods. *Complex PLDs* (CPLDs) are comprised of a collection of PLD-like blocks on a single chip, using a programmable interconnect. CPLD architecture is presented in the following section of this chapter.

Gate arrays offer more orders of magnitude, more hardware resources, greater logic-gate equivalents and more memory storage devices than do PLDs. These devices provide multiple logic gates, *configurable logic blocks* (CLBs) and *flip-flops*. In a gate-array product, the semiconductor layer is fixed and cannot be changed. The gate array can be customized by controlling how the logic gates, configurable logic blocks and flip-flops are interconnected and set up. Field-programmable gate arrays are customized after the integrated circuits leave the factory. The logic functions are implemented by using configurable logic blocks, and the interconnection of these configurable logic blocks and flip-flops is controlled by a programmable interconnect. FPGA architecture is presented in the following section of this chapter.

Factory-programmed gate arrays must be customized by the manufacturer. A designer determines how the logic gates and memory devices are connected and how the read-only memories should be programmed for specific applications. The designer provides this information to the manufacturer. The manufacturer can customize the gate array by adding a metal-interconnect layer over the semiconductor substrate used to connect the logic gates and flip-flops and program read-only memories. Since FPGAs are programmable by the customer, they do not incur the fixed costs associated with a custom step in IC manufacturing. Factory-programmed gate arrays, on the other hand, directly implement logic functions by using logic gates without the programmable interconnect. They typically offer higher performance and are more cost-effective in higher volumes compared to FPGAs. Since the factory-programmed gate arrays require an additional fixed cost to produce these ICs, they are often not cost-effective for lower production volumes.

Application specific integrated circuits (ASICs) provide another level of customization compared to gate arrays. The semiconductor layer that creates the logic functions, as well as the metal layers connecting the logic functions, are customizable in an ASIC. The manufacturer provides the designer with a library of standard logic gates that can be used. The designer can control which logic gate is used, where it is placed and how it is interconnected. This level of control allows the designer to optimize performance, power consumption and cost savings by controlling the type and placement of every logic gate. For instance, AND gates can be designed for high performance, large drive strength, low power consumption or low cost. The selection of the AND gate and the gate's placement affects the IC's performance, power consumption and cost. But ASICs also require additional fixed costs to customize the metal and semiconductor layers of the integrated circuit. High production volumes for the IC may offset the fixed costs and allow the ASIC to be a more cost-effective solution than a gate array. In addition, a digital system may need implementation in an ASIC to meet the performance or power-consumption requirements of the product.

Finally, a *full custom-integrated circuit* gives the design the highest level of flexibility and integration. The design is implemented at the transistor level. The size, location and interconnection of each transistor is determined by the designer. The designer is not limited to using predesigned logic gates from a library, as is the case with ASIC implementations. A full custom-integrated circuit gives the designer more flexibility to make tradeoffs for performance, power consumption and costs compared to an ASIC. The design time and fixed costs are large for full custom-integrated circuits. High-performance processors are typically designed using full custom-integrated circuits.

Programmable Elements

Designs are implemented on a PLD by programming the device's configurable elements. A device's *programming elements* can be classified as one-time programmable (e.g., fuse or antifuse), or reconfigurable. Reconfigurable programming elements are classified as either volatile (e.g., SRAM) or nonvolatile (e.g., fuse or EEPROM). They can physically connect two nets in a device or provide a type of memory cell that adds a logic value to a device component. PLDs contain circuitry allowing each programming element to be addressed during the configuration.

One-time programmable devices contain elements that can be configured only once. One-time programmable devices are nonvolatile, since they retain their configuration when power is turned off. Examples of one-time programmable elements are the *fuse* and the *antifuse*. Fuse-based PLDs come with all available programmable connections held in place by fused metal links. These fuses are *blown* during programming by passing a large current through the link. Antifuse-based devices come with no connections in place and are programmed by *growing* links that form a connection. Antifuse devices are more common than fuse-based devices because most potential connection points remain disconnected after configuration.

Reconfigurable devices can be programmed multiple times. Devices today can be programmed thousands of times. These devices have either volatile or nonvolatile programming elements. CPLDs typically contain nonvolatile programming elements, such as *electrically programmable read-only memory* (EPROM) and *electrically erasable **programmable read-only memory*** (EEPROM). EPROM-based devices are programmed electronically and erased by exposing the programmable elements to ultraviolet light through a window on the chip. EEPROM devices are programmed and erased by supplying electronic pulses to the programming element during configuration. EEPROM elements typically are larger in size than EPROM elements. Flash-based FPGAs are nonvolatile and use flash memory as the programming element. Similar to EEPROM elements, flash cells are electrically erasable and have a small cell size. SRAM cells are reconfigurable, volatile programming elements that lose their configuration when power is turned off. This programming element is used in the majority of commercially available FPGAs. These devices are configured during programming by storing the appropriate value (logic 0 or logic 1) in each SRAM cell. Flash and antifuse programming elements are commonly found in CPLD.

CPLD Architecture

As shown in Figure 3.53, CPLD architecture contains three standard components: *macroblocks*, IO blocks and a programmable interconnect. The makeup of each functional block, the number on a particular device and the number of available IO pins varies among CPLD families and manufacturers. In general, each macroblock consists of a number of SPLD-like *macrocells*.

A macrocell implements a logic function using a two-level AND-OR structure, in which the input signals to the AND gates are programmable. Figure 3.54 depicts a generic macrocell. The logic function is programmed on the device during configuration, corresponding to the device's programming element (EEPROM, antifuse). Additional elements are found in the macrocell that configures the logic function of the macrocell. The output of the macrocell function is routed to other parts of the CPLD or to an output pin by the programmable interconnect.

The CPLD macrocell provides additional hardware resources to customize the implemented function. An exclusive-OR and D flip-flop can be used to invert or store the sum-of-products output. The signal leaving the AND-OR array can be complemented by configuring the programmable input of the exclusive-OR gate to a logic 1. There is a D flip-flop in the macrocell registering the AND-OR-array output signal. A set of multiplexers (MUXs) selects product terms and configures the macrocell output.

There are four multiplexers depicted in the macrocell: a product-term MUX, an output MUX, a feedback MUX and an output-enable MUX. The product-term MUX can select a single product term within the macrocell or from another macrocell. The output MUX selects either the combinational output from the AND-OR (possibly inverted) or the registered output from the flip-flop. The feedback MUX selects the combinational signal, the registered signal or a signal from the programmable interconnect that is sent back to the AND-OR array. The enable MUX controls the output enable on the three-state buffer at the macrocell's output. The output enable can be controlled by a product term, a global output enable signal, or it can be always on or always off.

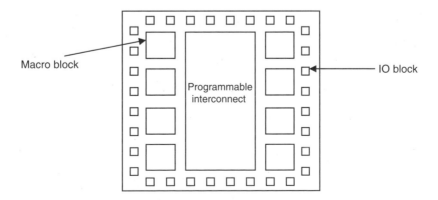

FIGURE 3.53   Functional blocks on a CPLD: macroblock, IO and interconnect.

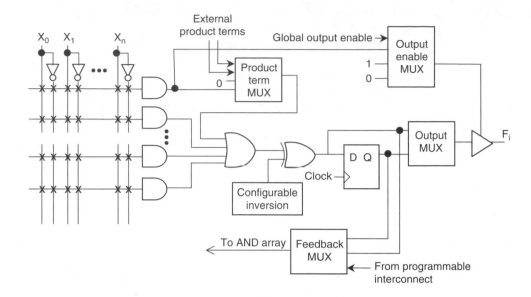

FIGURE 3.54   Macrocell of a generic GPLD.

The programmable interconnect permits signal routing from one macroblock to any other macroblock on the integrated circuit chip. The programmable interconnect will be discussed in more detail in the FPGA Architecture section. The IO blocks connect internal nets to IO pins and contain circuitry to configure pins to meet any one of a number of IO standards supported by the device.

FPGA Architecture

A standard FPGA *fabric* contains a number of configurable logic blocks and IO blocks and has a programmable interconnect. A configurable logic block mixes combinational and sequential logic at a more general level than does the AND-OR array in a CPLD. However, the configurable logic block can implement only a small part of a complex design. FPGAs that contain specialized dedicated blocks, such as *random access memory* (RAM), multipliers and on-chip processors are becoming more common and more powerful. These are often referred to as platform FPGAs. The devices are capable of implementing large, complex *system-on-chip* (SoC) designs. Figure 3.55 depicts a generic layout of the functional blocks on an FPGA.

The IO blocks determine the electrical characteristics of the IO pin for a particular IO standard and whether the IO pin is acting as in input, output or bidirectional pin. When configured as an output pin, additional logic in the IO block determines the signal routed to the pin through a three-state buffer. When acting as an input, the signal is buffered and routed through a programmable delay. It may also be registered.

FPGAs typically use SRAM cells as the programmable element for configuration. These are volatile memory elements that must be configured when power is turned on, either by downloading the configuration from a computer or from an external memory device. The values stored in the SRAM cells depend on the system implementation. These values are generated during the design flow process, discussed in the last section of this chapter. In general, SRAM cells control pass transistors, MUX select lines and look-up table entries. They are depicted in Figure 3.56.

SRAM cells can be configured to drive the select lines of a multiplexer, found in a configurable logic block or an IO block. A 4:1 MUX is shown in Figure 3.56 (a) where the SRAM cells control which of the four input signals is routed to the MUX output. The configurable logic distinguishing the FPGA in the spectrum of PLDs is called the look-up table (LUT). An LUT is a configurable type 0 MUX design that can implement a Boolean

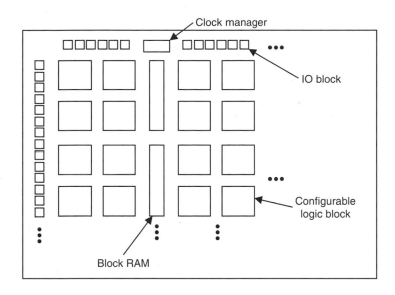

FIGURE 3.55 FPGA fabric.

function of *n* variables, where *n* is the number of control inputs on the MUX. The LUT must contain 2^n SRAM cells. In Figure 3.56 (a) *n* is two, there are two MUX control inputs and four SRAM cells and the LUT can implement any logic function consisting of two variables. Function variables are routed to the select lines of a multiplexer, and SRAM cells are connected to the data inputs of the MUX. An LUT can be configured to implement a logic function or it can be configured, with additional logic, to act as RAM. LUTs on FPGAs typically have four select inputs and can implement a logic function of four variables. A configurable logic block contains multiple LUTs that can be combined with multiplexers to implement logic functions in the tens of variables.

When used as a pass transistor control, as in Figure 3.56 (c), the SRAM cell drives the gate of an n-channel MOS transistor. This transistor behaves like a switch between two wires on the FPGA. To configure a connection between points A and B, a logic 1 is stored in the SRAM cell. Current flows between the two points through the transistor. To disconnect points A and B, a logic 0 is stored in the SRAM cell, and the transistor is turned off.

In the programmable interconnect, when a potential connection lies at the intersection of four nets, six pass transistors (one vertical, one horizontal and four diagonal) control the intersection's connectivity, as depicted in Figure 3.57. The programmable interconnect is configured as a switch matrix permitting the signal on a

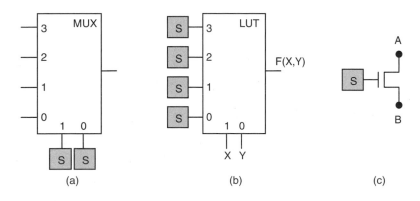

FIGURE 3.56 Programmable SRAM control: (a) MUX control, (b) look-up table, (c) pass transistor.

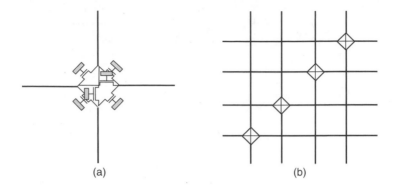

FIGURE 3.57 Programmable interconnect: (a) connectivity of four nets can be controlled by configuring six pass transistors, (b) switch matrix with SRAM cells and pass transistors depicted by one vertical, one horizontal and four diagonal segments at each intersection.

horizontal net to be routed to any vertical net. There generally are three types of interconnects on an FPGA: a global interconnect routing signals across the device, a local interconnect routing signals directly between logic blocks, and a timing interconnect permitting special handling of timing and control signals.

The configurable logic block is the primary configurable-logic element on an FPGA. It contains a number of look-up tables, flip-flops, multiplexers and specialized carry and control logic, as shown in Figure 3.58. The exact configuration varies among device families and manufacturers. Both combinational and sequential functions are implemented in the CLB. Neighboring CLBs have direct interconnects between them, while CLB input and output signals are routed across the device using the programmable interconnect.

Recent advances in FPGAs include the addition of specialized elements integrated into the fabric, as shown in Figure 3.59. Dedicated hardware multipliers, optimized for speed and size, have greatly improved the ability of an FPGA to perform signal and image processing. On-chip processors that access the fabric, called hard-core processors, enable embedded systems to be designed and implemented on a single chip. The advent of these processors has coincided with the development of embedded system-software packages for the design and integration of software running on the processor. Developments also include logic modules on the fabric and interfacing with external components, such as the PCI bus, to create powerful and diverse systems.

Design Example for CPLD and FPGA

In this section, a design example illustrates how programmable logic is configured to implement a useful function. An 8-bit, ripple-carry adder will be implemented on a CPLD and FGPA. A modular design technique is used to partition the 8-bit adder into smaller subcomponents that can be implemented with CPLD and

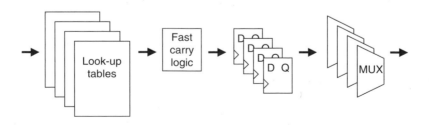

FIGURE 3.58 Configurable logic block.

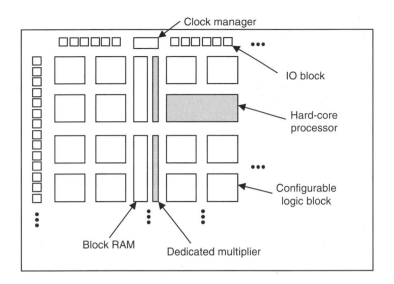

FIGURE 3.59 Platform FPGA with dedicated multipliers and on-chip processor.

FPGA building blocks. The adder subcomponent first is implemented using two-level logic suited for CPLD implementations. The adder subcomponent then is implemented using an FPGA look-up table structure.

A general adder has three input operands, A, B and C_{in}, and two output operands, S and C_{out}. The adder sums the inputs, A, B and C_{in}, and produces the output sum, S, and the carry-out, C_{out}. For an 8-bit adder, the input and output operands A, B and S are 8-bit binary numbers represented as $(a_7\ a_6\ a_5\ a_4\ a_3\ a_2\ a_1\ a_0)$, $(b_7\ b_6\ b_5\ b_4\ b_3\ b_2\ b_1\ b_0)$ and $(s_7\ s_6\ s_5\ s_4\ s_3\ s_2\ s_1\ s_0)$ respectively, where bit 0 is the least significant bit and bit 7 is the most significant. C_{in} is the carry into the least significant bit position of the adder and C_{out} is the carry-out of the adder's most significant bit position.

The bit-level-addition operation is represented using the diagram in Figure 3.60. In this figure, the carry in, C_{in}, is represented as c_0 and the intermediate carries c_1–c_7 are shown. Figure 3.60 illustrates the 8-bit addition, broken down into eight 1-bit binary additions starting from the least significant bit position. The first 1-bit addition computes $c_0 + a_0 + b_0$ and produces s_0 and c_1. The second 1-bit addition computes $c_1 + a_1 + b_1$ and produces s_1 and c_2. Generally, for the ith bit position a 1-bit full adder is used to compute $c_i + a_i + b_i$ and produce s_i and c_{i+1}. The task of designing an 8-bit, ripple-carry adder is partitioned into the task of designing a full adder and then connecting 8 instances of the full adder with cascaded carries, shown in Figure 3.61.

The full adder may be designed using standard combinational-logic design techniques. The minimum sum-of-products Boolean equation for the sum and carry out functions of a full adder follow:

$$s_i = a_i b'_i c'_i + a'_i b_i c'_i + a'_i b'_i c_i + a_i b_i c_i \tag{3.14}$$

$$c_{i+1} = a_i b_i + b_i c_i + a_i c_i \tag{3.15}$$

$$
\begin{array}{r}
C_{out}\quad c_7\ \ c_6\ \ c_5\ \ c_4\ \ c_3\ \ c_2\ \ c_1\ \ c_0 = C_{in} \\
a_7\ \ a_6\ \ a_5\ \ a_4\ \ a_3\ \ a_2\ \ a_1\ \ a_0 \\
+\qquad b_7\ \ b_6\ \ b_5\ \ b_4\ \ b_3\ \ b_2\ \ b_1\ \ b_0 \\
\hline
s_7\ \ s_6\ \ s_5\ \ s_4\ \ s_3\ \ s_2\ \ s_1\ \ s_0 \\
\end{array}
$$

FIGURE 3.60 Binary addition of $A + B + C_{in} = S$ and C_{out}.

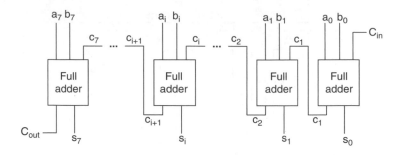

FIGURE 3.61 An 8-bit ripple-carry adder.

In the previous section of this chapter, the architecture for a CPLD was presented. A CPLD consists of an array of macroblocks on a single integrated circuit with programmable interconnect. An 8-bit ripple-carry adder will fit into one macroblock if the macroblock contains 16 or more macrocells. Each macrocell implements a different Boolean function using programmable product terms. For this example, it is assumed that the CPLD is constructed using 18 macrocells per macroblock. Each macrocell uses an AND-OR structure with five programmable product terms. Figure 3.62 indicates how a full adder can be implemented using the CPLD macrocell above. The Xs indicate where the interconnect has been programmed to be shorted together. For example, the Xs indicate that the inputs to the first programmable product term, p_0, are a_i, b_i' and c_i'. This product term is equivalent to the first product term in Equation (3.14), $p_0 = a_i \, b_i' c_i'$. Product terms 1 through 3 implement the remaining three product terms in Equation (3.14). Since there are only four product terms needed for the sum Boolean function, the last product term, p_4, has been programmed to produce a zero, $p_4 = a_i \, a_i' = 0$. The Boolean equation for the sum bit, s_i, is implemented by OR-ing together with the five

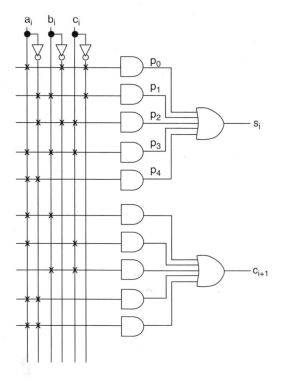

FIGURE 3.62 Full adder implementation using two CPLD macroblocks.

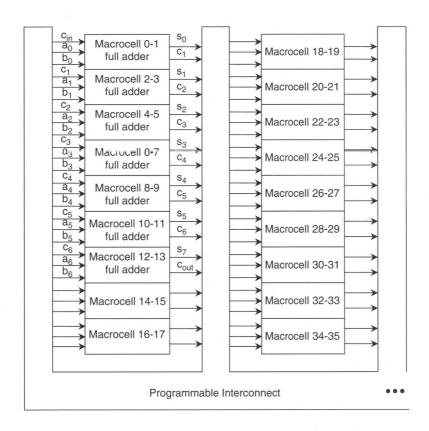

FIGURE 3.63 8-bit Ripple-carry adder mapped to a CPLD macroblock.

product terms.

$$s_i = p_0 + p_1 + p_2 + p_3 + p_4$$
$$= a_i b'_i c'_i + a'_i b_i c'_i + a'_i b'_i c_i + a_i b_i c_i + 0$$
$$= a_i b'_i c'_i + a'_i b_i c'_i + a'_i b'_i c_i + a_i b_i c_i$$

Similarly, the product terms to implement Equation (3.15) for the carry-out, c_i, are programmed in the bottom macrocell in Figure 3.62.

Each Boolean function is implemented using one CPLD macrocell. To implement an 8-bit adder, the carry chain must be connected, and the input and output operands must be routed into the appropriate macrocell. Figure 3.63 shows how one macroblock of a CPLD can implement the 8-bit adder, assuming the programmable interconnect is correctly configured to connect the input operands, the output operands and the carry chain.

In the rest of this section, an 8-bit ripple-carry adder will be implemented using FPGA architecture, based on look-up tables (LUTs) as described in the previous section. FPGA architectures use small, RAM-based LUTs to implement logic functions. The size and configuration of these LUTs vary depending on product family and manufacturer. A common CLB structure consists of two small LUTs that can be configured into a larger LUT. In this example, it is assumed that the CLB can be configured into a 32×1-bit LUT. Figure 3.64 (a) shows how three 32×1-bit LUTs can be accessed in parallel to effectively implement a 32×3-bit LUT. This LUT in Figure 3.64 (a) has five address inputs, $addr_4$ $addr_3$ $addr_2$ $addr_1$ $addr_0$, and three data outputs, d_2 d_1 d_0. Each of these three 32×1-bit LUTs lies in a different CLB. The programmable interconnect is configured to connect the common LUT address bits for all three LUTs. This 32×3-bit LUT implements the two-bit adder in

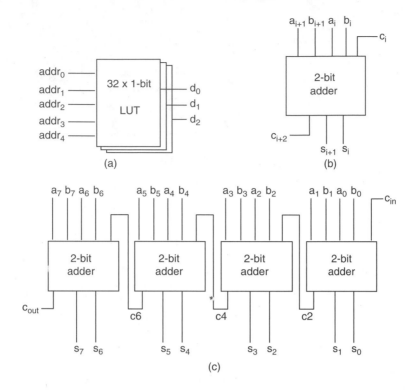

FIGURE 3.64 An 8-bit adder implemented for an FPGA architecture using look-up tables. (a) three 32×1-bit look-up tables accessed in parallel, (b) a 2-bit adder subcomponent, and (c) 8-bit adder.

Figure 3.64 (b). A 2-bit adder is equivalent to two cascaded full adders. Therefore the 2-bit adder adds the two 2-bit binary inputs A and B and the carry-in input, producing a 2-bit sum output and a carry-out. The a_i, a_{i+1}, b_i, b_{i+1} and c_i inputs are connected to the LUT address bits $addr_2$, $addr_4$, $addr_1$, $addr_3$, and $addr_0$, respectively. The LUT must be programmed so that the outputs d_0, d_1 and d_2 generate the 2-bit sum, s_i, s_{i+1} and the carry-out, c_{i+2}. Four 2-bit adders are cascaded, as shown in Figure 3.64 (c), to produce an 8-bit adder.

The truth table for the 2-bit adder subcomponent is shown in Figure 3.65 (a). The inputs are the same as those above. To construct the truth table, inputs are summed using a weighting of one for the ith column inputs and a weighting of two for the $i + 1$st column inputs. In the output, the s_i, s_{i+1} and c_{i+2} columns have a weighting of one, two and four, respectively. A MUX type 0 design is used to select the appropriate value from the SRAM cells for the LUT. Figure 3.65 (b) illustrates how three 32×1-bit SRAM LUTs are used with multiplexers to implement the 2-bit adder subcomponent. One 32×1-bit SRAM stores the output values for each of the three Boolean functions needed to implement the 2-bit adder. The independent input variables, on the left half of the truth table, control which SRAM cells are selected by the multiplexers. Since the truth table is listed in the order a_{i+1}, b_{i+1}, a_i, b_i and c_i, the most significant bit with respect to the truth-table ordering is a_{i+1} and the least significant bit is c_i. Therefore, the multiplexer lines should be connected as $select_4 = a_{i+1}$, $select_3 = b_{i+1}$, $select_2 = a_i$, $select_1 = b_i$ and $select_0 = c_i$.

Four 2-bit adder subcomponents implemented using LUTs, as described above, are cascaded to implement the 8-bit adder, as shown in Figure 3.64 (c). Figure 3.66 shows how the adder subcomponents might be mapped to an FPGA architecture with configurable logic blocks. Each row of three CLBs in Figure 3.66 represents a 2-bit adder subcomponent. The first CLB in the row computes the carry-out of the 2-bit adder, and the second and third CLBs in each row computes the s_i and s_{i+1} output of each 2-bit

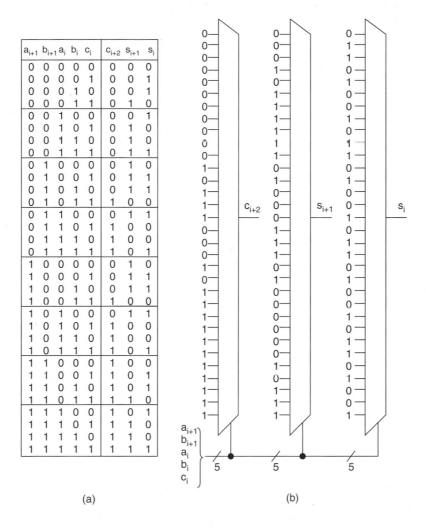

a_{i+1}	b_{i+1}	a_i	b_i	c_i	c_{i+2}	s_{i+1}	s_i
0	0	0	0	0	0	0	0
0	0	0	0	1	0	0	1
0	0	0	1	0	0	0	1
0	0	0	1	1	0	1	0
0	0	1	0	0	0	0	1
0	0	1	0	1	0	1	0
0	0	1	1	0	0	1	0
0	0	1	1	1	0	1	1
0	1	0	0	0	0	1	0
0	1	0	0	1	0	1	1
0	1	0	1	0	0	1	1
0	1	0	1	1	1	0	0
0	1	1	0	0	0	1	1
0	1	1	0	1	1	0	0
0	1	1	1	0	1	0	0
0	1	1	1	1	1	0	1
1	0	0	0	0	0	1	0
1	0	0	0	1	0	1	1
1	0	0	1	0	0	1	1
1	0	0	1	1	1	0	0
1	0	1	0	0	0	1	1
1	0	1	0	1	1	0	0
1	0	1	1	0	1	0	0
1	0	1	1	1	1	0	1
1	1	0	0	0	1	0	0
1	1	0	0	1	1	0	1
1	1	0	1	0	1	0	1
1	1	0	1	1	1	1	0
1	1	1	0	0	1	0	1
1	1	1	0	1	1	1	0
1	1	1	1	0	1	1	0
1	1	1	1	1	1	1	1

(a)　　　　　　　　　　(b)

FIGURE 3.65 2-bit adder subcomponent design based on FPGA architecture using look-up tables. (a) Truth-table for 2-bit adder subcomponent; (b) three 32×1-bit look-up table implementation of adder.

adder. The programmable interconnect routes the input operands and results to pass in and out of the configurable logic blocks. The carry outputs are passed to the neighbor, configurable logic blocks using the direct interconnects. The direct interconnects are much faster than the programmable local and global interconnects. Since the carry-chain is on the critical path, the direct interconnects improve the performance of the 8-bit adder.

Programmable Logic Design Flow Using a Hardware Description Language

Sophisticated computer-aided design (CAD) tools help facilitate hardware design. These tools allow the user to specify a hardware function using a hardware description language (HDL). HDLs are similar to high-level software programming languages since they allow the designer to work at a high level of abstraction and not be concerned with the underlying hardware implementation. Verilog and VHDL are hardware description languages commonly used for hardware design. Our adder function quickly can be specified using Boolean equations in an HDL. Figure 3.67 lists the VHDL code for an 8-bit ripple-carry adder. The entity section at the top of the VHDL code lists the input and output signals for the top-level design. In this case A, B and Cin are

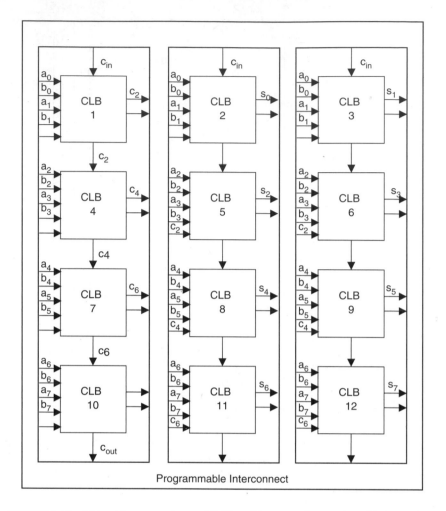

FIGURE 3.66   8-bit adder mapped to an FPGA architecture using configurable logic blocks.

the inputs and S and Cout are the outputs. The architecture section defines the functions of the hardware circuits. Since the 8-bit adder is designed with combinational logic, it can be specified using Boolean equations for each sum and carry bit, as shown in Figure 3.67.

A typical design flow that implements a hardware function using programmable logic begins with the designer writing the HDL code, such as the VHDL code in Figure 3.67. The behavioral code often is simulated to ensure that there are no errors in the HDL code. The designer also must assign the circuit's input and outputs signals to specific I/O pins on the programmable logic device. Other constraints may be entered, including timing requirements.

The rest of the design flow is executed automatically using CAD tools. A synthesis CAD tool generates the design's hardware representation by using standard logic gates and components. The hardware design is represented by a netlist, or a text file containing a list of the hardware gates and components. The nets represent signal names indicating how the hardware devices are interconnected. After the netlist has been generated, it must be translated, or mapped, to the hardware architecture of the programmable-logic device. For an FPGA, the netlist is mapped by a CAD tool to the LUT implementation structure in the configurable logic blocks. The next step is to execute the place and route function that assigns the logic functions to specific logic blocks on a programmable-logic device. The CAD tool must then determine how to route signals through the programmable interconnect. Once the placing and routing is complete, a programming file is

```
library IEEE;
use IEEE.STD_LOGIC_1164.ALL;

entity adder8 is
Port (
    a: in std_logic_vector(7 downto 0);
    b: in std_logic_vector(7 downto 0);
    cin: in std_logic;
    s: out std_logic_vector(7 downto 0);
    cout: out std_logic);
end entity adder8;

architecture behavioral of adder8 is
begin
    -- Carry Boolean Equations
    c(1) <= (a(0) and b(0)) or (a(0) and c(0)) or (b(0) and c(0));
    c(2) <= (a(1) and b(1)) or (a(1) and c(1)) or (b(1) and c(1));
    c(3) <= (a(2) and b(2)) or (a(2) and c(2)) or (b(2) and c(2));
    c(4) <= (a(3) and b(3)) or (a(3) and c(3)) or (b(3) and c(3));
    c(5) <= (a(4) and b(4)) or (a(4) and c(4)) or (b(4) and c(4));
    c(6) <= (a(5) and b(5)) or (a(5) and c(5)) or (b(5) and c(5));
    c(7) <= (a(6) and b(6)) or (a(6) and c(6)) or (b(6) and c(6));
    cout <= (a(7) and b(7)) or (a(7) and c(7)) or (b(7) and c(7));

    -- Sum Boolean Equations
    s(0)  <= a(0) xor b(0) xor c(0);
    s(1)  <= a(1) xor b(1) xor c(1);
    s(2)  <= a(2) xor b(2) xor c(2);
    s(3)  <= a(3) xor b(3) xor c(3);
    s(4)  <= a(4) xor b(4) xor c(4);
    s(5)  <= a(5) xor b(5) xor c(5);
    s(6)  <= a(6) xor b(6) xor c(6);
    s(7)  <= a(7) xor b(7) xor c(7);
end architecture adder8;
```

FIGURE 3.67 VHDL code for an 8-bit ripple carry adder.

generated that can be downloaded into the programmable logic device. The programming file contains the binary data to configure the programmable elements in the programmable-logic device. This file is downloaded into the device using a serial interface. Once the program has been successfully downloaded, the programmable-logic device will function as it has been programmed.

This chapter began with a discussion on various hardware implementation technologies and the advantages and disadvantages of each. Programmable logic-device architectures were presented, followed by a design example to illustrate how these programmable devices implement hardware functions. Finally, the typical design flow used for hardware design, including programmable logic design, was presented. As technologies improve, larger and more complex designs will become more feasible with programmable logic.

Defining Terms

Antifuse: An interconnect that originally is open-circuited but can become shorted-circuited by growing a link to make a connection.

Application specific integrated circuit (ASIC): A semi-custom, integrated circuit designed for a specific application using standard components found in a library.

Block RAM (BRAM): On-chip RAM that can be configured either as RAM or ROM in a user's design on an FPGA.

Complex programmable-logic device (CPLD):   A collection of programmable-logic devices with an interconnection structure on a single integrated circuit.

Configurable logic block (CLB):   The programmable-logic element of an FPGA used to implement logic functions.

Electrically programmable read-only memory (EPROM) cell: A programmable element that can be electronically programmed by applying a voltage or electric field across the programmable elements. An EPROM is erased by exposing the programmable elements to ultraviolet light through a window on the chip.

Electrically erasable programmable read-only memory (EEPROM) cell: Programmable elements that can be electronically programmed or erased by applying a voltage or electric field across the programmable elements.

Fabric: Generic term applying to the programmable architecture of an FPGA.

Field programmable gate array (FPGA):   A field programmable gate array has significantly more logic but a smaller programmable-logic block than a programmable-logic device and includes a more complicated programmable interconnection network to interconnect the programmable-logic blocks.

Flash cell: A memory cell that is electrically erasable, but with a small cell size, similar to an EEPROM element..

Flip-flop: A digital storage device used to store a binary value.

Full custom integrated circuit: A custom integrated circuit designed for a specific application that allows complete customization of the semiconductor and metal layers of the integrated circuit.

Fuse: An interconnect that is originally short-circuited but can be open-circuited, or blown, during programming by passing a large current through the link.

Gate arrays: An integrated circuit device with dedicated hardware resources, such as logic gates and memory storage devices, that is customized in the field or factory for a specific application.

Hardware description language (HDL):   A language type used to specify and design hardware circuits.

Integrated circuit (IC):   An electronic circuit that consists of one or more logic gates fabricated on a single chip.

Macroblock: A programmable-logic block containing several macrocells. The exact number of macrocells per macroblock varies by product and manufacturer.

Macrocell: The fundamental building block for programmable-logic devices that implements a single sum-of-products, or two-level AND-OR, logic function using programmable product terms.

Programmable array of logic array (PAL):   A logic-integrated circuit device with a two-level AND-OR structure, suitable to implement arbitrary logic functions. Only the AND gates can be programmed in a PAL.

Programming element: The device that is configured to implement logic design. Programming elements can be one-time programmable (e.g., antifuse) or reconfigurable (e.g., flash), and can be volatile (e.g., SRAM) or non-volatile (e.g., EEPROM).

Programmable-logic array (PLA):   An integrated circuit device with a two-level AND-OR structure suitable to implement arbitrary logic functions. Both the AND gates and OR gates can be programmed in a PLA.

Programmable logic device (PLD):   A programmable-logic device is a broad classification of programmable-integrated circuits that consist of 20 to 200 logic gates.

Random access memory (RAM):   Memory that stores digital information.

Simple programmable-logic device (SPLD):   A programmable-logic device that typically implements logic functions using programmable two-level logic.

SRAM cell: A single bit of SRAM that is the typical programmable element in a FPGA. The SRAM cells on an FPAG are stored with logic 0 or logic 1 values during programming to configure the chip.

System on chip (SoC):   A complete system on a single chip that may include a processor, memory and custom digital logic.

Very large scale integration (VLSI):   Single integrated circuit devices with more than a million transistors.

References

M.M. Mano and C.R. Kime, *Logic and Computer Design Fundamentals*, 3rd ed., Upper Saddle River, NJ: Pearson Prentice Hall, 2004.

M.M. Mano and C.R. Kime, *VLSI Programmable Logic Devices*, supplement to *Logic and Computer Design Fundamentals*, 3rd ed., Upper Saddle River, NJ: Pearson Prentice-Hall, 2004.

R.S. Sandige, *Digital Design Essentials*, Upper Saddle River, NJ: Prentice-Hall, 2002.

J.F. Wakerly, *Digital Design Principles & Practices*, 3rd ed., Upper Saddle River, NJ: Prentice-Hall, 2001.

W. Wolf, *FPGA Based System Design*, Upper Saddle River, NJ: Prentice-Hall, 2004.

Further Information

The monthly magazine *Xilinx Xcell* presents recent articles on programmable logic. One example is Z. A. Zamindar, "Implement an Embedded System with FPGAs," in the Fall 2004 issue, Volume 50, pp. 43–45.

There are a number of journals that publish articles related to PLDs and FPGAs. Those publications include the *IEEE Transactions on VLSI* and the *International Journal of Embedded Systems*.

P. Chow, S. Seo, J. Rose, K. Chung, I Rahardja and G. Paez, "The Design of an SRAM-Based Field-Programmable Gate Array: Part I: Architecture," *IEEE Transactions on VLSI*, Vol. 7 No. 2, June 1999, pp. 191–197.

P. Chow, S. Seo, J. Rose, K. Chung, I Rahardja, and G. Paez, "The Design of an SRAM-Based Field-Programmable Gate Array: Part II: Circuit Design and Layout," *IEEE Transactions on VLSI*, Vol. 7 No. 3, Sept. 1999, pp. 321–330.

4

Microprocessors

John Staudhammer
University of Florida

Phillip J. Windley
Brigham Young University

James F. Frenzel
University of Idaho

4.1   Practical Microprocessors

John Staudhammer

A microprocessor (μP) is a semiconductor die containing the components of a computer's central processor, complete with instruction processing unit, arithmetic, interrupt electronics and basic communication facilities. Such devices have been available since the early 1970s and have benefited from continuing improvements in electronics. As microelectronics technology allowed feature sizes of components to decrease, more powerful systems were put on single dies. Those developments have changed the functions of microprocessors, markedly among commercial electronic components. Enhanced electronics capabilities led to two major trends: an increase in circuit speed and a widening of the datapath. Data acquisition components, memory blocks and output ports now are commonly added to a basic μP to build a complete microcontroller (μC) on a single chip. This allows the μC to be used for a wide range of applications, whose function is merely a matter of programming. Current clock speeds can be set in the low GHz for personal computer chips but are usually found in the low MHz range for simple control applications. μP systems commonly have either 8, 16 or 32 bits, although original 4-bit and high-end 64-bit processors are available. μC systems form the heart of embedded devices performing specific control tasks in all applications. These devices are universal and highly adaptable for control applications. One of the early μPs, the Intel 8080, still is widely used. Similarly, the Motorola 6800 and its family, the HC11 and related units, are widely used in dedicated control applications. The original 4-bit μP, intended for calculators and containing a few thousand transistors, has been largely supplanted by 8-bit and 16-bit units, even when their increased capabilities are not utilized. This is due largely to economies of scale and design simplifications. High-end μP chips can contain several billion transistors. The chips are built with CMOS technology, minimizing power consumption.

  μP applications in system control and data processing involve adding memory for programs and data storage and both input and output circuitry. They also may require communication ports and analog–digital (**A/D**) converters. Several manufacturers offer these components on a basic μP chip, producing a capable microprocessor-based controller that needs only programming to be configured into a customized μC system. Manufacturers also offer CD-ROM software for the customization, design, simulation and configuration of a chip.

An excellent overview of microprocessors and systems is found in Peatman [2003], Gaonkar [2001], Gaonkar [2002] and Kheir [1997]. Many books are available for designing systems with μC chips. Most books are slanted toward a specific device, but the design process remains similar for all μP and μC chips.

The μC is the basic controller of an embedded system. A μC is a μP with memory and peripheral circuits on the same chip. These include various memory types, interrupt structures, arithmetic-logic units (including floating-point processing on some advanced chips), communication, timing and data acquisition circuits. As the feature sizes of switching circuits decrease and the speed improves in excess of Moore's Law, the manufacturer has three options:

1. Reduce the chip's die size — this results in a less expensive device with about the same capabilities.
2. Increase processing power at the same cost.
3. Add peripheral circuits to the processor, putting devices on the chip that normally would be added to the μP system. Adding the right combination of peripherals to the basic chip offers the most benefit for system cost.

The chip developments above occur simultaneously, resulting in a successful product line. For example, the Motorola 6800 processor was introduced about 25 years ago. Today, it is still available, but the manufacturer advises designers to use the successor chips 6809 MP and 68HC11 μC.

A μP chip communicates with its peripherals through three sets of busses: a bidirectional data bus, a unidirectional memory-address bus and a bidirectional control bus. The μP sends and receives data on communication lines; the number and types of these vary among chips. On the same chip, a μC will have a number of memory elements (typically, different kinds), timers, communication ports, buffers, counters and analog-digital converters.

All processors require a clock, typically a crystal or a ceramic resonator, that is directly connected to dedicated pins on the processor. The processor's internal circuits run at this speed, but external devices (and communication busses) run at a submultiple of this rate. The bus clock rate is usually 2 to 20 times slower than that of internal circuits. The advertised clock rate is the internal μP clock rate, not the bus speed.

There are more than 100 different μP and μC chips on the market. They each have peculiarities and may have advantages for specific applications. They differ in the types of data they handle (such as the number of basic bits they manage per operation), the amount of processing they do (for example, floating-point capabilities) and the software support available from the manufacturer or user groups. What makes a μP practical for any situation is not its claimed prowess (typically stated in MIPS, or millions of instructions per second), but the ease of use and cost for a given application. That is determined to a large degree by the kind and amount of software support from the manufacturer.

Types of Microprocessors and Microcontrollers

The annual compendium of μP/μC chips (Markowitz 2004) categorizes these chips by the width of their data path: 4, 8, 16, 32 or 64 bits. In addition, high-performance chips include bit/word slice chips. These implement the functions of a central processing unit for a limited number of bits (4 or 8 bits) and are concatenated for a full-width computer word. For a discussion of these chips, a widely used reference book comes from Mick and Brick [1980]. Additional references are in the IEEE-ACM Design Automation Conference archives, accessible through the Internet.

The most precious resource in a μP chip is the number of connection pins. Great efforts are made to use the pins chosen by the processor's various control lines and mode selection. A μP may have four or more modes of operation: it is cheaper to build and support a more flexible chip than to build several different ones.

The majority of μPs work with several data-access means (memory addressing modes); they support different modes of working with memory and external data items. They use complex instruction set computers (CSC.) These chips appear as single items in computer-controlled devices; most of these chips are used in embedded-systems applications, in household appliances and in toys.

Most μPs are 8-bit devices; they use 8-bit-wide datapaths but have memory address busses of 16 bits (with 64K of address space.) Often the data bus is time-multiplexed with lower address bits. The typical instruction execution speed is three to seven bus cycles. The 16-bit μP chips have 16-bit address and data lines. The typical execution speed is a few clock cycles.

μC chips contain a μP and the various components making up a μP system: a random-access memory (RAM) for holding volatile data; at least one kind of read-only memory (ROM), possibly written in a special configuration mode and holding the control program; communications peripherals, including parallel and serial interfaces; and timing and counting circuits to measure input pulses and digital-analog converters. Because processors use an external clock, usually a crystal clock, timing intervals can be determined with great precision. Many procedures have been developed for this feature: voltage-to-frequency converters bring pulses externally to the μP system, which then accurately counts them. These timing and counting capabilities give μP systems their ubiquitous applicability.

Software for μP/μC Systems

μP/μC device manufacturers have made special efforts to make their products attractive to system designers. Much software is offered with the chips to support their design. ROM-based monitor programs are available for most chips, so normal communication tasks can be easily accomplished. Programs for usual input/output tasks, data acquisition, timing and program examples are distributed throughout Internet bulletin boards, making relatively error-free software available to designers. The bulk of such software is in machine assembly code, but C-language procedures often are available. Programming then adapts these software snippets to the task the system is to perform. System task analysis and program system design, as well as validation and software validation, are still the designer's responsibilities. The processor carries out instructions that are a combination of 0 and 1s, which can be difficult to decipher. Programs are written in mnemonic assembly-language, text-readable form. The manufacturer supplies assemblers and loaders to set the program into the memory of the μP/μC system. These programs run on personal workstations. Programs that simulate and monitor the chip's operation through the PC ports are effective development tools. The typical development-software package includes an assembler, a loader and a monitor. More user-oriented software for these tasks is available from third-party vendors. Many of these software packages cost more money but can be cost-effective. Interactive assemblers and debugging tools are particularly good investments. To find effective software tools, the manufacturer's customer engineering group should be consulted.

Packaging and Cost

The simplest processors are housed in dual in-line packages (DIP) of 16 pins or more and resemble conventional TTL chips. Mid-range processors are housed in large DEP packages of 40 or more pins, but many reside in dense-pin sockets of 100 or more pins. High-performance chips are found in multi-row pin packages, can require more than 200 connections and sometimes use unusual chip sockets. These high-performance chips may require special cooling, typically from a fan clipped to the chip.

Virtually all processors are multi-sourced, making them available from more than one manufacturer. This is important for continued product support. Processors come in various speed grades. High-speed ones are three to ten times more expensive than slower ones. Low-end processors, containing no memory and few communication ports, cost less than a dollar in large quantities; high-end processors approach $1,000 in single units.

Programming of μPs

Many books explain the fundamentals of μP systems and their programming for various tasks. Some books are oriented towards specific chips; however, they present "how-to" examples that can be applied to other chips. They provide good training for all μP/μC systems. Two handy references are Kheir [1997], used in Motorola 68HC11, and Barnett [1995], used in Intel 8051.

Most μP systems are programmed in assembly language. Each instruction the processor executes is given an English-like name (ADD for adding two numbers), and translator programs can convert this English-like source code to the 0/1 sequence of bits used by the processor. These programs, used with PC workstations, are available from chip manufacturers, while convenient assemblers are available from third-party vendors. High-level language compilers also are available from third-party vendors. Usually, these are C-language compilers, but Pascal may be available also. The best guide again is the application engineering group of the chip manufacturer.

Usually, the functions that the μP and μC perform offer time-critical, real-time control. The programs can become intricate and require detailed knowledge of the the μP chip functions. To select a suitable chip for a given task, a logical design must be made of the control task for the entire hardware configuration, and a quantitative evaluation of the system architecture must be completed. This is seldom a simple task but can be performed in a simplified manner, especially if only order-of-magnitude evaluations are made. A standard reference for the quantitative analysis of computing systems is Hennessey and Patterson [2004].

μP system programming involves the creation of carefully tailored code to acquire input signals and create control signals that interact with the controlled actuators and receivers.

Development Support

Complex μP/μC systems are developed top-down. Task requirements are refined until they can be implemented with relatively simple subroutines. They then are coded and assembled into a control program loaded into the memory of the processor. The most important support for a designer is software for program checking, both for logical flow and for cycle-by-cycle activity. Subtle errors and data dependencies often will occur, and finding them may be a daunting task. Simulators are first-level tools for checking the program. Simulators do not use actual hardware but instead a software model of it. They calculate the internal action of the hardware and display the contents of all computer registers and selected memory locations, allowing an effective check of the internal operations. Many simulators come from third-party vendors.

The actual operation of the μP/μC system can be checked with an in-circuit emulator (ICE), an expensive but effective tool replacing the μP/μC pin-for-pin in the actual circuit. The ICE uses a powerful device to track signals, including brief transients, and can be used to show the behavior of the system and the response from the μP and its associated software.

Comparison of μP/μC chips

As with all computer devices, advertised speeds and performance figures must be interpreted cautiously. These numbers normally refer to manufacturer test cases and may not be directly comparable [Hennessey and Patterson, 2004]. The numbers may not apply to the user's requirements. Since the market for μP/μC chips is highly competitive, small differences become amplified in advertising. Unfortunately, finding valid comparison numbers for an application requires extensive benchmarking efforts. Advertised figures may be taken as a guide only if the user's task is similar to the advertised task. If the selected chip does not pass the comparison with a comfortable margin, a user should opt for a higher-performance version of the same chip if much program development has already been completed. Otherwise, another, superior chip should be chosen for a new design.

The μP issue of the magazine *EDN* is a handy starting place for a detailed overview of commercially available chips and a comparison of their characteristics.

Trends in μP/μC Developments

An entire μP can be used as a building-block in an application specific integrated circuit (ASIC), available from many vendors. ASIC and VLSI design tools are used to design such systems and are tailored to specific applications. μP manufacturers are developing μC chips, extending the use of their basic processors. For any application, the best procedure is to invite several vendors to propose alternate systems. Available chips are merely indicators of devices yet to come.

High performance chips are becoming more reduced instruction-set computer (RISC) oriented. The basic idea is to execute one instruction per clock cycle in a more regular processor structure rather than using the prevailing three to seven clock cycles in the normal complex instruction-set computer (CISC) processor. For users, the internal structure of the processor makes little difference; the support software is far more critical. The trend is to make processors simpler, speeding up process execution, even if some programs require more details to replace CISC "convenience" instructions.

High-end μP systems have such a wide data path (32 or 64 bits) that more than one instruction may be accessed at a time. These machines are termed very long instruction word (VLIW) machines. They may have higher throughputs but tend to be more complex internally. RISC and VLIW can provide increased performance but are not necessarily more cost-effective solutions.

Microelectronics can produce multi-million and few-billion transistor chips. While expensive, they can provide a cost-effective solution to complex control and computation problems. However, the humble 8- and 16-bit μC chips are the most cost-effective workhorses for the bulk of control applications.

References

R.H. Barnett, *The 8051 Family of Microcontrollers*, Englewood Cliffs, NJ: Prentice-Hall, 1995.

R.S. Gaonkar, *Microprocessor Architecture, Programming and Applications with the 8085*, Englewood Cliffs, NJ: Prentice-Hall, 2002.

R.S. Gaonkar, *Z 80 Microprocessor: Architecture, Interfacing, Programming and Design*, 3rd ed., Englewood Cliffs, NJ: Prentice-Hall, 2001.

J. Hennessey and D.A. Patterson, *Computer Architecture: A Quantitative Approach*, 3rd ed., Amsterdam, Holland: Elsevier, 2004.

M. Kheir, *The M68HC11 Microcontroller: Applications in Control, Instrumentation, and Communications*, Englewood Cliffs, NJ: Prentice-Hall, 1997.

I.S. MacKenzie, *The 8051 Microcontroller*, 2nd ed., Englewood Cliffs, NJ: Prentice-Hall, 1995.

M.C. Markowitz, "EDN's annual μP/μC chip directory," *EDN* magazine (yearly).

J. Mick and J. Brick, *Bit-Slice Microprocessor Design*, New York: McGraw-Hill, 1980.

J.B. Peatman, *Embedded Design with the PIC452 Microcontroller*, Englewood Cliffs, NJ: Prentice-Hall, 2003.

J.W. Valvano, *Introduction to Embedded Microcomputer Systems: Motorola 6811 and 6812 Simulation*, Pacific Grove, CA: Brooks/Cole/Thomson Learning, *2003*.

Further Information

Microprocessor and microcontroller chips are two of the fastest-changing (and improving) items in electronics. A large number of reference texts exist, but the latest changes are best tracked through technical magazines and direct contact with vendors. Contact information is best made through an Internet search engine.

The magazine *EDN* runs an annual issue on μP and μC developments (Markowitz, 1995). This review is published late in the year. This special issue is a good snapshot of the state of the industry.

The IEEE magazine *Micro* publishes detailed articles on device developments and applications of μC technology. Device specifics are found in the manufacturers' reference literature. A serious user must become familiar with the applicable manual and design notes. Even for a modest chip, the reference manual may run to 300 pages. The manufacturer's free-of-charge support software is usually a fully-loaded CD-ROM disk. The usual μC chip is meant to be configured with. flash memory typically loaded from a PC through a parallel or serial port. Required software is found on the CD-ROM or a bulletin board from the manufacturer.

4.2 Applications

Phillip J. Windley and James F. Frenzel

Microprocessors are cheap, small, and consume little power. In addition, in recent years their performance has increased at a greater rate than the performance of larger computers. These factors have led to an explosion in the application of microprocessors. A short section could never do justice to every application; therefore, we will view representative applications in three broad areas:

- Data collection, where microprocessors are used to monitor sensors and either record the collected information or communicate the information to some other computer.
- Control, where microprocessors have largely replaced analog electronics for controlling everything from manufacturing robots to home appliances.
- Computing, where microprocessors have transformed the concept of computer and made parallel processing possible.

Admittedly, these categories are not strictly disjoint. They do, however, represent the most pervasive uses for microprocessors at an abstract level.

Data Collection

In data collection the microprocessor-based system serves primarily as a low-cost data recorder. Basic functions include the polling of sensors, acceptance of data, data storage, and data transmission or display. Additional features might include preprocessing of the raw data. Such a classification spans a broad range of applications, from automotive diagnostics to space-born monitoring stations.

Microprocessors are well suited as the controller for such tasks because of their cost and flexibility. Sufficient numbers of processors may be used to allow real-time data acquisition. Because the microprocessor is programmable, sensors may be added, removed, or rearranged without major system impact. Finally, because the microprocessor is a computational device, calculations may be performed on the recorded data to produce useful information, such as calculating speed from distance and time. In the next section we will examine the components of one such system, the retail point-of-sale terminal [Hordeski, 1984].

Point-of-Sale Terminal

The function of a point-of-sale (POS) terminal is characteristic of the applications under the category of data collection. The microprocessor is not being used for intensive computations, nor for controlling a complex process, but rather to collect data, perform some processing, and then pass the results on to a central collector. The cost and flexibility of the microprocessor make it an excellent choice over special-purpose hardware.

System Components. In addition to the microprocessor and storage capability, the typical retail terminal has one or more input devices for entering prices (e.g., keyboard, bar code scanner) and one or more output devices for displaying totals (e.g., paper tape, display). Often these terminals are part of a large network of terminals and may support additional features beyond totaling purchases such as automated inventory control and credit checking. A complete system is shown in Figure 4.1, including magnetic tape for storing transactions and **a universal asynchronous receiver/transmitter (UART)** for communication with a central processing facility. Because of the high unit volume, it is desirable to keep the cost and complexity low. Typically, each terminal will have limited storage capability, relying on a central processor for maintaining store inventory and credit checks. In order to reduce communication traffic with the central processor, however, each terminal generally has in storage the current price for all items.

Universal Product Code. The use of the Universal Product Code (UPC) has enabled the development of intelligent POS terminals which can "read" the UPC symbol and determine the identity of the item. The UPC symbol consists of ten decimal digits, split into two fields of five digits each. Each digit is encoded using a 7bit binary number, represented by a group of 7 dark (binary 1) and light (binary 0) bars. The five left-hand digits

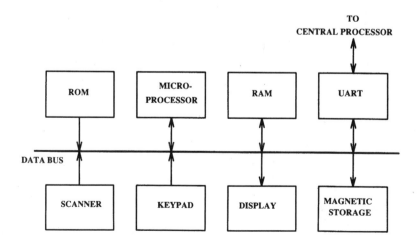

FIGURE 4.1   Point-of-sale terminal system.

are encoded using odd parity and the right-hand digits are encoded using even parity. This allows correct recognition of the symbol, independent of its orientation.

For groceries, the first five decimal digits identify the manufacturer and the second group of five digits identify the specific product. There are additional codes in use as well, such as the National Drug Code. By using a microprocessor-based system, a POS terminal can be quickly reconfigured to recognize a different code (or multiple codes) through a simple software change.

Operation. A typical sale might involve the following steps. The clerk inquires whether the sale is to be a cash purchase or charged to an account. If the latter, the clerk enters the necessary information and the terminal transmits a request to a central processor, inquiring as to the available credit. In the interim, items are entered, either through the bar code scanner or the keypad, and the price and running total are displayed. The identity of the items purchased is also stored for later transmission to the central processor responsible for inventory control. Finally, the terminal checks the available credit against the total and records the transaction for later transmission to the central processor.

Digital Tachometer

Another example of using a microprocessor for data collection is the implementation of a digital tachometer [Bonert, 1989]. The microprocessor samples the output of a shaft **encoder** and compares it with a reference signal to determine the rotational speed. The calculated value is passed to a digital-to-analog converter to generate an analog speed signal. The system is shown in Figure 4.2.

Speed Evaluation Methods. Various methods may be used to evaluate the speed value, all of which involve some combination of pulse counting and time measurement. The constant elapsed time (CET) method provides a good compromise between measurement accuracy and response time. The CET method records the number of encoder pulses observed during a fixed time interval. The rotational speed, n, is then given by

$$n = C_p/(C_t m/T_c)$$

where C_p is the number of encoder pulses, C_t is the number of clock pulses, m is the number of encoder marks per turn, and T_c is the clock pulse period.

Implementation. Rather than continuously stopping and resetting external counters, it is possible to take advantage of features often found in modern **microcontrollers**, microprocessors containing additional interface circuitry. Microcontrollers often contain counters, timers, and **capture registers**. Capture registers allow the storing of timer or counter values triggered by an external signal. At the start of evaluation, the rising edge of the next encoder pulse triggers the capture of the timer count and the pulse count. After a minimum evaluation time has elapsed, the next encoder pulse again triggers the capture of the current counter values. The rotational speed can then be computed using the difference between the captured values. A flowchart of the algorithm is shown in Figure 4.3.

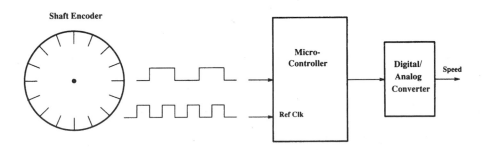

FIGURE 4.2 Digital tachometer.

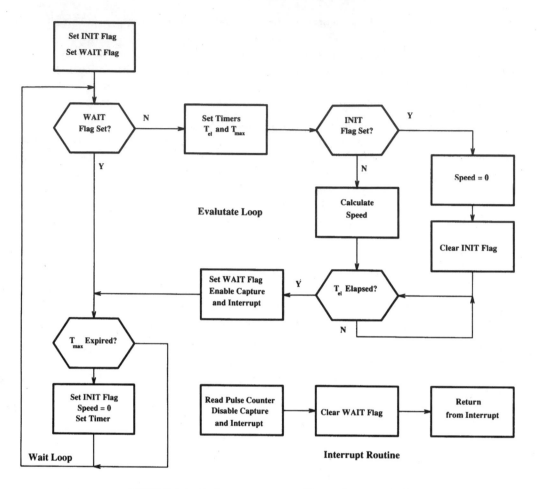

FIGURE 4.3   Tachometer program. (*Source:* Bonert, 1989.)

Performance.   Using an encoder with 1024 marks per revolution, a 2-MHz reference clock, and an evaluation period of 2.3 ms resulted in a measurable speed range of 25.5–4883 rpm. The maximum relative error was 0.123%, induced primarily by the encoder tolerance [Bonert, 1989].

Control

Microprocessors are ubiquitous in control applications. While some custom analog controllers are still built, the advantages of cost and flexibility inherent in microprocessors make them a natural choice. The advantages of microprocessors are particularly obvious in mass-produced goods where time-to-market can be a significant driving force.

Microcontrollers

Microprocessors designed especially for use in control applications are called *microcontrollers*. Typically, the major difference between a microcontroller and a standard microprocessor is the presence of scratchpad RAM, input and output ports, timers, and even analog-to-digital (**A/D**) and digital-to-analog (**D/A**) converters on-chip.

  Figure 4.4 shows a simplified microcontroller architecture. The process to be controlled is monitored by means of sensors. The outputs from the sensors are fed to A/D converters which convert the analog signals from the sensors to digital signals appropriate for use in the microprocessor. The microprocessor reads the digital signal from the A/D converter and uses it for input to a control program stored in the

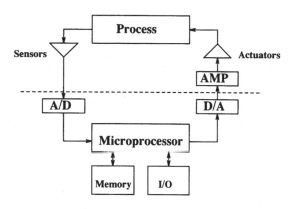

FIGURE 4.4 Typical microcontroller design.

microprocessor memory. The program produces digital outputs which are fed to D/A converters. The analog outputs from the D/A converters (which are typically low power) are fed to amplifiers, and the amplified signal is used to control actuators that affect the process being controlled.

Control Applications

Consumer Electronics.   A survey of the typical home will show numerous microprocessors where 10 years ago, there were none. Microprocessors are used for controlling VCRs, TVs, stereo equipment, microwave ovens, sprinkler systems, telephone equipment, heating systems, and virtually every other appliance using electricity.

Manufacturing.   Microprocessors have found numerous applications in manufacturing. Perhaps none is better known than the robot. Microprocessor technology has made the modern robot possible. Robot arms used in manufacturing typically have five or six joints. Current practice is to treat each joint in the robot arm as a separate servomechanism with its own control system. For example, the PUMA 560 robot arm, manufactured by Unimation, has six rotating joints. Each joint is controlled by an individual microcontroller system. Another computer calculates paths and sends individual joint motion information to the six joint servomechanisms [Fu et al., 1987].

The servomechanism system shown in Figure 4.5 consists of an 8-bit Rockwell 6503 microprocessor, a D/A converter, an amplifier, a joint motor, and an encoder. The 6503 microprocessor receives joint position information from the supervisory computer every 28 ms. The microprocessor calculates the joint error information by comparing the current position to the desired joint position using the PID (proportional-integral-derivative) control method. The error is converted to an analog signal by the D/A converter and amplified before going to the joint motor. The encoder is connected to the motor shaft and provides a digital signal to the microprocessor.

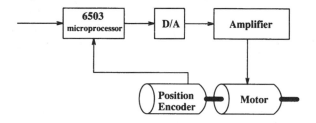

FIGURE 4.5   Microprocessor-controlled servomechanism from a PUMA 560 robot.

The microprocessor performs the following functions:

1. Receives the desired joint position from the supervisory computer every 28 ms
2. Reads the position signal from the encoder every 0.875 ms
3. Calculates the error every 0.875 ms
4. Sends the error to the D/A converter

The microprocessor calculates joint error and sends the correction signal to the joint motor 32 times for every joint position received from the supervisory computer.

Transportation.   Microprocessors are used for control applications in every facet of the transportation industry. Microprocessors are used to control the operation of the vehicles themselves such as controlling engines, air surfaces in aircraft, antilock brakes in automobiles, and rudders in ships. Microprocessors are also used in wide-area applications such as traffic control.

In controllers for motor traffic, the microprocessor has replaced hardwired logic and analog systems to provide systems which are much more capable and typically more reliable [Hordeski, 1984]. A typical traffic light controller is shown in Figure 4.6. The microprocessor provides the CPU, memory, and I/O ports. The system includes a real-time clock for timing external events and a power-fail restart unit which restarts the system after a power failure (including restoring volatile data). The system monitors traffic at the intersection through the use of loop detectors and controls the traffic by changing the traffic lights. Other components of the system monitor and control pedestrian traffic and provide an interface to the system for human operators.

The loop detectors are paired coils of wire placed under the pavement. The impedance of the loop detectors changes in response to the presence of a car on the roadway. The change in impedance changes the frequency of an RC oscillator, which is converted to a digital signal reported to the microprocessor. Loop detectors can be used to monitor the presence of a car at a traffic light, the length of a line of cars, and the speed of traffic.

The function of the traffic controller is to optimize traffic flow. For example, during busy periods of the day, the goal may be to optimize flow through an intersection. Another goal may be to ensure that traffic flows smoothly in certain directions to effectively feed larger roads. Traffic lights can be synchronized to provide a highway through a busy network of roads by ensuring that a car that enters the roadway and maintains a recommended speed can travel along the entire length without stopping at a traffic light. On the other hand, during periods of low use, such as night and early morning, the system may monitor for the presence of a car at an intersection and immediately switch the light to let it pass.

Microprocessors offer advantages in traffic control situations in addition to optimized traffic flow. When properly designed, the system can provide a certain degree of fault tolerance. When a loop detector is giving a

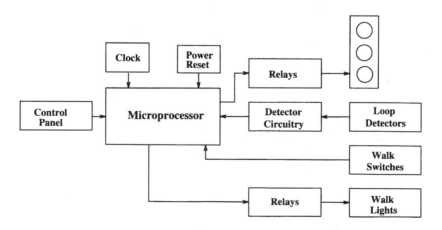

FIGURE 4.6   Traffic control system. (*Source:* M. Hordeski, *Microprocessors in Industry*, New York: Van Nostrand Reinhold, 1984, p. 398. With permission.)

faulty value, the system can be programmed to ignore its value and use values from adjoining lanes. An error report can be forwarded to a central traffic facility and after repairs are made, the loop brought automatically online. The system can also monitor feedback information from the traffic light to ensure that the lights are actually lit. When a problem is detected, the system can enter an emergency mode and report the problem.

Social Issues

The explosive growth in the use of microprocessors in control applications has caused discussion about the utility and safety of such devices.

An issue many people can identify with is feature overload. The advent of cheap microprocessors has turned design upside-down. Designers can add additional features for very little additional increased manufacturing cost. Competition spurs even more features until even the simplest of consumer items come with thick instruction manuals. Naturally, consumers become frustrated with features that are difficult to use.

Perhaps more important are the safety hazards that may be engendered by replacing analog control systems with digital control systems. Most analog systems are based on physical properties with continuous behavior. Digital systems, on the other hand, are discrete and are thus much more prone to problems where small errors can result in large changes in behavior due to the digital representation of value; a single bit change can result in a large change in magnitude. Digital control systems are becoming more and more prevalent in systems controlling aircraft, automobiles, nuclear power plants, and other safety-critical systems. Engineers who design the systems and officials charged with ensuring their safety are still coming to grips with the implications of this trend. New techniques for analyzing computer system designs for errors are being developed which promise to alleviate some of these concerns [Windley, 1995].

Computing

While microprocessors have been put to a plethora of interesting special-purpose uses such as data collection and control, perhaps the most visible use of microprocessors has been in the area of general-purpose computing.

Microcomputers

The advent of microprocessors has resulted in a personal computer on virtually every desktop. Even the slowest of these computers rival the performance of the largest computers available 15 years ago.

Figure 4.7 shows the major hardware components of a simple microcomputer. The central processing unit (CPU) is the execution engine of the microcomputer and is most often a microprocessor. One popular family of microprocessors used as the CPU in microcomputers is manufactured by Intel. These chips, with names such as the 8088, 80286, 80386, 80486, and Pentium are used in microcomputers such as the IBM personal computer. Another important family of microprocessors is the Motorola Power PC series, which is used in microcomputers manufactured by Apple Computer [Matloff, 1992].

In addition to the CPU, there are a number of other components in a microcomputer. General-purpose memory is not typically part of the microprocessor but must be added as a separate component. In simple microcomputers, the memory may be directly attached to the microprocessor. In more complex designs, the

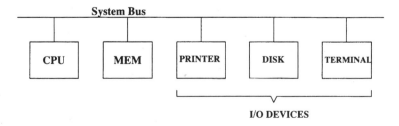

FIGURE 4.7 Major components of a simple microcomputer.

memory is attached to the microprocessor by a system bus that allows system components other than the microprocessor to access memory as well. In addition, the memory may have its own controller, called a memory management unit.

Other components in the system include input/output (I/O) interfaces to devices such as printers, terminals, disks, mice, and so on. The common feature of all of these devices is that they interface the microprocessor to the outside world. All of the components in the microcomputer are connected together by a system bus. The bus is a set of parallel wires that carry information from one component to another.

Multiprocessing

The desire for greatly increased computer performance has fueled research in using microprocessors as the computing engines in multiprocessors which would achieve performance gains over single-processor computers through the use of numerous low-cost microprocessors.

There are numerous multiprocessor architectures. An example architecture that is well suited to using large numbers of microprocessors is the hypercube. The hypercube architecture was originally developed by Charles Seitz and others at California Institute of Technology in the early 1980s. The hypercube depends on using large numbers of commodity microprocessors, each with private memory, in a hypercube network [Bell, 1989].

In a hypercube network, N microprocessors are arranged in an n-dimensional cube, where $N = 2^n$. Each processor is connected to n other processors and the longest communications path from any processor to any other is n links. For example, a three-dimensional hypercube contains eight processors and is arranged as a standard cube, where the nodes are the processors and the edges of the cube are the communication paths.

Figure 4.8 shows a four-dimensional hypercube represented as a tesseract. A four-dimensional hypercube has 16 processors, each is connected to 4 other processors, and the longest path between any two processors (shown in bold in Figure 4.8) is 4. Thus, doubling the number of processors results in a unit increase in the communications path length. This logarithmic relationship results in the great advantage of the hypercube: it scales well. A system with 1024 processors has a maximum communications path length of just 10.

There are several manufacturers of hypercube systems including NCUBE and Intel. Most of these systems have between 32 and 1024 processors. NCUBE has a hypercube architecture with 8192 nodes operating at 2.4 megaflops each.

Digital Signal Processing

Digital signal processing (DSP) may be considered a specific example belonging to the category of computation. Specialized microprocessors are finding widespread application in many areas of digital signal processing such as telecommunications, speech processing, medical imaging, and radar [Aliphas and Feldman, 1987]. These microprocessors are designed for very high data rates and contain specialized circuitry to accelerate computations that are specific to signal processing. Figure 4.9 illustrates the architectural differences between digital signal processors and conventional microprocessors.

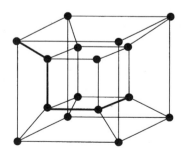

FIGURE 4.8   A four-dimensional hypercube.

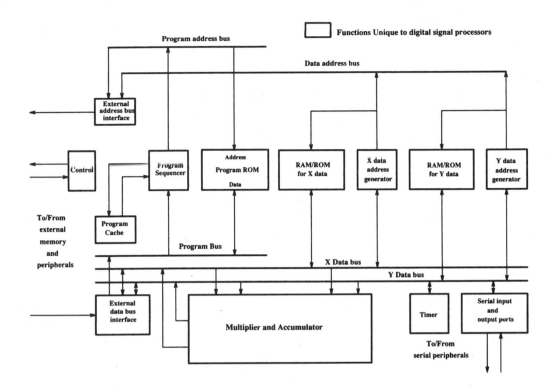

FIGURE 4.9 Digital signal processor architecture. (*Source:* Aliphas and Feldman, 1987.)

Architectural Features. A common task among most signal processing algorithms is the summation of multiple products. The most notable distinction between general-purpose microprocessors and digital signal processors is the existence of a high-speed multiplier-accumulator [Allen, 1985]. This circuitry can complete a multiply-add operation in one cycle, as opposed to roughly 25 cycles for a conventional microprocessor. Traditionally, only fixed-point arithmetic was available, but newer DSP chips provide floating-point arithmetic with 32 bits of precision.

The second most noticeable feature on DSP chips is the existence of multiple data buses and memories. Many chips have two data memories, each with a data bus, allowing the simultaneous fetch of two operands for the multiply-accumulate operation. Furthermore, most chips use the Harvard architecture, characterized by separate program and data memories, so that instructions and data can be fetched simultaneously. Others use a modified Harvard architecture, where data can be stored in slower, cheaper program memory and moved to the faster data memory as needed.

Finally, DSP chips typically have separate arithmetic-logic units (ALU) for data arithmetic and address calculations. This serves two purposes: (1) data calculations can proceed unhindered by address calculations, maintaining a high throughput, and (2) each unit can be specialized for its particular task. For example, the data ALU may have additional circuitry to support saturation arithmetic, whereas the ALU used for address calculations may provide indexing, auto-increment, or even bit-reversal, an operation required for the fast Fourier transform (FFT).

Dedicated digital signal processors offer an excellent alternative or supplement to general-purpose microprocessors for signal processing applications. As a slave to a conventional processor, the DSP chip is freed from communicating with peripherals, increasing throughput. For additional performance, DSP chips may be operated in a multiprocessor configuration, controlled by a central processor. Such an arrangement would be appropriate for applications such as phased-array radar, where the volume of data and uniformity of the calculations lend themselves to distributed processing.

Defining Terms

A/D: Analog to digital. Usually a device that changes an analog signal to a digital signal of corresponding magnitude.

Capture registers: Internal registers which, triggered by a specified internal or external signal, store or "capture" the contents of an internal timer or counter.

D/A: Digital to analog. Usually a device that changes a digital signal to an analog signal of corresponding magnitude.

Encoder: A sensor that directly creates a digital signal for use in a control application. An example is a shaft encoder that turns an angular shaft position into a digital signal.

Interrupts: Special hardware on a computer that suspends the executing program so that another procedure can be run to service an external device.

Microcontroller: A special-purpose microprocessor with scratchpad RAM, input and output ports, timers, and even analog to digital (A/D) and digital-to-analog (D/A) converters on-chip used in control applications.

Universal asynchronous receiver/transmitter (UART): Circuitry (often a separate module), which provides all of the interface functions necessary for a microprocessor to communicate with a serial device.

References

A. Aliphas and J. Feldman, "The versatility of digital signal processing chips," *IEEE Spectrum*, vol. 24, no. 6, pp. 40–45, June 1987.

J. Allen, "Computer architecture for digital signal processing," *Proceedings of the IEEE*, vol. 73, no. 5, pp. 852–873, May 1985.

G. Bell, "The future of high performance computers in science and engineering," *Communications of the ACM*, 32(9), pp. 1091–1099, September 1989.

R. Bonert, "Design of a high performance digital tachometer with a microcontroller," *IEEE Transactions on Instrumentation and Measurement*, vol. 38, no. 6, pp. 1104–1108, December 1989.

K.S. Fu, R.C. Gonzalez, and C.S.G. Lee, *Robotics: Control, Sensing, Vision, and Intelligence*, New York: McGraw-Hill, 1987.

M. Hordeski, *Microprocessors in Industry*, New York: Van Nostrand Reinhold, 1984.

N.S. Matloff, *IBM Microcomputer Architecture and Assembly Language*, Englewood Cliffs, N.J.: Prentice-Hall, 1992.

P.J. Windley, "Formal modeling and verification of microprocessors," *IEEE Trans. on Computers*, vol. 44, no. 1, January 1995.

Further Information

Byte magazine is a good resource for entry-level articles on microprocessor applications. For subscriptions contact: BYTE, One Phoenix Mill Lane, Petersborough, NH 03458.

The Institute of Electrical and Electronics Engineers (IEEE) publishes several magazines and journals that frequently contain articles concerning microprocessor applications. *IEEE Micro* is a bimonthly magazine which addresses the design and use of microprocessors and minicomputers. *IEEE Computer* is a monthly magazine covering all aspects of computing. Three pertinent journals published bimonthly by the IEEE are *Transactions on Industry Applications*, *Transactions on Industrial Electronics*, and *Transactions on Instrumentation and Measurement*. The address for the IEEE Service Center is 445 Hoes Lane, Piscataway, NJ 08855.

5
Displays

James E. Morris
Portland State University

Larry F. Weber
The Society for Information

5.1   Light-Emitting Diodes

James E. Morris

The light-emitting diode (LED) has found a multitude of roles as the field of optoelectronics has bloomed. Infrared devices are used in conjunction with spectrally matched phototransistors in optoisolation couplers; hand-held remote controllers; interruptive, reflective and fiber-optic sensing techniques; and low-cost IR sources for fiber optic communications. Visible spectrum applications include simple status indicators and dynamic power level bar graphs on a stereo or tape deck. This section will concentrate on digital display applications of visible output devices.

Semiconductor Device Principles

The operation of an LED is based on the recombination of electrons and holes in a semiconductor. As an electron carrier in the conduction band recombines with a hole in the valence band, it loses energy ΔE equal to the bandgap E_g with the emission of a photon of frequency

$$v = c/\lambda = \Delta E/h \tag{5.1}$$

where λ is the radiation wavelength, c is the velocity of light, and h is Planck's constant.

The incidence of recombination under equilibrium conditions is insufficient for practical applications, but can be enhanced by increasing the minority carrier density. In an LED this is accomplished by forward biasing the diode, the injected minority carriers recombining with the majority carriers within a few diffusion lengths of the junction edge. Figure 5.1 illustrates the process. The potential barrier eV_o is reduced by forward bias eV leading to net forward current and the minority carrier distributions shown on either side of the depletion layer. As the carriers diffuse away from the junction edges, these distributions decay exponentially because of recombination with the majority carriers. Each recombination

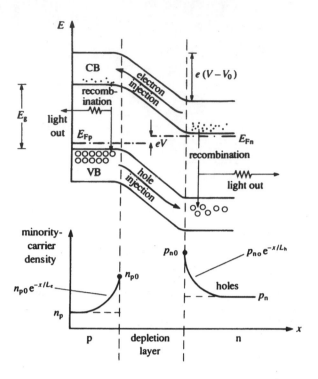

FIGURE 5.1   Light emission due to radiative recombination of injected carriers in a forward-biased pn junction. (*Source*: J. Allison, *Electronic Engineering Semiconductors and Devices*, 2nd ed., London: McGraw-Hill, 1990, p. 302. With permission.)

event shown on either side of the junction gives off a photon. This process is called *injection electroluminescence*.

Equation (5.1) implies that the radiation emitted will be monochromatic, but in practice $\Delta E > E_g$, and there is a spectral distribution corresponding to the energy distributions of the carriers in the conduction and valence bands.

Semiconductor Materials

Silicon is the most common material used in current semiconductor technologies, but it is not at all suitable for an LED. The reason is that silicon has an indirect bandgap, and a direct bandgap is required for process efficiency. Direct and indirect bandgaps are compared in Figure 5.2, where carrier energy is plotted versus momentum for both cases. The photon momentum

$$p = h\lambda = h\upsilon/c \tag{5.2}$$

is very small, and conservation of momentum can be readily accommodated by small deviations from the vertical transition shown in Figure 5.2(a). For the indirect case illustrated in Figure 5.2(b), the energy change ΔE defines the photon energy and momentum, again according to Equation (5.1) and Equation (5.2), but conservation of momentum additionally requires that the much greater electron momentum on the order of $h/2a$ be accounted for. For lattice dimensions, a, are on the order of 10^{-10} m and wavelengths, λ, are on the order of 10^{-6} m; it is clearly not possible for both conservation criteria to be met without the participation of a third body, i.e., a phonon. The two consequences of this result are that the indirect transition is inefficient

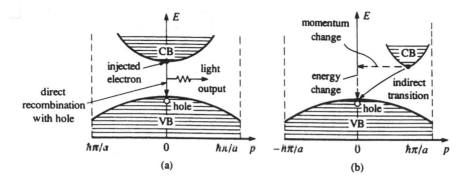

FIGURE 5.2 (a) Interband recombination in a direct-bandgap semiconductor; (b) recombination in an indirect-gap semiconductor also involves a momentum change. (*Source*: J. Allison, *Electronic Engineering Semiconductors and Devices*, 2nd ed., London: McGraw-Hill, 1990, p. 303. With permission.)

(in that it must transfer momentum and hence thermal energy to the lattice) and less likely to occur than the direct transition (because of the requirement for all three particles to simultaneously meet the energy and momentum conditions). Indirect bandgaps therefore lead to long diffusion lengths and recombination times, which produce good transistors but poor LEDs.

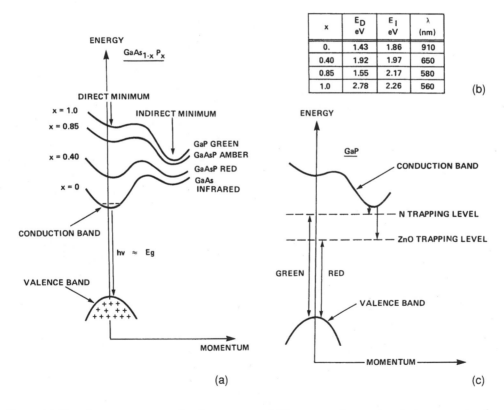

x	E_D eV	E_I eV	λ (nm)
0.	1.43	1.86	910
0.40	1.92	1.97	650
0.85	1.55	2.17	580
1.0	2.78	2.26	560

FIGURE 5.3 (a) Plot of momentum versus bandgap energy, and (b) corresponding semiconductor parameters for various compounds of the GaAs/GaP system; (c) plot of momentum versus bandgap energy for indirect GaP materials showing special trapping levels. (*Source*: S. Gage et al., *Optoelectronics/Fiber-Optics Applications Manual*, 2nd ed., New York: Hewlett-Packard/McGraw-Hill, 1981, pp. 1.3–4. With permission.)

The most common direct-bandgap semiconductor is GaAs, but the photon wavelength calculated for $E_g = E_D = 1.43$ eV as listed in Figure 5.3(b) is in the infrared. Such a material may be ideal for communications and sensory optoelectronic applications but is unsuitable for display purposes. The bandgap may be adjusted, however, by the substitution of phosphorus for arsenic in the lattice as shown in Figure 5.3(a). The color range listed corresponds to the range of LED colors commonly available: red, yellow and green. The direct and indirect bandgaps, E_D and E_λ, of $GaAs_{1-x}P_x$ vary with x as

$$E_D = 1.441 + 1.091x + 0.210x^2 \tag{5.3}$$

and

$$E_\lambda = 1.977 + 0.144x + 0.211x^2 \tag{5.4}$$

(Wang, 1989), enabling one to design the material to produce the required LED color.

Note the continuous transition from the direct GaAs to the indirect GaP. The materials have an indirect bandgap for $x > 0.4$, and have the same problems as light emitters as silicon. The efficiency of an indirect-gap emitter can be greatly enhanced by the introduction of appropriate impurity recombination centers, as shown in Figure 5.3(c). In the process shown, an injected minority carrier electron (in p-type material) is first trapped by the localized impurity (which is itself electrically neutral but which introduces a local potential to the lattice, which attracts electrons). The center is then negatively charged and attracts a hole to complete the recombination process, which produces the photon. The recombination center solves the momentum transfer problem, because the trapped electron is localized to the impurity lattice site, and has a momentum range according to the Heisenberg Uncertainty Principle of

$$\Delta p \sim h/2\pi a \tag{5.5}$$

that is, sufficient to include the processes shown in the diagram at $p \sim 0$. In the cases used as examples, a nitrogen atom substitutes for a phosphorus, or a zinc–oxygen pair substitutes for adjacent gallium–phosphorus atoms in the $GaAs_{1-x}P_x$ lattice.

The $GaAs_{1-x}P_x$ system is well established, but can only produce wavelengths defined by the range of energy gap widths, i.e., down to green. Blue LEDs require higher band-gap materials:

1. SiC technology is well developed for high temperature semiconductor applications, but it has an indirect band gap, so its emission efficiency is very poor [Pierret, 1996].
2. GaN is a direct band gap material system producing successful blue and blue-green devices [Jiles, 1994; Nakamura, 1995; Pierret, 1996]. AlGaInN alloys cover the spectrum from red ($\sim$1.9 eV from InN) to deep UV ($\sim$6.2 eV from AlN).
3. II-IV compounds such as ZnS and ZnSe possess direct band gaps in the 1.5–3.6 eV range, offering the possibility of full spectrum LEDs within the single materials system [Jiles, 1994].

In principle, white light LEDs can be constructed by balancing different wavelengths of different intensities from different LEDs of a single material system. More commonly, however, part of the output of a blue LED is absorbed by a yellow-green phosphor coating, with the combined spectrum yielding near-white light.

The discovery and development of n-type and p-type organic materials has led to the growth of a new field of organic electronics. Juxtaposition of n- and p-type materials produces diode characteristics, directly analogous to the semiconductor device, and metal-oxide semiconductor field-effect transistor (MOSFET) devices have also been made. An organic light emitting diode (OLED) requires one of the semi-conducting polymers to be electro-luminescent, since this is the source of the emitted light rather than junction photo-emission. The material systems offer the potential for very low cost printed devices, possibly on flexible substrates. This manufacturing flexibility suggests applications for monochromatic liquid crystal display (LCD) back-lighting, flexible displays, etc. The basic device might be monochromatic with a wide variety of colors available by electro-luminescent material selection, but the addition of a red dye, for example, to a blue-green emitter can yield a direct white-light device. A wide-area white OLED array can form the basis of a color display with the addition of sub-pixel RGB filters, as for LCD displays [Howard and Prache, 2001].

Device Efficiency

In considering LED efficiencies, it is convenient to consider the emission process to consist of three distinct steps: (a) excitation, (b) recombination, and (c) extraction. These will be discussed with reference to Figure 5.4 and semiconductor devices.

(a) Photons created by minority electron recombination on the p-type side of the junction are more likely to be successfully emitted from the surface of the device, for the structure shown in Figure 5.4(a) and Figure 5.4(b) if the p-type region is a thin surface layer. For a given total LED current, I, made up of electron, hole and space-charge region recombination components, I_n, I_p and I_r, respectively, the electron injection efficiency (which provides the excitation) is

$$\gamma_n = I_n/(I_n + I_p + I_r) \tag{5.6}$$

In principle, all the physical processes described above apply equally to both electrons and holes. However, the electron mobility, μ_n, is greater than that of a hole, μ_p, and since

$$I_n/I_p = N_d\,\mu_n/N_a\,\mu_p \tag{5.7}$$

(where N_d, N_a are n-type donor and p-type acceptor doping densities, respectively) greater γ_n is attainable for a given doping ratio than hole injection efficiency, γ_p. Consequently, LEDs are usually p-n$^+$ diodes constructed as in Figure 5.4, with the p-layer at the surface.

(b) Some of the recombinations undergone by the excess electron distribution, Δn, in the p-type region will lead to radiation of the photon desired, but others will not because of the existence of doping and various impurity levels in the bandgap. The total recombination rate, R, can be written in terms of the radiative and non-radiative rates R_r and R_{nr}, as

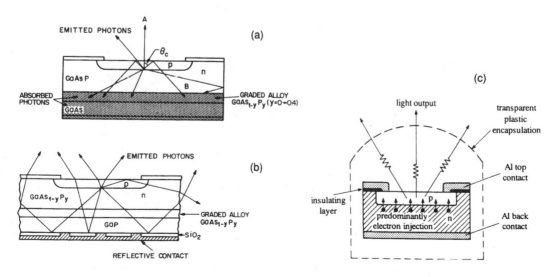

FIGURE 5.4 Effect of (a) opaque substrate, (b) transparent substrate, and (c) encapsulation on photons emitted at the *pn* junction. (*Source*: (a and b) S.M. Sze, *Semiconductor Devices: Physics and Technology*, New York: Wiley, 1985, p. 262. Reprinted by permission of John Wiley & Sons, Inc. (c) J. Allison, *Electronic Engineering Semiconductors and Devices*, 2nd ed., London: McGraw-Hill, 1990, p. 307. With permission.)

$$R = R_r + R_{nr} \tag{5.8}$$

where

$$R_r = \Delta n / \tau_r, \quad R_{nr} = \Delta n / \tau_{nr}, \quad R = \Delta n / \tau \tag{5.9}$$

and where τ_r and τ_{nr} are the minority carrier lifetimes associated with the radiative and non-radiative recombination processes, and τ is the effective lifetime. The radiative efficiency is defined as

$$\eta = R_r / (R_r + R_{nr}) = \tau / \tau_r \tag{5.10}$$

and the *internal quantum efficiency* is

$$\eta_i = \eta \gamma \tag{5.11}$$

(c) It is clear from Figure 5.4 that many of the photons generated on either side of the junction will pass through sufficient bulk semiconductor to be reabsorbed. In fact the photon energy may be ideally suited to reabsorption if it exceeds the semiconductor direct bandgap. It is obvious, then, why GaAs is opaque and GaP transparent to photons from Ga(As:P) junctions. Clearly, a greater efficiency might be expected from the transparent substrate with reflecting contact [Figure 5.4(b)].

The photon must strike the LED surface at an angle less than the critical angle for total internal reflection, θ_c, where

$$\sin \theta_c = n_{ext} / n_{LED} = 1/n \tag{5.12}$$

and n_{ext}, n_{LED} are the external and internal refractive indices, respectively. For air, $n_{ext} = 1$, but critical angle loss can be reduced by encapsulating the device in an epoxy lens cap [Figure 5.4(c)] to increase both $n_{ext} > 1$ and the angle of incidence at the air interface.

Even within angles less than θ_c, there is Fresnel loss, with transmission ratio

$$T = 4n / (1 + n)^2 \tag{5.13}$$

The total *external quantum efficiency* is then the fraction of photons emitted [Neamen, 1992], given by

$$\eta_e = 1 / (1 + \alpha v_o / AT) \tag{5.14}$$

where α is the average absorption coefficient, v_o is the LED volume, and A is the emitting area [Yang, 1988].

In considering LED effectiveness for display purposes, one must also include radiation wavelength in relation to the spectral response of the human eye [Sze, 1985]. Although the GaP green LED is intrinsically less efficient than the GaAsP red LED, the eye compensates for the deficiency with a greater sensitivity to green.

More recently developed heterojunction LEDs (Figure 5.5) offer two mechanisms to improve LED efficiencies [Yang, 1988]. The electron injection efficiency can be enhanced, but, in addition, absorption losses through the wider 2.1-eV bandgap n-type layer are essentially eliminated for photons emitted by recombination in the lower 2.0-eV bandgap p-type region.

Improving efficiencies have led to extensive use of LEDs in traffic lights and automotive applications (e.g., in hazard, brake, turn and running lights).

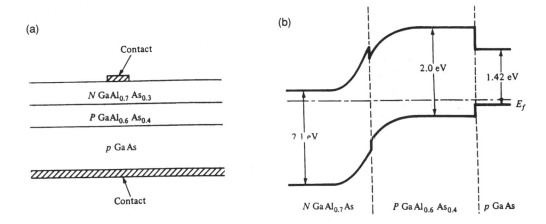

FIGURE 5.5 A GaAlAs heterojunction LED: (a) cross-sectional diagram; (b) energy-band diagram. (*Source*: E.S. Yang, *Microelectronic Devices*, New York: McGraw-Hill, 1988, p. 401. With permission.)

Interfacing

In circuit design applications the LED may be treated much as a regular diode, but with a much greater forward voltage, V_F. Since one usually seeks maximum brightness from the device, it is usually conducting heavily and V_F approaches the contact potential. As one moves from GaAs to GaP [Figure 5.3(a)], V_F varies from about 1.5 to around 2.0 V. The variation in V_F with temperature (at constant current) follows similar rules as apply to conventional diodes, but radiant power and wavelengths also change [Gage et al., 1981].

Single LEDs are commonly driven by logic gates, perhaps as status indicators, and some of the simplest interface circuits are shown in Figure 5.6. In many cases, the gate output will not be able to source or sink sufficient current for visibility, and an amplifier will be required, as in Figure 5.7. Bar graph displays are commonly used to indicate signal level on audio equipment, with a modification of the position indicator seen in Figure 5.8 to guide fine tuning. Matrix LED arrays can be used for flexible, high-density panel displays [Figure 5.9(a)], and are conventionally controlled by row or column strobing [Figure 5.9(b)] controlled by a microprocessor interface.

Multiple LEDs are commonly packaged together in a single integrated device, organized in one of the standard display fonts [Figure 5.10(a)], with decoding often included within the package [Figure 5.10(b)].

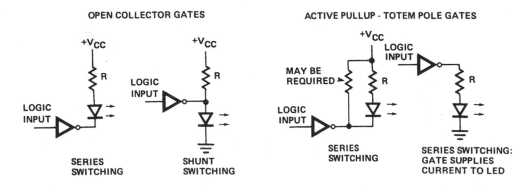

FIGURE 5.6 Digital logic can interface directly to LED lamps. (*Source*: S. Gage et al., *Optoelectronics/Fiber-Optics Applications Manual*, 2nd ed., New York: Hewlett-Packard/McGraw-Hill, 1981, p. 2.20. With permission.)

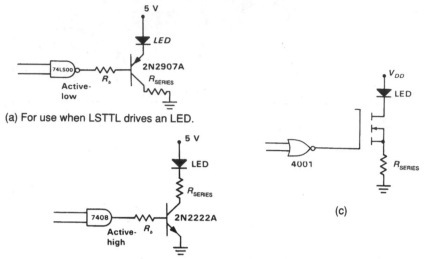

(a) For use when LSTTL drives an LED.

(b) For use when a logic high is needed to drive an LED.

FIGURE 5.7 LED interfacing for (a) low-power transistor-transistor logic, (b) logic high drive, and (c) CMOS. (*Source*: M. Forbes and B.B. Brey, *Digital Electronics*, Indianapolis: Bobbs-Merrill, 1985, p. 242. With permission.)

The 7-segment display is adequate for hexadecimal applications, but the 16-segment display is required for alphanumerics. To limit pin-out requirements the LEDs of a single package are connected in either the common anode or common cathode configuration [Figure 5.11(a)] with multiple display digits multiplexed, as illustrated in Figure 5.11(b).

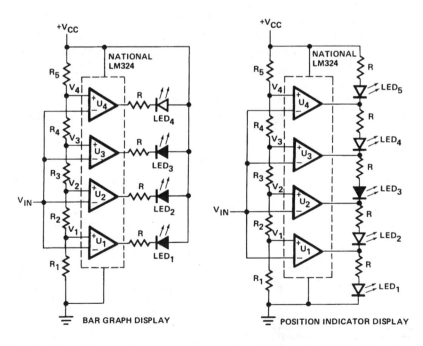

FIGURE 5.8 Operational amplifiers or voltage comparators used to decode an analog signal into a bar graph or position indicator display. (*Source*: S. Gage et al., *Optoelectronics/Fiber-Optics Applications Manual*, 2nd ed., New York: Hewlett-Packard/McGraw-Hill, 1981, p. 23.3. With permission.)

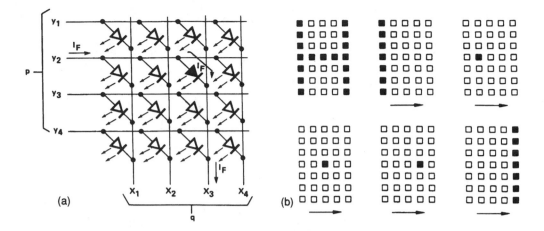

FIGURE 5.9 Matrix displays. (a) One LED will be turned on by applying the proper signal to one *x* axis and one *y* axis. (b) Character generation using column strobe methods. (*Source*: S. Gage et al., *Optoelectronics/Fiber-Optics Applications Manual*, 2nd ed., New York: Hewlett-Packard/McGraw-Hill, 1981, pp. 2.25, 5.44. With permission.)

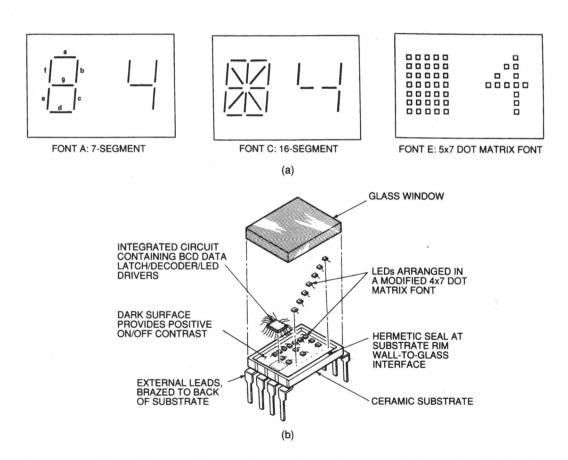

FIGURE 5.10 (a) Display fonts used in LED displays. (b) Construction features of a hermetic LED display. (*Source*: S. Gage et al., *Optoelectronics/Fiber-Optics Applications Manual*, 2nd ed., New York: Hewlett-Packard/McGraw-Hill, 1981, pp. 5.3, 5.6. With permission.)

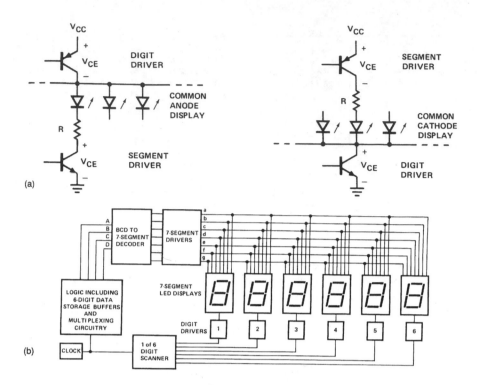

FIGURE 5.11   (a) Generalized drive circuits for strobed operation. (b) Block diagram of a strobed (multiplexed) six-digit LED display. (*Source*: S. Gage et al., *Optoelectronics/Fiber-Optics Applications Manual*, 2nd ed., New York: Hewlett-Packard/ McGraw-Hill, 1981, pp. 5.25, 5.23. With permission.)

Defining Terms

External quantum efficiency: The proportion of the photons emitted from the pn junction that escape the device structure (but sometimes alternatively defined as $\eta_i\eta_e$).

Injection electroluminescence: Electroluminescence is the general term for optical emission resulting from the passage of electric current; injection electroluminescence refers to the case where the mechanism involves the injection of carriers across a pn junction.

Internal quantum efficiency: The product of injection efficiency and radiative efficiency corresponds to the ratio of power radiated from the junction to electrical power supplied.

Acknowledgment

Anjani Kulkarni's assistance with the preparation of this chapter is recognized with thanks.

References

J. Allison, *Electronic Engineering Semiconductors and Devices*, 2nd ed., London, England: McGraw-Hill, 1990.

M. Forbes and B.B. Brey, *Digital Electronics*, Indianapolis, IN: Bobbs-Merrill, 1990.

S. Gage, D. Evans, M. Hodapp, H. Sorensen, R. Jamison, and R. Krause, *Optoelectronics/Fiber-Optics Applications Manual*, 2nd ed., New York, NY: Hewlett-Packard/McGraw-Hill, 1981.

W.E. Howard and O.F. Prache, "Microdisplays based upon organic light-emitting diodes," *IBM J. Res. & Dev.*, vol. 45, no. 1, pp. 115–127, 2001.

D. Jiles, *Introduction to the Electronic Properties of Materials*, London, England: Chapman & Hall, 1994.

S. Nakamura, "A bright future for blue/green LEDs," *IEEE Circ. & Devices*, vol. 11, no. 3, pp. 19–23, 1995.

D.A. Neamen, *Semiconductor Physics and Devices: Basic Principles*, Boston, MA: Irwin, 1992.

R.F. Pierret, *Semiconductor Device Fundamentals*, New York, NY: Addison-Wesley, 1996.

S.M. Sze, *Semiconductor Devices: Physics and Technology*, New York, NY: Wiley, 1985.

S. Wang, *Fundamentals of Semiconductor Theory and Device Physics*, Englewood Cliffs, NJ: Prentice-Hall, 1989.

E.S. Yang, *Microelectronic Devices*, New York, NY: McGraw-Hill, 1988.

Further Information

More extensive semiconductor device treatments of the LED are contained in *Semiconductor Devices and Integrated Electronics* by A.G. Milnes (Van Nostrand Reinhold, New York, 1980) and in *Introduction to Optical Electronics* by K.A. Jones (Harper and Row, New York, 1987). E. Uiga provides more interfacing and design detail for the LED as a circuit element in optoelectronics (Prentice-Hall, Englewood Cliffs, NJ, 1995). Wang [1989] considers second-order effects extensively. In *Semiconductor Optoelectronics* by J. Singh (McGraw-Hill, New York, 1996), the emphasis is on communications applications, but the temperature dependence and frequency response issues covered there are also relevant to displays.

Chapter 2 of Gage et al. [1981] contains detailed information on the optical and thermal design constraints on the LED package and on LED back-lit display systems. Chapter 6 considers filtering and other techniques for the contrast enhancement required for direct sunlight viewing.

Professional society magazines are good sources of up-to-date information at the non-specialist level, especially the occasional special issues devoted to topic reviews. *IEEE Spectrum* is a good example, as is the *IEEE Circuits & Devices* magazine.

The Howard and Prache paper [2001] and S.J.M. Connor et al. ("Towards Full Colour LEP Displays," *Proc. SPIE*, 4105, 2001, 9–17) describe different materials systems and optical characteristics for OLED devices. The March/April 2002 issue of the *IEEE Journal on Selected Topics in Quantum Electronics* (8, 2) is devoted to high-efficiency light-emitting diodes, with papers collected into sections on LEDs for communications, lighting, and UV emission, and on OLEDs.

5.2 Liquid-Crystal Displays

James E. Morris

In a low-power CMOS digital system the dissipation of a light-emitting diode (LED) or other comparable display technology can dominate the total system's power requirements. In such circumstances the low-power dissipation advantage of CMOS technology can be completely lost. This is the situation in which liquid-crystal display (LCD) technology is commonly used. The LED (or other active system, such as a plasma or vacuum fluorescent display) emits optical power supplied (comparatively inefficiently) by the system battery or other source. The passive LCD is fundamentally different in that the optical power is supplied externally (by sunlight or room lighting typically), and the system source need supply only the relatively minute amount of power (microwatts per square centimeter) required to change the device's reflective optical properties.

Principle of Operation

Materials classed as liquid crystals are typically liquid at high temperatures and solid at low temperatures, but in the intermediate temperature range they display characteristics of both. Although there are many different types of liquid crystals used, we will concentrate here on the use of *nematic* crystals in *twisted nematic* devices, the most common by far.

The essential feature of a liquid crystal is the long rod-like molecule. In a nematic crystal the molecules align as shown in Figure 5.12. If the container surface is microscopically grooved, the interface molecules will be aligned by the grooves and intermolecular forces will maintain that orientation across the liquid crystal

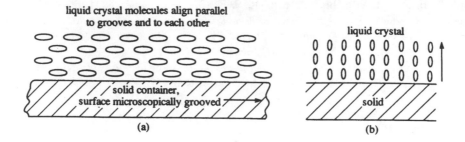

FIGURE 5.12 Liquid-crystal/grooved interface (a) with no field applied, and (b) with an electric field $\varepsilon >$ a critical value. (*Source*: J. Allison, *Electronic Engineering Semiconductors and Devices*, 2nd ed., London: McGraw-Hill, 1990, p. 308. With permission.)

[Figure 5.12(a)]. The molecules will align in an electric field, and beyond a critical value the field may be sufficient to overcome the alignment with the grooves [Figure 5.12(b)]. (In practice, the transition is not so abrupt, and groove alignment persists at the interface itself [Figure 5.13].)

The process of alignment in the electric field is the result of the anisotropic dielectric constant characteristic of liquid crystals. For the electric field parallel to the molecular alignment, $\varepsilon_r = \varepsilon_{\parallel}$, and for a perpendicular field, $\varepsilon_r = \varepsilon_{\perp}$. In a "positive" liquid crystal, $\varepsilon_{\parallel} > \varepsilon_{\perp}$, and the molecules align parallel to the field as described above in order to minimize the system's potential energy.

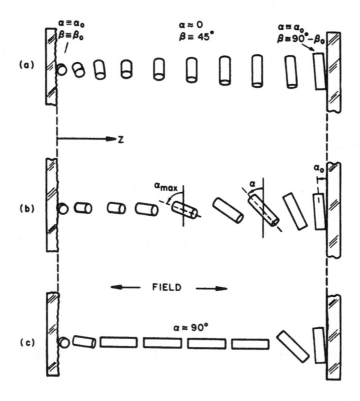

FIGURE 5.13 Diagram of the orientation of the liquid-crystal axis in a cell (a) with no applied field, (b) with about twice the critical field, and (c) with several times the critical field. Note slight permanent tilt (α_0) and turn (β_0) at the surfaces. (*Source*: G. Baur, in *The Physics and Chemistry of Liquid Crystal Devices*, G.J. Sprokel, Ed., New York: Plenum, 1980, p. 62. With permission.)

The principle of the twisted nematic cell is illustrated in Figure 5.14. The confining plates, typically 10 μm apart, are grooved orthogonally, forcing the molecular orientation to spiral through 90 degrees [Figure 5.14(a)]. In the LCD, two polarizers and a mirror are added as shown in Figure 5.14(b). Incident ambient light is polarized and enters the liquid-crystal cell with the plane of polarization parallel to the molecular orientation. As the light traverses the cell, the plane of polarization is rotated by the twist in the liquid crystal, so that it reaches the opposite face with a polarization 90° to the original direction but now parallel to the

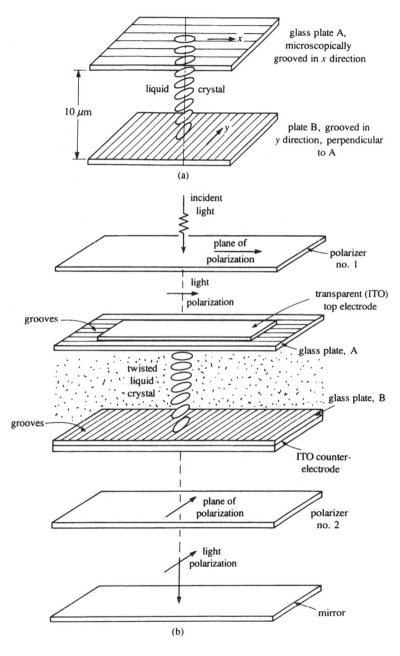

FIGURE 5.14   (a) Twisted nematic cell, $\varepsilon = 0$. (b) Liquid-crystal display element. (*Source*: J. Allison, *Electronic Engineering Semiconductors and Devices*, 2nd ed., London: McGraw-Hill, 1990, p. 309. With permission.)

direction of the second polarizer, through which it may therefore pass. The light is then reflected from the mirror and passes back through the cell, reversing the prior sequence.

When an electric field (greater than the critical field) is applied between the transparent electrodes, usually conductive *indium–tin oxide* (ITO) thin films, the 90° twist in the crystal is destroyed as the molecules align parallel to the field, so that the rotation of the light's plane of polarization cannot be sustained. Consequently, the crossed polarizers effectively block reflection of the incident light from the backing mirror, and the surface appears to be dark, with excellent contrast to the light gray color of the device in the reflecting mode. The contrast ratio can be further enhanced by the use of the *super twisted nematic* crystal, where the molecular orientation is rotated through 270 degrees rather than 90 degrees.

Transmission LCDs function very similarly to the devices just described, but without the mirror, which is replaced by a powered backlighting source. Obviously, the low-power advantage of the passive device is lost in this active alternative, but monochromatic backlighting does provide one means of constructing displays with varied background colors.

Another form of color display is provided by *cholesteric* crystals. The three main types of liquid crystals, nematic, cholesteric and smectic, are distinguished by the different types of molecular ordering they display. In the cholesteric crystal the direction of molecular alignment rotates in each successive parallel plane (Figure 5.15). The spatial period of the rotation, p, is called the pitch, and Bragg reflections occur when the wavelength of incident light meets the condition

$$\lambda = p/n \tag{5.15}$$

where n is an integer. The liquid crystal can thus appear to be colored in incident white light. In practice, the color is strongly temperature dependent and the effect is more appropriate to temperature-sensing applications than to digital displays.

Large area color LCD displays have become commonplace for computer and HDTV monitors, where the color pixels are organized in the traditional television RGB format, with each pixel made up of three separate red, green and blue sub-pixels. If each subpixel intensity can be modulated to 256 (8-bit) levels, the pixel has a range of 256^3 (about 16 million) colors. The subpixel colors can be defined by external filters or internal dyes [Braithwaite and Weaver, 1995]. In the latter case, the dye molecules align with the LCD molecules and absorb correctly polarized light.

Uiga [1995] discusses some LCD problems, in particular the slow response times and the variation of effective critical voltages with limited viewing angles and the temperature dependences of both.

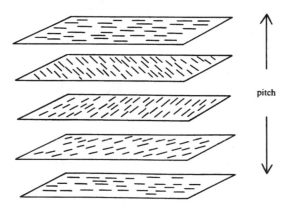

FIGURE 5.15   Cholesteric ordering: a large number of planes of nematic ordering are formed where the directors rotate as we move along a direction perpendicular to the planes. (*Source*: J. Wilson and J.F.B. Hawkes, *Optoelectronics: An Introduction*, London: Prentice-Hall, 1989, p. 145. With permission.)

Interfacing

LCDs can be organized in all the ways available to competing technologies—e.g., LEDs (see Section 5.1), including seven-segment, alphanumeric, and dot matrix. The LCD differs from LED displays where each pixel or segment must be a separate device, because the LCD segment or pixel areas are defined by transparent electrodes separated from a common overlapping backplane by a single liquid crystal [Figure 5.16(a)]. In a large matrix array it may take a significant period to scan all pixels, and the simple addressing scheme of Figure 5.16(b) may lead to noticeable flicker. The high off-resistance of the MOSFETs of Figure 5.16(c) can reduce this problem by increasing the discharge time to hold the LCD on after the address pulse has gone. The MOSFETs in this *active matrix* technology are implemented in practice in the form of polysilicon or hydrogenated amorphous silicon (a-Si:H) thin film transistors (TFT) [Shur, 1990; Braithwaite and Weaver, 1990].

The interfacing requirements, which are otherwise similar in multiplexing techniques, etc., are complicated by the requirement for zero net DC bias across the cell in order to avoid electrochemical degradation of the material. LCDs require AC drive signals, and square waves of frequency between 25 Hz and 1 kHz are typically used [Wilson and Hawkes, 1989]. A square wave is applied to the backplane (Figure 5.17), with in-phase and

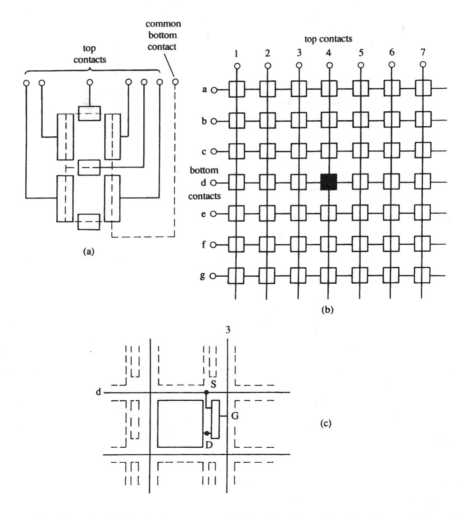

FIGURE 5.16 LCD addressing: (a) simple (seven-segment) addressing; (b) matrix addressing; and (c) matrix addressing with MOSFETs. (*Source*: J. Allison, *Electronic Engineering Semiconductors and Devices*, 2nd ed., London: McGraw-Hill, 1990, p. 312. With permission.)

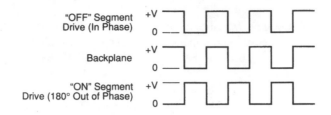

FIGURE 5.17 Drive signals from a direct connect LCD driver. (*Source*: R. Lutz, Application Note 350, in *Interface Databook*, Santa Clara, Calif.: National Semiconductor Corporation, 1990, p. 4–109. With permission.)

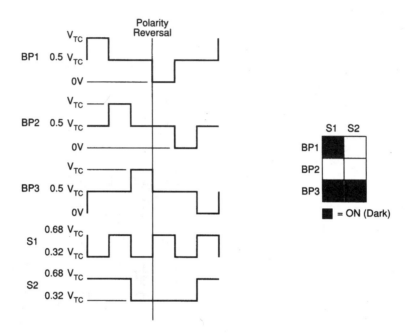

FIGURE 5.18 Example of backplane and segment patterns. (*Source*: R. Lutz, Application Note 350, in *Interface Databook*, Santa Clara, Calif.: National Semiconductor Corporation, 1990, p. 4–109. With permission.)

antiphase signals to the counter electrode determining whether the given pixel or segment is on or off. In practice the state is determined by the root mean square (rms) value of the differential voltage applied.

Figure 5.18 illustrates the additional complexity that would be required by even a simple multiplexed addressing system. The backplane and segment drivers might correspond to rows and columns of a dot matrix, as implied in the diagram, or the backplanes may identify specific characters of an alphanumeric display. Calculating the rms values of the difference voltages shown gives $0.42\ V_{tc}$ for the *on* pixels and $0.24\ V_{tc}$ for *off*, from which V_{tc} can be calculated for reliable operation if the critical voltage is known for the LCD to be used.

Defining Terms

Active matrix: Each pixel in a high density display matrix, such as for flat-screen television, requires its own active (switching element) driver (e.g., a TFT).

Cholesteric: In the cholesteric liquid crystal successive layers of aligned molecules are rotated naturally.

Indium–tin oxide (ITO):   A mixture of the semiconducting oxides SnO_2 and In_2O_3; the most common transparent conductor.

Nematic: The type of liquid crystal in which the molecular chains align; such alignment can be controlled across the liquid crystal if it can be constrained at the boundaries.

Twisted nematic: The alignments of the nematic planes are rotated through 90 degrees across the crystal by constraining alignments to be orthogonal at the boundaries.

References

J. Allison, *Electronic Engineering Semiconductors and Devices*, 2nd ed., London, England: McGraw-Hill, 1990.

G. Baur, "Optical characteristics of liquid crystal displays," in *The Physics and Chemistry of Liquid Crystal Devices*, G. J. Sprokel, Ed., New York, NY: Plenum, 1980.

N. Braithwaite and G. Weaver, Eds., *Electronic Materials*, Milton Keynes: The Open University/Butterworths, England, 1990.

R. Lutz, "Designing an LCD dot matrix display interface, Application note 350," in *Interface Databook*, Santa Clara, CA: National Semiconductor Corporation, 1990.

M. Shur, *Physics of Semiconductor Devices*, Englewood Cliffs, NJ: Prentice-Hall, 1990.

E. Uiga, *Optoelectronics*, Englewood Cliffs, NJ: Prentice-Hall, 1995.

J. Wilson and J.F.B. Hawkes, *Optoelectronics: An Introduction*, London: Prentice-Hall, 1989.

Further Information

Nematic liquid-crystal molecules typically incorporate two separated benzene rings in a complex chain molecule (Wilson and Hawkes, 1989). The organic chemistry of liquid-crystal compounds will lie outside the interests of most readers but is briefly reviewed in "Liquid Crystal Materials for Display Devices," by J.A. Castellano and K.J. Harrison, in *The Physics and Chemistry of Liquid Crystal Devices*, edited by G.J. Sprokel [Plenum, 1980].

One technique used in liquid-crystal color switches requires the use of electrically controlled birefringence. This topic is covered at an elementary level by Wilson and Hawkes [1989].

An interesting historical perspective on the development of LCD technology is provided by the extensive reviews of 150 patents in the field contained in *Liquid Crystal Devices*, edited by T. Kallard (*State of the Art Review*, Vol. 7, Optosonic Press, New York, 1973). The book also contains a bibliography of more than 1100 entries.

The various professional societies' magazines are excellent sources of material for recent developments in this field (and others). These publications regularly devote a special issue to research developments in a single field at a level intended for the non-specialist. A good example in the LCD area is provided by two articles on TFT silicon for active matrix displays contained in the *Materials for Flat-Panel Displays* issue of the *MRS Bulletin*, 21(3), March 1996 (Materials Research Society), which cover the transition from a-Si:H to polysilicon and the prospects for single crystals.

5.3   Plasma Displays

Larry F. Weber

Introduction

The high image quality and sleek thin profile of large-diagonal plasma displays has generated great consumer excitement for high definition digital home theatre. This has created explosive growth in manufacturing capacity and interest in full color plasma displays. This is fueled by the realization that plasma displays can fulfill the long sought-after goal of consumer-affordable hang-on-the-wall flat-panel television displays with diagonals in the range of 32 to 80 in. Color plasma displays operate on the same physical principle as fluorescent lamps. A gas discharge generates ultraviolet light which excites a phosphor layer that fluoresces visible light. Differing phosphors are used for the red, green, and blue primaries and a full color moving image is obtained by modulating each primary color subpixel to one of typically 256 or more intensity levels at 60 times a second.

Color Plasma Display Markets

The plasma display manufacturers have adopted the strategy of a strong attack on the greater than 40-in diagonal NTSC television and high definition television (HDTV) markets. Display diagonals smaller than 30 in are specifically avoided in this strategy. Plasma displays have found their proper place in the market by evading the fierce competition from the smaller diagonal liquid crystal displays and CRTs since both of these technologies have difficulty with 40- to 80-in diagonals. However, recently the very large market for large-screen flat-panel TV is attracting strong interest from the liquid crystal manufacturers. Liquid crystals are likely to dominate for diagonals less than 30 in while the plasma displays will dominate for diagonals larger than 40 in. In this larger diagonal range, projection displays will also play a strong role because they have lower cost than both the plasma displays and the liquid crystals. The plasma display will enjoy a strong place in the flat TV market due to its superior image quality compared to the liquid crystal or projection displays.

The potential market for large plasma displays is quite large. The 2004 sales for plasma display modules is expected to be $5 billion on 2 million units. This is projected to increase to over $10 billion on 9 million units in 2007. These numbers are for modules sold at the OEM level. There is of course considerable value added to the final TV set which increases the final selling price to the consumer. In 2004 the 42-in diagonal size had the most sales. As prices decline in future years the most popular size is expected to increase to 50-in diagonal and beyond. The largest product available in 2004 was a 71-in diagonal, and this will surely increase as the technology and the market mature.

Color Plasma Display Attributes

Table 5.1 shows some of the attributes of color plasma displays which make them successful. The following reviews each attribute.

1. The electrical characteristics of the gas discharge allow plasma displays to be made with diagonals in the 32- to 80-in range. Such large diagonals are facilitated by very strong nonlinearity and inherent memory of the discharge, as discussed in items 3 and 4 below, which present no practical limitations to the number of lines that can be multiplexed. Also, the high impedance characteristic (covered below in item 11) coupled with the ability to use highly conductive opaque electrodes greatly reduces electrode loss limitations to size. Color plasma display products are available with sizes as great as 71-in diagonal

TABLE 5.1   Color Plasma Display Attributes

1.	Diagonals of 32 to 80 in
2.	1 billion colors
3.	Very strong nonlinearity
4.	Inherent memory
5.	Long lifetime
6.	Very wide viewing angle
7.	Instant update time
8.	Good luminance and luminous efficiency
9.	CRT-like manufacturing model
10.	Tolerant to shock, vibration and temperature extremes
11.	Reasonable impedance characteristics
12.	Precise digital grayscale
13.	CRT-like color gamut
14.	Excellent dark room contrast ratio

having over 6 million subpixels. In 2004 the largest prototype demonstrated was an 80-in diagonal with the full 1920 × 1080 HDTV resolution.

2. The all-digital grayscale technique used in color plasma displays allows each primary subpixel to display as many as 1024 intensity levels. This allows full 30-bit color or as many as 1 billion colors. These 1024 intensity levels are important for reducing the low-level contouring artifacts that are visible for digital grayscale having fewer intensity levels.

3. The plasma display has a very large nonlinearity due to the electrical characteristic of the gas discharge used in all plasma displays. This is an electrical nonlinearity, meaning that below a certain threshold voltage the gas discharge will emit no light. Of course, above that threshold voltage the gas discharge fires and emits a desired color. Very sharp nonlinearity allows plasma displays to be multiplexed without limit, which makes very large plasma displays practical. This is demonstrated by a number of recently developed 1920 × 1080 × 3 subpixel color plasma displays. This is a considerable advantage when compared to other display technologies such as the liquid crystal. The liquid crystal display does not have a very good nonlinearity, and therefore some other nonlinear element, such as a thin-film transistor, is frequently added in series with each liquid crystal element to increase the display nonlinearity. Of course, this greatly complicates and adds cost to this active matrix liquid crystal.

4. All color plasma displays have inherent memory which is stored directly in the glass plasma panel. Memory is very desirable for flat-panel displays because it allows the display to be very bright, even for very large sizes. This is because a display with memory has a pixel duty cycle of one. Displays without memory have a pixel duty cycle of one divided by the number of scanned lines. Thus, as the nonmemory displays get bigger and the number of scanned lines increases, the duty cycle and therefore the brightness of the display decrease. An additional value of memory is the elimination of flicker because the pixels are on all of the time.

5. The lifetime of color plasma displays can be very long. Plasma display television products are frequently delivered with guaranteed specified lifetimes to half luminance of 60,000 h and measured lifetimes in excess of 100,000 h. While this is better than most CRT products, there is considerable effort to extend this lifetime further. The failure mode is usually a slow degradation in the phosphor that gradually decreases the display luminance. In general the blue phosphor degrades the most and the red phosphor degrades the least. If all of the pixels are aged uniformly with the usual randomly moving television image, the display will still be usable after the specified lifetime but at reduced luminance. However, displays used for computer images require a much tougher life specification because images such as icons may be left on the same screen location for long periods and burn in an image. These problems are very similar to the phosphor degradation observed on a CRT.

6. One of the major advantages of all plasma displays over liquid crystal and projection displays is the very wide viewing angle. Plasma displays have a nearly Lambertian light emission distribution in both the vertical and the horizontal directions, which means that the brightness appears the same in all viewed directions.

7. Gas discharges switch in microseconds and so plasma displays can be updated instantly. Speed is especially important for full motion high definition television images.

8. The luminance and luminous efficiency of color plasma displays are quite good. Display module products having 1000 candelas per meter squared at 1.8 lumens per watt are available [Sato et al., 2002]. Unlike other technologies such as liquid crystal or projection displays, the plasma display is a power-on-demand technology. This means that power is dissipated by the display in proportion to the luminance of a given image region. In contrast, the liquid crystal power is mostly in the backlight and the projector power is mostly in the lamp, so these technologies take the same power independent of image luminance. This makes the plasma display the lowest power technology for television images since the average pixel level of a television image is 20% or less. The potential for the luminous efficiency of plasma displays to increase in the future is substantial, since the currently available 1.8 lumens per watt is much less than the 80 lumens per watt of a common fluorescent lamp that uses the same fundamental light generation technology.

9. Color plasma displays are manufactured in a plant that has many features in common with CRT manufacturing plants. This contrasts sharply with the semiconductor-like manufacturing plant of active matrix liquid crystal displays (AMLCDs). Therefore, the plasma display plant will cost typically one-third the cost of the equivalent sized AMLCD plant.

10. The structure can withstand very high levels of shock and vibration when properly mounted. Military plasma displays have been designed for in excess of 150 Gs of shock. Plasma displays can easily operate at both high and low temperature extremes. Ac plasma displays have a temperature limit dependent almost solely on the drive circuit characteristics.

11. Plasma displays have a high input impedance characteristic that makes them easy to drive. The dielectric constant of the gas is equal to one, which means that plasma displays have virtually the lowest possible electrode capacitance. This is 1000 times smaller than electroluminescent displays and about 100 times smaller than liquid crystal displays. This translates to lower current requirements and therefore smaller drive circuit silicon area for the plasma displays. While plasma displays do require 80-volt address drivers, it is frequently easier to design high voltage circuits than high current circuits. Also, the larger displays can be designed with little power dissipation in panel electrodes.

12. The grayscale technique used in color plasma displays is 100% digital, which allows design of an all digital image system having reduced noise and increased stability in color representations. This is very important as signal sources with very high quality digital signals, such as those from digital video disks and HDTV, become widely available.

13. The color gamut of the available plasma display phosphors is very good. While the color coordinates do not yet exactly match those used in the CRT, future process adjustments are expected to produce the desired close match.

14. Plasma display products are available with dark room contrast ratios in excess of 5000:1 at all viewing angles. This is a major advantage over liquid crystal TVs which generally specify contrast ratio only at the normal viewing direction since the liquid crystal contrast ratio degrades rapidly for off-axis angles. Dark room contrast is very important for home theatre television displays since these are frequently viewed in low ambient light conditions.

Gas Discharge Physics

A brief account of gas discharge physics will be covered below. A more detailed discussion of this material is presented in Weber [1985] and Boeuf [2003].

Figure 5.19 shows the important reactions that occur in a gas discharge for the monochrome gas mixture of neon and argon. Color plasma displays use a neon-xenon gas mixture, but the fundamental discharge physics is the same. The reactions in the gas volume include ionization (I), excitation (E), metastable generation (M), and Penning ionization (P). The three surface reactions that occur at the cathode cause ejection of electrons from the cathode by a bombarding neon ion, a neon metastable atom or by a high energy photon. For simplicity, only the neon reactions are shown in Figure 5.19. The most important volume reaction is ionization (I), which can cause the generation of an avalanche in the gas volume as shown in Figure 5.19. This avalanche is started by an electron near the cathode, and as it grows toward the anode it generates a large number of electron-ion pairs. The number of electron-ion pairs increases with increasing applied voltage across the gas. Ions, photons, or metastable atoms that are transported to the cathode can then eject electrons with a cathode surface-dependent probability, and these ejected electrons will initiate further avalanches. These mechanisms act as a positive feedback system that becomes unstable when the loop gain is greater than 1. The onset of the unstable condition is defined as the gas firing voltage. Above this firing voltage the discharge current will continue to grow without bounds if the initial avalanche is primed with at least a single electron.

Figure 5.20 shows the *I-V* characteristic of a typical gas discharge found in plasma displays. Note that the current is plotted on a log scale over nine orders of magnitude. The most striking feature is the very strong nonlinearity at the firing voltage, which is a major attribute of gas discharges that allows matrix addressing.

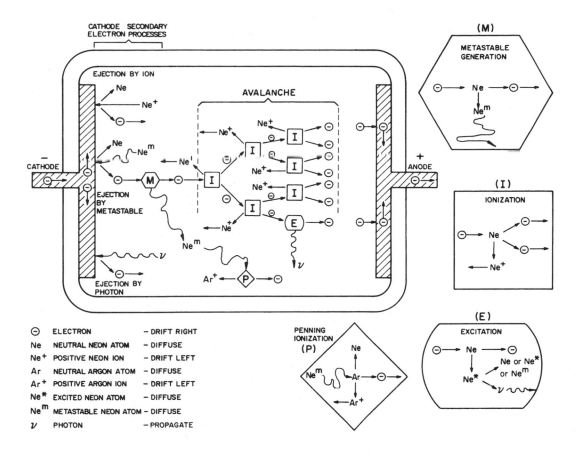

FIGURE 5.19 Model of important gas discharge reactions.

When the discharge current has sufficient magnitude, space charge distortion sets in and the characteristic achieves a negative resistance region. Most plasma displays operate near the junction of the normal and the abnormal glow regions of the characteristic.

One critical aspect of gas discharges is the requirement for external priming as shown in the lower part of Figure 5.20. The avalanche process shown in Figure 5.19 needs at least one electron to start the discharge growth. Without this first electron the discharge will not start at any voltage. Priming electrons can come from a number of different active particles created either by a prior discharge or by neighboring discharging pixels. Active particles include free electrons, free ions, metastable atoms, and ultraviolet photons.

Figure 5.21 shows the characteristics of the glow discharge commonly found in operating plasma displays. The light comes from two luminous regions: the negative glow and the positive column. Plasma displays on the market today use light from both regions. These regions are caused by the space charge distribution of the electrons and ions that distort the electric field and voltage distribution. The positive column generates light with a much higher efficiency than the negative glow, and so intense research is now underway to exploit this advantage of the positive column [Oversluizen, et al., 2004].

Current Limiting for Plasma Displays

To avoid a catastrophic arc, the current in a gas discharge must be limited by some means. There are a number of ways of accomplishing this, but only two, shown in Figure 5.22, have achieved commercial success. DC plasma displays use a resistor, a semiconductor current source, or a short applied voltage pulse to limit the current and have the electrodes in intimate contact with the gas discharge. AC plasma displays limit the

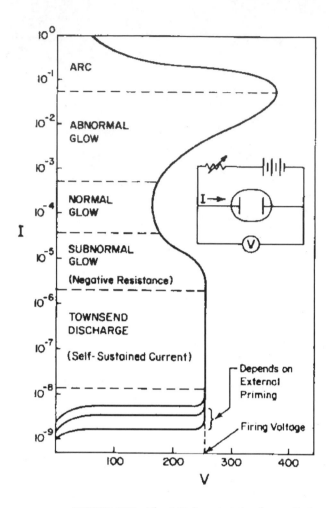

FIGURE 5.20   The *I-V* characteristic of a gas discharge.

current with an internal glass dielectric that couples the electrodes capacitively to the gas discharge. Most current color plasma displays use AC current limiting. DC current limiting is important to understand for historical and possible future research reasons.

DC displays can have the resistors or current sources connected to a display electrode external to the panel, which allows only one discharge to be ignited along that electrode at any one time. This works well for scanned displays. Multiple discharges and DC memory require placing the resistor internal to the panel in series with each subpixel. Materials and process advances have allowed practical DC color displays with memory to be made having a resistor per subpixel [Koike et al., 1995].

AC displays can achieve memory and the necessary current limiting with a simple dielectric layer that forms a capacitor in series with each pixel. When a voltage pulse is applied to an AC panel, the discharge deposits a charge on the wall that reduces the voltage across the gas. After a short time, the discharge will extinguish and the light output will end until the applied voltage reverses polarity and a new discharge pulse occurs. This wall charge allows the AC plasma displays to operate in a memory mode, which greatly increases the brightness of large displays.

AC Plasma Displays

Figure 5.23 shows the AC monochrome structure that was developed in the late 1960s and is still manufactured for limited special applications today [Criscimagna and Pleshko, 1980]. These panels are made

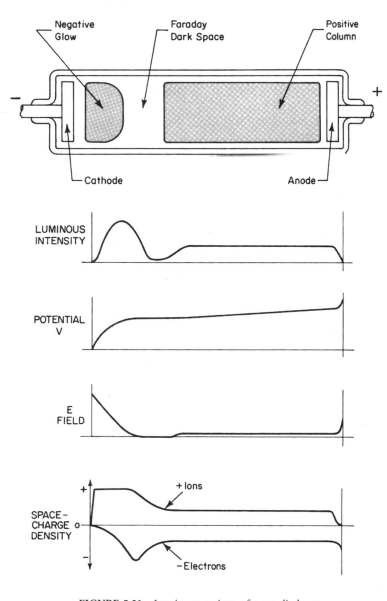

FIGURE 5.21   Luminous regions of a gas discharge.

by depositing thin film electrodes on the front and back substrates and then covering those electrodes with a thin dielectric glass. Recall from Figure 5.22 that this dielectric glass makes a capacitor that is used to limit the discharge current. This dielectric also is used to store the charge that gives these panels inherent memory. The two substrates are then sealed together around the perimeter and filled with neon gas. The AC monochrome panels have a very simple structure that allows the pixels to be isolated simply by the action of electric fields.

The AC plasma panels require the inner surface of the dielectric that is in contact with the gas to have a special protection layer of magnesium oxide. This MgO layer is necessary for the panel to have low operating voltages and long life. Being a refractory oxide, MgO sputters away at a very low rate, and it is also well known for its high secondary electron emission.

Ac displays require that an AC signal, called the sustain voltage, be applied during operation as shown on the right side of Figure 5.22. The typical sustain frequency for monochrome displays is 50 kHz. For high luminance color displays it is typically 250 kHz. Figure 5.24 shows the details of this operation for a pixel in

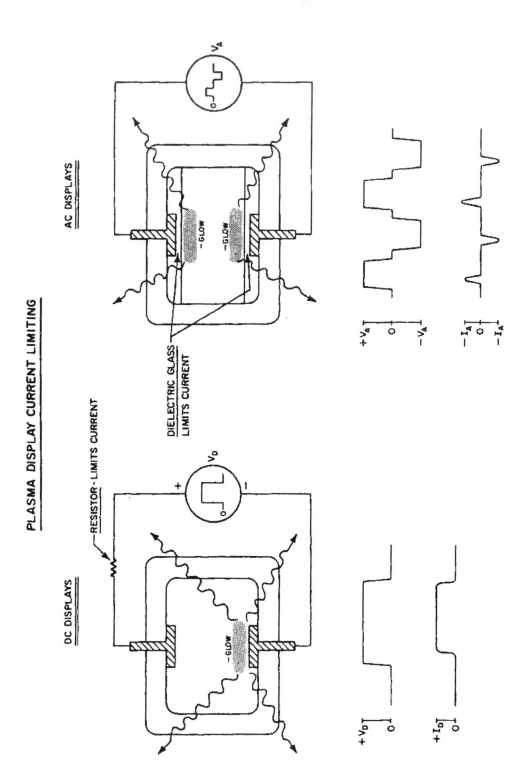

FIGURE 5.22 The two current limiting techniques used in plasma display products.

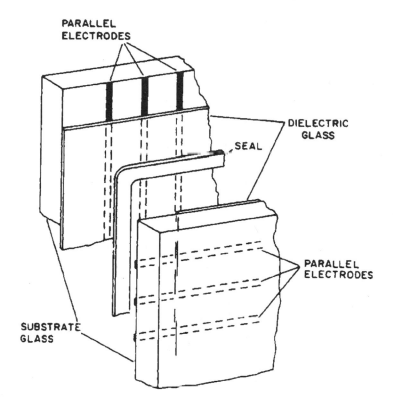

FIGURE 5.23 Monochrome AC plasma display structure.

both the on and off states. When a pixel is discharging, charge collects on the dielectric glass walls and influences the voltage across the gas. The component of voltage due to this charge is called the wall voltage. When a pixel is on, the wall voltage changes for each polarity reversal of the sustain voltage. This change in wall voltage coincides with a pulse of light due to the gas discharge. When the pixel is off, there are no light pulses, and the wall voltage remains at a zero level.

Pixel addressing is achieved through a partial discharge by introducing an address pulse timed between the sustain pulses. A write pulse causes the wall voltage to transit from zero volts to the final equilibrium wall voltage level. Likewise, an erase pulse causes the wall voltage to return to zero.

Color Plasma Display Devices

Color is achieved by placing phosphors in the plasma panel and then exciting those phosphors with the ultraviolet light of the gas discharge. This is the same principle as that used in the fluorescent lamp. Xenon is the active UV generating gas, which provides atomic resonance radiation at 147 nm and a molecular band centered at 173 nm. Neon or helium buffer gases are always mixed with the xenon in order to lower the operating voltage.

Figure 5.25 shows the two fundamental structures for achieving color. The opposed discharge structure is very similar to the monochrome AC structure shown in Figure 5.23. The surface discharge structure (also called coplanar or single-substrate) separates the discharge cathode areas from the phosphor by applying the sustain voltage only to the lower electrodes while the phosphor is on the top. The surface discharge approach achieves longer phosphor life because it is not directly sputtered by the energetic ions that are directed toward the cathodes. For this reason all modern color plasma displays use the surface discharge structure.

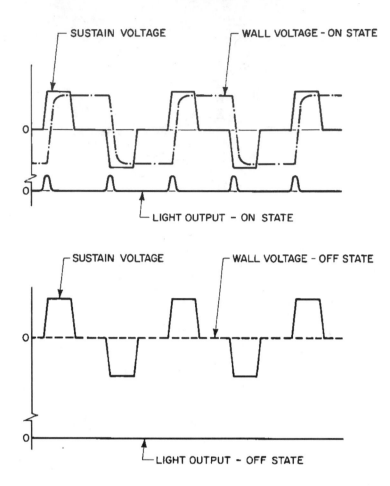

FIGURE 5.24   Sustain voltage, wall voltage, and light output for AC plasma pixels in the on and off states.

The detailed structure for the surface discharge AC plasma displays is shown in Figure 5.26 [Shinoda et al., 1993]. Note that the structures placed on the front and back substrates each have simple one-dimensional features. Since the structures on the two substrates are positioned orthogonally, there is no critical alignment between the two substrates because the pixels will automatically occur wherever the orthogonal electrodes intersect. This allows for straightforward manufacture of large panels.

The phosphors are placed on the rear substrate of the panel in Figure 5.26 and are excited by the ultraviolet light generated by the electrodes on the top substrate. This is shown as a single subpixel cross section in Figure 5.27. Note that the sustain and scan electrodes of Figure 5.27 have been schematically rotated 90° from the correct positions shown in Figure 5.26 in order to illustrate the action of the discharge. In this design, both sets of AC sustain electrodes are on the front plate. The AC voltage is applied to these electrodes in the normal way, and the fringing fields from these electrodes reach into the gas and create a discharge. Note that the structure in Figure 5.26 and Figure 5.27 has glass barrier rib separators between each subpixel. This is necessary to reduce crosstalk between the different colors, which will reduce the color purity. These barrier ribs do not transmit the 147-nm or 173-nm radiation generated by the xenon gas used in color plasma displays. The phosphors are placed on all walls of the subpixel channel except for the front plate, which has phosphor-damaging sputtering activity at the cathodes. This nearly complete phosphor coverage of the walls maximizes luminance while minimizing sputtering damage.

Other important features of the surface discharge structure in Figure 5.26 and Figure 5.27 are the address electrodes buried beneath the phosphors of the rear plate. These are the column electrodes that are selectively

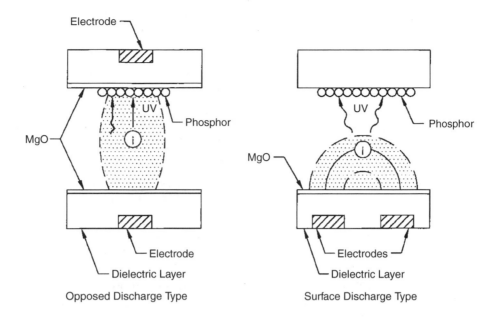

FIGURE 5.25   Two structural designs of color AC plasma displays.

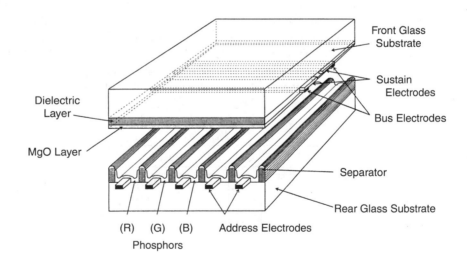

FIGURE 5.26   Structure of surface discharge color AC panel.

pulsed depending on the input image data. While these address operations do create discharge activity that could potentially sputter damage the phosphor, the address pulse frequency is orders of magnitude lower than the sustain frequency and so the amount of address damage is minimal.

The sustain and scan electrodes shown in Figure 5.26 are made of a conductive transparent material such as indium–tin oxide. Unfortunately, the resistance is orders of magnitude too high. To correct this problem, narrow bus electrodes of high conductivity materials, such as silver or chrome-copper-chrome, are placed over the tin oxide to reduce the electrode resistance to values on the order of 100 ohms.

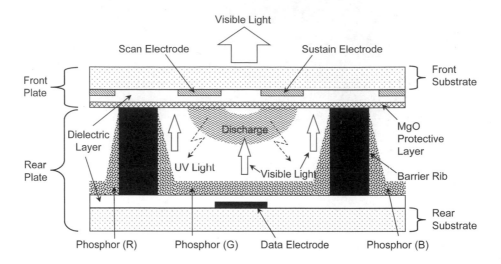

FIGURE 5.27   Structure details of surface discharge color AC panel.

Grayscale

The memory displays cannot use pulse intensity or pulse width modulation for grayscale because the pixels in memory mode are either on or off, and such pulse perturbations would in many cases have the undesirable effect of changing the state of the pixel. Instead these memory displays achieve grayscale by modulating the percentage of time that the pixel is on in a given frame. This means that the pixels must be addressed multiple times per frame. In the sequence shown in Figure 5.28 for 256 intensity levels, each frame is divided into eight subfields and each subfield consists of an address period and a sustain period [Yoshikawa, 1992]. During a given address period, address pulses are applied to all pixels in the panel according to the subfield image data. Each of the eight subfields has a sustain period with a different number of sustain cycles which emits an

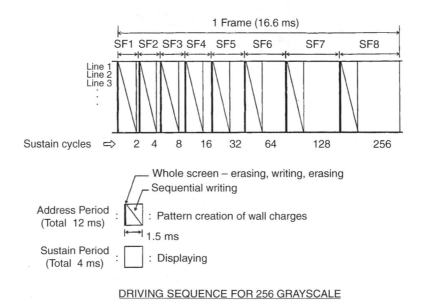

FIGURE 5.28   Addressing sequence for 256 intensity grayscale in AC memory plasma displays.

amount of light proportional to the number of sustain cycles. If each data bit of a given pixel intensity word is allowed to control one of the subfields, then the total number of sustain cycles (and light) per frame will be proportional to the 8-bit intensity value. This results in a precise inherently digital grayscale.

Defining Terms

AC plasma displays: These employ an internal capacitive dielectric layer to limit the gas discharge current.

DC plasma displays: These employ an internal or external resistor to limit the gas discharge current.

Luminous efficiency: The measure of the display output light luminance for a given input power usually measured in lumens per watt, which is equivalent to the nit.

Memory: The property of a display pixel that allows it to remain stable in an initially established state of luminance. Memory gives a display high luminance and absence of flicker.

Metastable atom: An atom in a temporary but long lived excited state in which photon emission is forbidden by electrodynamic theory. Metastables can give up their energy by ionizing other atoms or through wall collisions.

Plasma: The fourth state of matter comprised of positive ions and negative electrons of equal and sufficiently high density to nearly cancel out any applied electric field. Not to be confused with blood plasma.

Priming: The mechanism whereby particles such as ions, electrons, photons, or metastable atoms provide the one electron needed to start the gas discharge.

Sputtering: The physical process whereby an ion with kinetic energy in the gas collides with a solid surface ejecting an atom from the solid into the gas.

References

J.P. Boeuf, "Plasma display panels: physics, recent developments and key issues," *J. Phys. D: Appl. Phys.*, 36, R53–79, 2003.

T.N. Criscimagna and P. Pleshko, "AC plasma display," in *Topics in Applied Physics, Display Devices*, vol. 40, Berlin, Germany: Springer-Verlag, 1980, pp. 91–150.

J. Koike et al., "Long-life, high luminance 40-in. color DC PDP for HDTV," Intl. Display Res. Conf., Hamamatsu, Japan, pp. 943–944, 1995.

G. Oversluizen, et al., "High-Xe-content high-efficacy PDPs," *J. of SID*, 12, 51–55, 2004.

Y. Sato et al., "A 50-in. Diagonal plasma display panel with high luminous efficiency and high display quality," 2002. SID Int. Symposium, Boston, MA, pp. 1060–1063, 2002.

T. Shinoda et al., "Development of technologies for large-area color AC plasma displays," SID Intl. Symp., Seattle, WA, pp. 161–164, 1993.

H. Tolner, "Color-plasma-display manufacturing," 2003 *SID Seminar Lecture Notes*, vol. 1, Baltimore, MD, p. M-9/1, 2003.

L.F. Weber, "Plasma displays," in *Flat-Panel Displays and CRTs*, L.E. Tannas Jr., Ed., New York, NY: Van Nostrand Reinhold, 1985, pp. 332–414.

K. Yoshikawa et al., "A full color AC plasma display with 256 gray scale," Intl. Display Res. Conf., Hiroshima, Japan, pp. 605–608, 1992.

Further Information

A more detailed account of the material presented in this section is provided in Weber [1985] and Boeuf [2003]. The details of the manufacturing process can be found in Tolner [2003].

The Society for Information Display (SID) annual International Symposium publishes a digest of technical papers which is the best source for new display developments. Tutorial material can be found in the annual SID Seminar Lecture Notes. More research-oriented papers can be found in the technical digest of the

International Display Research Conference which rotates annually among Europe, Japan, and North America. In addition, SID publishes the *Journal of the SID*, which contains more detailed archival papers. These materials can be obtained from the SID at 610 S. 2nd Street, San Jose, CA 95112 or see the web site http://www.sid.org. Many of these can be found by searching Scitation at http://scitation.aip.org.

The International Electrotechnical Commission (IEC) has a very active plasma display standards group. Many of the terms and their fundamental usage can be found in the document: Terminology and Letter Symbols IEC 61988–1 Ed. 1.0. IEC standards are available at http://www.iec.ch/.

6

Data Acquisition

Dhammika
Kurumbalapitiya
Harvey Mudd College

6.1 Introduction

Data acquisition includes everything from gathering data, to transporting it, to storing it. The term *data acquisition* is described as the "phase of data handling that begins with sensing of variables and ends with a magnetic recording of raw data, may include a complete telemetering link" (McGraw-Hill, *Dictionary of Scientific and Technical Terms,* Second Edition, 1978). Here, the term *variables* refers to those physical quantities that are associated with a natural or artificial process. A data acquisition phase involves a real-time computing environment where the computer must be keyed to the time scale of the process. Figure 6.1 gives a simplified block diagram of a data acquisition system current in the early 1990s.

The path the data travels through the system is called the data acquisition channel. Data are first captured and subsequently translated into usable signals using transducers. In this discussion, usable signals are assumed to be electrical voltages, either unipolar (that is, single ended, with a common ground so that we need just one lead wire to carry the signal) or bipolar (that is, common mode, with the signal carried by a wire pair, so that the reference of the rest of the system is not part of the output). These voltages can be either analog or digital, depending on the nature of the measurand (the quantity being captured). When there is more than one analog input, they are subsequently sent to an analog **multiplexer** (MUX). Both the analog and the digital signals are then conditioned using signal conditioners. There are two additional steps for those conditioned analog signals. First they must be sampled and next converted to digital data. This conversion is done by **analog-to-digital converters** (ADC).

Once the analog-to-digital conversion is done, the rest of the steps have to deal with digital data only. The calendar/clock block shown in Figure 6.1 is used to add the time-of-date information, an important parameter of a real-time processing environment, into the half-processed data. The digital processor performs the overall system control tasks using a software program, which is usually called system software. These control tasks also include display, printer, data recorder, and communication interface management. A well-regulated **power supply unit** (PSU) and a stable clock are essential components in many data acquisition systems. There are systems where massive amounts of data points are produced within a very short period of time, and they are equipped with *on-board memory* so that a considerable amount of data points can be stored locally. Data are transmitted to the host computer once the local storage has reached its full capacity. Historically, data acquisition evolved in modular form, until monolithic silicon came along and reduced the size of the modules.

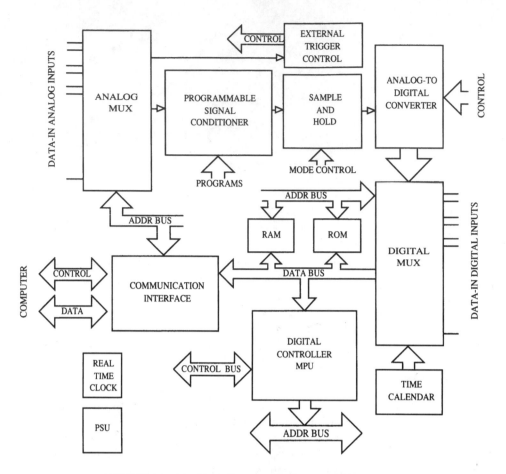

FIGURE 6.1 The block diagram of a data acquisition system.

The analysis and design of data acquisition systems is a discipline that has roots in the following subject areas: signal theory, transducers, analog signal processing, noise, sampling theory, quantizing and encoding theory, analog-to-digital conversion theory, analog and digital electronics, data communication, and systems engineering. Cost, accuracy, bit resolution, speed of operation, on-board memory, power consumption, stability of operation under various operating conditions, number of input channels and their ranges, on-board space, supply voltage requirements, compatibility with existing bus interfaces, and the types of data recording instruments involved are some of the prime factors that must be considered when designing or buying a data acquisition system. Data acquisition systems are involved in a wide range of applications, such as machine control, robot control, medical and analytical instrumentation, vibration analysis, spectral analysis, correlation analysis, transient analysis, digital audio and video, seismic analysis, test equipment, machine monitoring, and environmental monitoring.

6.2   The Analog and Digital Signal Interface

The data acquisition system must be designed to match the process being measured as well as the end-user requirements. The nature of the process is mainly characterized by its speed and number of measuring points, whereas the end-user requirement is mainly the flexibility in control. Certain processes require data acquisition with no interruption where computers are used in controlling. On the other hand, there are cases where the acquisition starts at a certain instance and continues for a definite period. In this case the acquisition cycle is repeated in a periodic manner, and it can be controlled manually or by software. Controllers access the process via the analog and digital interface submodules, which are sometimes called analog and digital front ends.

Many applications require information capturing from more than one channel. The use of the analog MUX in Figure 6.1 is to cater to multiple analog inputs. A detailed diagram of this input circuitry is shown in Figure 6.2 and the functional description is as follows. When the MUX is addressed to select an input, say, $x_i(t)$, the same address will be decoded by the decoding logic to generate another address, which is used in addressing the programmable register. The programmable register contains further information regarding how to handle $x_i(t)$. The outcome of the register is then used in subsequent tuning of the signal conditioner. Complex programmable control tasks might include automatic gain selection for each channel, and hence the contents of this register are known as the channel gain list. The MUX address generator could be programmed in many ways, and one simple way is to scan the input channels in a cyclic fashion where the address can be generated by means of a binary counter. Microprocessors are also used in addressing MUXs in applications where complex channel selection tasks are involved. Multiplexers are available in integrated circuit form, though relay MUXs are widely used because they minimize errors due to cross talk and bias currents. Relay MUX modules are usually designed as plugged-in units and can be connected according to the requirements.

There are applications where the data acquisition cycle is triggered by the process itself. In this case an analog or digital trigger signal is sent to the unit by the process, and a separate external trigger interface circuitry is supplied. The internal controller assumes its duties once it has been triggered. It takes a finite time to settle the signal $x_i(t)$ through the MUX up to the signal conditioner once it is addressed. Therefore, it is possible to process $x_{i-1}(t)$ during the selection time of $x_i(t)$ for greater speeds. This function is known as pipelining and will be illustrated in Section 6.3.

In some data acquisition applications the data acquisition module is a plugged-in card in a computer, which is installed far away from the process. In such cases, transducers—the process sensing elements—are connected to the data acquisition module using transmission lines or a radio link. In the latter case a complete demodulating unit is required at the input. When transmission lines are used in the interconnection, care must be taken to minimize electromagnetic interference since transmission lines pick up noise easily. In the case of a single-ended transducer output configuration, a single wire is adequate for the signal transmission, but a common ground must be established between the two ends as given in Figure 6.3(a). For the transducers that have common mode outputs, a shielded twisted pair of wires will carry the signal. In this case, the shield, the transducer's encasing chassis, and the data acquisition module's reference may be connected to the same ground as shown in Figure 6.3(c). In high-speed applications the transmission line impedance should be

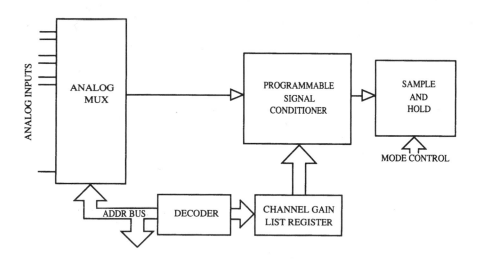

FIGURE 6.2   Analog input circuitry—the analog front end.

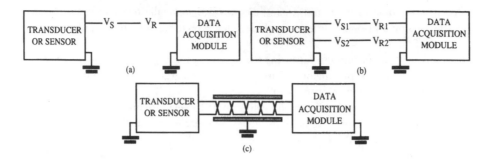

FIGURE 6.3 (a) Connecting transducers to the data acquisition unit, (b) single-ended (unipolar) output, and (c) common-mode (bipolar) output.

matched with the output impedance of the transducer in order to prevent reflected traveling waves. If the transducer output is not strong enough to transmit for a long distance, then it is best to amplify it before transmission.

Transducers that produce digital outputs may be first connected to Schmitt trigger circuits for pulse shaping purposes, and this can be considered as a form of digital signal conditioning. This becomes an essential requirement when such inputs are connected through long transmission lines where the line capacitance significantly affects the rising and falling edges of the incoming wave. Opto-isolators are sometimes used in coupling when the voltage levels of the two sides of the transducer and the input circuit of the data acquisition unit do not match each other. Special kinds of connectors are designed and widely used in interconnecting transmission lines and data acquisition equipment in order to screen the signals from noise. Analog and digital signal grounds should be kept separate where possible to prevent digital signals from flowing in the analog ground circuit and including spurious analog signal noise.

6.3 Analog Signal Conditioning

The objective of an analog signal conditioner is to increase the quality of the transducer output to a desired level before analog-to-digital conversion. A signal conditioner mainly consist of a preamplifier, which is either an instrumentation amplifier or an operational amplifier and/or a filter. Coupling more and more circuits to the data acquisition channel has to be done taking great care that these signal conditioning circuits do not add more noise or unstable behavior to the data acquisition channel. General purpose signal conditioner modules are commercially available for applications. Some details were given in the previous section about program-mable signal conditioners and the discussion is continued here.

Figure 6.4 shows an instrumentation amplifier with programmable gain where the programs are stored in the channel-gain list. The reason for having such sophistication is to match transducer outputs with the maximum allowable input range of the ADC. This is very important in improving accuracy in cases where transducer output voltage ranges are much smaller than the full-scale input range of an ADC, as is usually the case. Indeed, this is equally true for signals that are larger than the full-scale range, and in such cases the amplifier functions as an attenuator. Furthermore, the instrumentation amplifier converts a bipolar voltage signal into a unipolar voltage with respect to the system ground. This action will reduce a major control task as far as the ADC is concerned; that is, the ADC is always sent unipolar voltages, and hence it is possible to maintain unchanged the mode control input which toggles the ADC between the unipolar and bipolar modes of an ADC.

Values of the **signal-to-noise ratio**

$$\text{SNR} = \left[\frac{\text{RMS signal}}{\text{RMS noise}} \right]^2 \tag{6.1}$$

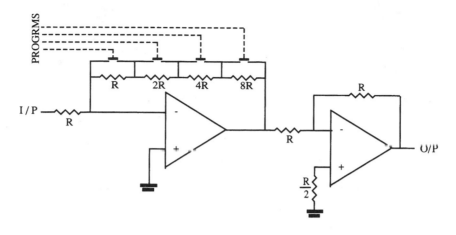

FIGURE 6.4 Programmable gain instrumentation amplifier.

at the input and the output of the instrumentation amplifier are related to its **common-mode rejection ratio** (CMRR) given by

$$CMRR = \sqrt{\frac{SNR_{output}}{SNR_{input}}} \qquad (6.2)$$

Hence, higher values of SNR_{output} indicate low noise power. Therefore, instrumentation amplifiers are designed to have very high CMRR figures. The existence of noise will result in an error in the ADC output. The allowable error is normally expressed as a fraction of the **least significant bit** (LSB) of the code such as $\pm(1/X)$LSB. The amount of error voltage (V_{error}) corresponding to this figure can be found considering the bit resolution (N) and the ADC's maximum analog input voltage (V_{max}) as given in

$$V_{error} = \pm\left[\frac{V_{max}}{2^N - 1} \times \frac{1}{X}\right] \text{volts} \qquad (6.3)$$

Other specifications of amplifiers include the temperature dependence of the input offset voltage (V_{offset}, $\mu V/^\circ C$) and the current (I_{offset}, $pA/^\circ C$) associated with the operational amplifiers in use. High slew rate ($V/\mu s$) amplifiers are recommended in high-speed applications. Generally, the higher the bandwidth, the better the performance.

Cascading a filter with the preamplifier will result in better performance by eliminating noise. Active filters are commonly used because of their compact design, but passive filters are still in use. The cut-off frequency, f_c, is one of the important performance indices of a filter that has to be designed to match the channel's requirements. The value f_c is a function of the preamplifier bandwidth, its output SNR, and the output SNR of the filter.

6.4 Sample-and-Hold and A/D Techniques in Data Acquisition

Sample-and-hold systems are primarily used to maintain a constant magnitude representing the input, across the input of the ADC throughout a precisely known period of time. Such systems are called **sample-and-hold amplifiers** (SHA), and their characteristics are crucial to the overall system accuracy and reliability of digital data. The SHA is not an essential item in applications where the analog input does not vary more than $\pm(1/2)$LSB of voltage. As the name indicates, a SHA operates in two different modes, which are digitally

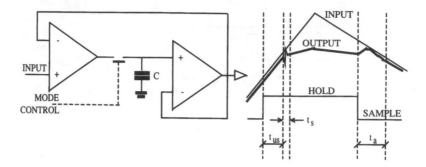

FIGURE 6.5 · Sample-and-hold circuit diagram and switching waveforms.

controlled. In the sampling mode it acts as an input voltage follower, where, once it is triggered into its hold mode, it should ideally retain the signal voltage level at the time of the trigger. When it is brought back into the sampling mode, it instantly assumes the voltage level at the input.

Figure 6.5 shows the simplified circuit diagram of a monolithic sampling-and-hold circuit and the associated switching waveforms. The differential amplifiers function as input and output buffers, and the capacitor acts as the storage mechanism. When the mode control switch is at its *on* position, the two buffers are connected in series and the capacitor follows the input with minimum time delay, if it is small. Now, if the mode control is switched *off*, the feedback loop is interrupted, and the capacitor ideally retains its terminal voltage until the next sampling signal occurs. Leakage and bias currents usually cause the capacitor to discharge and or charge in the hold mode and the fluctuation of the hold voltage is called *droop*, which could be minimized by having a large capacitor. Therefore, the capacitance has to be selected such that the circuit performs well in both modes. Several time intervals are defined relative to the switching waveform of SHAs. The *acquisition time* (t_a) is the time taken by the device to reach its final value after the sample command has been given. The *setting time* (t_s) is the time taken to settle the output. The *aperture uncertainty* or *aperture jitter* (t_{us}) is the range of variation of the aperture time. It is important to note here that the sampling techniques have a well-formulated theoretical background.

ADCs perform a key function in the data acquisition process. The application of various ADC technologies in a data acquisition system depends mainly on the cost, bit resolution, and speed. Successive approximation types are more common at high resolution at moderate speeds (<1 MHz). This kind of ADC offers the best trade-offs among bit resolution, accuracy, speed, and cost. Flash converters, on the other hand, are best suited for high-speed applications. Integrating-type converters are suitable for high-resolution and -accuracy applications.

Many techniques have been developed in coupling sample-hold circuits and ADCs in data acquisition systems because no single ADC or sampling technology is able to satisfy the ever increasing requirements of data acquisition applications. Figure 6.6 illustrates the various sampling and ADC configurations used in practice. It can be seen that the sampling frequencies are increased because of pipelining, parallelism, or concurrent architecture. The increase in the sampling frequency improves the bandwidth, improving in turn the SNR in the channel.

6.5   The Communication Interface of a Data Acquisition System

The communication interface is the module through which the acquired data are sent as well as other control tasks are established between the data acquisition module and the host computer (Figure 6.1). There are basically two different ways of establishing a data link between the two. One way is to use interrupts and the other is through **direct memory access** (DMA). In the case of an interrupt-driven mode, an interrupt-request signal is sent to the computer. Upon receiving it, the computer will first finish the execution of the current instruction, suspend the next, and then send an interrupt-acknowledge signal asking the module to send data.

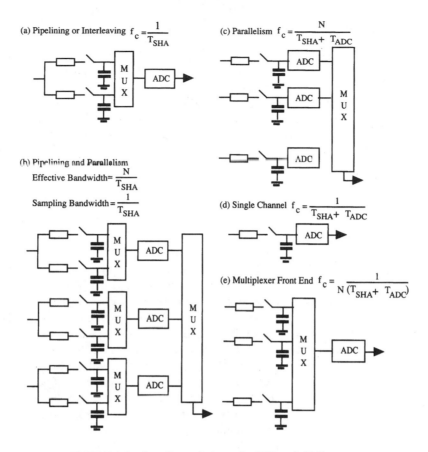

FIGURE 6.6 Coupling techniques for SHA and ADC systems.

The operation is asynchronous since the sender sends data when it wants to do so. Getting the computer ready to receive data is known as handshaking. In the case of a DMA transfer, the DMA controller is given the starting address of the memory location where the data have to be written. The DMA controller asks the computer to freeze its operations until it has finished writing data directly into the memory. The operation does not need any waiting time and therefore it is fast.

Data acquisition systems are usually designed to couple with existing computer systems, and many computer systems provide standard bus architecture, allowing users to connect various peripherals that are compatible with its bus. Data acquisition systems are computer peripherals that follow the above description. Since ADCs produce parallel data, many data acquisition systems provide outputs compatible with parallel instrument buses such as the IEEE-488 (HP-IB or GPIB) or the VMEbus. Personal computer-based data acquisition boards must have communication interfaces compatible with the computer bus in order to share resources. The RS-232 standard communication interfaces are widely used in serial data transfer. Communication interfaces for data acquisition systems are normally designed to satisfy the electrical, mechanical, and protocol standards of the interface bus. Electrical standards include power supply requirements, methods of supply, the data transfer rate (baud rate), the width of the address, and the line terminating impedance. Mechanical requirements are the type, size, and the pin assignments of the connectors. The data transfer protocol determines the procedure of data transfer between the two systems. A definition of the timing and input–output philosophy—whether the transfer is in synchronous, asynchronous, or quasi-synchronous mode and how errors are detected and handled—are important factors to be considered.

6.6   Data Recording

It is important to provide storage media to cater to large streams of data being produced. Data acquisition systems use graph paper, paper tapes, magnetic tapes, magnetic floppy disks, hard disks, or any combination of these as their data recorders. Paper and magnetic tape storage schemes are known as sequential access storage, whereas disk storage is called direct access storage. Tapes are cost-effective media compared to disk drives and are still in wide use. In many laboratory situations it will be much more cost effective to network a number of systems to a single, high-capacity hard drive, which acts as a file server. This adoption of digital recording provides the ultimate in signal-to-noise ratio, accuracy of signal waveform and freedom from tape transfer flutter. Data storage capacity, access time, transfer rate, and error rate are some of the performance indices that are associated with these devices.

6.7   Software Aspects

So far the discussion has been mainly on the hardware side of the data acquisition system. The other most important part is the software system associated with a data acquisition system, which can generally be divided into two—the system software and the user-interface program. Both must be designed properly in order to achieve the maximum use of the system. The system software is mainly written in assembly language with many lines of code, whereas the user interface is built using a high-level software development tool. One main part of system software is written to handle the input–output (I/O) operations. The use of assembly language results in the fast execution of I/O commands. The I/O software has to deal with how the basic input–output programming tasks such as interrupt and DMA handling are done. The other aspects of system software are to perform the internal control tasks such as providing trigger pulses for the ADC and SHA, addressing the input multiplexer, the accessing and editing of the channel-gain list, transferring data into the on-board memory, and the addition of the clock/calendar information into data. Multitasking software programs are best suited for many data acquisition systems because it may be necessary to read data from the data acquisition module and display and print it at the same time. Menu-driven user interfaces are common and have a variety of functions built into them.

Defining Terms

Analog-to-digital converter (ADC):   A device that converts analog input voltage signals into digital form.

Common-mode rejection ratio (CMRR):   A measure of quality of an amplifier with differential inputs and defined as the ratio between the common-mode gain and the differential gain.

Direct memory access (DMA):   The process of sending data from an external device into the computer memory with no involvement of the computer's central processing unit.

Least significant bit (LSB):   The 20th bit in a digital word.

Multiplexer (MUX):   A combinational logic device with many input channels and usually just one output. The function performed by the device is connecting one and only one input channel at a time to the output. The required input channel is selected by sending the channel address to the MUX.

Power supply unit (PSU):   The one that generates the necessary voltage levels required by a system.

Sample-and-hold amplifier (SHA):   A unity gain amplifier with a mode control switch where the input of the amplifier is connected to a time-varying voltage signal. A trigger pulse at the mode control switch causes it to read the input at the instance of the trigger and maintain that value until the next trigger pulse.

Signal-to-noise ratio (SNR):   The ratio between the signal power and the noise power at a point in the signal traveling path.

References

For further reading consult the following texts, which were used along with the authors' experience and other sources as a basis for this article:

Analog Devices, *Data Conversion Handbook,* Analog Devices, Inc., 1989/90.

R. Annino and R. Driver, *Scientific and Engineering Applications with Personal Computers*, New York: Wiley Interscience, 1986.

D.L. Feucht, *Handbook of Analog Circuit Design*, San Diego: Academic Press, 1990.

D.G. Fink and D. Christiansen (Eds.), *Electronics Engineers' Handbook*, 3rd ed., New York: McGraw-Hill, 1989.

P.M. Garrett, *Analog Systems for Microprocessor and Minicomputers*, Reston, Va.: Reston Publishing Company, 1978.

P. Holloway, "Technology focus interview," *Electronic Engineering*, December 1990.

F. Jorgensen, *The Complete Handbook of Magnetic Recording*, 4th ed., Blue Ridge Summit, Penn.: Tab Books, 1995.

F.F. Mazda, *Electronic Instruments and Measurement Techniques*, New York: Cambridge University Press, 1987.

D.A. Mellichamp (Ed.), *Real-Time Computing With Applications to Data Acquisition and Control*, New York: Van Nostrand Reinhold, 1983.

M. Tatkow and J. Turner, "New techniques for high-speed data acquisition," *Electronic Engineering*, September 1990.

Further Information

To probe further in the subject area, refer to the *Data Acquisition Handbook,* published by Data Translation, Marlboro, Mass., 1990, and the *Data Acquisition Handbook,* published by Rector Press, 1995.

7
Testing

Michaela Serra
University of Victoria

Bulent I. Dervisoglu
Silicon Graphics, Inc.

7.1 Digital IC Testing

Michaela Serra

In this chapter, an overview is provided of digital testing techniques, with reference to material containing details of methodologies and algorithms. First, a general introduction of terminology and a taxonomy of testing methods and fault models is presented. A discussion is presented of main approaches for the generation of test patterns, both algorithmically and pseudo-randomly. The article concludes with an introduction to signature analysis and to built-in self-tests.

Taxonomy and Definitions

Technology advances are causing the density of circuits to continually increase, while the number of I/O pins remains small in comparison. This causes a serious escalation in complexity, and testing is becoming a major cost burden (estimated to run as high as 40 percent of total production cost). Moreover, digital devices now are ubiquitous in many portable electronics for safety-critical applications, such as biomedical devices. Ensuring reliability through proper testing is crucial. Devices should be tested before and after packaging, after mounting them on a board and periodically during operation. Different methods may be necessary for each case, but it is important that defective units be detected as soon as possible in the production chain. The *ruleoften* is often invoked: a detection cost C at the component level becomes a cost of 10 C at the board level, a cost of 100 C for a system and as high as 1,000 C in the field. *Testing* is generally assumed to be the means by which some qualities or attributes are determined to be fault-free or faulty. The main purpose of testing is to detect malfunctions (in a go/no-go test), and to increase yield or a change in the production process. Testing might also involve fault diagnosis to narrow down the actual location of the malfunction.

Evaluating the reliability and quality of a digital device is commonly called testing, yet it comprises distinct phases usually kept separate both in the research community and in industrial practice.

1. *Verification* is the initial phase at design time, ensuring the match of specifications to functionality. A digital device needs to be verified for correctness. Verification checks that all design rules are followed. Generally, this type of functional testing checks that the circuit (a) implements what it is supposed to do and (b) does not implement what it is not supposed to do. Both conditions are necessary. This type of

evaluation uses a variety of techniques, including logic verification with hardware description languages, full-functional simulations and the generation of functional test vectors. Verification techniques will not be discussed.

2. *Testing* refers to the phase of ensuring that only defect-free production devices are shipped and of detecting manufacturing faults. Testing methods must (a) be fast enough to be applied to large numbers of devices during production; (b) take into consideration possible access to large, expensive external tester machines; and (c) consider whether the implementation of built-in self-tests (BISTs) proves advantageous. In BIST, the circuit is designed to include self-testing extra circuitry and can signal its possible failure status directly during testing and in the field. This involves a certain amount of area overhead, and trade-offs must be considered. The development of appropriate testing algorithms and tool support can require a large engineering effort but it may need to be done only once per design. The application speed of this method to many copies of the devices can be of great importance.

If many defects are found in the manufacturing process, the final yield is lowered. Estimates can be derived for the relationship between manufacturing yield, testing effectiveness (fault coverage) and the defect level remaining after testing [Williams, 1986]. Let Y denote the yield, a value between 1 (100 percent defect-free production) and 0 (all circuits faulty after testing), that is approximated by the ratio of "good devices" over "total devices." Let FC be fault coverage, calculated as the percentage of detected faults over the total number of detectable modeled faults (see below for fault models). The value of FC ranges from 1 (all possible faults detected) to 0 (no useful testing done). The goal is to estimate the final defect level (DL) after testing, defined as the probability of shipping a defective product. DL measures the number of bad devices passing all tests; its value is expressed as number of defects per million (DPM). It has been shown that tests with high fault coverage for certain fault models also have high defect coverage. The empirical equation is $DL = 1 - Y^{(1-FC)}$. Plotting this equation provides interesting and practical results. For example, if a value of $DPM = 300$ (that is, $DL = 0.0003 = 0.03\%$) and a value of $Y = 0.5$ are desired, then it must follow that $FC = 1 - (\log(1-DL)/\log Y) = 0.999567$, which is 99.957 percent. Conversely, with a similar desired yield $Y = 0.5$ and $FC = 0.9$, then $DL = 1 - 0.5^{(1-0.9)} = 0.06697$, implying that about 6.7 percent of shipped devices are defective. It can be concluded that high fault coverage must be achieved to obtain an acceptable defect-level value, and that manufacturing yield must be continually improved to maintain reliability for shipped products.

Most testing techniques focus only on combinational circuits. This is a realistic assumption, based on the design of sequential circuits by partitioning the memory elements from the control functionality in the combinational modules. This general approach is used as a method in design for testability (DFT) (see the next chapter). DFT can encompass any design strategy to enhance a circuit's testability. Scan design is the best-known implementation for separating the latches from combinational modules. Some of the latches can be reconfigured and used as either tester units or as input generator units. These are essential for built-in testing.

Figure 7.1 shows the taxonomy for testing methods. The main division is between **online** and **offline testing**. In the former, each input/output word from a circuit can be tested during normal operation, normally implying that the circuit has been augmented to contain an embedded coding scheme that provides for error detection. In the latter, the circuit must suspend normal operation and enter a test mode, when the appropriate testing method is applied. Offline testing can be performed either through external testing, possibly by using a tester machine external to the circuitry, or through the use of BIST.

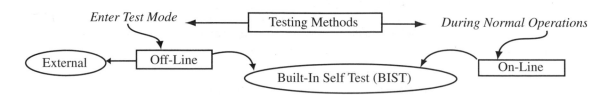

FIGURE 7.1   Taxonomy of testing methods.

Fault Models

At the defect level, an enormous number of failures could be present. It is infeasible to analyze all of them. Failures are grouped by their *logical* fault effect on the circuit's functionality, and this leads to the construction of logical fault models to test algorithms and evaluate fault coverage [Abramovici et al., 1992; Jha et al., 2003]. A *fault* denotes the physical failure mechanism; the *fault effect* denotes the logical effect of a fault on a signal-carrying net; and an *error* is defined as the condition (or state) of a system containing a fault (deviation from correct state). Faults also can be divided into classes according to their characteristics, as shown in Figure 7.2. Here, only *permanent* faults, or faults in existence long enough to be observed at test time, are discussed. This is different from *temporary* faults (transient or intermittent), which appear and disappear in short time intervals, and *delay* faults, which affect the circuit's operating speed.

The fundamental fault model is a **stuck-at fault**, implying the fault effect as a line segment stuck at logic 0 or 1 (*stuck-at 0* or *stuck-at 1*). Testing may consider single or multiple stuck-at faults, and Figure 7.3 shows an example for a simple circuit. The fault-free function is shown as Z, while the faulty output functions, occurring as single stuck-at faults of either line a stuck-at 0 (a/0) or of line b stuck-at 1 (b/1), are shown as Z^*. A stuck-at fault requires a single input pattern to stimulate the fault by controlling the inputs and making the faulty output observable. The goal of test-pattern generation algorithms is to find such settings for all detectable faults. On the other hand, a delay fault defines the effect to be *slow-to-rise* (from 0 to 1) or *slow-to-fall* (from 1 to 0). The final value may be correct, but it is outside the timing parameters that were expected. To detect such faults, two patterns are applied as stimuli. The first pattern sets a line at a certain value, and the second changes that value. This increases the complexity of a fault-detection algorithm.

Even considering only single stuck-at faults, not all possible faults need to be explicitly tested. Many have the same fault effect and are indistinguishable. For example, an input stuck-at-0 fault for an AND gate has the same logical effect as the output stuck-at-0, and it is necessary to test for only one fault. *Fault collapsing* is the process of reducing the number of faults to be examined, using fault-equivalence classes based on fault effects. Table 7.1 shows the main fault equivalence classes for gates.

Test-Pattern Generation

Test-pattern generation is the process of generating an appropriate (but minimal) subset of all input patterns, stimulating the circuit inputs so that a desired percentage of detectable faults can be exercised and detected [Abramovici et al., 1992; Jha et al., 2003]. The process can be divided into two phases: (1) derivation of a test and (2) test application. For the former, one first must select appropriate circuit models (at the gate or

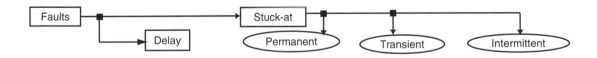

FIGURE 7.2 Fault characteristics.

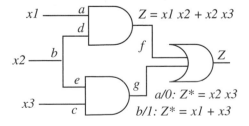

FIGURE 7.3 Stuck-at faults.

TABLE 7.1   Fault Equivalence

Gate	Fault		Equivalent to:
AND	any input/0	← →	output/0
OR	any input/1	← →	output/1
NAND	any input/0	← →	output/1
NOR	any input/1	← →	output/0
NOT	input/0	← →	output/1
	input/1	← →	output/0

transistor lcvcl) and the models for faults. A test is then constructed so that the output signal from a faulty circuit differs from that of a good circuit. This can be computationally expensive, but the process is performed only once during the design stage. The generation of a test set can be obtained either by algorithmic methods (with or without heuristics), or by pseudo-random methods. For the second phase, a test is subsequently applied many times to each device and must be efficient both in space (for pattern storage requirements) and in time. Often such a set is not minimal, but near minimality may be sufficient. The considerations in evaluating a test set are (a) the time needed to construct a minimal test set; (b) the size of the test-pattern generator, in terms of the software or hardware module used to stimulate the circuit tested; (c) the size of the test set itself; (d) the time to load the test patterns; and (e) the equipment required (if external), or the BIST overhead (see Figure 7.4).

Most algorithmic test-pattern generators are based on the concept of *sensitized paths,* through controllability and observation. Given a line in a circuit, it is necessary to find a sensitized path to stimulate a possible fault and carry its logical effect to an observable output. For example, to sensitize a path passing through an AND gate, all other gate inputs must be set to logic 1 to permit the sensitized signal to carry forward. Figure 7.5 summarizes the principles of constructing a test pattern for each fault through path sensitization, which is not a full algorithm but underlies the overall logic. In Figure 7.5(a) possible input patterns are found to test the fault online $g/0$. Line g must be set to 1 and a faulty signal is expected to be 0. This leads to the notation g: 1/0. In forward propagation, line h must be set to 0 for the fault to become observable at output Z: 1/0. In Figure 7.5(b), the second phase of backward propagation is shown, where

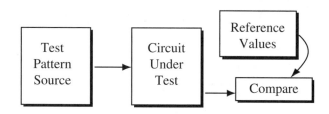

FIGURE 7.4   Test sets.

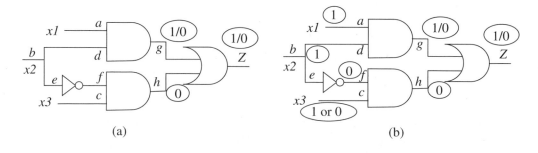

(a)   (b)

FIGURE 7.5   Path sensitization.

possible controlling inputs are set. Lines *a* and *d* must be set to 1 to control line *g* being 1. This assigns $x1 = x2 = 1$, forcing line *f* to be 0. This leaves *x3* with a choice of either 0 or 1 to maintain $h = 0$. Two test patterns are found: $(x1\ x2\ x3) = \{(1\ 1\ 0)\ \text{or}\ (1\ 1\ 1)\}$.

The best-known algorithms are the D-algorithms (precursors to others), PODEM and FAN. Many algorithmic variations are used with heuristics and optimizations, often tailored to certain classes of devices [Abramovici, 1992; Jha et al., 2003]. Major steps can be identified in most automatic test pattern generation (ATPG) programs: (1) listing the signal on the line where a fault should be detected; (2) sensitizing the path from that line to a primary output so that the fault can be observed; (3) determining the primary input conditions needed to set the testing signal (using back propagation); and (4) repeating this procedure until all detectable faults in a given fault set have been covered. Powerful heuristics speed the steps by helping with the sequential selection of faults examined and by reducing the amount of back and forward propagation needed. The process is summarized graphically in Figure 7.6.

Except for heuristics, algorithmic test pattern generation is computationally expensive and can lead to numerous difficulties, especially in certain types of networks. Newer alternatives are based on pseudo-random pattern generation [Bardell et al., 1987] and fault simulation. With this strategy, a large set of patterns is generated pseudo-randomly with the help of an inexpensive hardware or software generator. Typical choices for these are linear feedback shift registers (LFSR) and linear cellular automata registers (LCAR) (see examples below). The pseudo-random set stimulates a circuit. Using a fault simulator, the number of faults covered by this set can be evaluated. An algorithmic test-pattern generator is then applied to find coverage for the remaining faults — hopefully, a small number — and the pseudo-random set is augmented. However, the resulting set is large, and fault simulation is computationally expensive. Still, this method presents an alternative for circuits where the application of deterministic algorithms for all faults is not feasible. Pseudo-random pattern generation is the main alternative when BIST is introduced, because the overhead costs of storing even a minimal test set would be impractical in circuit testing.

Output Response Analysis and Built-In Self-Test

Output response analysis uses methods focusing on the output stream, with the assumption that the circuit is stimulated either by an exhaustive or a pseudo-random set of input combinations. When designing a circuit with BIST, a decision must be made on how to check the correctness of the circuit's responses [Bardell et al., 1987]. It is not feasible to store all expected responses on a chip. A common solution is to reduce circuit responses to relatively short sequences. This process is called *data compaction* or *signature analysis,* and the short, compacted resulting sequence is called a *signature.* The normal configuration for data-compaction testing is shown in Figure 7.7. The circuit is stimulated by an input pattern generator (pseudo-random or even exhaustive if $n < 20$); the resulting long output vectors are compacted to a short signature of length *k*, where *k* is usually 16 to 32 bits. The signature is compared to a known good value. The advantages of this method are that (1) the testing can be done at circuit speed by the appropriate choice of the pseudo-random generator;

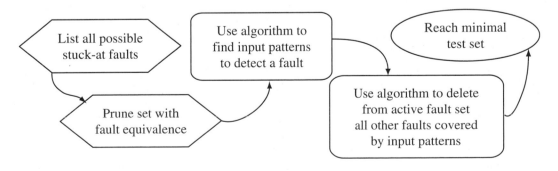

FIGURE 7.6 Process for test pattern generation.

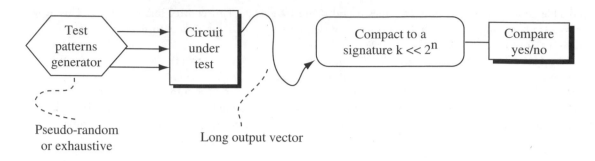

FIGURE 7.7   Data compaction.

(2) there is no need to generate algorithmic test patterns; and (3) testing circuitry involves a very small area, especially if the circuit has been designed using scan techniques (see the next chapter). The issues involve the design of efficient input generators and compactors.

The main disadvantage of this method is the possibility of aliasing. When a short signature is formed, a loss of information occurs. This can be caused by a faulty circuit producing the same signature as a fault-free circuit. The design method for data compaction aims at minimizing the aliasing probability. Using the compactors described below, the probability of aliasing has been theoretically proven to be 2^{-k}, where k is the length of the compactor and thus the length of the signature. The result is asymptotically independent of the size and complexity of the circuit under test. For example, for $k = 16$, the probability of aliasing is about 10^{-6}. The empirical results show that, in practice, this method is even more effective. This is the chosen methodology when BIST is required for its effectiveness, speed and small overhead area.

A secondary issue in data compaction is the determination of the expected "good" signature. The best method is to use fault-free simulation for both the circuit and the compactor. The appropriate comparator then can be built as part of the testing circuitry [Bardell et al., 1987; Abramovici, 1992; Jha et al., 2003].

The most important issues are in the choices of a pseudo-random generator and a compactor. Although no "perfect" compactor can be found, several have been shown to be effective. Several compaction techniques have been researched. Counting techniques, including 1s count, syndrome testing, transition count and Walsh spectra coefficients have been used. Other techniques are based on LFSRs and LCARs. Only these latter methods are discussed in this article. LFSRs and LCARs are preferred for implementation of the input pattern generators.

Pseudo-Random Pattern Generators

An autonomous LFSR (ALFSR) is a clocked synchronous shift register, augmented with appropriate feedback taps and receiving no external input [Bardell et al., 1987]. This provides an example of a general, linear, finite-state machine, where memory cells are simple D flip-flops and the next state operations are implemented only by EXOR gates. Figure 7.8(a) shows an ALFSR of length $k = 3$. An ALFSR of length k can be described by a polynomial with binary coefficients of degree k, where the nonzero coefficients of the polynomial denote the positions of the respective feedback taps. In Figure 8(a), the high-order coefficient for x^3 is 1, and there is a feedback tap from the right-most cell s_2 (reading from right to left); the coefficient for x^2 is 0, and no feedback tap exists after cell s_1; however, taps are present from cell s_0 and to the left-most stage, since x^1 and x^0 have nonzero coefficients. Since this is an autonomous LFSR, there is no external input to the left-most cell.

The state of the ALFSR is denoted by the binary state of its cells. In Figure 7.8(a), the next state of each cell is determined by the implementation provided by its polynomial and can be summarized as follows: $s_0^+ = s_2, s_1^+ = s_0 \oplus s_2$ and $s_2^+ = s_1$, where s_i^+ denotes the next state of cell s_i at each clock cycle. If the ALFSR is initialized in a nonzero state, it cycles through a sequence of states and eventually comes back to the initial state. This follows the functionality of next-state rules implemented by its polynomial description. If the

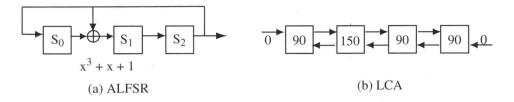

(a) ALFSR   (b) LCA

FIGURE 7.8   Autonomous linear generator.

polynomial chosen to describe the ALFSR is primitive, the ALFSR of length k cycles through all possible 2^{k-1} nonzero states in a single cycle (see the theory of Galois fields for the definition of primitive). These polynomials can be found from tables [Bardell et al., 1987].

By connecting the output of each cell to each input of a tested circuit, the ALFSR provides an ideal input generator. It is inexpensive in its implementation and provides the stimuli in pseudo-random order.

An alternative to an ALFSR is an LCA, a one-dimensional array of two cell types: rule 150 and rule 90 cells [Cattell et al., 1996]. Each cell is composed of a flip-flop that saves the current state of the cell and an EXOR gate that computes the next cell state. A *rule 150* cell computes its next state as the EXOR of its present state and of the states of its two (left and right) neighbors, as in $s_i^+ = s_{i-1} \oplus s_i \oplus s_{i+1}$, while a *rule 90* cell computes its next state as the EXOR of the states of its two neighbors only, as in $s_i^+ = s_{i-1} \oplus s_{i+1}$. In Figure 7.8(b), all connections in an LCA are near-neighbor connections, saving routing area and delays. There are advantages of using LCAs instead of ALFSRs: the localization of all connections, the ease of concatenation to obtain various sizes and, most importantly, the concept that LCAs are better pseudo-random pattern generators when used in autonomous mode. In that mode, they do not show the correlation of bits from the shifting of ALFSRs. The better pattern distribution provided by LCA as input stimuli provides better detection for delay faults, requiring a two-pattern stimuli.

As with ALFSRs, LCAs are described by a characteristic polynomial. Through the polynomial, any linear finite state machine can be built either as an ALFSR or as an LCA. However, it is more difficult to derive the corresponding LCA with a polynomial, and tables are now used. The main disadvantage of LCA is in the area overhead costs absorbed by the extra EXOR gates needed for the implementation of the cell rules.

Data Compaction or Signature Analysis

If the left-most cell of an LFSR is connected to an external input, as shown in Figure 7.9, the LFSR can be used as a data compactor [Bardell et al., 1987]. The underlying operation of such an LFSR is to compute polynomial division over a finite field. The effectiveness of the signature analysis is based on this functionality, with the fundamentals coming from the theory of cyclic codes. The divisor polynomial describes the LFSR implementation. The binary input stream can represent the coefficients (high order first) of a dividend polynomial. For example, if the input stream is 1001011 (bits are input left to right in time), the dividend polynomial is represented as $x^6 + x^3 + x + 1$. After seven clock cycles are performed for the input bits to have entered the LFSR, the binary output stream exiting from the right denotes the quotient polynomial. The last cell state in the LFSR denotes the remainder polynomial.

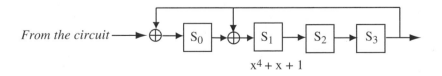

FIGURE 7.9   LFSR as signature analyzer.

In computing a signature for testing a circuit, the input stream to the LFSR, used as a compactor, is the output stream from the circuit under test. At the end of the testing cycles, only the last LFSR state is examined and considered to be the compacted signature of the circuit. In most cases, circuits have many outputs, and the LFSR is converted into a multiple-input shift register (MISR). An MISR is constructed by adding EXOR gates to the input of some or all the flip-flop cells; the circuit outputs are fed through these gates into the compactor. The probability of aliasing for an MISR is the same as that for an LFSR; however, some errors are missed due to cancellation. This occurs when an error in one output at time t is canceled by the EXOR operation, with the error in another output at time $t + 1$. Assuming equal probabilities for the different errors, the probability of error cancellation has been shown to be 2^{1-m-N}, where m is the number of outputs compacted and N is the length of the output streams.

Given that the normal signature length varies between $k = 16$ and $k = 32$, the probability of aliasing is minimal and acceptable in practice. In MISR, the length of the compactor also depends on the number of outputs tested. If the number of outputs is greater than the length of the MISR, algorithms or heuristics exist to combine outputs with EXOR trees before feeding them to the compactor. If the number of outputs is much smaller, choices can be evaluated. The amount of aliasing that occurs in a particular circuit can be computed by full fault simulation by injecting each possible fault into a simulated circuit and computing the resulting signature. Changes in aliasing can be performed by changing the polynomial that defines the compactor. Primitive polynomials, essential for the generation of exhaustive input generators (see above), possess good aliasing characteristics. All implementations of such linear finite-state compactors, be it LFSR or LCA, possess the same aliasing properties.

Summary

Accessibility to internal dense circuitry is becoming a greater problem. It is essential that a designer consider how a device will be tested and incorporate structures into the design to help that testing. Formal DFT techniques provide testing access points. As test-pattern generation grows more prohibitively expensive, probabilistic solutions based on data compaction and using fault simulation will become more widespread, especially if they are supported by DFT techniques and can avoid the major expense of dedicated external testers. Any chosen technique must be incorporated within the framework of a powerful CAD system that provides semiautomatic analysis and feedback.

Defining Terms

Aliasing: This occurs if the faulty output produces the same signature as a fault-free output.

Built-in self-test (BIST):   The inclusion of on-chip circuitry to provide testing.

Fault coverage: The percentage of detected faults over all possible detectable faults.

Fault simulation: An empirical method to determine how faults affect the operation of a circuit or how much testing is required to obtain a desired fault coverage.

Linear feedback shift register (LFSR):   A shift register formed by flip-flops and EXOR gates, chained together with a synchronous clock, used either as an input pattern generator or as a signature analyzer.

MISR: Multiple-input LFSR.

Offline testing: A testing process carried out while the tested circuit is not in use.

Online testing: Concurrent testing to detect errors while the circuit is operating.

Pseudo-random pattern generator: This generates a binary pattern sequence where the patterns appear to be random in the local sense but are deterministically repeatable.

Random testing: A testing process using a set of pseudo-randomly generated patterns.

Signature analysis/data compaction: A test where the output responses of a device over time are compacted into a characteristic value called a *signature*, which is then compared to a known good response.

Stuck-at fault: A fault model represented by a signal stuck at a fixed logic value (0 or 1).

Test pattern (test vector): An input vector where the faulty output is different from the fault-free output (the fault is stimulated and detected).

References

M. Abramovici, M.A. Breuer and A.D. Friedman, *Digital Systems Testing and Testable Design*, Rockville, MD.: IEEE Press, 1992.

P.H. Bardell, W.H. McAnney and J. Savir, *Built-In Test for VLSI: Pseudorandom Techniques*, New York, NY: John Wiley and Sons, 1987.

K. Cattell and J.C. Muzio, "Synthesis of one-dimensional linear hybrid cellular automata," *IEEE Trans. Computer Aided Design*, vol. 15, no. 3, pp. 325–335, 1996.

N. Jha and S. Gupta, *Testing of Digital Systems*, Cambridge University Press, Cambridge, NY: 2003.

T.W. Williams, Ed., *VLSI Testing*, Amsterdam, The Netherlands: North-Holland, 1986.

Further Information

Books by Abramovici et al. [1992] and Jha et al. [2003] give the most comprehensive view of testing methods and design for testability. More information on deterministic pattern generation also can be found in *Fault Tolerant Computing*, edited by D.K. Pradhan, Englewood Cliffs, NJ: Prentice-Hall, 1986. For newer approaches to random testing, the book by Bardell et al. [1987] contains basic information. The latest state-of-the-art research is found mainly in proceedings of the IEEE International Test Conference.

7.2 Design for Test

Bulent I. Dervisoglu

Testing of electronic circuits, which has long been pursued as an activity that follows the design and manufacture of (at least) the prototype product, has currently become a topic of up-front investigation and commitment. Today, it is not uncommon to list the *design for testability* (DFT) features of a product among the so-called *functional* requirements in the definition of a new product to be developed. Just how such a major transformation has occurred can be understood by examining the testability problems faced by manufacturing organizations and considering their impact on time to market (TTM).

The Testability Problem

The primary objective of testing digital circuits at chip, board, or system level is to detect the presence of hardware failures induced by faults in the manufacturing processes or by operating stress or wearout mechanisms. Furthermore, during manufacturing, a secondary but equally important objective is to accurately determine which component or physical element (e.g., connecting wire) is faulty so that quick diagnosis/repair of the product becomes possible. These objectives are necessary due to imperfections in the manufacturing processes used in building digital electronic components/systems. All digital circuits must undergo appropriate level testing to avoid shipping faulty components/systems to the customer. Analog circuits may have minimum and maximum allowable input signal values (e.g., input voltage) as well as infinitely many values in between these that the component has to be able to respond to. Testing of analog circuits is often achieved by checking the circuit response at the specified upper and lower bounds as well as observing/quantifying the change of the output response with varying input signal values. On the other hand, the behavior of a digital system is characterized by discrete (as opposed to continuous) responses to discrete operating state/input signal permutations such that testing of digital circuits may be achieved by checking their behavior under every operating mode and input signal permutation. In principle this approach is valid. However, in practice, most digital circuits are too complex to be tested using such a brute force technique. Instead, test methods have been developed to test digital circuits using only a fraction of all possible test conditions without sacrificing test coverage. Here, *test coverage* is used to refer to the ratio of faults that can be detected to all faults which are taken into consideration, expressed as a percentage. At the present time the most popular *fault model* is the so-called *stuck-at* fault model that refers to individual nets being considered to be fault-free (i.e., *good network*) or considered to be permanently stuck at either one of the logic 1 or logic 0 values. For example, if the *device under test* (DUT) contains several components (or building blocks), where the sum of all input and output terminals (*nodes*) of the components is k, there are said to be $2k$ possible stuck-at faults, corresponding to each of the circuit nodes being permanently stuck at one of the two possible logic states. In general, a larger number of possible stuck-at faults leads to increased difficulty of testing the digital circuit.

For the purpose of *test pattern* (i.e., input stimulus) generation it is often assumed that the *circuit under test* (CUT) is either fault-free or it contains only one node which is permanently stuck at a particular logic state. Thus, the most widely used fault model is the so-called *single stuck-at fault* model. Using this model each fault is tested by applying a specific test pattern that, in a good circuit, drives the particular node to the logic state which has the opposite value from the state of the fault assumed to be present in the faulty circuit. For example, to test if node v is stuck at logic state x (denoted by v/x or $v–x$), a test pattern must be used that would cause node v to be driven to the opposite of logic state x if the circuit is not faulty. Thus, the test pattern attempts to show that node v is not stuck at x by driving the node to a value other than x, which for a two-valued digital circuit must be the opposite of x (denoted by $\sim x$). This leads to the requirement that to detect any stuck-at fault v/x, it is necessary to be able to control the logic value at node v so that it can be set to $\sim v$. If the signal value at node v can be observed directly by connecting it to a test equipment, the particular fault v/x can be detected readily. However, in most cases, node v may be an *internal* node, which is inaccessible for direct observation from outside the component package. In that case, it is necessary to create a condition where the value of the signal on an externally observable node, say node t, will be different for each of the two possible values that node v can take on, that is, node t shall be driven to logic state y or $\sim y$ depending upon whether node v is at logic state x or $\sim x$, respectively. Note that x and y may represent the same or different logic states.

The external pins of a component are the only means of applying the stimuli and observing the behavior of that component. During testing, a test pattern is used as the stimulus to detect the presence of a particular fault by causing at least one output pin of the component to take on a different value depending upon whether the targeted fault is present or not. Thus, a test pattern is used for *controlling* the circuit's nodes so that the presence of a fault on a circuit node can be *observed* on at least one of the circuit's external pins. Solving the dual problems of *controllability* and *observability* is the primary objective of all test methods. The *logic-to-pin ratio* of a digital circuit is a relative measure of the ratio of possible faults in the circuit to the number of signal pins (i.e., not including the constant power/ground pins) of that component. A large-value logic-to-pin ratio implies that logic states of a large number of circuit nodes must be controlled using a small number of external pins. As a result, conflicting requirements for controllability and observability become harder to satisfy, and the circuit is considered to be more difficult to test.

Consider Figure 7.10, which depicts a single (hypothetical) *integrated circuit* (IC) component and shows its internal circuitry which uses four NAND gates. The nodes of the circuit are numbered 1 through 12 and the external pins of the component are labeled *A, B,* and *C*. To detect if node 7 is stuck at logic 0 (i.e., 7/0), a test

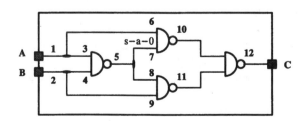

FIGURE 7.10   Example logic circuit with internal node 7 stuck at 0 (7/0).

TABLE 7.2   Test Pattern for Node 7/0 for the Circuit in Figure 7.9

A	B	1	2	3	4	5	6	7	8	9	10	11	12	C
1	0	1	0	1	0	1	1	1	1	0	0	1	1	1 good circuit
1	0	1	0	1	0	1	1	0	1	0	1	1	0	0 with fault 7/0

pattern must be found that sets node 7 (and hence, node 5) to the logic 1 state. This can be achieved by setting either or both of the external pins A and B to the logic 0 state. Furthermore, to observe (or deduce) the value of node 7 at the only externally visible circuit pin, C, it is necessary to create a condition where the logic state of node 12 becomes dependent on the value of node 7. The only path from node 7 to node 12 passes through node 10, and since node 10 is the output of a NAND gate the second input to that gate (i.e., node 6) must be set to the logic 1 state by setting input pin A to the logic 1 state. Therefore, the only possible test pattern for 7/0 is $A = 1$ and $B = 0$. At this point, we must still continue the analysis to see if indeed node 12 will reflect the value of node 7. With input terminals A and B set to logic 1 and logic 0, respectively, node 9 will be set to logic 0, which causes node 11 to become logic 1. With these settings, the value at node 12 will be determined by the value at node 10 and the test pattern is valid. Table 7.2 shows the values of all circuit nodes when this test pattern is applied to the circuit of Figure 7.10.

It should be evident from the simple example of a *combinational circuit* described above that test pattern generation for digital circuits can be very difficult and involved. The problem becomes much more complex when dealing with *sequential circuits,* where the *internal state variables* (i.e., bistable memory storage elements such as latches and flip-flops) must be treated as *pseudo-inputs* and *pseudo-outputs* that must be controlled and observed using the external pins of the component. In this case test patterns become *test sequences* that must be applied in precise order, and outputs must be observed only at prescribed times. Thus, the testing of sequential circuits is much harder to achieve compared to the testing of combinational circuits. Computer programs, called automatic test pattern generation (ATPG) programs, have been developed for generating test patterns for combinational or sequential circuits. By far, the generation of test patterns for combinational circuits is better understood and automated than doing the same for sequential circuits.

Before discussing the various techniques that may be used to improve testability of digital circuits, it is necessary to mention the related problem of determining test effectiveness. A typical digital system contains a very large number of possible stuck-at faults. This and the logical complexity of the circuits make it unacceptable to "guess" how effective the test patterns (or the diagnostic program) will be in detecting all possible faults. This problem is often approached in a formal manner by using a class of test tool called a *fault simulator* program. A fault simulator uses the given set of test patterns to simulate the given circuit first when there are no faults assumed present (i.e., good circuit simulation). Next, the circuit is simulated with the same set of test patterns, but this time the effects of each possible stuck-at fault are considered one at a time. For a given test pattern, and given stuck-at-type fault, if the output of the good circuit simulation differs from the output obtained during fault simulation, then the given fault will be detected by the given test pattern. This way, it is possible to determine the percentage of all possible stuck-at faults that may be present in a digital circuit which will be covered by the given set of test patterns.

Most ATPG programs operate by picking a possible fault from among the possible faults, generating a specific test pattern that covers it, simulating the logic circuit with the newly generated test pattern to determine which other faults are incidentally covered by the same pattern, and continuing the process until all faults have been considered. Of the two related processes of *test pattern generation* and *fault simulation*, the latter is by far the more time-consuming one.

A different approach is taken in some testability analysis tools whereby rather than determining which faults are covered by a given test pattern, the analysis program assigns a numeric value to indicate the degree of difficulty of controlling and observing the digital circuit's nodes. This analysis, which can be done much more quickly compared to performing fault simulation, should be done prior to attempting to generate the test patterns for a circuit so that time will not be spent unnecessarily on digital circuits which are likely to present difficulties for the ATPG/fault-simulation process to deal with.

Design for Testability

Low-cost/high-volume manufacturing requires that product testability be considered up front since a product which is inherently hard to test will cost both time and money to achieve a desired level of quality. There are many steps that can be taken to improve the testability of digital circuits and systems. The following subsections describe some of the techniques that can be used.

Ad-Hoc Techniques [Abramovici et al., 1990; Bardell et al., 1978]

Circuit/System Reset Requirements.   A simple and straightforward mechanism for resetting a digital circuit to a known state is an essential requirement for testability. It should be noted that the requirement is not only for having the reset function provided but further that it should be simple to execute. For example, applying a defined sequence of external signals to a circuit which must be synchronized with a free-running clock signal would not be considered a simple reset mechanism. Instead, keeping an external signal at some logic value for a minimum duration is a much more desirable approach. It is very desirable that the reset function be asynchronous (i.e., not require system clock pulses to execute) since during power-up a circuit may need to be reset even before free-running clock pulses can be started.

Clock Control Requirements.   Another very important requirement for implementing DFT is the ability to control the clocking of the internal logic of the digital circuit. If the external clock signal is gated with some other signals such that it is necessary to determine how to set these other signals to their required values to allow the externally applied clock pulse to reach the internal flip-flop clock terminals, then the ATPG program has another level of constraints to resolve in generating the test patterns. Furthermore, some of these additional requirements may pose difficulties in satisfying them during component and/or system testing. Most ATPG programs assume that once the test pattern has been applied to the pins of the component, the system's response to that pattern can be captured by applying an external clock pulse which enables the internal flip-flops to respond to the test pattern. Thus, the ATPG programs assume that the internal flip-flop clock inputs are controlled directly from an external pin of the component. This very desirable characteristic is often expressed by stating that *externally applied clock pulses are not allowed to be gated by other signals before these reach the clock terminals of the internal flip-flops.* A side benefit of this design rule is that it prevents glitches (i.e., undesirable pulses) which might be generated at the flip-flop clock terminals due to changing the other inputs to the clock gating circuit while the clock pulse is present.

Managing "Unused" Inputs of Components.   When designing digital systems from existing components there may be inputs of those components that, for the current implementation, are not needed. For example, if a two-input AND gate is needed to implement a logic circuit on a printed circuit board, it may be possible to use one of the unused three-input AND gate elements from an IC package already present on that board. In this case, the unused third input of that AND gate must be connected to the logic 1 level in order that a three-input AND function may be implemented using the other two inputs to that gate. Thus, the unused input to the AND gate may be connected directly to the V_{cc} (i.e., power supply) signal. Similarly, if a flip-flop contains unused *preset* or *clear* terminals, these may be tied off to their respective deasserted states. In many cases printed circuit boards are tested using an *in-circuit tester* which uses a *bed-of-nails* test fixture to make physical contact with selected nets on the board so that their values can be observed or controlled by the tester. For the in-circuit tester to control the value of a net it has to backdrive the output of the component which normally drives that net. Since IC components have limited output drive capabilities, the in-circuit tester can overcome the electrical drive from that component and can force that net to a value opposite the value which the driving IC is trying to achieve. By keeping such backdriving conditions to last only a very short period, damage to the opposing IC component is prevented. However, if the net is driven not by an IC but directly from the V_{cc} or ground (*Gnd*) signals, then the in-circuit tester may not be able to overcome their drive. Furthermore, backdriving the V_{cc} or *Gnd* levels would prevent the other IC components from being able to perform their normal functions. Instead, if the logic signals to such unused terminals are applied using *pull-up* or *pull-down* resistors when connecting these to the V_{cc} or *Gnd* levels, respectively, these signals may be controlled by the in-circuit tester. For example, this way it becomes possible to set/reset a flip-flop value by using the normally "unused" preset/clear terminal of that flip-flop. Note that if the flip-flop contains both a preset and a clear input which are unused, these must be pulled up (or pulled down) through separate resistors so that each can be controlled by the in-circuit tester independent of the other. This is illustrated in Figure 7.11.

Synchronous versus Asynchronous Design Style. More than any other issue, discussions concerning synchronous versus asynchronous design style create the most disagreements concerning design for testability. Many logic designers who are experienced in using SSI and MSI IC chips have adapted a design style where synchronous (e.g., clocked) and asynchronous (e.g., self-timed) designs are freely mixed together. Using clocked flip-flops with asynchronous preset/clear inputs is a typical example of this design style. Similarly, building latches out of, say, cross-coupled NAND gates and using these as state variables in implementing finite-state machines used to be a very common technique. However, concerns about system initialization and pattern generation have made this style undesirable for implementing DFT. Indeed, most of the so-called *structured* design styles described below make it a requirement that all internal storage elements be constructed from clocked flip-flops, and feedback loops in combinational circuits are broken with the insertion of such flip-flops, along the feedback paths. Asynchronous circuits suffer from combinational circuit hazards that are glitches created as a result of delay differences along circuit paths. Some hazards may be prevented by constraining the manner (i.e., sequence) in which circuit inputs are allowed to be changed. Whereas such constraints may be met during regular system operation, often test pattern generation algorithms cannot take such constraints into account. Therefore, asynchronous logic may create severe problems during testing.

Avoiding Redundant Logic. Technically speaking, redundancy is the only reason why a given stuck-at fault might not be detectable by any test. For example, if an INVERTER function is implemented by tying both inputs of a two-input NAND gate together, then a stuck-at 1 fault on either one of the inputs becomes undetectable since the output signal can still be determined correctly by the remaining nonfaulty input signal. This creates two problems. First, conventional ATPG programs might spend a lot of time trying to generate a test pattern for such a fault before they declare the fault untestable. Second, the presence of an undetectable fault can cause a detectable fault to become undetectable (it may also cause an undetectable fault to become detectable). For example, consider a parity checking circuit in which an existing stuck-at fault may cause the wrong parity to be generated, and the existence of a second fault may correct the parity and hence hide both failures. The remedy for these situations is to try to avoid redundancy in the first place, and when this is not possible provide additional circuit modes where the redundant circuits might be isolated. Alternately (or in addition) it may be useful to provide additional test points, as described below.

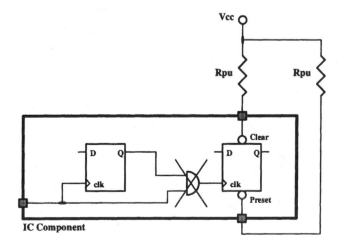

FIGURE 7.11 Using pull-up resistor to tie off unused preset/clear inputs of flip-flops.

Providing Test Points. A test point is an input or output signal to control or observe intermediate signals in a logic circuit. For example, if triple redundancy has been used to implement a fault-tolerant circuit, additional output signals might be provided so that signal values from the identical functional units become individually observable, improving the testability of the overall circuit. Similarly, control signals might be provided so that, during testing, outputs from some functional units may be forced into certain states which allow easier observation of the outputs from other circuits. Recommended sites for inserting test points include redundant nets, nets with large fan-outs, preset and clear inputs of flip-flops, nets that carry system clock signals, (at least some of the) inputs to logic circuit gates with large number of inputs (i.e., large fan-in), data and/or address lines of bus lines, as well as intermediate points in cascaded circuits (such as long ripple counters, shift registers).

Logic Partitioning. Traditionally logic partitioning has been used as a strategy when the circuit is too large/complex for the test generation tools to handle. Thus, its objective is to reduce the number of circuit nodes that must be considered jointly in order to generate test patterns. The partitioning process identifies the *logic cones*, which are sections of logic receiving inputs from multiple input sources and generating a single output. Thus, a digital circuit would be broken into as many individual logic cones as there are individually observable output signals. Obviously, the logic cones may (and often do) overlap with each other since they share common input signals or intermediate signals generated from inside one partition and used in another partition. This is illustrated in Figure 7.12(a), where two overlapping cones of logic are shown. Here, logic cones O_1 and O_2 contain primary inputs I_1, I_2, I_3, I_4 and I_3, I_4, I_5, I_6, respectively. When either partition is dependent on more inputs than what the ATPG tools or the tester can accommodate, it is possible to insert an additional gate, controlled by a tester input in order to test each partition independently of the other. This is illustrated in Figure 7.12(b), where an additional input pin I_t has been added such that with I_t set to logic 0 by the tester, it is possible to test either partition without requiring to control shared inputs I_3 or I_4. Logic partitioning has become more important as a result of increased use of *pseudo-exhaustive testing* (to be described later).

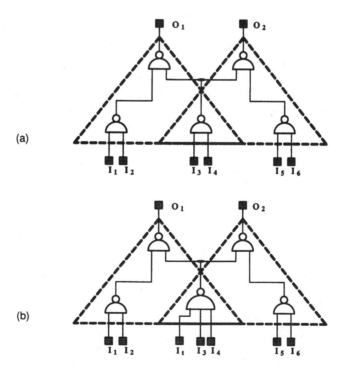

FIGURE 7.12 (a) Logic partitioning with overlapping logic cones. (b) Adding an additional test point to reduce dependence on primary inputs.

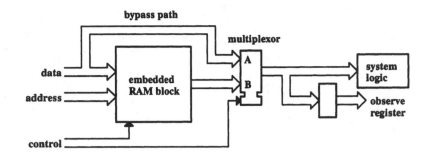

FIGURE 7.13   Providing testability in a design containing an embedded memory block.

Testing Embedded Memory Blocks.   A major testability problem arises when a regular-structure memory block such as random-access memory (RAM) or read-only memory (ROM) is embedded into a logic circuit. This creates three problems:

1. Testing logic that is downstream from the RAM block (i.e., output of RAM block drives the downstream logic) is difficult since this requires setting the test pattern at the RAM outputs. This problem is usually solved by providing a bypass mode where data inputs to the RAM (or ROM) block are channeled directly to the RAM (or ROM) outputs without (or in addition to) being stored inside the RAM block. This way the RAM data outputs can be controlled by controlling the data inputs as desired.
2. Testing logic that is upstream from the RAM block (i.e., outputs from logic circuit are captured by the RAM block) is difficult since the observation point is the RAM block. That is, it is necessary to access the RAM block in order to observe the test results. This problem might be solved by improving the observability of the RAM inputs and/or making the RAM outputs more easily observable as well as providing the *bypass* capability. This way, inputs to the RAM might be bypassed directly to the RAM outputs where they may be observed. This may require adding an *observe-only* register to capture the RAM outputs.
3. Testing of the RAM block itself is difficult since controlling its inputs and observing its outputs require manipulating the upstream and downstream logic circuit blocks, which may be difficult to achieve. Solution to this problem involves providing adequate control of the RAM block inputs (data, address, and read/write control) as well as providing observability of the RAM outputs. In effect, the embedded RAM block can be made testable as if it was a stand-alone block where established memory test algorithms can be applied [Breuer and Friedman, 1976].

Figure 7.13 illustrates how to improve testability of an embedded RAM structure.

Structured Techniques

An alternate approach to improving the testability of digital circuits is to carry out the circuit design by following certain rules that, by construction, assure high testability of the resulting circuits. Since the main problem in achieving testability of a digital circuit is achieving adequate controllability/observability of its internal nodes, structured DFT approaches [Bardell and McAnney, 1978] follow strict design rules that are aimed at achieving this goal. Furthermore, most structured DFT approaches require/recommend additional design rules aimed at preventing incorrect circuit operation as a result of signal races and hazards.

Level-Sensitive Scan Design (LSSD).   Level-sensitive scan design [Eichelberger and Williams, 1978] imposes strict rules on clock signal usage and allows implementing sequential behavior to be implemented only using the shift-register latch (SRL). In the first place, by not allowing any feedback involving combinational circuit elements alone, the LSSD approach prevents timing failures that might be present in purely asynchronous designs. Furthermore, rigid clocking rules are stated in order to prevent SRL data inputs from changing while the clock pulse(s) is (are) transitioning. Hence, the digital circuit is separated into two sections: (1) a robust (i.e., level-sensitive) multi-input/multi-output combinational circuit and (2) a set of SRL elements with which

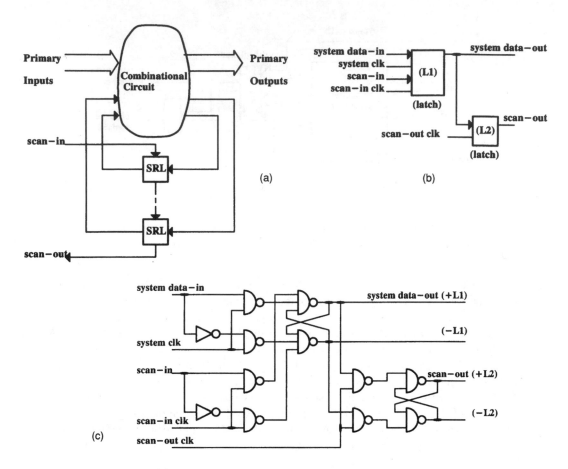

FIGURE 7.14   (a) LSSD circuit model. (b) SRL block diagram. (c) SRL logic diagram.

sequential behavior is implemented. In addition to their normal system interconnections each SRL is also connected to its two neighboring SRLs to form a shift-register structure. The serial shift input and shift output signals are labeled *scan-in* and *scan-out*, respectively, and treated as primary input/output terminals. Figure 7.14 shows an LSSD circuit model and the general form of an SRL. The significance of the shift-register (often referred to as the *scan-register*) structure is that, during testing, it allows each SRL's value to be individually controllable and observable by shifting (i.e., scanning) a serial vector into/out of the scan register. Hence, the SRLs can be treated as *pseudo-input/output terminals,* and the testing of the digital circuit is reduced to that of a combinational circuit only. Figure 7.14(a) shows an LSSD circuit model, and the general form of an SRL is given in Figure 7.14(b). A possible gate-level circuit implementation of an SRL is shown in Figure 7.14(c).

Among the most important LSSD design rules are the following:

1. All internal storage is implemented using SRLs. Each SRL operates such that the L1 latch accepts one or the other of the system data-in or the scan-in data values depending upon whether the system clk or the scan-in clk clock pulse is applied, respectively. The L2 latch accepts the L1 latch value when the scan-out clk clock pulse is applied. The L1 and L2 latches are stable (i.e., cannot change) when the clocks are off.
2. The SRL clocks system clk, scan-in clk, and scan-out clk must be controlled from primary circuit terminals and must be operated in nonoverlapping fashion. This eliminates dependency on minimum circuit delay and assures hazard-free (i.e, level-sensitive) operation.
3. System data-out from SRL_1 may feed the system data-in terminal of SRL_2 only if the system clk which feeds SRL_1 does not overlap with the system clk which feeds SRL_2. This rule prevents the data input to a latch from changing while its clock signal is transitioning.

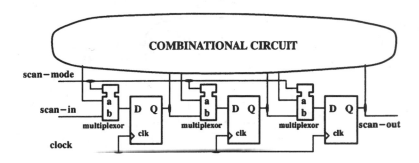

FIGURE 7.15 Model of a digital circuit with scan path.

4. All SRLs are interconnected into one or multiple shift registers by connecting the scan-out terminal from one SRL to the scan-in terminal of the next one in series. If multiple shift registers are implemented, each must be capable of being shifted simultaneously with the others and must have its own scan-in and scan-out primary terminals.

Scan Path. The *scan-path* [Funatsu et al., 1975] approach can be seen as a generalization of the LSSD approach since it follows the same principles but uses standard *D*-type flip-flops as the storage elements instead of the SRLs. The scannable flip-flops can be implemented using dual-ported latches (similar to the L1 latch in the SRL) or using a multiplexor to select between the scan-in and system data-in signals to feed the *D* input of a standard *D*-type flip-flop, as shown in Figure 7.15.

Scan/Set Logic. Scan/set [Stewart, 1977] is another form of implementing scan technology whereby the sequential circuit structure is separated from its accompanying scan/set register. This is illustrated in Figure 7.16. A variation on this scheme is the so-called shadow-register concept that has been implemented in some off-the-shelf IC components [AMDI, 1987].

Random-Access Scan. Random-access scan [Ando, 1980] uses a technique akin to addressing locations in a memory (e.g., RAM) block in order to make the states of all storage elements controllable and observable from primary input/output terminals. Using this approach, each storage element is made individually addressable (i.e., accessible) so that in order to control and/or observe the value of an individual storage element it is not necessary to shift in/shift out all other storage elements as well. Figure 7.17(a) shows the general model of a digital circuit employing the random-access scan approach. A possible gate-level circuit implementation of an addressable latch is given in Figure 7.17(b).

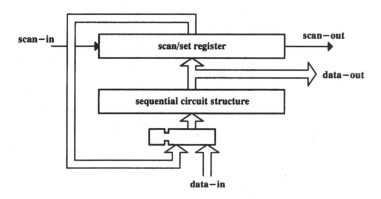

FIGURE 7.16 Generic scan/set circuit design.

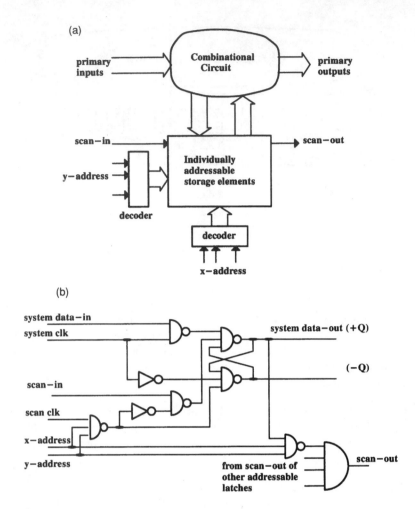

FIGURE 7.17   (a) General model for digital circuit implementing random-access scan. (b) Logic diagram for addressable latch.

Using this approach, each storage element in the circuit is given a unique x/y address and the decoded address signals are connected to the x/y address inputs of the latches. As seen in the circuit of Figure 7.17(b), each latch can then be individually written into using the *scan-in* terminal or its output can be observed using the *scan-out* terminal, provided that the pair of x/y address lines connected to the current latch are both asserted (i.e., set to logic 1). Furthermore, whereas it is also necessary to apply the *scan-in clk* in order to write into the latch, no clock is necessary to observe the latch output. This is a convenient feature that allows the latch values to be selectively observable even while the regular system operations are being executed. The *scan-out* values from the individual latches are combined together into a single AND gate and brought out to a primary output terminal of the circuit. This arrangement works since for any given address only one of the addressable latches will be selected and the scan-out from all other latches will be forced to the logic 1 state. On the other hand, a disadvantage of this approach is that before addressing each latch its proper address must first be applied to the circuit.

Boundary Scan.   Unlike the other scan-based techniques described above, boundary scan [IEEE, 1990] is intended primarily for testing the board-level interconnections among the IC components on a printed circuit board (PCB). In effect, boundary scan is a special form of scan path that is implemented around every *I/O* pin of an IC component in order to provide controllability and observability of the *I/O* pin values during testing.

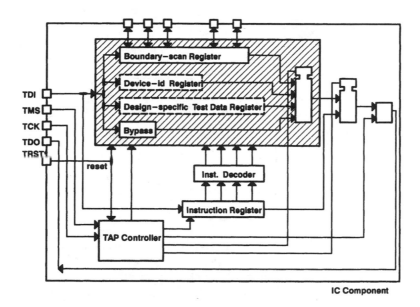

FIGURE 7.18   Architecture of IEEE 1149.1 boundary-scan standard.

Test control signals provided by an on-chip controller are used to disable the boundary-scan cells during regular system operation so that signal values can flow in/out of the IC component without interference from the test circuits. During testing, *output* pin values can be controlled using values preloaded into the boundary-scan register. Similarly, signal values received on the *input* pins can be captured into the boundary-scan register and subsequently shifted out to be observed on an external tester.

Boundary scan has become an important tool in achieving design for testability following the adoption of the IEEE 1149.1 Test Access Port and Boundary-Scan Architecture in 1990. The IEEE 1149.1 Standard defines a mandatory four-pin (plus an optional fifth pin) test access port (TAP) for providing the interface between the IC component and a digital tester. TAP signals comprise test data input (TDI), test data output (TDO), test clock (TCK), and test mode select (TMS) plus an optional asynchronous tap reset (TRST*) signal. The overall IEEE 1149.1 test architecture (see Figure 7.18) includes:

- The TAP
- The TAP controller
- The instruction register (IR)
- A group of mandatory and optional test data registers (TDRs)

The TAP controller is characterized by a 16-state finite-state machine (FSM) whose behavior is defined by the IEEE 1149.1 Standard. State transitions of the TAP FSM are controlled by the TMS input line and the dedicated test clock, TCK. Figure 7.19 shows the state-transition diagram for the TAP FSM.

A most important test data register defined by the IEEE 1149.1 Standard is the boundary-scan register that has individual cells associated with each *I/O* pin of the IC component. Mandatory and permissible features of the boundary-scan register cells are defined by the standard. In addition, a special single-bit register called the BYPASS register has been provided to furnish a more efficient way to shift data through IC components when multiple ICs are chained together by connecting the TDO output from one component to the TDI input of another.

Another mandatory feature of the IEEE 1149.1 Standard is the instruction register and an associated list of mandatory/permissible instructions that govern the behavior of the IC component during testing. The three mandatory instructions are called SAMPLE/PRELOAD, BYPASS, and EXTEST. SAMPLE allows taking a snapshot of the normal operation of the IC, whereas PRELOAD is used for shifting the captured values out while new values are loaded into the boundary-scan register. BYPASS allows shortening the (electrical) distance

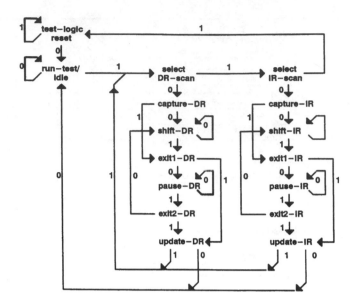

FIGURE 7.19   State-transition diagram for the TAP FSM.

between the TDI and TDO pins by providing a single-bit register as a shortcut during scan operations involving multiple IC components that are connected in series. EXTEST is the "workhorse" instruction that allows driving the signal values on the component's output pads from the boundary register while capturing the input values into their respective cells in the boundary register. This is followed by shifting the captured values out (using the TDO output) while simultaneously shifting in the new driving values (using the TDI input).

An alternative to using boundary scan is to use a "traditional" in-circuit tester that uses a special "bed-of-nails" fixture. In this approach [Parker, 1987], every net on a PCB would be probed using a tester pin which comes in physical contact with that net such that the current signal value of the net can be observed by the tester. The tester can also be used to control the signal values of the individual nets by injecting appropriate currents through the tester pins. However, since each net is already connected to an output pin of a component on the PCB, this approach amounts to *backdriving* the output drivers of IC components and therefore poses a potential risk of damage to the IC components. This approach is becoming more difficult and/or costly to implement as the number of nets goes up and IC pin spacing is reduced. Furthermore, due to fixturing difficulties, double-sided PCBs cannot be tested in this manner. The IEEE 1149.1 boundary-scan standard [IEEE, 1990] helps solve these problems by providing convenient direct access to the *I/O* pins of an IC component without requiring the traditional bed-of-nails fixture.

The "CrossCheck" Technique.   The CrossCheck approach [Gheewala, 1989] uses cells with built-in test points to observe critical signal values. The test points are connected to an underlining grid structure using very small FETs called *cross-point switches*. An on-chip test control circuit generates the necessary signals to address the individual probe lines and capture the results in a *multi-input signature register* (MISR). Test patterns can be generated externally or by using an on-chip pattern generator, and the final test signature (i.e., contents of MISR) can be accessed using dedicated test pins, such as by providing an IEEE 1149.1 TAP (see previous subsection). Figure 7.20 shows how the CrossCheck technique is implemented on an ASIC.

CrossCheck methodology provides a high degree of observability of the ASIC. Since it is not possible to provide observability of all signals of a design, careful analysis must be performed to determine the most effective points for inserting the cross-point switches. Similarly, the size of the grid structure for the probe lines might be chosen to be design-dependent. However, in many instances it may be better to implement the probe lines as part of the IC master slice in order to reduce the amount of customization to a minimum.

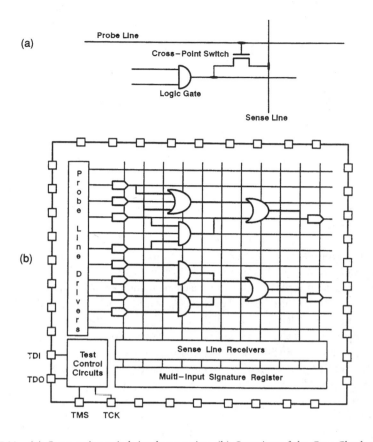

FIGURE 7.20 (a) Cross-point switch implementation. (b) Overview of the CrossCheck technique.

The benefit offered by the CrossCheck technique is due to the potential for the reduced number of test patterns necessary to test an ASIC. This is due to the fact that as observability of internal nodes is increased it becomes easier to generate efficient test patterns which can detect many faults simultaneously. Furthermore, increased observability of internal nodes also improves diagnosability and may help determine the root cause of a failure sooner. On the negative side, the CrossCheck technique does not help improve controllability of internal nodes as achieved using scan-path techniques. Also, a primary disadvantage of the CrossCheck methodology is area penalty due to routing channels that must be set aside for the grid structure. Furthermore, added capacitance of the cross-point switches may affect performance, especially in high-speed applications. In addition, since the technique offers very good observability but no controllability of the internal nodes, it lacks the advantage offered by scan-based approaches for system debug and internal path-delay testing [Dervisoglu and Stong, 1991]. However, recent advances have been made that improve the controllability of internal nodes using the CrossCheck technique in gate-array ICs.

Built-in Self-Test (BIST) Techniques.   The term built-in self-test (or BIST) is a generic name given to any test technique in which an external test resource (e.g., component tester) is not needed to apply test patterns and check a circuit's response to those patterns. This implies that the test patterns must be preloaded into the target device or be generated by the target device itself, in real time. For example, dedicating a section of an IC component for implementing a ROM-based sequencer to apply prestored patterns to test another section of that IC would be classified as a BIST technique. It is often more cost effective to generate the test patterns in real time (i.e., during testing), but in general it is not possible to develop real-time test pattern generation techniques that generate arbitrarily selected test patterns without additionally generating unnecessary ones. Note that whereas storing the test patterns in a ROM might be acceptable in some cases,

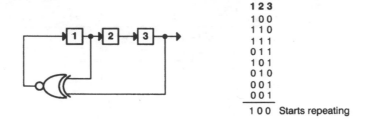

```
1 2 3
1 0 0
1 1 0
1 1 1
0 1 1
1 0 1
0 1 0
0 0 1
0 0 1
─────
1 0 0   Starts repeating
```

FIGURE 7.21   Three-bit maximal-length LFSR.

the size of ROM necessary to store the test patterns prevents this technique being used for implementing BIST in large/complex digital circuits.

One approach to test vector generation is to ignore the specifics of the target circuit and enumerate all possible permutations of inputs. Thus, using *exhaustive* testing, an *n*-input combinational logic cone would be tested by checking its response to all 2**n permutations of input values. In this case, a binary counter can be used as the test pattern generator (TPG). Other, more efficient counter forms (such as a *maximal-length linear feedback shift register,* LFSR) may also be used as the TPG. An LFSR is a special kind of circular-shift register where the serial data input is determined by an EXCLUSIVE-OR function of some of the bit positions. Bit positions which are included in the feedback EXCLUSIVE-OR function are referred to as the tap positions. For any given *degree* (i.e., number of bits) *n* of LFSR there is at least one set of tap positions that result in the LFSR going through all nonzero *n*-bit permutations when it is started in any nonzero state. An LFSR that can go through all 2**n states is called a maximal-length LFSR. Figure 7.21 shows a 3-bit maximal-length LFSR and the state sequence that it produces. Exhaustive testing guarantees that all *detectable* faults which do not transform a combinational circuit into a sequential circuit will be detected. Depending upon the clock frequency, this approach becomes impractical to apply when the number of input variables goes up (usually above 22) [McCluskey, 1984].

In cases where the number of test patterns necessary to achieve exhaustive testing is too large to be applicable, a related technique, called pseudo-random testing, may be used. Pseudo-random testing achieves many of the benefits of exhaustive testing but requires much fewer test patterns. This is achieved by generating the test patterns in random fashion from among the 2**n possible patterns. However, the random generation of test patterns is done using a deterministic algorithm that produces test patterns in repeatable sequence. Before pseudo-random testing is chosen, it is necessary to examine the pseudo-random test resistance of the circuit. For example, if 500,000 pseudo-random test patterns are applied to a 20-input AND gate, there is only a 0.00004% probability that an essential test pattern (which sets all 20 inputs to logic 1) will be included among them.

Yet another related technique is to use *pseudo-exhaustive* testing that aims at breaking a circuit into separate partitions and testing each partition exhaustively [Barzilai et al., 1985; Dervisoglu, 1985; Bardell and McAnney, 1984]. Pseudo-exhaustive testing uses the same techniques used in exhaustive testing for testing the individual partitions without generating test patterns that cover the entire circuit. Mathematical considerations for pseudo-random/pseudo-exhaustive testing are too complex to describe here. The following example is presented for illustration purposes only. Figure 7.22 depicts the combinational portion of a digital circuit consisting of a number of overlapping logic cones that each produce a single output signal. All inputs are assumed to be connected to scannable flip-flops (i.e., pseudo-inputs) or to primary input pins of the component such that all inputs are 100% controllable either by controlling the values in the flip-flops or the primary input pins. All flip-flops are assumed to be scannable and are arranged into a single *scan path* such that the logic cones have *n* or fewer inputs all of which lie within *k* consecutive bits along the scan path. Outputs from the individual logic cones connect (not shown here) to the inputs of flip-flops and/or primary output pins. Thus, all logic cone outputs are also 100% observable. Now, assume that the serial output from the LFSR shown in Figure 7.21 is connected as the "scan-in" input to the scan-path register shown in Figure 7.22. In this case any *consecutive* 3-bit

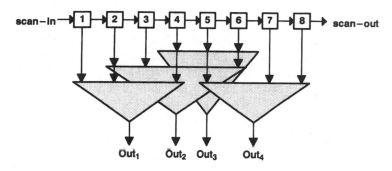

FIGURE 7.22 Overlapping logic cones connected to a common scan path.

partition of the scan-path register will go through the same state sequence as the LFSR itself, delayed from it by the number of flip-flops between that partition and the output bit of the LFSR. For example, the third logic cone that has inputs from flip-flops 4, 5, and 6 will see all input permutations except the all-zeros case which can be applied separately as a special case. On the other hand, the first logic cone, with inputs from flip-flops 1, 2, and 4, will not receive all possible nonzero permutations of three input variables. This is because the first logic cone receives its three inputs from three *nonconsecutive* positions of the scan-path register. In this case only input permutations that have even parity across positions 1, 2, and 4 will be received by the first logic cone. Furthermore, the fourth logic cone that also receives inputs from three nonconsecutive bit positions which are 4 bits apart will receive all 3-bit nonzero input permutations. Analysis of which set of input permutations may be generated across nonconsecutive n bits of a scan-path register which receives the outputs from an mth degree ($m \geq n$) LFSR is based on *linear dependence* and is outside the scope of this section. However, the problem may also be approached statistically by choosing the degree of the LFSR to be higher than n but smaller than k which is the largest span of inputs to any logic cone. For example, in Figure 7.22 the degree of the LFSR may be chosen as 4. In this case, the probability that a logic cone which has 4 or fewer inputs separated by k bits (here, $k = 5$) may be calculated [Lempel and Cohn, 1985]. It should be noted that a logic cone may be tested in full even when it has not received all $2^{**}n$ input permutations.

BIST also requires ability to capture the test results without the need for an external tester. This is often achieved by using a *multi-input signature register* (MISR) to capture individual test results and compress these into an overall value called the test *signature*. Figure 7.23 shows a sample signature register that can compress test results captured from four separate outputs into a single 4-bit signature. Provided that the test circuit has deterministic behavior, a signature register can be started in a given starting state, and its final value may be compared to a known good signature to determine pass/fail status. However, compressing test results into a single overall signature may prevent proper fault detection if multiple erroneous outputs (which may result from the same fault being detected on multiple test vectors) causes the final test signature to be correct even though interim signatures were wrong. The probability that a faulty circuit signature will be the same as the good circuit signature is known as aliasing probability. It can be shown that if the test length is sufficiently long, aliasing probability diminishes toward 2^{-t}, where t is the number of bits of the signature register [Dervisoglu, 1985].

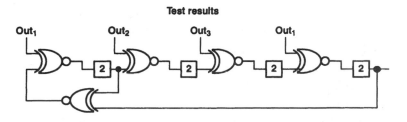

FIGURE 7.23 A four-bit parallel-input signature register.

The two constructs of LFSR and the MISR can be merged into a single multipurpose register in a *built-in logic block observation* (BILBO) approach [Konemann et al., 1979] where each register can have multiple modes of operation including the LFSR mode, MISR mode, SCAN mode, and NORMAL mode. In this case an on-chip test-control circuit may be used to control the modes of operation of the BILBO registers so that, in turn, each register is used as a test pattern generator or signature register to test a digital component. Figure 7.24 illustrates how to use the BILBO scheme in a stepwise fashion to test a large digital circuit.

Path-Delay Testing

Path-delay testing is aimed at testing whether a given component/system operates at a specified performance level that is often measured as the maximum system clock frequency. For example, the lower bound for the maximum clock frequency which a microprocessor IC is specified that it can reach needs to be verified. However, due to the very large number of different operations that a microprocessor can perform it is not practical to verify correct behavior of such a component operating at maximum clock frequency for every possible single operation or sequence of operations that it is designed to perform. On the other hand, it may be possible to examine the structure of the design to discover its *logic paths* and verify that signals can be propagated along these paths within a specified propagational delay time between the initiation of a signal transition at the beginning of the path and the arrival of the final values at the end of that path. This is called *path-delay* testing. A modern IC component with typical complexity would contain many hundreds of

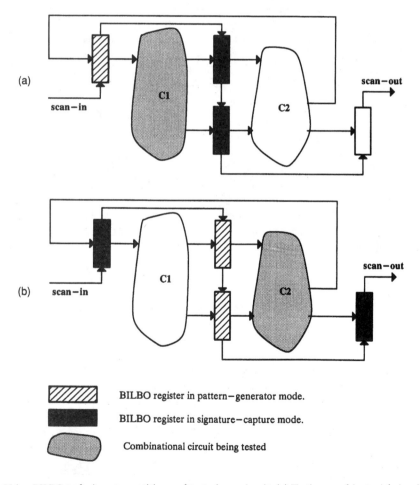

FIGURE 7.24   Using BILBO technique to partition and test a large circuit. (a) Testing combinatorial circuit C_1. (b) Testing combinatorial circuit C_2.

thousands of logic paths, so that it becomes impractical to test all of them for at-speed operation. All *synchronous* digital circuits are designed so that there is a fixed clock period resulting from the use-constant frequency clock signals to time their operation. Obviously, the clock period constitutes an upper bound for the propagational delay through any logic path, since otherwise clock pulses may arrive at the flip-flops while their data input signals may still be transitioning. On the other hand, propagational delay through some logic paths may be very close to this upper bound (i.e., clock period) value whereas others may have more slack in them. It is therefore important to identify the *critical* paths and perform path-delay testing on these. Hence path-delay testing can be broken into the two phases of critical-path selection and path-delay test pattern generation.

Several different approaches can be used in identifying the critical paths, including:

1. Select sufficiently large number of paths selected at random from a list of all logic paths.
2. Calculate worst-case timing for all logic paths and select a certain percentage of the slowest paths.
3. First identify certain key nodes and then select paths that pass through those nodes using either of the two approaches listed in (1) and (2) above.

The more challenging problem is to generate the test patterns to verify that none of the signal propagations along a given logic path require longer than the clock-period time to complete. A path-delay test pattern is a pair of patterns that generates the desired signal transition(s) and provides the sensitization of the signal paths whereby the generated transition(s) is (are) sensitized through the combinational circuit to the input of a flip-flop where it will be captured when the system clock is applied. For example, Figure 7.25 shows a combinational circuit and identifies a specific signal path for which the path delay is to be measured. To determine the appropriate path-delay test patterns, a dummy AND gate is first added to the circuit as shown. An input to the AND gate is derived from the output of the combinational circuit through which the input signal transition is to be propagated. This signal is used in its true or complemented form depending upon whether the final value of the signal transition is a logic 1 or logic 0, respectively. Other inputs to the dummy AND gate come from all remaining inputs of gates through which the desired signal transitions must flow. If the desired signal transition is flowing through an AND or NAND gate, the remaining inputs of these gates are also fed to the inputs of the dummy AND gate, whereas if the desired signal transitions flow through OR or NOR gates, their remaining inputs are inverted and then connected to the inputs of the dummy AND gate. The dummy AND gate is not actually implemented as part of the combinational logic but rather acts as

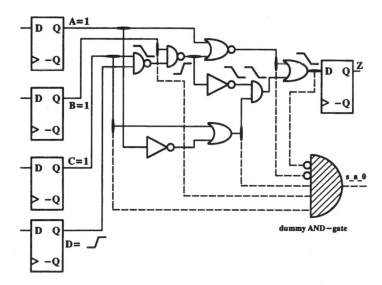

FIGURE 7.25 Circuit example to illustrate path-delay test pattern generation (all flip-flops are clocked using a common clock signal that has not been shown).

a convenient place to collect all the necessary conditions for sensitizing the transitions. For example, in the example given above the first pattern requires input flip-flops A, B, and C all to be set to the logic 1 value in order to sensitize a *low-to-high* transition at the D input, whereas the second test pattern requires A, B, and C all to remain at logic 1 while D is changed from logic 0 to the logic 1 value. This way the transitions created on input D will travel through the identified signal path to reach the destination flip-flop Z.

Path-delay test patterns become much easier to generate and also apply to a circuit if the circuit is designed using scannable flip-flops that are additionally capable of storing two arbitrarily selected values in them. This can be done in such a fashion that the initial value available at the flip-flop output will be replaced by the second value when a first clock pulse is applied, and the flip-flop will revert to its normal mode of operation before the second clock pulse is applied. This way the pair of test patterns that form a path-delay test are first loaded into the flip-flops (using scan) and then two clock pulses are applied at speed. The final result captured by the second clock pulse is then scanned out and examined to determine pass/fail status. It is also possible to get an actual measurement of the path delays by repeating the same test over and over again while systematically reducing the time distance between the two clock pulses to determine the minimum separation of the two clock pulses required for proper operation.

Figure 7.26 shows a modified LSSD latch design [Malaiya and Narayanaswamy, 1983] that can be used to enable path-delay testing as described above. Using this design, it is possible to load any two arbitrary test vectors to the combinational circuit in rapid succession. First, test vector Q_1, Q_2,...,Q_n would be scanned into the L1 latches outputs by using clocks C_3 and C_2. Next, the test vector would be moved into the L2 latches by applying a single C clock. This way the flip-flop outputs would be set to their initial values defined by Q_1, Q_2,..., Q_n. Following this, the second test vector Y_1, Y_2,...,Y_n would be scanned into the L1 latches using clock signals C_3 and C_2. Now applying the C clock causes the first test vector (Q_i) to be replaced by the second test vector (Y_i), and if the C_1 clock is applied next, the response of the combinational circuit will be captured in the L1 latches. This way, the minimum delay between the clock signals C and C_1 that is necessary to allow the signals to propagate through the combinational circuit can be determined. Other flip-flop designs with built-in features to support *double-strobe* testing are also possible [Dervisoglu and Stong, 1991].

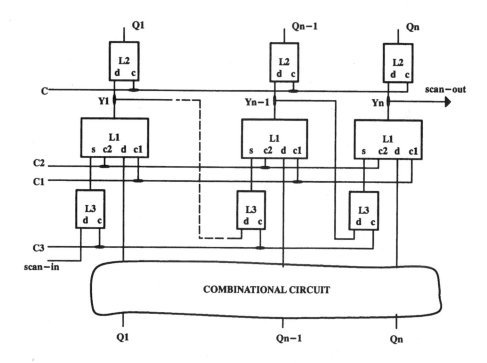

FIGURE 7.26   Using a three-latch flip-flop design to enable path-delay testing.

A different and more difficult-to-use approach for generating test patterns for path-delay measurement is to perform scan-in to load the internal flip-flops with a special pattern that prior circuit analysis will have determined will be transformed into the actually intended test pattern when the first functional clock pulse is applied. The circuit analysis required to use this approach amounts to performing simulation in reverse time flow to determine what state the device under test should be placed in (using scan) so that its next state corresponds to the desired test pattern.

Future for Design for Test

Present-day trends for striving to achieve shorter time to market while at the same time meeting competitive cost demands are going to continue into the foreseeable future. Design for testability is one of several areas that manufacturers from IC components to complete systems are paying increased emphasis to in order to meet their product goals. Twenty years ago some product managers considered testing as being necessary to weed out the bad from the good but did not consider DFT to be adding value to a product. However, since testing is essential, the value of DFT is seen in reducing the cost of an essential item. Hence DFT adds value to a product at least by an amount equal to the savings in test costs that it brings about. Furthermore, DFT improves time to market by making it possible to identify initial production problems at an earlier point in time. For example, initial productions of high-performance ASIC components may contain flaws that prevent their at-speed operation under certain circumstances. If these flaws are not discovered in a timely manner, they may turn into "showstopper" issues causing serious delays in revenue shipments of products. Whereas no "guaranteed" solutions exist to prevent and/or find a solution for all types of problems, design for testability is a rapidly maturing field of digital design.

Defining Terms

Boundary scan: A technique for applying scan design concepts to control/observe values of signal pins of IC components by providing a dedicated boundary-scan register cell for each signal I/O pin.

Built-in self-test (BIST): Any technique for applying prestored or real-time-generated test cases to a subcircuit, IC component, or system and computing an overall pass/fail signature without requiring external test equipment.

Path-delay testing: Any one of several possible techniques to verify that signal transitions created by one clock event will travel through a particular logic/path in a subcircuit, IC component, or system and will reach their final steady-state values before a subsequent clock event.

Pseudo-random testing: A technique that uses a linear feedback shift register (LFSR) or similar structure to generate binary test patterns with statistical distribution of values (0 and 1) across the bits; these patterns are generated without considering the implementation structure of the circuit to which they will be applied.

Scan design: A technique whereby storage elements (i.e., flip-flops) in an IC are connected in series to form a shift-register structure that can be entered into a test mode to load/unload data values to/from the individual flip-flops.

References

M. Abramovici, M.A. Breuer, and A.D. Friedman, *Digital Systems Testing and Testable Design*, Rockville, Md.: Computer Science Press, 1990.

Advanced Micro Devices Inc. [AMDI], "Am29C818 CMOS Pipeline Register with SSR Diagnostics," product specification, Bus Interface Products Data Book, 1987, pp. 47–55.

H. Ando, "Testing VLSI with random access scan," in digest of papers, COMPCON, February 1980, pp. 50–52.

P.H. Bardell and W.H. McAnney, "Parallel pseudorandom test sequences for built-in test," *in Proc. International Test Conference*, October 1984, pp. 302–308.

P.H. Bardell, W.H. McAnney, and J. Savir, *Built-In Test for VLSI. Pseudorandom Techniques,* New York: Wiley, 1978.

Z. Barzilai, D. Coppersmith, and A.L. Rosenberg, "Exhaustive generation of bit patterns with applications to VLSI self-testing," *IEEE Trans. on Computers,* vol. C-32, no. 2, pp. 190–194, February 1985.

M.A. Breuer and A.D. Friedman, *Diagnosis and Reliable Design of Digital Systems,* Rockville, Md.: Computer Science Press, 1976, pp. 139–146, 156–160.

B.I. Dervisoglu, "VLSI self-testing using exhaustive bit patterns," in Proc. IEEE International Conference on Computer Design, October 1985, pp. 558–561.

B.I. Dervisoglu and G.E. Stong, "Design for testability: Using scanpath techniques for path-delay test and measurement," in Proc. International Test Conference, October 1991, pp. 364–374.

E.B. Eichelberger and T.W. Williams, "A logic design structure for LSI testability," *Journal of Design Automation and Fault-Tolerant Computing,* vol. 2, no. 2, pp. 165–178, 1978.

S. Funatsu, N. Wakatsuki, and T. Arima, "Test generation systems in Japan," in Proc. 12th Design Automation Symposium, June 1975, pp. 114–122.

T. Gheewala, "CrossCheck: A cell based VLSI testability solution," in Proc. 26th Design Automation Conference, 1989, pp. 706–709.

"IEEE Standard Test Access Port and Boundary-Scan Architecture," IEEE Std. 1149.1–1990, May 1990.

B. Konemann, J. Mucha, and G. Zwiehoff, "Built-in logic block observation technique," in digest of papers, International Test Conference, October 1979, pp. 37–41.

A. Lempel and M. Cohn, "Design of universal test sequences for VLSI," *IEEE Trans. on Information Theory,* vol. IT-31, no. 1, pp. 10–17, 1985.

Y.K. Malaiya and R. Narayanaswamy, "Testing for timing faults in synchronous sequential integrated circuits," in Proc. International Test Conference, 1983, pp. 560–571.

E.J. McCluskey, "Verification testing. A pseudoexhaustive test technique," *IEEE Trans. on Computers,* vol. C33, no. 6, pp. 541–546, June 1984.

K.P. Parker, *Integrating Design and Test,* New York: IEEE Computer Society Press, 1987.

J.H. Stewart, "Future testing of large LSI circuit cards," in Proc. Semiconductor Test Symposium, Cherry Hill, N.J., October 1977, pp. 6–15.

Further Information

An excellent treatment of design for testability topics is found in Abramovici et al. [1990]. Also, Breuer and Friedman [1976] provide a very good treatment of pseudo-random test topics.

C.M. Maunder and R.E. Tulloss (*The Test Access Port and Boundary-Scan Architecture,* IEEE Computer Society Press Tutorial, 1990) provide a user's guide for boundary-scan and the IEEE 1149.1 Standard.

B.I. Dervisoglu ("Using Scan Technology for Debug and Diagnostics in a Workstation Environment," in Proc. International Test Conference, 1988, pp. 976–986) provides a very good example of applying DFT techniques all the way from the IC component level to the system level. Also, B.I. Dervisoglu ("Scan-Path Architecture for Pseudorandom Testing," *IEEE Design & Test of Computers,* vol. 6, no. 4, pp. 32–48, August 1989) describes using pseudo-random testing at the system level. Similarly, P.H. Bardell and M.J. Lapointe ("Production Experience with Built-in Self-Test in the IBM ES/9000 System," in Proc. International Test Conference, October 1991, pp. 28–36) describe application of BIST for testing a commercial product at the system level.

Computer Engineering

8
Organization

Richard F. Tinder
Washington State University

S.N. Yanushkevich
University of Calgary

Carl Hamacher
Queen's University

Zvonko Vranesic
University of Toronto

Safwat Zaky
University of Toronto

Jacques Raymond
University of Ottawa

8.1 Number Systems

Richard F. Tinder

Number systems provide the basis for conveying and quantifying information. Weather data, stocks, pagination of books, weights and measures—these are just a few examples of the use of numbers that affect our daily lives. For this purpose we find the decimal (or arabic) number system to be reliable and easy to use. This system evolved presumably because early humans were equipped with a crude type of calculator, their ten fingers. A number system that is appropriate for humans, however, may be intractable for use by a machine such as a computer. Likewise, a number system appropriate for a machine may not be suitable for human use.

Before concentrating on those number systems that are useful in computers, it will be helpful to review the characteristics that are desirable in any number system. There are *four* important characteristics in all:

- Distinguishability of symbols
- Arithmetic operations capability
- Error control capability
- Tractability and speed

To one degree or another the decimal system of numbers satisfies these characteristics for hard-copy transfer of information between humans. Roman numerals and **binary** are examples of number systems that do not satisfy all four characteristics for human use. On the other hand, the binary number system is preferable for use in digital computers. The reason is simply put: current digital electronic machines recognize only two identifiable states physically represented by a high voltage level and a low voltage level. These two physical states are logically interpreted as the binary symbols 1 and 0.

A fifth desirable characteristic of a number system to be used in a computer should be that it have a minimum number of easily identifiable states. The binary number system satisfies this condition. However, the digital computer must still interface with humankind. This is done by converting the binary data to a decimal and character-based form that can be readily understood by humans. A minimum number of identifiable characters (say 1 and 0, or true and false) is not practical or desirable for direct human use. If this is difficult to understand, imagine trying to complete a tax form in binary or in any number system other than decimal. On the other hand, use of a computer for this purpose would not only be practical but, in many cases, highly desirable.

Positional and Polynomial Representations

The *positional form* of a number is a set of side-by-side (juxtaposed) digits given generally in *fixed-point* form as

$$
\begin{array}{c}
\text{MSD} \qquad\qquad \text{Radix Point} \qquad\qquad \text{LSD} \\[4pt]
Nr = (a_{n-1} \ldots a_3 a_2 a_1 a_0 \cdot a_{-1} a_{-2} a_{-3} \ldots a_{-m})_r \\[4pt]
\underbrace{\qquad\qquad}_{\text{Integer}} \qquad \underbrace{\qquad}_{\text{Fraction}}
\end{array}
\tag{8.1}
$$

where the **radix** (or base) r is the total number of digits in the number system and a is a digit in the set defined for radix r. Here, the radix point separates n integer digits on the left from m fraction digits on the right. Notice that a_{n-1} is the most significant (highest-order) digit, called MSD, and that a_{-m} is the least significant (lowest-order) digit, denoted by LSD.

The *value* of the number in Equation (8.1) is given in *polynomial form* by

$$
\begin{aligned}
N_r &= \sum_{i=-m}^{n-1} a_i r^i \\
&= \begin{pmatrix} a_{n-1} r^{n-1} + \cdots + a_2 r^2 + a_1 r^1 + a_0 r^0 \\ + a_{-1} r^{-1} + a_{-2} r^{-2} + \cdots + a_{-m} r^{-m} \end{pmatrix}_r
\end{aligned}
\tag{8.2}
$$

where a_i is the digit in the ith position with a *weight* r^i.

Application of Equation (8.1) and Equation (8.2) follows directly. For the decimal system $r = 10$, indicating that there are 10 distinguishable characters recognized as decimal numerals $0, 1, 2, \ldots, r - 1(= 9)$. Examples of the positional and polynomial representations for the decimal system are

$$
\begin{aligned}
N_{10} &= (d_3 d_2 d_1 d_0 . d_{-1} d_{-2} d_{-3})_{10} \\
&= 3017.528
\end{aligned}
$$

and

$$
\begin{aligned}
N_{10} &= \sum_{i=-3}^{n-1} d_i 10^i \\
&= 3 \times 10^3 + 0 \times 10^2 + 1 \times 10^1 + 7 \times 10^0 + 5 \times 10^{-1} + 2 \times 10^{-2} + 8 \times 10^{-3} \\
&= 3000 + 10 + 7 + 0.5 + 0.02 + 0.008
\end{aligned}
$$

where d_i is the decimal digit in the ith position. Exclusive of possible leading and trailing zeros, the MSD and LSD for this number are 3 and 8, respectively. This number could have been written in a form such as $N_{10} = 03017.52800$ without altering its value but implying greater accuracy of the fraction portion.

Unsigned Binary Number System

Applying Equation (8.1) and Equation (8.2) to the binary system requires that $r = 2$, indicating that there are two distinguishable characters, typically 0 and $(r - 1) = 1$, that are used. In positional representation these characters (numbers) are called *binary digits* or *bits*. Examples of the positional and polynomial notations for a binary number are

$$N_2 = (b_{n-1} \ldots b_3 b_2 b_1 b_0 . b_{-1} b_{-2} b_{-3} \ldots b_{-m})_2$$

$$= 101101.101_2$$

$$\text{MSB} \longrightarrow \qquad \qquad \longleftarrow \text{LSB}$$

and

$$N = \sum_{i=-m}^{n-1} b_i 2^i$$
$$= 1 \times 2^5 + 0 \times 2^4 + 1 \times 2^3 + 1 \times 2^2 + 0 \times 2^1 + 1 \times 2^0 + 1 \times 2^{-1} + 0 \times 2^{-2} + 1 \times 2^{-3}$$
$$= 32 + 8 + 4 + 1 + 0.5 + 0.125$$
$$= 45.625_{10}$$

where b_i is the bit in the ith position. Thus, the bit positions are weighted $\ldots$, 16, 8, 4, 2, 1, $\frac{1}{2}$, $\frac{1}{4}$, $\frac{1}{8}$, $\ldots$ for any number consisting of integer and fraction portions. Binary numbers so represented are sometimes referred to as *natural* binary. In positional representation the bits on the extreme left and extreme right are called the MSB (most significant bit) and LSB (least significant bit), respectively. Notice that by obtaining the value of a binary number a conversion from binary to decimal has been performed. The subject of radix (base) conversion will be dealt with more extensively later.

For reference purposes Table 8.1 provides the binary-to-decimal conversion for two-, three-, four-, five-, and six-bit binary. The six-bit binary column is only halfway completed for brevity.

In the natural binary system the number of bits in a unit of data is commonly assigned a name. Examples are:

- 4-data-bit unit: nibble (or half-byte)
- 8-data-bit unit: byte
- 16-data-bit unit: two bytes (or half-word)
- 32-data-bit unit: word (or four bytes)
- 64-data-bit unit: double-word, etc.

The word size for a computer is determined by the number of bits that can be manipulated and stored in registers. The foregoing list of names would be applicable to a 32-bit computer.

Unsigned Binary-Coded Decimal, Hexadecimal, and Octal Systems

While the binary system of numbers is most appropriate for use in computers, it has several disadvantages when used by humans who have become accustomed to the decimal system. For example, binary machine code is long, difficult to assimilate, and tedious to convert to decimal. However there exist simpler ways to

TABLE 8.1   Binary-to-Decimal Conversion

Two-Bit Binary	Decimal Value	Three-Bit Binary	Decimal Value	Four-Bit Binary	Decimal Value	Five-Bit Binary	Decimal Value	Six-Bit Binary	Decimal Value
00	0	000	0	0000	0	10000	16	100000	32
01	1	001	1	0001	1	10001	17	100001	33
10	2	010	2	0010	2	10010	18	100010	34
11	3	011	3	0011	3	10011	19	100011	35
		100	4	0100	4	10100	20	100100	36
		101	5	0101	5	10101	21	100101	37
		110	6	0110	6	10110	22	100110	38
		111	7	0111	7	10111	23	100111	39
				1000	8	11000	24	101000	40
				1001	9	11001	25	101001	41
				1010	10	11010	26	101010	42
				1011	11	11011	27	101011	43
				1100	12	11100	28	101100	44
				1101	13	11101	29	101101	45
				1110	14	11110	30	101110	46
				1111	15	11111	31	101111	47
								.	.
								.	.
								.	.

represent binary numbers for conversion to decimal representation. Three examples, commonly used, are natural binary-coded decimal (NBCD), binary-coded **hexadecimal** (BCH), and binary-coded **octal** (BCO). These number systems are useful in applications where a digital device, such as a computer, must interface with humans. The NBCD code representation is also useful in carrying out computer arithmetic.

The NBCD Representation

The BCD system as used here is actually an 8, 4, 2, 1 weighted code called *natural* BCD or NBCD. This system uses patterns of four bits to represent each decimal position of a number and is one of several such weighted BCD code systems. The NBCD code is converted to its decimal equivalent by polynomials of the form

$$N_{10} = b_3 \times 2^3 + b_2 \times 2^2 + b_1 \times 2^1 + b_0 \times 2^0$$
$$= b_3 \times 8 + b_2 \times 4 + b_1 \times 2 + b_0 \times 1$$

for any $b_3 b_2 b_1 b_0$ code integer. Thus, decimal 6 is represented as $(0 \times 8) + (1 \times 4) + (1 \times 2) + (0 \times 1)$, or 0110 in NBCD code. Like natural binary, NBCD code is also called "natural" because its bit positional weights are derived from integer powers of 2^n. Table 8.2 shows the NBCD bit patterns for decimal integers 0 through 9.

The NBCD code is currently the most widely used of the BCD codes. There are many excellent sources of information on BCD codes. One, in particular, provides a fairly extensive coverage of both weighted and unweighted BCD codes [Tinder, 1991].

Decimal numbers greater than 9 or less than 1 can be represented by the NBCD code if each digit is given in that code and if the results are combined. For example, the number 63.98 is represented by (or converted to) NBCD code as

$$
\begin{array}{ccccc}
& 6 & 3 & \cdot & 9 & 8 \\
63.98_{10} = & 0110 & 0011 & \cdot & 1001 & 1000)_{NBCD}
\end{array}
$$
$$= 11100011.10011_{NBCD}$$

TABLE 8.2 NBCD Bit Patterns and Decimal Equivalent

NBCD Bit Pattern	Decimal	NBCD Bit Pattern	Decimal
0000	0	1000	8
0001	1	1001	9
0010	2	1010	NA
0011	3	1011	NA
0100	4	1100	NA
0101	5	1101	NA
0110	6	1110	NA
0111	7	1111	NA

NA = not allowed.

Here, the code weights are 80, 40, 20, 10; 8, 4, 2, 1; 0.8, 0.4, 0.2, 0.1; and 0.08, 0.04, 0.02, 0.01 for the tens, units, tenths, and hundredths digits, respectively, representing four decades. Conversion between binary and NBCD requires conversion to decimal as an intermediate step. For example, to convert from NBCD to binary requires that groups of four bits be selected in both directions from the radix point to form the decimal number. If necessary, zeros are added to the leftmost or rightmost ends to complete the groups of four bits as in the above example. Negative NBCD numbers can be represented either in sign-magnitude notation or 1's or 2's **complement** notation as discussed later.

Another BCD code that is used for number representation and manipulation is called excess 3 BCD (or XS3 NBCD, or simply XS3). XS3 is an example of a *biased-weighted* code (a bias of 3). This code is formed by adding 0011_2 ($= 3_{10}$) to the NBCD bit patterns in Table 8.2. Thus, to convert XS3 to NBCD code, 0011 must be subtracted from XS3 code. In four-bit quantities the XS3 code has the useful feature that when adding two numbers together in XS3 notation a carry will result and yield the correct value any time a carry results in decimal (i.e., when 9 is exceeded). This feature is not shared by either natural binary or NBCD addition.

The Hexadecimal and Octal Systems

The hexadecimal number system requires that $r = 16$ in Equation (8.1) and Equation (8.2), indicating that there are 16 distinguishable characters in the system. By convention, the permissible hexadecimal digits are 0, 1, 2, 3, 4, 5, 6, 7, 8, 9, A, B, C, D, E, and F for decimals 0 through 15, respectively. Examples of the positional and polynomial representations for a hexadecimal number are

$$N_{16} = (h_{n-1} \ldots h_3 h_2 h_1 h_0 . h_{-1} h_{-2} h_{-3} \ldots h_{-m})_{16}$$
$$= (\text{AF3.C8})_{16}$$

with a decimal value of

$$N = \sum_{i=-m}^{n-1} h_i 16^i$$
$$= 10 \times 16^2 + 15 \times 16^1 + 3 \times 16^0 + 12 \times 16^{-1} + 8 \times 16^{-2}$$
$$= 2803.78125_{10}$$

Here, it is seen that a hexadecimal number has been converted to decimal by using Equation (8.2).

The octal number system requires that $r = 8$ in Equation (8.1) and Equation (8.2), indicating that there are eight distinguishable characters in this system. The permissible octal digits are 0, 1, 2, 3, 4, 5, 6, and 7, as one might expect. Examples of the application of Equation (8.1) and Equation (8.2) are

$$N_8 = (o_{n-1} \ldots o_3 o_2 o_1 o_0 . o_{-1} o_{-2} o_{-3} \ldots o_{-m})_8$$
$$= 501.74_8$$

with a decimal value of

$$N = \sum_{i=-m}^{n-1} o_i 8^i$$
$$= 5 \times 8^2 + 0 \times 8^1 + 1 \times 8^0 + 7 \times 8^{-1} + 4 \times 8^{-2}$$
$$= 321.9375_{10}$$

When the hexadecimal and octal number systems are used to represent bit patterns in binary, they are called binary-coded hexadecimal (BCH) and binary-coded octal (BCO), respectively. These two number systems are examples of *binary-derived radices*. Table 8.3 lists several selected examples showing the relationships between BCH, BCO, binary, and decimal.

What emerges on close inspection of Table 8.3 is that each hexadecimal digit corresponds to four binary digits and that each octal digit corresponds to three binary digits. The following example illustrates the relationships between these number systems:

$$
\begin{aligned}
10110111111.11011_2 &= \overset{5}{0101}\ \overset{B}{1011}\ \overset{F}{1111} : \overset{D}{1101}\ \overset{8}{1000} \\
&= 5BF.D8_{16} \\
&= \overset{2}{010}\ \overset{6}{110}\ \overset{7}{111}\ \overset{7}{111} : \overset{6}{110}\ \overset{6}{110} \\
&= 2677.66_8 \\
&= 1471.84375_{10}
\end{aligned}
$$

To separate the binary digits into groups of four (for BCH) or groups of three (for BCO), counting must begin from the radix point and continue outward in both directions. Then, where needed, zeros are added to the leading and trailing ends of the binary representation to complete the MSDs and LSDs for the BCH and BCO forms.

Conversion between Number Systems

It is not the intent of this section to cover all methods for radix (base) conversion. Rather, the plan is to provide general approaches, separately applicable to the integer and fraction portions, followed by specific examples.

TABLE 8.3   The BCH and BCO Number Systems

Binary	BCH	BCO	Decimal	Binary	BCH	BCO	Decimal
0000	0	0	0	1010	A	12	10
0001	1	1	1	1011	B	13	11
0010	2	2	2	1100	C	14	12
0011	3	3	3	1101	D	15	13
0100	4	4	4	1110	E	16	14
0101	5	5	5	1111	F	17	15
0110	6	6	6	10000	10	20	16
0111	7	7	7	11011	1B	33	27
1000	8	10	8	110001	31	61	49
1001	9	11	9	1001110	4E	116	78

Conversion of Integers

Since the polynomial form of Equation (8.2) is a geometrical progression, the integer portion can be represented in *nested radix* form. In source radix s, the nested representation is

$$N_s = (a_{n-1}s^{n-1} + a_{n-2}s^{n-2} + \cdots + a_1 s^1 + a_0 s^0)_s$$
$$= a_0 + s(a_1 + s(a_2 + \cdots + a_{n-1}))_s$$
$$= a_0 + s\left(\sum_{i=1}^{n-1} a_i s^{i-1}\right) \tag{8.3}$$

for digits a_i having integer values from 0 to $s-1$. The nested radix form not only suggests a conversion process but also forms the basis for computerized conversion.

Consider that the number in Equation (8.3) is to be represented in nested radix r form

$$N_r = b_0 + r(b_1 + r(b_2 + \cdots + b_{m-1}))_r$$
$$= b_0 + r\left(\sum_{i=1}^{m-1} b_i r^{i-1}\right) \tag{8.4}$$

where, in general, $m \neq n$. Then, if N_s is divided by r, the results are of the form

$$\frac{N_s}{r} = Q + \frac{R}{r} \tag{8.5}$$

where Q is the integer quotient rearranged as $Q_0 = b_1 + r(b_2 + \ldots + b_{m-1})_r$ and R is the remainder $R_0 = b_0$. A second division by r yields $Q_0/r = Q_1 + R_1/r$, where Q_1 is arranged as $Q_1 = b_2 + r(b_3 + \ldots + b_{m-1})_r$ and $R_1 = b_1$. Thus, by repeated division of the integer result Q_i by r, the remainders yield $(b_0, b_1, b_2, \ldots, b_{m-1})_r$ in that order.

The conversion method just described, called the *radix divide method*, can be used to convert between any two integers of different radices. However, the requirement is that *the arithmetic required by N_s/r must be carried out in source radix, s*. Except for source radices 10 and 2, this poses a severe problem for humans. Table 8.4 provides the recommended procedures for integer conversion. The radix divide method is suitable for computer conversion providing, of course, that the computer is programmed to carry out the arithmetic in different radices.

The integer conversion methods of Table 8.4 can be illustrated by the following simple examples:

Example 8.1 $139_{10} \rightarrow N_2$

N/r	Q	R
$139/2 =$	69	1
$69/2 =$	34	1
$34/2 =$	17	0
$17/2 =$	8	1
$8/2 =$	4	0
$4/2 =$	2	0
$2/2 =$	1	0
$1/2 =$	0	1

$139_{10} = 10001011_2$

Example 8.2 $10001011_2 \rightarrow N_{10}$. By positional weights,

$$N_{10} = 128 + 8 + 2 + 1 = 139_{10}$$

TABLE 8.4 Summary of Recommended Methods for Integer Conversion by Noncomputer Means

Integer Conversion	Conversion Method
$N_{10} \rightarrow N_r$	Radix division by radix r using Equation (8.5)
$N_s \rightarrow N_{10}$	Equation (8.2) or Equation (8.3)
$N_s)_{s \neq 10} \rightarrow N_r)_{r \neq 10}$	$N_{10} \rightarrow N_r$ by Equation (8.2) or (8.3)
	$N_{10} \rightarrow N_r$ radix division by r using Equation (8.5)

Special Cases for Binary Forms

$N_2 \rightarrow N_{10}$	Positional weighting
$N_2 \rightarrow N_{BCH}$	Partition N_2 into groups of four bits starting from radix point, then apply Table 8.3
$N_2 \rightarrow N_{BCO}$	Partition N_2 into groups of three bits starting from radix point, then apply Table 8.3
$N_{BCH} \rightarrow N_2$	Reverse of $N_2 \rightarrow N_{BCH}$
$N_{BCO} \rightarrow N_2$	Reverse of $N_2 \rightarrow N_{BCO}$
$N_{BCH} \rightarrow N_{BCO}$	$N_{BCH} \rightarrow N_2 \rightarrow N_{BCO}$
$N_{BCO} \rightarrow N_{BCH}$	$N_{BCO} \rightarrow N_2 \rightarrow N_{BCH}$
$N_{NBCD} \rightarrow N_{XS3}$	Add $0011_2 (= 3_{10})$ to N_{NBCD}
$N_{XS3} \rightarrow N_{NBCD}$	Subtract $0011_2 (= 3_{10})$ from N_{NBCD}

Example 8.3 $139_{10} \rightarrow N_8$

$$
\begin{array}{lll}
N/r & Q & R \\
139/8 = & 17 & 3 \\
17/8 = & 2 & 1 \\
2/8 = & 0 & 2 \quad 139_{10} = 213_8
\end{array}
$$

Example 8.4 $10001011_2 \rightarrow N_{BCO}$

$$
\begin{array}{ccc}
2 & 1 & 3 \\
010 & 001 & 011 = 213_{BCO}
\end{array}
$$

Example 8.5 $213_{BCO} \rightarrow N_{BCH}$

$$
\begin{array}{ccccc}
2 & 1 & 3 & 8 & B
\end{array}
$$
$$
213_{BCO} = 010\,001\,011 = 10001011_2 = 1000\,1011 = 8B_{16}
$$

Example 8.6 $213_8 \rightarrow N_5$

$$
213_8 = 2 \times 8^2 + 1 \times 8^1 + 3 \times 8^0 = 139_{10}
$$

$$
\begin{array}{lll}
N/r & Q & R \\
139/5 = & 27 & 4 \\
27/5 = & 5 & 2 \\
5/5 = & 1 & 0 \\
1/5 = & 0 & 1 \quad\quad 213_8 = 1024_5
\end{array}
$$

Check: $1 \times 5^3 + 2 \times 5^1 + 4 \times 5^0 = 125 + 10 + 4 = 139_{10}$

Conversion of Fractions

By extracting the fraction portion from Equation (8.2) one can write

$$.N_s = (a_{-1}s^{-1} + a_{-2}s^{-2} + \cdots + a_{-m}s^{-m})_s$$
$$= s^{-1}(a_{-1} + s^{-1}(a_{-2} + \cdots + a_{-m}))_s$$
$$= s^{-1}\left(a_{-1} + \sum_{i=2}^{m} a_{-i}s^{-i+1}\right)_s \tag{8.6}$$

in radix s. This is called the *nested inverse radix* form that provides the basis for computerized conversion.

If the fraction in Equation (8.6) is represented in nested inverse radix r form, then

$$.N_r = r^{-1}(b^{-1} + r^{-1}(b^{-2} + \cdots + b^{-p}))_r$$
$$= r^{-1}\left(b_{-1} + \sum_{i=2}^{p} b_{-1}r^{-i+1}\right)_r \tag{8.7}$$

for any fraction represented in radix r. Now, if N_s is multiplied by r, the result is of the form

$$.N_s \times r = I \times F \tag{8.8}$$

where I is the product integer, $I_1 = b_{-1}$, and F_0 is the product fraction arranged as $F_1 = r^{-1}(b_{-2} + r^{-1}(b_{-3} + \ldots + b_{-p}))_r$. By repeated multiplication by r of the remaining fractions F_i, the resulting integers yield $(b_{-1}, b_{-2}, b_{-3}, \ldots b_{-m})_r$, in that order.

The conversion just described is called the *radix multiply method* and is perfectly general for converting between fractions of different radices. However, as in the case of integer conversion, the requirement is that *the arithmetic required by* $.N_s \times r$ *must be carried out in source radix, s.* For noncomputer use by humans, this procedure is usually limited to fraction conversions $N_{10} \rightarrow N_r$, where the source radix is 10 (decimal). The recommended methods for converting between fractions of different radices are given in Table 8.5. The radix multiply method is well suited to computer use.

For any integer of radix s, there exists an exact representation in radix r. This is not the case for a fraction whose conversion is a geometrical progression that never converges. Terminating a fraction conversion at n digits (to the right of the radix point) results in an error or uncertainty. In decimal, this error is given by

$$\epsilon_{10} = a_{-n}r^{-n} + a_{-(n+1)}r^{-(n+1)} + a_{-(n+2)}r^{-(n+2)} + \cdots$$
$$= r^{-n}\left[a_{-n} + \sum_{i=1}^{\infty} a_{-(n+i)}r^{-(n+i)}\right]_r$$

where the quantity in brackets approaches the value of $a_{-n} + 1$. Therefore, terminating a fraction conversion at n digits from the radix point results in an error with bounds

$$0 < \epsilon_{10} \leqslant r^{-n}(a_{-n} + 1) \tag{8.9}$$

in decimal. Equation (8.9) is useful in deciding when to terminate a fraction conversion.

Often, it is desirable to terminate a fraction conversion at $(n + 1)$ digits and then round off to n from the radix point. A suitable method for rounding to n digits in radix r is: Perform the fraction conversion to $(n + 1)$

TABLE 8.5   Summary of Recommended Methods for Fraction Conversion by Noncomputer Means

Fraction Conversion	Conversion Method
$.N_{10} \rightarrow .N_r$	Radix multiplication by using Equation (8.8)
$.N_s \rightarrow .N_{10}$	Equation (8.2) or Equation (8.6)
$.N_r)_{s \neq 10} \rightarrow .N_r)_{r \neq 10}$	$N_s \rightarrow N_{s10}$ by Equation (8.2) or Equation (8.6)
	$N_{10} \rightarrow N_r$ radix multiply by Equation (8.8)
Special Cases for Binary Forms	
$.N_2 \rightarrow .N_{BCH}$	Partition $.N_2$ into groups of four bits from radix point, then apply Table 8.3
$.N_2 \rightarrow .N_{BCO}$	Partition $.N_2$ into groups of three bits from radix point, then apply Table 8.3
$.N_{BCH} \rightarrow .N_2$	Reverse of $.N_2 \rightarrow .N_{BCH}$
$.N_{BCO} \rightarrow .N_2$	Reverse of $.N_2 \rightarrow .N_{BCO}$
$.N_{BCH} \rightarrow .N_{BCO}$	$.N_{BCH} \rightarrow .N_2 \rightarrow .N_{BCO}$
$.N_{BCO} \rightarrow .N_{BCH}$	$.N_{BCO} \rightarrow .N_2 \rightarrow .N_{BCH}$

digits from the radix point, then drop the $(n+1)$ digit if $a_{-(n+1)} < r/2$, or add $r^{-(n-1)}$ to the result if $a_{-(n+1)} \geqslant r/2$.

After rounding off to n digits, the maximum error becomes the difference between the rounded result and the smallest value possible. By using Equation (8.9), this difference is

$$\epsilon_{max} = r^{-n}(a_{-n} + 1) - r^{-n}(a_{-n} + a_{-(n+1)}/r)$$
$$= r^{-n}(1 - a_{-(n+1)}/r)$$

Then, by rounding to n digits, there results an error with bounds

$$0 < \epsilon_{10} \leqslant r^{-n}(1 - a_{-(n+1)}/r) \tag{8.10}$$

in decimal. If $a_{-(n+1)} < r/2$ and the $(n+1)$ digit is dropped, the maximum error is r^{-n}. Note that for $N_s \rightarrow N_{10} \rightarrow N_r$ type conversions, the bounds of errors aggregate.

The following examples illustrate the fraction conversion methods of Table 8.5.

Example 8.7 $0.654_{10} \rightarrow N_2$ rounded to eight bits

$.N_s \times r$	F	I	
0.654×2	0.308	1	
0.308×2	0.616	0	
0.616×2	0.232	1	
0.232×2	0.464	0	
0.464×2	0.928	0	
0.928×2	0.856	1	$0.654_{10} = 0.10100111_2$
0.856×2	0.712	1	
0.712×2	0.424	1	
0.424×2	0.848	0	$\epsilon_{max} = 2^{-8}$

Example 8.8 $0.654_{10} \rightarrow N_8$ terminated at four digits

$.N_s \times r$	F	I
0.654×8	0.232	5
0.232×8	0.856	1
0.856×8	0.848	6
0.848×8	0.784	6

$0.654_{10} = 5166_8$
with error bounds
$0 < \epsilon_{10} \leqslant 7 \times 8^{-4} = 1.71 \times 10^{-3}$

Example 8.9 $0.5166_8 \rightarrow N_2$ rounded to eight bits and let $0.5166_8 \rightarrow N_{10}$ be rounded to four decimal places.

$$0.5166_8 = 5 \times 8^{-1} + 1 \times 8^{-2} + 6 \times 8^{-3} + 6 \times 8^{-4}$$
$$= 0.625000 + 0.015625 + 0.011718 + 0.001465$$
$$= 0.6538 \text{ rounded to four decimal places; } \epsilon_{10} \leqslant 10^{-4}$$

$.N_s \times r$	F	I
0.6538×2	0.3076	1
0.3076×2	0.6152	0
0.6152×2	0.2304	1
0.2304×2	0.4608	0
0.4608×2	0.9216	0
0.9216×2	0.8432	1
0.8432×2	0.6864	1
0.6864×2	0.3728	1
0.3728×2	0.7457	0

$0.5166_8 = 0.10100111_2$ (compare with Example 7)

$\epsilon_{10} \leq 10^{-4} + 2^{-8} = 0.0040$

Example 8.10 $0.10100111_2 \rightarrow N_{BCH}$

$$0.10100111_2 = 0.\underset{A}{1010} \quad \underset{7}{0111} = 0.A7_{BCH}$$

Signed Binary Numbers

To this point only unsigned numbers (assumed to be positive) have been considered. However, both positive and negative numbers must be used in computers. Several schemes have been devised for dealing with negative numbers in computers, but only four are commonly used:

- Signed-magnitude representation
- Radix complement representation
- Diminished radix complement representation
- Excess (offset) code representation

Of these, the radix 2 complement representation, called 2's complement, is the most widely used system in computers.

Signed-Magnitude Representation

A signed-magnitude number consists of a magnitude together with a symbol indicating its sign (positive or negative). Such a number lies in the decimal range of $-(r^{n-1} - 1)$ through $+(r^{n-1} - 1)$ for n integer digits in radix r. A fraction portion, if present, would consist of m digits to the right of the radix point.

The most common examples of signed-magnitude numbers are those in the decimal and binary systems. The sign symbols for decimal ($+$ or $-$) are well known. In binary it is established practice to use $0 = $ plus and $1 = $ minus for the sign symbols and to place one of them in the MSB position for each number. Examples in eight-bit binary are

$$+45.5_{10} = 0 \quad \overset{\text{Magnitude}}{\overbrace{101101.1_2}} \quad +0_{10} = 0\ 0000000_2$$

Sign bit

$$-123_{10} = 1 \quad \overset{\text{Magnitude}}{\overbrace{1111011_2}} \quad -0_{10} = 1\ 0000000_2$$

Sign bit

Although the sign-magnitude system is used in computers, it has two drawbacks. There is no unique zero, as indicated by the examples, and addition and subtraction calculations require time-consuming decisions regarding operation and sign as, for example, (-7) minus (-4). Even so, the sign-magnitude representation is commonly used in **floating-point** number systems.

Radix Complement Representation

The *radix complement* of an n-digit number N_r is obtained by subtracting it from r^n, that is $r^n - N_r$. The operation $r^n - N_r$ is equivalent to complementing the number and adding 1 to the LSD. Thus, the radix complement is $\overline{N}_r + 1_{\text{LSD}}$ where $\overline{N}_r = r^n - 1 - N_r$ is the complement of a number in radix r. Therefore, one may write

$$\begin{aligned} \text{Radix complement of } N_r &= r^n - N_r \\ &= \overline{N}_r + 1 \end{aligned} \tag{8.11}$$

The complements $\overline{N}_r$ for digits in three commonly used number systems are given in Table 8.6. Notice that the complement of a binary number is formed simply by replacing the 1's with 0's and 0's with 1's as required by $2^n - 1 - N_2$.

With reference to Table 8.6 and Equation (8.11), the following examples of radix complement representation are offered.

Example 8.11   The 10's complement of 47.83 is

$$\overline{N}_{10} + 1_{\text{LSD}} = 52.17$$

Example 8.12   The 2's complement of 0101101.101 is

$$\overline{N}_2 + 1_{\text{LSB}} = 1010010.011$$

Example 8.13   The 16's complement of A3D is

$$\overline{N}_{16} + 1_{\text{LSD}} = 5C2 + 1 = 5C3$$

The decimal value of Equation (8.11) can be found from the polynomial expression

$$N_{\text{radix compl.}})_{10} = -(a_{n-1}r^{n-1}) + \sum_{i=-m}^{n-2} a_i r^i \tag{8.12}$$

for any n-digit number of radix r. In Equation (8.11) and Equation (8.12) the MSD is taken to be the position of the sign symbol.

TABLE 8.6 Complements for Three Commonly Used Number Systems

| | Complement $(-N_r)$ | | |
Digit	Binary	Decimal	Hexadecimal
0	1	9	F
1	0	8	E
2		7	D
3		6	C
4		5	B
5		4	A
6		3	9
7		2	8
8		1	7
9		0	6
A			5
B			4
C			3
D			2
E			1
F			0

TABLE 8.7 Examples of Eight-Bit 2's and 1's Complement Representations (MSB = Sign Bit)

Decimal Value	2's Complement	1's Complement
− 128	10000000	
− 127	10000001	10000000
− 31	11100001	11100000
− 16	11110000	11101111
− 15	11110001	11110000
− 3	11111101	11111100
− 0	00000000	11111111
+ 0	00000000	00000000
+ 3	00000011	00000011
+ 15	00001111	00001111
+ 16	00010000	00010000
+ 31	00011111	00011111
+ 127	01111111	01111111
+ 128		

2's Complement Representation. The radix complement for binary is the 2's complement representation. In 2's complement the MSB is the sign bit, 1 indicating a negative number or 0 if positive. The decimal range of representation for n-integer bits in 2's complement is from $-(2^{n-1})$ through $+(2^{n-1})$. From Equation (8.11), the 2's complement is formed by

$$N_2)_{2\text{'s compl.}} = 2^n - N_2 = \overline{N}_2 + 1 \tag{8.13}$$

A few examples in eight-bit binary are shown in Table 8.7. Notice that application of Equation (8.13) changes the sign of the decimal value of a binary number (+ to − , and vice versa) and that only one zero representation exists.

Application of Equation (8.12) gives the decimal value of any 2's complement number, including those containing a radix point. For example, the pattern $N_{2\text{'s compl.}} = 11010010.011$ has a decimal value

$$N_{2\text{'s compl.}})_{10} = -1 \times 2^7 + 1 \times 2^6 + 1 \times 2^4 + 1 \times 2^1 + 1 \times 2^{-2} + 1 \times 2^{-3}$$
$$= -128 + 64 + 16 + 2 + 0.25 + 0.125$$
$$= -45.625_{10}$$

The same result could have easily been obtained by first applying Equation (8.13) to $N_{2\text{'s compl.}}$ followed by the use of positional weighting to obtain the decimal value. Thus,

$$N_{2\text{'s compl.}} = 00101101.101$$
$$= 32 + 8 + 5 + 0.5 + 0.125$$
$$= 45.625_{10}$$

which is known to be a negative number, -45.625_{10}.

Negative NBCD numbers can be represented in 2's complement. The foregoing discussion on 2's complement applies to NBCD with consideration of how NBCD is formed from binary. As an example, -59.24_{10} is represented by

$$0101\ 1001.0010\ 0100)_{\text{NBCD}} = 10100110.11011100)_{2\text{'s compl. NBCD}}$$

In a similar fashion, negative NBCD numbers can also be represented in 1's complement following the procedure given in the next paragraph. Sign-magnitude representation of a negative NBCD number simply requires the addition of a sign bit to the NBCD magnitude.

Diminished Radix Complement Representation

The diminished radix complement of a number is obtained by

$$N_r)_{\text{dim .rad.compl.}} = r^n - N_r - 1$$
$$= \overline{N_r} \tag{8.14}$$

Thus, the complement of a number is its diminished radix complement. It also follows that the radix complement of a number is the diminished radix complement with 1 added to the LSD as in Equation (8.13). The range of representable numbers is $-(r^{n-1} - 1)$ through $+(r^{n-1} - 1)$ for radix r.

In the binary and decimal number systems, the diminished radix complement representations are the 1's complement and 9's complement, respectively. Examples of 1's complement are shown in Table 8.7 for comparison with those of 2's complement. Notice that in 1's complement there are two representations for zero, one for $+0$ and the other for -0. This fact limits the usefulness of the 1's complement representation for computer arithmetic.

Excess (Offset) Representations

Other systems for representing negative numbers use *excess* or *offset* codes. Here, a bias B is added to the true value N_r of the number to produce an excess number N_{xs} given by

$$N_{xs} = N_r + B \tag{8.15}$$

When $B = r^{n-1}$ exceeds the usable bounds of negative numbers, N_{xs} remains positive. Perhaps the most common use of the excess representation is in floating-point number systems—the subject of the next section.

Two examples are given below in eight-bit excess 128 code.

Example 8.14

$$
\begin{array}{rll}
-43_{10} & 11010101 & N_{\text{2's compl.}} \\
+128_{10} & 10000000 & B \\
\hline
85_{10} & 01010101 & N_{xs} = -43_{10} \text{ in excess 128 code}
\end{array}
$$

Example 8.15

$$
\begin{array}{rll}
27_{10} & 00011011 & N_{\text{2's compl.}} \\
+128_{10} & 10000000 & B \\
\hline
155_{10} & 10011011 & N_{xs} = 27_{10} \text{ in excess 128 code}
\end{array}
$$

The representable decimal range for an excess 2^{n-1} number system is -2^{n-1} through $+(2^{n-1} - 1)$ for an n-bit binary number. However, if $N_2 + B > 2^{n-1} - 1$, *overflow* occurs and 2^{n-1} must be subtracted from $(N_2 + B)$ to give the correct result in excess 2^{n-1} code.

Floating-Point Number Systems

In fixed-point representation [Equation (8.1)], the radix point is assumed to lie immediately to the right of the integer field and at the left end of the fraction field. The fixed-point system is the most commonly used system for representing bounded orders of magnitude. For example, with 32 bits a binary number could represent decimal numbers with upper and lower bounds of the order of $\pm 10^{10}$ and $\pm 10^{-10}$. However, for greatly expanded bounds of representation, as in scientific notation, the *floating-point* representation is needed.

A floating-point number (FPN) in radix r has the general form

$$\text{FPN})_r = F \times r^E \tag{8.16}$$

where F is the *fraction* (or **mantissa**) and E is the *exponent*. Only fraction digits are used for the mantissa! Take, for example, Planck's constant $h = 6.625 \times 10^{-34}$ J · s. This number can be represented many different ways in floating point notation:

$$\text{Planck's constant } h = 0.625 \times 10^{-33}$$
$$= 0.625 \times 10^{-32}$$
$$= 0.00625 \times 10^{-31}$$

All three adhere to the form of Equation (8.16) and are, therefore, legitimate floating-point numbers in radix 10. Thus, as the radix point *floats* to the left, the exponent is *scaled* accordingly. The first form for h is said to be *normalized* because the MSD of F is nonzero, a means of standardizing the radix point position. Notice that the sign for F is positive while that for E is negative.

In computers the FPN is represented in binary where the normalized representation requires that the MSB for F always be 1. Thus, the range in F in decimal is

$$0.5 \leqslant F < 1$$

Also, the mantissa F is represented in sign-magnitude from. The normalized format for a 32-bit floating-point number in binary, which agrees with the IEEE standard (IEEE, 1985), is shown in Figure 8.1. Here, the sign bit (1 if negative or 0 if positive) is placed at bit position 0 to indicate the sign of the fraction. Notice that the radix point is assumed to lie between bit positions 8 and 9 to separate the E bit-field from the F bit-field.

Before two FPNs can be added or subtracted in a computer, the E fields must be compared and equalized and the F fields adjusted. The decision-making process can be simplified if all exponents are converted to positive numbers by using the excess representation given by Equation (8.15). For a q-digit number in radix r, the exponent in Equation (8.16) becomes

$$E_{xs} = E_r + r^{q-1} \tag{8.17}$$

where E is the actual exponent augmented by a bias of $B = r^{q-1}$. The range in the actual exponent E_r is usually taken to be

$$-(r^{q-1} - 1) \leqslant E_r \leqslant +(r^{q-1} - 1)$$

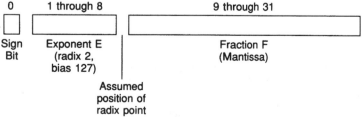

FIGURE 8.1   IEEE standard bit format for normalized floating-point representation.

In the binary system, required for computer calculations, Equation (8.17) becomes

$$E_{xs} = E_2 + 2^{q-1} \tag{8.18}$$

with a range in actual exponent of $-(2^{q-1} - 1) \leqslant E_2 \leqslant +(2^{q-1} - 1)$. In 32-bit normalized floating-point form, the exponent is stored in excess 128 code, while the number is stored in sign-magnitude form.

There still remains the question of how the number 0 is to be represented. If the F field is zero, then the exponent can be anything and the number will be zero. However, in computers the normalized FPN_2 limits F to $0.5 \leqslant F < 1$ since the MSB for F is always 1. The solution to this problem is to assume that the number is zero if the exponent bits are all zero regardless of the value of the mantissa. This leads, however, to a discontinuity in normalized FPN_2 representation at the low end.

The IEEE standard for normalized FPN_2 representation attempts to remove the problem just described. The IEEE system stores the exponent in excess $2^{q-1} - 1$ code and limits the decimal range of the actual exponent to

$$-(2^{q-1} - 2) \leqslant E_2 \leqslant +(2^{q-1} - 1)$$

For 32-bit FPN representation, the exponent is stored in excess 127 code as indicated in Figure 8.1. Thus, the allowable range of representable exponents is from

$$-126_{10} = 00000001_2 \quad \text{through} \quad +127_{10} = 11111110_2$$

This system reserves the use of all 0's or all 1's in the exponent for special conditions [IEEE, 1985; Pollard, 1990]. So that the F field magnitude can diminish linearly to zero when $E = -126$, the MSB $= 1$ for F is not specifically represented in the IEEE system but is implied.

The following example attempts to illustrate the somewhat confusing aspects of the IEEE normalized representation:

The number 101101.11001_2 is to be represented in IEEE normalized FPN_2 notation.

$$101101.11001_2 = .10110111001 \times 2^6$$
$$\text{Sign bit} = 0 \, (\text{positive})$$
$$E_{xs} = 6 + 127 = 133_{10} = 10000101_2$$
$$F = 0110111001\ldots00(\text{the MSB} = 1 \text{ is not shown})$$

Therefore, the IEEE normalized FPN is

$$FPN_2 = 0 \quad 10000101 \quad 0110111001\ldots0$$

Still other forms of FPNs are in use. In addition to the IEEE system, there are the IBM, Cray, and DEC systems of representation, each with its own single- and double-precision forms.

Defining Terms

Binary: Representation of quantities in base 2.
Complement: Opposite form of a number system.
Floating point: Similar to "scientific notation" except used to represent binary operations in a computer.
Hexadecimal: Base 16 number system.
Mantissa: Fraction portion of a floating-point number.
Octal: Base 8 number system.
Radix: Base to which numbers are represented.

References

H.L. Garner, "Number systems and arithmetic," in *Advances in Computers*, vol. 6, F.L. Alt et al., Eds., New York: Academic, 1965, pp. 131–194.

IEEE, *IEEE Standard for Binary Floating-Point Arithmetic*, ANSI/IEEE Std. 754–1985.

D.E. Knuth, *The Art of Computer Programming: Seminumerical Algorithms*, vol. 2, Reading, Mass: Addison-Wesley, 1969.

C. Tung, "Arithmetic," in *Computer Science*, A.F. Cardenas et al., Eds., New York: Wiley-Interscience, 1972, chap. 3.

Further Information

K. Hwang, *Computer Arithmetic*, New York: Wiley, 1978.

L.H. Pollard, *Computer Design and Architecture*, Englewood Cliffs, N.J.: Prentice-Hall, 1990.

R.F. Tinder, *Digital Engineering Design: A Modern Approach*, Englewood Cliffs, N.J.: Prentice-Hall, 1991.

8.2 Computer Arithmetic

S.N. Yanushkevich

Basics of Computing Arithmetic

State-of-the-art computing arithmetic includes:

Binary arithmetic (adders, multipliers, and dividers),
Residue number systems (RNS) arithmetic, and
Stochastic arithmetic.

There are also specialized arithmetics such as *multi-valued logic*, *fuzzy logic*, and *threshold* arithmetic for neural networks.

Binary arithmetic is prevalent in today's computers. The addition and subtraction of numbers using binary arithmetic are fundamental operations performed frequently in any computation. The speed with which these operations are performed has a strong impact on the overall performance of a computer. The speed of arithmetic circuit is limited by the longest signal delay in the circuit, often referred to as the *critical-path delay*. A commonly used indicator of the value of an arithmetic circuit is its *price/performance* ratio.

The computational reliability is a key problem of the new generation of computer systems based on nanotechnology, and can be achieved in various ways, in particular by applying stochastic arithmetic and residue techniques.

Number Representation and Addition of Binary Numbers

In the decimal system the sign of a number is indicated by a special symbol, " $+$ " or " $-$ ". In the binary system the sign of a number is denoted by the left-most bit. Positive numbers are represented using the positional number representation. Negative numbers can be represented in three different ways (Figure 8.2):

Sign-and-magnitude, such that the sign symbol distinguishes a number as being positive or negative. While performing addition, the magnitudes are added, and the resulting sum is given the sign of the operands. If the operands have opposite signs, it is necessary to subtract the smaller number from the larger one (logic circuits that compare and subtract numbers are needed). The range of signed integers is $-(2^{n-1} - 1) \leq x \leq 2^{n-1} - 1$. In such a system zero has two representations: positive zero $00...0$ and negative zero $10...0$.

1's complement; in that system, the negative numbers are defined according to a subtraction operation involving positive numbers. An n-bit negative number K is obtained by subtracting its equivalent positive number P from 2^n-1; i.e., $K = (2^n - 1) - P$. An advantage of 1's complement representation is

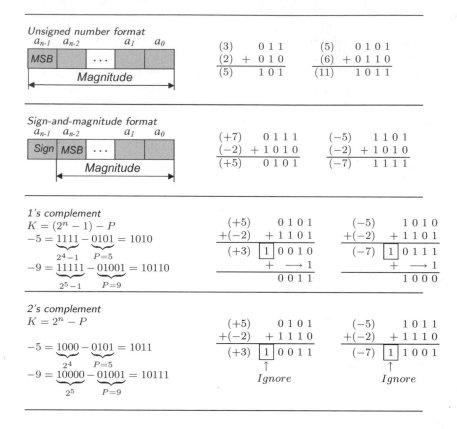

FIGURE 8.2 Examples of representation and addition of binary numbers: unsigned, sign-and-magnitude, 1's and 2's complemented.

that a negative number is generated by complementing all bits of the corresponding positive number. The addition of 1's complement numbers may require a correction, so that must be performed, and the time needed to add two 1's complement numbers may be twice as long as the time needed to add two unsigned numbers.

2's complement; in this system, the negative numbers are defined according to a subtraction operation involving positive numbers. An n-bit negative number K is obtained by subtracting its equivalent positive number P from 2^n; i.e., $K = 2^n - P$. An advantage of 2's complement representation is that when the numbers are added, the result is always correct. If there is a carry-out from the sign-bit position, it is simply ignored.

The technique of performing a subtraction operation by addition of a complement of the subtrahend can be generalized for the decimal number system.

Residue Number Systems

The hardware implementation of an arithmetic algorithm is largely affected by the choice of a specific numbering system. The attractive properties of RNS include carry-free, fault isolating and modular characteristics; these are widely used particularly in high-speed digital signal processing. The most attractive property of RNS is that there is no carry propagation inside the set. In an RNS-based system conversion procedures, from conventional binary representation to residue format and vice versa, are used.

In RNS an integer is represented as a set of residues with respect to a set of relatively prime integers called *moduli*. An RNS is defined in terms of a set of relatively prime moduli $\{r_1, r_2, \ldots, r_s\}$ where the greatest common divisor is equal to 1 for each pair of the moduli. The set of integers $\{r_1, r_2, \ldots, r_s\}$ is called a *complete residue system modulo m* if

1. $r_i \neq r_j$ (mod m) whenever $i \neq j$, and
2. for each integer n there corresponds an r_i such that $n = r_i$ (mod m).

Definitions. If a and b are integers and m is a natural number, the statement $a \equiv b$ (mod m) (a is *congruent* to b modulo m) means that the difference, $a - b$, is exactly divisible by the positive integer m. If a and b are two integers and $a \equiv b$ (mod m), then b is a *residue* of a modulo m.

If s different integers $\{r_1, r_2, \ldots, r_s\}$ form a complete residue system modulo m, then $s = m$. For example, the sets $\{1, 2, 3\}$, $\{-1, 0, 1\}$, and $\{1, 7, 9\}$ are all complete residue systems modulo 3. The set $\{0, 1, 2, 3, 4, 5\}$ is a complete residue system modulo 6. Note this set can be reduced to $\{1, 5\}$.

Properties. The following properties are useful in design of modular arithmetic circuits.

Property of addition: If $a \equiv b$ (mod m) and $c \equiv d$ (mod m), then $a + c \equiv b + d$ (mod m) and $a - c \equiv b - d$ (mod m).

The first property of multiplication: If $a \equiv b$ (mod m) and $c \equiv d$ (mod m), then $ac \equiv bd$ (mod m).

Note that properties of addition and multiplication of two congruences can be extended to an arbitrary number of congruences with the same modulus.

The second property of multiplication (squaring): If $a \equiv b$ (mod m), then $a^n \equiv b^n$ (mod m) for every positive integer n.

The transitive property: If $a \equiv b$ (mod m) and $b \equiv c$ (mod m), then $a \equiv c$ (mod m).

The proof of the above properties is omitted. A simple way of proving properties of congruences is to write the congruence $a \equiv b$ (mod m) as the equality $a = b + km$, where k is some integer.

Addition, Subtraction, Multiplication, and Division. It follows that two or more congruences may be added, subtracted, and multiplied provided the same modulus is used throughout; i.e., congruences behave like ordinary equations in algebra. However, the division of congruences is based on several specific rules, which derive from the property that it is not always possible to divide both sides of a congruence by the same integer and obtain a true congruence with the same modulus. For example, it does not follow from $76 \equiv 28$ (mod 8) that $76/4$ is congruent to $28/4$ modulo 8.

While in ordinary arithmetic there is an infinite number of integers $0, 1, 2, \ldots$, in the modular arithmetic there is essentially only a finite number of integers (Figure 8.3). The operation of division is defined as solution $x \equiv b/a$ (mod m) of the congruence $ax \equiv b$ (mod m). For example, $13/8$ (mod 9) means the solution x of $8x \equiv 13$ (mod 9), which may be written as $-x \equiv 13$ or $x \equiv -13 = -4 \equiv 5$ (mod 9). Thus, $13/9 \equiv 5$.

Any operation consisting of integers or fractions in ordinary arithmetic will remain true in modular arithmetic. For example, calculate

$$11 \cdot \left(\frac{3}{2} - \frac{2}{3} \right) = 11 \cdot \frac{9 - 4}{6} = \frac{55}{6}$$

In modular arithmetic (mod 7) the fraction $3/2$ means the solution of $2x \equiv 3$ (mod 7) or $x = 5$, since $10 \equiv 3$ (mod 7), and $2/3$ means the solution of $3x \equiv 2$ (mod 7) or $x = 3$, since $9 \equiv 2$ (mod 7). The above operation becomes

$$4 \cdot (5 - 3) = 4 \cdot 2 = 8 \equiv 1 \ (\text{mod } 7)$$

On the other hand, $55/6$ means the solution of $6x \equiv 55$ (mod 7) or $6x \equiv 6$ or $x \equiv 1$, which is the same answer as before.

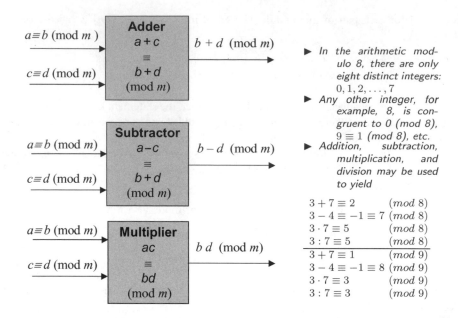

The block diagram at the top shows:

Adder $a+c \equiv b+d$ (mod m): inputs $a \equiv b$ (mod m) and $c \equiv d$ (mod m), output $b+d$ (mod m).

Subtractor $a-c \equiv b+d$ (mod m): inputs $a \equiv b$ (mod m) and $c \equiv d$ (mod m), output $b-d$ (mod m).

Multiplier $ac \equiv bd$ (mod m): inputs $a \equiv b$ (mod m) and $c \equiv d$ (mod m), output bd (mod m).

▶ In the arithmetic modulo 8, there are only eight distinct integers: $0, 1, 2, \ldots, 7$

▶ Any other integer, for example, 8, is congruent to 0 (mod 8), $9 \equiv 1$ (mod 8), etc.

▶ Addition, subtraction, multiplication, and division may be used to yield

$$3 + 7 \equiv 2 \qquad (mod\ 8)$$
$$3 - 4 \equiv -1 \equiv 7\ (mod\ 8)$$
$$3 \cdot 7 \equiv 5 \qquad (mod\ 8)$$
$$3 : 7 \equiv 5 \qquad (mod\ 8)$$
$$3 + 7 \equiv 1 \qquad (mod\ 9)$$
$$3 - 4 \equiv -1 \equiv 8\ (mod\ 9)$$
$$3 \cdot 7 \equiv 3 \qquad (mod\ 9)$$
$$3 : 7 \equiv 3 \qquad (mod\ 9)$$

FIGURE 8.3   Examples of addition, subtraction, multiplication, and division in residue number system, mod 8.

The choice of moduli sets and the conversion of the residue to binary numbers are important issues to residue arithmetic. The RNS based on the set moduli $\{2^n - 1,\ 2^n,\ 2^n + 1\}$ is popular in digital signal processing. These converters are using $2n$-bit or n-bit adders.

Binary Adders

The operands of addition are the *addend* and *augend*. The addend is added to the augend to form the sum. In most arithmetic circuits the augmented operand (the augend) is replaced by the sum, whereas the addend is unchanged. High speed adders are not only for addition but also for subtraction, multiplication, and division. The speed of a digital processor depends heavily on the speed of adders. The adders add vectors of bits, and the principal problem is to speed up the carry signal.

An *n-bit binary adder* is a combinational circuit that has two n-bit inputs $A = a_{n-1}, \ldots, a_0$ and $B = b_{n-1}, \ldots, b_0$ representing the operands A and B, respectively, and n-bit output $S = s_{n-1}, \ldots, s_0$, and performs binary addition of the input operands. Additional input and output signals, *carry-in* C_i and *carry-out* C_{i+1} are used to implement module-based architecture—i.e., the design of larger adders.

Full-Adder

A logic network that performs addition at a single bit position is the generic cell used not only to perform addition but also arithmetic multiplication division and filtering operations.

The truth table for this network (1-bit adder) is given in Figure 8.4(a). Such a network is referred to as a binary *full-adder* (FA). The optimized functions of the full adder outputs, sum S_i and C_{i+1}, are illustrated in Figure 8.4(b). The OR and two-level AND-OR combination circuits to implement S_i and C_{i+1}, correspondingly, are shown in Figure 8.4(c), and an alternative circuit formed of two half-adders (HAs), the subcircuits that compute p_i and g_i, is given in Figure 8.4(d).

Figure 8.5 illustrates the *reduced ordered binary decision diagram* (ROBDD), a graph-based representation of the two-output function of the full adder. In this graph, Shannon expansion (S) is used to represent switching functions s_i and c_i by decision diagrams. Then, the nodes of the minimized decision diagrams, ROBDD, are represented by multiplexers (MUX), so that a multiplexer-based full adder is built [Figure 8.5(b)].

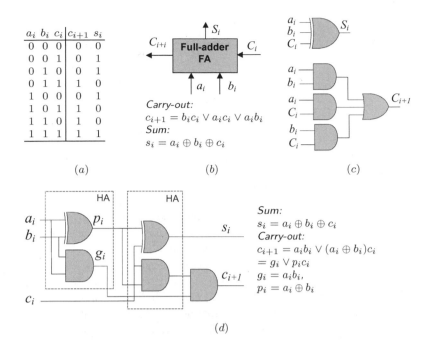

FIGURE 8.4 Full-adder: the truth table (a), the formal description (b), the logic network over library of AND, OR and EXOR gates (c) and the half-adder (HA) based design (d).

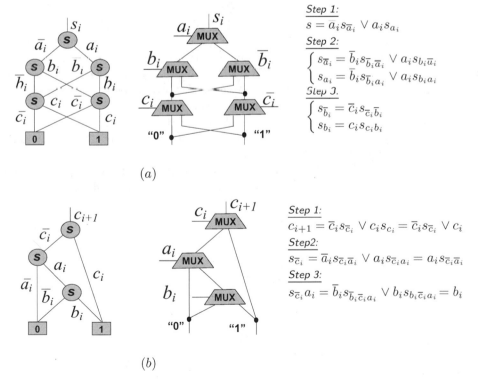

FIGURE 8.5 Multiplexer-based synthesis of full adder using Shannon decision diagrams of functions sum s_i (a) and carry c_i (b).

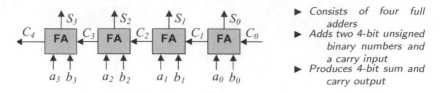

FIGURE 8.6   Ripple-carry 4-bit adder.

Ripple-Carry Adder

The *ripple-carry adder* is a multilevel network designed by the connection of full-adders (Figure 8.6). For the multilevel adder the total time required is calculated as the delay from carry-in C_i to carry-out C_{i+1}.

Depending on the position at which a carry signal has been generated, the propagation time can be variable. In the best case, when there is no carry generation, the addition time will only take into account the time to propagate the carry signal. With a ripple-carry adder, if the input bits A_i and B_i are different for all positions i, then the carry signal is propagated at all positions (thus never generated) and the addition is completed when the carry signal has propagated through the whole adder. In this case the ripple-carry adder is as slow as it is large. Actually, ripple-carry adders are fast only for some configurations of the input words where carry signals are generated at some positions. They can be divided into blocks where a special circuit detects quickly if all the bits to be added are different. These *carry-skip adders* take advantage both of the generation or the propagation of the carry signal.

Carry-Lookahead Adder

A nonoptimized 4 bit adder can be made by the use of the generic 1-bit adder cell connected one to the other (ripple-carry adder). In this case the sum resulting at each stage need to wait for the incoming carry signal to perform the sum operation. The carry propagation can be sped up in two ways. The first and most obvious way is to use a faster logic circuit technology. The second way is to generate carries by means of forecasting logic that does not rely on the carry signal being rippled from stage to stage of the adder. This, a faster alternative to the carry-ripple adder, can be obtained at the cost of more gates with a larger number of inputs. To reduce the delay, the carry-propagation path is broken so that a compromise among speed and cost is obtained by performing addition as a two-step process. First, the values of all carries into the full-adder modules are determined, and then simultaneously all result bits are computed (Figure 8.7). This adder is called a *carry-lookahead adder*. Formally, it is possible to express a carry as a function of all the preceding low order carries by using the recursivity of the carry function.

A *carry generator* determines the values of all intermediate carries before the corresponding sum bits are computed. A *sum generator* uses precomputed carriers to determine the value of the sum bits.

The cost of addition of two numbers of m and n digits is $O(\min\{m, n\})$.

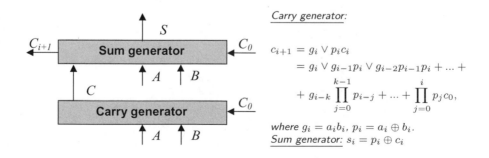

Carry generator:

$$c_{i+1} = g_i \vee p_i c_i$$
$$= g_i \vee g_{i-1} p_i \vee g_{i-2} p_{i-1} p_i + \dots +$$
$$+ g_{i-k} \prod_{j=0}^{k-1} p_{i-j} + \dots + \prod_{j=0}^{i} p_j c_0,$$

where $g_i = a_i b_i$, $p_i = a_i \oplus b_i$.
Sum generator: $s_i = p_i \oplus c_i$

FIGURE 8.7   Carry-lookahead adder.

Multipliers

Combinational multipliers for positive integers are used in floating-point processors and signal processing applications. An $n \times m$-bit combinational multiplier is a combinational circuit that produces the multiplication $0 \leqslant A \times B \leqslant (2^n - 1)(2^m - 1)$ (product) of two integer numbers: $0 \leqslant A \leqslant 2^n - 1$ (multiplicand) and $0 \leqslant B \leqslant 2^m - 1$ (multiplier):

$$A = \sum_{i=0}^{m-1} a_i 2^i, \quad B = \sum_{j=0}^{n-1} b_j 2^j, \quad m \geq n$$

$$A \times B = \sum_{h=0}^{m+n-2} \left(\sum_{i+j=h} a_i b_j \right) 2^h$$

When the operands are interpreted as integers, the product is generally twice the length of the operands in order to preserve the information content.

Multiplication can be considered as a series of repeated additions. This repeated addition method that is suggested by the arithmetic definition is slow, and is almost always replaced by an algorithm that makes use of positional number representation.

It is possible to decompose multipliers into two parts. The first part is dedicated to the generation of partial products, and the second one collects and adds them. As for adders, it is possible to enhance the intrinsic performances of multipliers. Acting in the generation part, the Booth (or modified Booth) algorithm is often used because it reduces the number of partial products (see Further Reading on page 8-29). The collection of the partial products can then be made using a regular array, a Wallace tree or a binary tree.

The simplest multiplication can be viewed as repeated shifts and adds (one adder, a shift register, and a small amount of control logic). The disadvantage is that it is slow. One fairly simple improvement to this is to form the matrix of partial products in parallel, and then use a 2-dimensional array of full adders to sum the rows of partial products. This 8×6 structure, shown in Figure 8.8, is known as an *array multiplier*. The multiplier consists of $n - 1 = 5$, $m = 8$-bit carry-ripple adders and $n = 6$ arrays of m AND gates. The delay of the multiplier is defined as the critical path equal to the sum of the delay of the buffer circuit connecting the input signal and AND gates, the delay of the AND gate, and the delay of the adders.

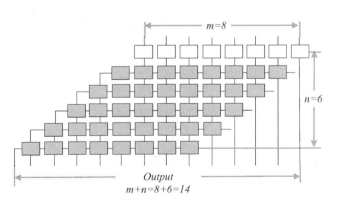

Architecture of $n \times m$ bits multiplier:
$n - 1 = 5$ $m = 8$-bit carry-ripple adders;
m arrays of n 2-input AND gates

Formal description

$$A \times B = A \left(\sum_{i=0}^{m-1} b_i 2^i \right)$$

$$= \sum_{i=0}^{m-1} A b_i 2^i$$

The multiplication is performed by adding the integers $A b_i 2^i$. Because b_i is either 0 or 1, we get

$$A b_i = \begin{cases} 0, & \text{if } b_i = 0; \\ A, & \text{if } b_i = 1. \end{cases}$$

$$
\begin{array}{cccccccc}
a_7 b_0 & a_6 b_0 & a_5 b_0 & a_4 b_0 & a_3 b_0 & a_2 b_0 & a_1 b_0 & a_0 b_0 \\
a_7 b_1 & a_6 b_1 & a_5 b_1 & a_4 b_1 & a_3 b_1 & a_2 b_1 & a_1 b_1 & a_0 b_1 \\
a_7 b_2 & a_6 b_2 & a_5 b_2 & a_4 b_2 & a_3 b_2 & a_2 b_2 & a_1 b_2 & a_0 b_2 \\
a_7 b_3 & a_6 b_3 & a_5 b_3 & a_4 b_3 & a_3 b_3 & a_2 b_3 & a_1 b_3 & a_0 b_3 \\
a_7 b_4 & a_6 b_4 & a_5 b_4 & a_4 b_4 & a_3 b_4 & a_2 b_4 & a_1 b_4 & a_0 b_4 \\
a_7 b_5 & a_6 b_5 & a_5 b_5 & a_4 b_5 & a_3 b_5 & a_2 b_5 & a_1 b_5 & a_0 b_5 \\
\end{array}
$$

FIGURE 8.8 An 8×6 multiplier: topology of architecture and multiplication scheme.

The advantage of the array multiplier is that it is a regular structure and a local interconnect: each cell is connected only to its neighbors. The disadvantage is that the worst case delay path goes from the upper left corner diagonally down to the lower right corner and then across the ripple carry adder; i.e., the delay is linearly proportional to the operand size. The method, which can be employed to decrease the delay of the array multiplier, is to replace the ripple-carry adder with a carry-lookahead adder. Another approach to collection of the partial products is based on the so-called *Wallance tree multiplier* (see section on further reading). The number of operations occurring in multiplication is at most $O(mn)$.

Arithmetic-Logic Units

Arithmetic-logic units (ALUs) are modules capable of realizing a set of arithmetic and logic functions. ALU performs the specific operations selected dynamically by the control unit of the processor. In Figure 8.9 a 4-bit ALU has two 4-bit data inputs A and B, a carry-in input C_0, a 4-bit data output S, and also P and G outputs that can be used for computing carry-out signal $C_4 = G \vee P \cdot C_0$. Generic arithmetic and logic operations performed by ALU are given in Figure 8.9. This 4-bit ALU module can be used to construct larger ALUs.

Other Number Representations

Besides the positional (radix-2 2's or 1's complement) number system, positive integers can be represented in *binary-coded decimal* (BCD) notation, *2-1-2-4* code, *excess-3* code, or *2-out-of-5*. In this section the BCD format is considered.

Numbers with the fraction part can be represented in the fixed-point and floating-point format. Floating-point format is dominant in today's processors.

Binary-Coded-Decimal Format

Let two decimal digits (operands) be denoted by the binary codes $A = a_3a_2a_1a_0$ and $B = b_3b_2b_1b_0$. Note that 4-bit A and B cannot take values larger than 9 (e.g., 1010,1011,1100,1101,1110, and 1111). A carry-in and carry-out are denoted by C_0 and C_3. The data output (sum) is $S = s_3s_2s_1s_0$. This sum must be corrected. These cases are shown in Figure 8.10 along with the circuit of BCD adder that consists of the binary adder and the correction circuit. The correction circuit implements the addition of number 0110 in the indicated cases. The combinational circuit is implemented with the optimal representation of the function c_{out}, z_3, z_2, z_1, and z_0. Alternatively, MUX-based design can be used: formally, the circuit can be represented by a ROBDD and by the diagram in Figure 8.11.

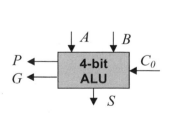

Control	Function, S
Transfer	A
Complement	$\overline{A}$
AND	AB
OR	$A \vee B$
EXOR	$A \oplus B$
Increment	$A + 1$
ADD	$A + B + 1$
	$A + B$
1C subtraction	$A + \overline{B}$
2C subtraction	$A + \overline{B} + 1$
Decrement	$A - 1$

FIGURE 8.9   Example of 4-bit arithmetic-logic units.

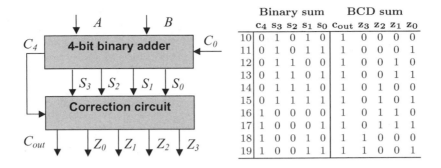

FIGURE 8.10 Decade of BCD adder and the part of a truth table where correction is needed.

	Binary sum					BCD sum				
	c_4	s_3	s_2	s_1	s_0	c_{out}	z_3	z_2	z_1	z_0
10	0	1	0	1	0	1	0	0	0	0
11	0	1	0	1	1	1	0	0	0	1
12	0	1	1	0	0	1	0	0	1	0
13	0	1	1	0	1	1	0	0	1	1
14	0	1	1	1	0	1	0	1	0	0
15	0	1	1	1	1	1	0	1	0	1
16	1	0	0	0	0	1	0	1	1	0
17	1	0	0	0	1	1	0	1	1	1
18	1	0	0	1	0	1	1	0	0	0
19	1	0	0	1	1	1	1	0	0	1

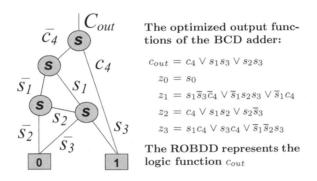

The optimized output functions of the BCD adder:

$$c_{out} = c_4 \vee s_1 s_3 \vee s_2 s_3$$
$$z_0 = s_0$$
$$z_1 = s_1 \bar{s}_3 \bar{c}_4 \vee \bar{s}_1 s_2 s_3 \vee \bar{s}_1 c_4$$
$$z_2 = c_4 \vee s_1 s_2 \vee s_2 \bar{s}_3$$
$$z_3 = s_1 c_4 \vee s_3 c_4 \vee \bar{s}_1 \bar{s}_2 s_3$$

The ROBDD represents the logic function c_{out}

FIGURE 8.11 Decision diagram of the correction function of the BCD adder.

Fixed-Point Format

A *fixed-point* number consists of *integer* and *fraction* parts [Figure 8.12(a)]. The position of the radix point is assumed to be fixed. Logic circuits that deal with fixed-point numbers are essentially the same as those used for integers. Fixed-point numbers have a range that is limited by the significant digits used to represent the number. The integer numbers are fixed-point numbers without a fraction part.

Floating-Point Format

A real number is represented by a *mantissa M* comprising the significant digits and an *exponent E* of the radix R $M \times R^E$. In the IEEE 754 standard two sizes of floating-point format are specified: a *single-precision* 32-bit format and *double-precision* 64-bit format. In Figure 8.12(b) the single-precision format is shown, where S is the sign (so, this representation is *sign-and-magnitude*), M is the fraction or mantissa, E is the exponent (positive or negative), and the radix (base) is 2 for computer arithmetic. The 8-bit exponent is specified in the *excess-127* format, which is convenient for adding and subtracting floating-point numbers. The hidden 1 is added in the formula of the number representation; that is the implicit leading bit in the normalized binary number 1. $M \times 2^{E-127}$, where 127 is the exponent bias for the single precision (it is 1023 for double precision). Thus, the 32-bit 00...00 in the IEEE 754 standard represents 0, while 000000000...01 represents

$$(-1)^0 \times (1 + .0)_2 \times 2^{0-127} = 2^{-127}$$
$$= 2_{10} \times 2^{-128} \approx 2_{10} \times 10^{-38}$$

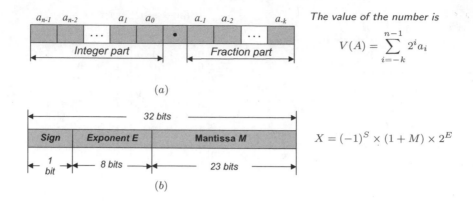

FIGURE 8.12   Fixed-point format (a) and single-precision floating-point format (b).

(the smallest fraction that can be represented by this format), and 11...11 represents

$$(-1)^1 \times (1 + .11111111111111111111111_2) \times 2^{255-127}$$
$$= -1.11111111111111111111111_2 \times 2^{128} \approx -2_{10} \times 10^{38}$$

(the largest negative number to be represented by this format).

Floating-Point Addition, Multiplication and Division

Two floating-point numbers must have equal exponents before they are added. This is implemented as follows:

- The number with the smaller exponent is chosen, and its mantissa is shifted right a number of steps equal to the difference in exponents.
- The exponent of the results is set equal to the larger exponent.
- Addition on mantissas is performed, and the sign of the result is determined.
- The resulting value is normalized.

Two numbers are multiplied by adding their exponents and multiplying their mantissas:
- The exponents are added and the exponent bias (127) is subtracted.
- The mantissas are multiplied, and the sign of the result is determined.
- The resulting value is normalized.

Division is implemented in three steps as is multiplication, except that the exponents are subtracted instead of added, and the mantissas are divided instead of multiplied.

Low Power Computing Arithmetic

Low power arithmetic circuits are in the critical application domain, which demands high-speed computations and complex functionalities with low consumption. The design for low power arithmetic circuits can be achieved by optimization tools, by using a power-efficient gate, and by module libraries. On the arithmetic level choice of the number representation is essential to reduce power consumption (often at the expense of lower speed). For that purpose the hybrid signed digit (HSD) representation is deployed. It combines both signed and unsigned digits, and renders the maximum length of carry propagation equal to the maximum distance between two consecutive signed digits. This representation allows choices from 2's complement representation where there are no signed digits, to the conventional fully signed digit representation wherein every digit is signed.

In radix-2 ($R = 2$) HSD representation, the signed digits can take any value from the set $\{-1, 0, +1\}$. The unsigned digits can assume any of the two values $\{0, 1\}$. The addition of two HSD numbers enables a signed digit to stop an incoming carry from propagating further. Consequently, the carries propagate between the signed digits and the maximum length of carry propagation equals the distance between the signed digits. Thus, addition in such a representation requires the carry in between all digit positions (signed or unsigned) to assume any value in the generic signed-digit system. The operations in a signed-digit position are the same as those in the signed-digit case. First, the signed-digit positions generate a carry-out and an intermediate sum, based on the two input signed digits and the two bits at the neighboring lower-order, unsigned digit position. In the second step, the carries generated out of the signed digit positions ripple through the unsigned digits up to the next higher-order, signed digit position where the propagation stops. All the carry propagation chains between the signed digit positions run simultaneously.

The most significant digit in the HSD representation is a signed digit, and all the other digits can be unsigned. For example, given 32 digits, the most significant digit is a signed digit. The remaining digits can unsigned or signed; for example, the 1st, 2nd, 4th, 8th, and 16th (and 32nd) digits can be signed, and all the remaining digits can be unsigned. The addition time for such a representation is determined by the longest possible carry propagation chain between consecutive signed digit positions.

Stochastic Arithmetic

Stochastic arithmetic is applied, in particular, to the implementation of fault tolerance systems, which are characterized by the ability to recover from transient errors during computing. An example of such systems is artificial neural networks.

Basics of Stochastic Computing

Stochastic arithmetic is based on special coding of numbers, so-called *stochastic encoding*. The input numbers x_1 and x_2 are encoded by stochastic pulse streams. Stochastic streams are independent (technically this means that independent generators of random pulse are used with some additional tools for decorrelation of signals) with mean $E(x_1) = P_{x_1}$ and $E(x_2) = P_{x_2}$. Operations with stochastic streams correspond to operations with probabilities P_{x_1} and P_{x_2}. The result is decoded to the values in the range [0,1].

The information in a pulse stream is contained in the *primary statistic of the bit stream*, or the probability of any given bit in the stream being a logic 1. Statistical characteristics of these streams are known; i.e., they can be measured. The errors are in the form of *random variance*. The main feature that distinguishes classical arithmetic computation and stochastic computation is that arithmetic operations are performed via the completely random data. Its actual value is

1. A random event which cannot be predicted, and
2. Repetition of a computation will result in a different sequence of logic levels.

In a conventional computer logic levels represent data change deterministically from value to value as the computation proceeds. If the computation is repeated, the same sequence of logic levels will occur.

Stochastic Encoding

A binary number X is compared with a uniform random number generated by a generator of random numbers (Figure 8.13). The upper limit of numbers is X_{max}. The firing probability P_f of the comparator output is equal to X/X_{max}, so the output value $\hat{X}$ which is obtained by accumulating the pulse times follows the binomial distribution.

Stochastic Adder

The averages of the stochastic pulse stream of two inputs x_1 and x_2 and the output signal f are $E(x_1)$, $E(x_2)$ and $E(f) = E(x_1) + E(x_2) = P_{x_1} + P_{x_2}$, respectively [Figure 8.14(a)], where autocorrelation function is defined as $K_f(\tau) = E[[f(t) - E[f]][f(t - \tau)]] - E[(f)]]$. The simplest stochastic adder is implemented by a single OR gate. The output signal probability is approximately $P_{x_1} + P_{x_2}$, except in the high rate region where collisions are too frequent.

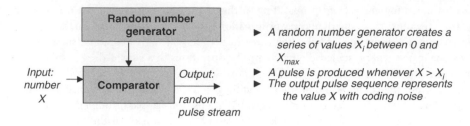

FIGURE 8.13   A coding circuit for generating random pulse sequences.

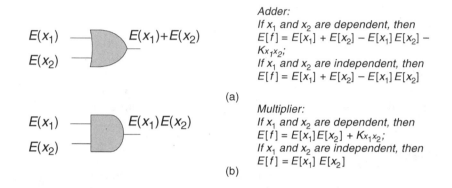

FIGURE 8.14   Simplest stochastic adder (a) and stochastic multiplier (b) based on pulse model of computing.

Stochastic Multiplier

The averages of the stochastic pulse stream of input and output signals are $E(x_1)$, $E(x_2)$ and $E(f) = E(x_1)E(x_2) = P_{x_1}P_{x_2}$; i.e., the output signal is equal to the product of probabilities P_{x_1} and P_{x_2}, and implemented by a single AND gate [Figure 8.14(b)].

Threshold Logic for Massively Parallel Systems

Neural networks are massively parallel systems. For many years arithmetic circuit design based on threshold gates has been considered an alternative to traditional logic. The threshold gate design is based on majority, or threshold logic gates, a fundamental paradigm of decision making in biological systems. Based on this principle, an arbitrary logic function can be implemented by a set of threshold gates.

Threshold and Majority Gates.   A *threshold gate* (Figure 8.15) is a multiple terminal device that calculates the weighted sum of the inputs x_i, $i = 1, 2, \ldots, n$. Afterwards this sum is converted into a output y by comparing the sum with a given threshold level T. A *majority gate* is defined similarly to a threshold gate but the output y takes values -1 and $+1$ that correlate to 1 and 0 values, respectively. The class of all threshold functions is equivalent to the class of all majority functions.

Threshold Networks for Arithmetic Functions.   Multiple addition, multiplication and division have been shown to be computable by small constant-depth, polynomial-size threshold networks. The efficient threshold networks for arithmetic functions rely in many cases on the new computational algorithms. These networks are characterized by a cost and delay comparable to that of logic gates.

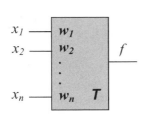

A threshold gate is a logic gate with

▶ *n inputs* $x_1, x_2, \ldots, x_n$, *which can take values 0,1, and*
▶ *a set of* $(n+1)$ *real numbers* $w_1, w_2, \ldots, w_n$ *and* T *called weights and threshold respectively,*

such that the output f *of the threshold gate is described by the equation*

$$f = \begin{cases} 1, \; for \; \sum_{i=1}^{n} w_i x_i \succeq T; \\ 0, \; for \; \sum_{i=1}^{n} w_i x_i \succeq T. \end{cases}$$

FIGURE 8.15   Threshold gate.

Computing Arithmetic of Nanostructures

Conventional electronics emphasized the development of logic families consisting of gates that are networked. This approach can be adopted for nanoelectronics. While this approach exists, the design of large circuits is problematic due to reliability issues and interconnection limitations. The alternative approach is massive parallel computation on locally interconnected networks of computing elements.

Three aspects that are critical in nanocomputing are:

1. *Fault tolerance*, the ability to recover from transient errors during computing.
2. *Robustness to errors*, the ability to operate correctly in the presence of errors.
3. *Defect tolerance*, the ability to operate correctly in the presence of permanent hardware errors that emerged in the manufacturing process.

To satisfy the above criteria, various methods, design strategies and architectures can be chosen, in particular using:

- Homogeneous architectures utilizing principles of massive parallelism
- Computational methods that are fault tolerant by nature, for example, stochastic computing
- Error correction coding
- Redundancy

Acknowledgment

Remarks and suggestions of Professor V. Shmerko are acknowledged.

Further Information

Basics of logic design can be found in many textbooks, for example, in Refs. [6,7]. Details on addition and multiplication algorithms including floating-point examples, can be found in Ref. [17]. A Wallace tree multiplier is considered in Ref. [10], and an advanced binary tree based multiplier design has been proposed in Ref. [11].

Number systems. In Ref. [13], RNS floating-point arithmetic (addition, subtraction, multiplication, division and square root) for an interval number is discussed with the goal to achieve reliable computation when hardware representations of numbers have inadequate precision. For example, a double-base representation ($m_1 = 2, m_2 = 3, n = 2$) is $x = \sum_{i,j} w_{i,j} 2^i 3^j$, where i and j are positive integers. For $j = 0$ and $i = 0$, this equation becomes a binary and ternary system representation, respectively. Adder based residue-to-binary number converters have been reported in Ref. [24]. In Ref. [25], the residue-to-binary number converters for the RNS $\{2^n - 1, \; 2^n, \; 2^n + 1\}$ were designed using $2n$-bit adders or n-bit adders that are twice as fast as generic ones, and achieved improvement in area and dynamic range as well.

Particular modulo arithmetic involves Galois fields. Addition and multiplication in Galois fields, $GF(2^n)$, plays an important role in coding theory and is widely used in digital computers and data transmission or storage systems. The group theory is used to introduce algebraic system called a *field*. A field is a set of elements in which we can do addition, subtraction, multiplication and division without leaving the set.

Stochastic computing.   The original motivation for using stochastic arithmetic was the simplicity of the computational elements involved. Stochastic arithmetic provides the possibility of carrying out computations with simple hardware, with the following properties: fault tolerance, simple interconnections, capability to trade off computation time and accuracy without architecture changes [8]. Fault-tolerance systems and neural networks that are massively parallel systems can benefit from stochastic arithmetic [27]. In Ref. [1], stochastic arithmetic was implemented by computational elements in neural network. In Ref. [12], the massively parallel stochastic computing architecture for generating products and additions has been introduced.

Threshold logic.   The common model of neural networks is the *feedforward multilayer network* in which the basic processing unit is a linear threshold gate. In threshold logic a linear threshold gate computes a Boolean function f [16]. Effectiveness of threshold logic as an alternative for contemporary logic gates is determined by the availability cost, and capabilities of the basic building blocks [22,26].

Low power arithmetic circuits design.   The hybrid signed digit (HSD) representation, which employs both signed and unsigned digits, and renders the maximum length of carry propagation equal to the maximum distance between two consecutive signed digits, was introduced in Ref. [18]. Low power strategy for VLSI circuit design is outlined in Ref. [19]. In Ref. [2], low-power design style for arithmetic circuits (adders and multipliers) is studied. Deployment of binary decision diagrams in the low power design is considered in Ref. [15].

Verification and testing of arithmetic circuits.   The verification problem is formulated as follows: given two circuits C_1 and C_2 with the same number of inputs and outputs, verify that C_1 and C_2 produce for each input assignment the same output sequence. There are two approaches: a complete ordered decision tree can be generated from the circuit and compared for equivalency; alternatively, ROBDDs can be generated and compared. This approach is possible due to the fact that both complete ordered decision tree and ROBDD are canonical forms. The second approach is more compact: the space of global properties of circuits C_1 and C_2 is reduced to local properties [29].

Decision diagram technique for arithmetic circuits design.   An overview of the technique can be found in Ref. [29] and this handbook.

Multi-valued arithmetic circuits.   Multi-valued logic provides new possibilities for design and implementation of adders, multiplies and dividers. In multi-valued logic m-level signals are used to carry information instead of two levels, 0 and 1, in conventional computers [5]. Arithmetic transforms of a multi-valued function generate pseudo-arithmetic polynomials, which can be linearized using bit manipulation in word-level representations of multi-output functions [30].

Various aspects of the theory and application of arithmetic circuit designs based on multi-valued gates can be found in the Proceedings of the Annual IEEE Symposia on multiple valued logic, started in 1970.

Arithmetic computing and new technologies.   Research in arithmetic computing is generally technologically dependent. Novel devices are being investigated, as are the algorithms for arithmetic computations. The features of the logic and arithmetic implementation for *nanodevices* are considered in Refs. [9,28] and [23].

The homogenous, highly parallel arithmetic circuits, in particular systolic structures [14–21], cellular automata [3], and error correction coding [20] developed in the last decade, are the focus in nanotechnology. The logarithmic multiplier can be considered a good candidate for nanotechnology. This multiplier computes the product of two terms. The property used is $\log(A \times B) = \log(A) + \log(B)$. To obtain the logarithm of a number, the look-up tables, recursive algorithms or the segmentation of the logarithmic curve can be used [4].

The reliability problem is the problem of the design of a reliable machine from unreliable elements formulated by von Neumann. Reliability is actively investigated with respect to nanodevice design.

References

1. B.D. Brown and H.C. Card, "Stochastic neural computation I: computational elements," *IEEE Trans. Comput.*, vol. 50, no. 9, pp. 891–905, 2001.

2. T.K. Callaway, E.E. Swartzlander, "Low-power arithmetic components," in *Low Power Design Methodologies*, J.M. Rabaey and M. Pedram, Eds., Dordrecht, Holland: Kluwer Academic Publishers, 2001, pp. 161–200 (Chapter 7).

3. A. Clementi, G.A. De Biase, and A. Massini, "Fast parallel arithmetic on cellulat automata," *Complex Syst.*, vol. 8, no. 6, pp. 435–441, 1994.

4. J.N. Coleman, C.I. Chester, C.I. Softley, and J. Kaldec, "Arithmetic on the European logarithmic microprocessor," *IEEE Trans. Comput.*, vol. 49, no. 7, pp. 702–715, 2000.

5. G. Epstein, *Multi-Valued Logic Design*, London, UK: Institute of Physics Publishing, 1993.

6. M.D. Ercegovac, T. Lang, and J.H. Moreno, *Introduction to Digital Systems*, New York, NY: Wiley, 1999.

7. M.D. Ercegovac and T. Lang, *Digital Arithmetic*, Los Altos, CA: Morgan Kaufmann, 2003.

8. B.R. Gaines, "Stochastic computing systems," in *Advances in Information Systems Science*, vol. 2, New York, NY: Plenum, vol. 2, pp. 37–172, 1969 (Chapter 2).

9. T. Gramß, S. Bornholdt, M. Groß, M. Mitchell, and T. Pellizzari, *Non-Standard Computation*, New York, NY: Wiley VCH Verlag, 1998.

10. D. Goldberg, "Computer arithmetic, appendix A of J. L. Hennesy, and D. A. Patterson," *Computer Architecture: A Quantitive Approach*, Los Altos, CA: Morgan Kaufmann Publishers, 1990.

11. Y. Harata, Y. Nakamura, H. Nagase, M. Takigawa, and N. Takagi, "A high-speed multiplier using a redundant binary adder tree," *IEEE J. Solid States Circ.*, vol. 22, no. 1, pp. 28–34, 1987.

12. C.L. Janer, J.M. Quero, J.G. Ortega, and L.G. Franquelo, "Fully parallel stochastic computation architecture," *IEEE Trans. Signal Process.*, vol. 44, no. 8, pp. 2110–2117, 1996.

13. E. Kinoshita and Ki-Ja. Lee, "A residue arithmetic extension for reliable scientific computation," *IEEE Trans. Comput.*, vol. 46, no. 2, pp. 129–138, 1997.

14. S.Y. Kung, *VLSI Array Processors*, Englewood Cliffs, NJ: Prentice-Hall, 1988.

15. P. Lindgren, M. Kerttu, M. Thornton, and R. Drechsler, "Low-power optimization techniques for BDD mapped circuits," in *Proc. Asia-Pacific Design Autom. Conf.*, pp. 615–621, 2001.

16. S. Muroga, *Threshold Logic and its Applications*, New York, NY: Wiley-Interscience, 1971.

17. D. Patterson and J. Hennessy, *Computer Organization and Design: The Hardware/Software Interface*, San Mateo, CA: Morgan Kaufmann, 1994.

18. D.S. Phatak and I. Koren, "Hybrid signed–digit number systems: a unified framework for redundant number representations with bounded carry propagation chains," *IEEE Trans. Comput. Spec. Issue Comp. Arith.*, vol. TC–43, no. 8, pp. 880–891, 1994.

19. J.M. Rabaey, M. Pedram, and P.E. Landman. "Introduction," in *Low Power Design Methodologies*. J.M. Rabaey and M. Pedram, Eds., Dordrecht, Holland: Kluwer Academic Publishers, pp. 2–18, 2001.

20. T.R.N. Rao, *Error Coding for Arithmetic Processors*, New York, NY: Academic Press, 1974.

21. B.P. Sinha and P.K. Srimani, "Fast parallel algorithms for binary multiplication and their implementation on systolic architectures," *IEEE Trans. Comput.*, vol. 38, no. 3, pp. 424–431, 1989.

22. K.Y. Siu, and J. Bruck, "Neural computation of arithmetic functions," *Proc. IEEE*, 78, 1669–1675, 1990.

23. V.P. Shmerko and S.N. Yanushkevich, "Three-dimensional feedforward neural networks and their realization by nano-devices," *Artif. Intelligence Rev., An Int. Science Eng. J.*, vol. 20, no. 3–4, pp. 473–494, (Special Issue on "Artificial Intelligence in Logic Design," UK), 2003.

24. B. Vinnakota and V.V.B. Rao, "Fast conversion technique for binary-residue number systems," *IEEE Trans. Circ. Sys. I*, vol. 41, no. 12, pp. 927–929, 1994.

25. Y. Wang, "Residue-to-binary converters based on new Chinese remainder theorems," *IEEE Trans. Circ. Sys. II*, 197–206, 2000 (March).

26. J. Webster, Ed., *Encyclopedia of Electrical and Electronics Engineering. Threshold Logic*, Vol. 22, New York, NY: Wiley, pp. 178–190, 1999.

27.   S. Winograd and J.D. Cowan, *Reliable Computation in the Presence of Noise*, Cambridge, MA: MIT Press, 1963.

28.   S.N. Yanushkevich, V.P. Shmerko, and S.E. Lyshevski, *Logic Design of NanoICs*, Boca Raton, FL: CRC Press, 2005.

29.   S.N. Yanushkevich, D.M. Miller, V.P. Shmerko, and R.S. Stankovic, *Handbook on Decision Diagram Technique for Electrical Engineer*, Boca Raton, FL: CRC Press, 2005.

30.   S.N. Yanushkevich, P. Dziurzanski, and V. P. Shmerko, "Word-level models for efficient computation of multiple-valued functions. Part 1: LAR based models," in *Proc. IEEE 32th Int. Symp. Multiple-Valued Logic*, Boston, MA, pp. 202–208, 2002 (May).

8.3   Architecture*

Carl Hamacher, Zvonko Vranesic, and Safwat Zaky

Computer architecture can be defined here to mean the functional operation of the individual hardware units in a computer system and the flow of information and control among them. This is a somewhat more general definition than is sometimes used. For example, some articles and books refer to instruction set architecture or the system bus architecture.

The main functional units of a single-processor system, a basic way to interconnect them, and features that are used to increase the speed with which the computer executes programs will be described. Following this, a brief introduction to systems that have more than one processor will be provided.

Functional Units

A *digital computer*, or simply a computer, accepts digitized input information, processes it according to a list of internally stored *machine instructions,* and produces the resultant output information. The list of instructions is called a *program*, and internal storage is called *computer memory.*

A computer·has five functionally independent main parts: input, memory, arithmetic and logic, output, and control. The input unit accepts digitally encoded information from human operators, through electromechanical devices such as a keyboard, or from other computers over digital communication lines. The information received is usually stored in the memory and then operated on by the arithmetic and logic unit circuitry under the control of a program. Results are sent back to the outside world through the output unit. All these actions are coordinated by the control unit. The arithmetic and logic unit, in conjunction with the main control unit, is referred to as the **processor.**

Input and output equipment is usually combined under the term **input-output unit** (*I/O unit*). The simplest example is the video terminal consisting of a keyboard for input and a cathode-ray tube or flat panel display for output.

The **memory unit** stores programs and data. There are two main classes of memory devices called *primary* and *secondary* memory. Primary storage, or main memory, is an electronic storage device, constructed from integrated circuits that consist of many millions of semiconductor storage cells, each capable of storing one bit of information. These cells are accessed in groups of fixed size called *words*. A word is typically four or eight *bytes* long, with a byte consisting of eight bits. The main memory is organized so that the contents of one word can be stored or retrieved in one basic operation called a *memory cycle*.

To provide direct access to any word in the main memory in a short and fixed amount of time, a distinct address number is associated with each word location. A given word is accessed by specifying its address and issuing a control command that starts the storage or retrieval process. Small machines such as notebook computers or personal computers typically have or few hundred megabytes of main memory, while large

---

*Adapted from V.C. Hamacher, Z.G. Vranesic, and S.G. Zaky, *Computer Organization*, 4th ed., New York, NY: McGraw-Hill, 1996. With permission.

server machines can have significantly larger main memories. The time required to access a word for reading or writing is less than 100 ns.

Although primary memory is essential, it tends to be expensive and volatile. Thus cheaper, more permanent, secondary storage is used for files of information that contain programs data. A wide selection of suitable devices is available, including magnetic disks, drums, diskettes, tapes and optically accessed CDs.

Execution of most operations within a computer takes place in the **arithmetic and logic unit** (ALU) of a processor. Consider a typical example. Suppose that two numbers located in the main memory are to be added, and the sum is to be stored back into the memory. Using a few instructions, each consisting of a few basic steps, determined by the **control unit**, the operands are first fetched from the memory into the processor. They are then added in the ALU, and the result is stored back in memory. Processors contain a number of high-speed storage elements called *registers,* which are used for temporary storage of operands. Each register contains one word of data and its access time is about 100 times faster than main memory access time. Large-scale microelectronic fabrication techniques allow processors to be implemented on a single semiconductor chip containing many millions of transistors.

Basic Operational Concepts

To perform a given computational task, an appropriate program consisting of a set of machine instructions is stored in the main memory, usually one instruction per word. Individual instructions are brought from the memory into the processor for execution. Data used as operands are also stored in the memory. A typical instruction may be

$$\text{MOVE} \quad \text{MEMLOC}, Ri$$

This instruction loads a copy of the operand at memory location MEMLOC into the processor register Ri. The instruction requires a few basic steps to be performed. First, the instruction must be transferred from the memory into the processor, where it is decoded. Then the operand at location MEMLOC must be fetched into the processor. Finally, the operand is placed into register Ri. After operands are loaded into the processor registers in this way, instructions such as

$$\text{ADD} \quad Ri, Rj, Rk$$

can be used to add the contents of registers Ri and Rj, and then place the result into register Rk.

Instruction set design has been intensively studied to determine the effectiveness of the various alternatives. See Patterson and Hennessey [2005] for a thorough discussion.

The connection between the main memory and the processor that allows for the transfer of instructions and operands is called the **bus**, as shown in Figure 8.16. A bus consists of a set of address, data, and control lines. The bus also allows program and data files to be transferred from their long-term location on magnetic disk or CD storage to the main memory. Long distance digital communication with other computers is also enabled by transfers over the bus to the communication line interface, as shown in the figure. The bus interconnects a number of devices, but only two devices (a sender and a receiver) can use it at any one time. Therefore, some control circuitry is needed to manage the orderly use of the bus when a number of devices wish to use it.

Normal execution of programs may sometimes be preempted if some I/O device requires urgent control action or servicing. For example, a monitoring device in a computer-controlled industrial process may detect a dangerous condition that requires the execution of a special service program dedicated to the device. To cause this service program to be executed, the device sends an *interrupt* signal to the processor. The processor temporarily suspends the program that is being executed and executes the special *interrupt service routine.* After providing the required service, the processor switches back to the interrupted program. To appreciate the complexity of the computer system software programs needed to control such switching from one program task to another and to manage the general movement of programs and data between primary and secondary storage, consult Tanenbaum [1999].

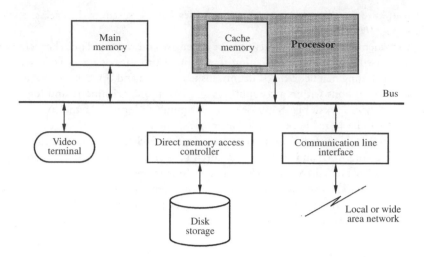

FIGURE 8.16   Interconnection of major components in a computer system.

The need often arises during program loading and execution to transfer blocks of data between the main memory and a disk or other secondary storage I/O device. Special control circuits are provided to manage these transfers without detailed control actions from the main processor. Such transfers are referred to as *direct memory access* (DMA). Assuming that accesses to the main memory from both I/O devices (such as disks) and the main processor can be appropriately interwoven over the bus, I/O-memory transfers and computation in the main processor can proceed in parallel, and performance of the overall system is improved.

Performance

A major performance measure for computer systems is the time, T, that it takes to execute a complete program for some task. Suppose N machine instructions need to be executed to perform the task. A program is typically written in some high-level language, translated by a compiler program into machine language, and stored on a disk. An operating system software routine then loads the machine language program into the main memory, ready for execution. Assume that, on average, each machine language instruction requires S basic steps for its execution. If basic steps are executed at the rate of R steps per second, then the time to execute the program is

$$T = (N \times S)/R$$

The main goal in computer architecture is to develop features that minimize T.

We will now give an outline of main memory and processor design features that help to achieve this goal. The first concept is that of a **memory hierarchy**. We have already noted that access to operands in processor registers is significantly faster than access to the main memory. Suppose that when instructions and data are first loaded into the processor, they are stored in a small, fast **cache memory** on the processor chip itself. If instructions and data in the cache are accessed repeatedly within a short period of time, as happens often with program loops, then program execution will be speeded up. The cache can only hold small parts of the executing program. When the cache is full, its contents are replaced by new instructions and data as they are fetched from the main memory. A variety of *cache replacement algorithms* are in use. The objective of these algorithms is to maximize the probability that the instructions and data needed for program execution are found in the cache. This probability is known as the cache *hit ratio*. A higher hit ratio means that a larger percentage of the instructions and data are

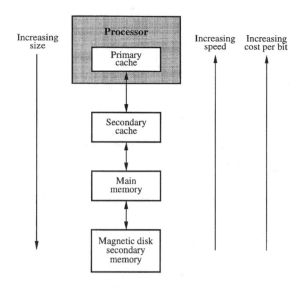

FIGURE 8.17   Memory hierarchy.

being found in the cache, and do not require access to the slower main memory. This leads to a reduction in the memory access basic step time components of S, and hence to a smaller value of T.

The basic idea of a cache can be applied at different points in a computer system, resulting in a hierarchy of storage units. A typical memory hierarchy is shown in Figure 8.17. Some systems have two levels of cache to take the best advantage of size/speed/cost tradeoffs. The main memory is usually not large enough to contain all of the programs and their data. Therefore, the highest level in the memory hierarchy is usually magnetic disk storage. As the figure indicates, it has the largest capacity, but the slowest access time. Segments of a program, often called *pages*, are transferred from the disk to the main memory for execution. As other pages are needed, they may replace the pages already in the main memory if the main memory is full. The orderly, automatic movement of large program and data segments between the main memory and the disk, as programs execute, is managed by a combination of operating system software and control hardware. This is referred to as **memory management**.

We have implicitly assumed that instructions are executed one after another. Most modern processors are designed to allow the execution of successive instructions to overlap, using a technique known as **pipelining**. In the example in Figure 8.18, each instruction is broken down into 4 basic steps—fetch, decode, operate and write—and a separate hardware unit is provided to perform each of these steps. As a result, the execution of successive instructions can be overlapped as shown, resulting in an instruction completion rate of one per basic time step. If the execution overlap pattern shown in the figure can be maintained for long periods of time, the effective value of S tends toward 1.

When the execution of some instruction I depends on the results of a previous instruction, J, which is not yet completed, instruction I must be delayed. The pipeline is said to be *stalled*, waiting for the execution of instruction J to be completed. While it is not possible to eliminate such situations altogether, it is important to minimize the probability of their occurrence. This is a key consideration in the design of the instruction set of modern processors and the design of the compilers that translate high-level language programs into machine language.

Now, imagine that multiple functional units are provided in the processor so that more than one instruction can be in the operate stage. This *parallel* execution capability, when added to pipelining of the individual instructions, means that execution rates of more than one instruction completion per basic step time can be achieved. This mode of enhanced processor performance is called **superscalar processing**.

The rate, R, of performing basic steps in the processor is usually referred to as the processor clock rate; and it is of the order of a billion steps per second in current high-performance VLSI processors. This rate is

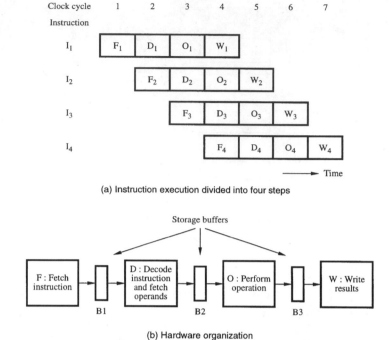

(a) Instruction execution divided into four steps

(b) Hardware organization

FIGURE 8.18   Pipelining of instruction execution.

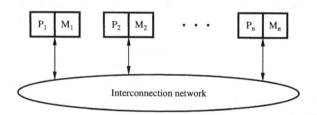

FIGURE 8.19   A multiprocessor system.

determined by the technology used in fabricating the processors, and is strongly related to the size or area occupied by individual transistors. This size feature, which is currently a small fraction of a micron, has been steadily decreasing as fabrication techniques improve, allowing increases in R to be achieved.

Multiprocessors

Physical limits on electronic speeds and power dissipation prevent single processors from being speeded up indefinitely. A major design trend has seen the development of systems that consist of a large number of processors. Such multiprocessors can be used to speed up the execution of large programs by executing subtasks in parallel. The main difficulty in achieving this type of speedup is in being able to decompose a given task into its parallel subtasks and assign these subtasks to the individual processors in such a way that communication among the subtasks can be done efficiently. Figure 8.19 shows a block diagram of a multiprocessor system, with the interconnection network needed for data sharing among the processors *Pi*. Parallel paths are needed in this network in order for parallel activity to proceed in the processors as they access the global memory space represented by the multiple memory units *Mi*.

Defining Terms

Arithmetic and logic unit: The logic gates and register storage elements used to perform the basic operations of addition, subtraction, multiplication and division of numeric operands, and the comparison, shifting and alignment operations on more general forms of numeric and nonnumeric data.

Bus: The collection of data, address and control lines that enables exchange of information, usually in word-size quantities, among the various computer system units. In practice, a large number of units can be connected to a single bus. These units contend in an orderly way for the use of the bus for individual transfers.

Cache memory: A high-speed memory for temporary storage of copies of the sections of program and data from the main memory that are currently active during program execution.

Computer architecture: The functional operation of the individual hardware units in a computer system and the flow of information and control among them.

Control unit: The circuits required for sequencing the basic steps needed to execute machine instructions.

Input-output unit (I/O):   The equipment and controls necessary for a computer to interact with a human operator, to access mass storage devices such as disks and tapes, or to communicate with other computer systems over communication networks.

Memory hierarchy: The collection of cache, primary, and secondary memory units that comprise the total storage capability in the computer system.

Memory management: The combination of operating system software and hardware controls that is needed to access and move program and data segments up and down the memory hierarchy during program execution.

Memory unit: The unit responsible for storing programs and data. There are two main types of units: primary memory, consisting of billions of bit storage cells fabricated from electronic semiconductor integrated circuits, used to hold programs and data during program execution; and secondary memory, based on magnetic disks, CDs and tapes, used to store permanent copies of programs and data.

Multiprocessor: A computer system comprising multiple processors and main memory unit modules, connected by a network that allows parallel activity to proceed efficiently among these units in executing program tasks that have been sectioned into subtasks and assigned to the processors.

Pipelining: The overlapped execution of the multiple steps of successive instructions of a machine language program, leading to a higher rate of instruction completion than can be attained by executing instructions strictly one after another.

Processor: The arithmetic and logic unit combined with the control unit needed to sequence the execution of instructions. Some cache memory is also included in the processor.

Superscalar processing: The ability to execute instructions at a completion rate that is faster than the normal pipelined rate, by providing multiple functional units in the pipeline to allow a small number of instructions to proceed through the pipeline process in parallel.

References

C. Hamacher, Z. Vranesic, and S. Zaky, *Computer Organization*, 5th ed., New York, NY: McGraw-Hill, 2002.

D.A. Patterson and J.L. Hennessey, *Computer Organization and Design—The Hardware/Software Interface*, 3rd. ed. San Francisco, CA: Morgan Kaufman, 2005.

A.S. Tanenbaum, *Structured Computer Organization*, 4th ed., Upper Saddle River, NJ: Prentice-Hall, 1999.

Further Information

The IEEE magazines *Computer*, *Micro*, and *Software* all have interesting articles on subjects related to computer architecture, including software aspects.

8.4   Microprogramming

Jacques Raymond

Since the 1950s when Wilkes et al. [1958] defined the term and the concept, microprogramming has been used as a clean and systematic way to define the instruction set of a computer. It has also been used to define a virtual architecture out of a real hardwired one.

Levels of Programming

In Figure 8.20, we see that a computer application is usually realized by programming a given algorithm in a high-level language. A system offering a high-level language capability is implemented at the system level via a compiler. The operating system is (usually) implemented in a lower-level language. The machine instruction set can be hardwired (in a hardware implementation) or implemented via microprogramming (Figure 8.21).

Therefore, microprogramming is simply an extra level in the general structure. Since it is used to define the machine instruction set, it can be considered at the hardware level. Since this definition is done via a program at a low level, but is still eventually modifiable, it can also be considered to be at the software level. For these reasons, the term firmware has been coined to name sets of microprograms. In short, microinstructions that specify hardware functions (microoperations such as Open a path, Select operation) are used to form a more complex instruction (Convert to binary, Add decimal). The machine instruction set is defined via a set of microprogram routines and a microprogrammed instruction decoder.

In a microprogrammed machine, the hardware is designed in terms of its capabilities (ALU, data paths, I/O, processing units) with little concern for how these capabilities will have to be accessed by the programmers. The microoperations are decoded from microinstructions. The way programmers view the machine is defined at the microprogramming level.

This approach offers some advantages over the hardwired approach. The advantages are that it is more systematic in implementation, modifiable, economical on most designs, and easier to debug. The disadvantages are that it is uneconomical on simple machines, slower, and needs support software. Like all programs, microprograms reside in memory. The term "control memory" is commonly used for microprograms.

Microinstruction Structure

On a given hardware, many processing functions are available. In general a subset O of these functions can be performed in parallel, for example, carrying on an addition between two registers while copying a register on an I/O bus. These functions are called microcommands.

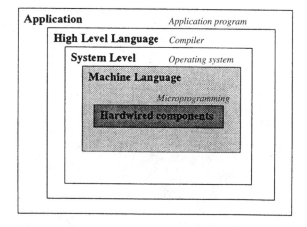

FIGURE 8.20   Levels of programming in a computer system.

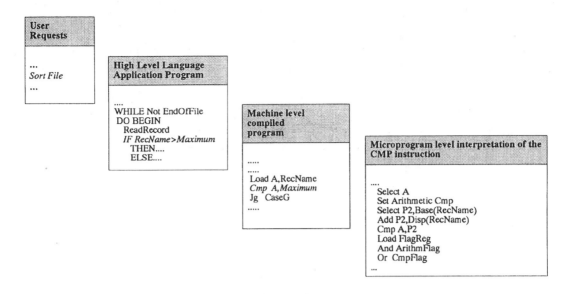

FIGURE 8.21 A view of computer system levels.

Horizontal Microinstructions

Each of the fields f of a microinstruction specifies a microcommand. If the format of the microinstruction is such that all possible microcommands can be specified, the instruction is called horizontal. Most of the time, it is wasteful in memory as, in a microprogram, not every possible microcommand is specified in each microinstruction. However, it permits the microprogrammer to fully take advantage of all possible parallelisms and to build faster machines.

For example, the horizontal specification of an ALU operation,

ALUOperation	SourcePathA	SourcePathB	ResultPath	CvtDecimal

specifies both operands, which register will contain the result, whether or not the result is to be converted to decimal, and the operation to be performed. If this instruction is, for example, part of a microprogram defining a 32-bit addition instruction, assuming a 16-bit path, it is wasteful to specify twice the source and result operands.

In some cases, it is possible to design a microinstruction that specifies more microcommands than can be executed in parallel. In that case, the execution of the microcommand is carried out in more than one clock cycle. For this reason they are called polyphase microinstructions (as opposed to monophase).

Schemes have been used to optimize the size of the word—for example, bit steering, where the value of one bit determines how a field is to be interpreted, reduces word size by combining nonparallel function specifications in one field

Vertical Microinstructions

At the other extreme, if the microinstruction allows only the specification of a single microcommand at a time, the instruction is then called vertical. In that case, only the necessary commands for a particular program are specified, resulting in smaller control memory requirements. However, it is not possible to take advantage of possible parallelism offered by the hardware, since only one microcommand is executed at a time. For example, the vertical specification of an ALU operation is as follows:

SourceA	Reg#	1st Operand
SourceB	Reg#	2nd Operand
Result	Reg#	Result
ALU	Op	Operation

Diagonal Microinstructions

Most cases fit in between these two extremes (see Figure 8.22). Some parallelism is possible; however, microcommands pertaining to a given processing unit are regrouped. This results in shorter microprograms than in the vertical case and may still allow some optimization. For example, a diagonal specification of an ALU operation is as follows:

SelectSources	RegA#	RegB#	Select Operands
SelectResult	Reg#	Dec/Bin	Result place and format
Select ALU Operation			Perform the operation

Optimization

Time and space optimization studies can be performed before designing the microinstruction format. The reader is referred to Das et al. [1973] and Agerwala [1976] for details and more references.

Microprogram Development

Microassemblers

The first level of specification of microinstructions is, just like its counterpart at the machine level, the assembler. Although the process and philosophy is exactly the same, it is traditionally called a microassembler. A microassembler is a software program (it is not relevant to know in which language it is written) whose function is to translate a source program into the binary code equivalent. Obviously, to write a source program, a language has to be designed. At assembly level, languages are usually very close to the hardware structure, and the objects defined are microregisters, gate level controls, and paths. Operations are the microoperations (sometimes slightly more sophisticated with a microassembler with macrofacilities).

This level provides an easily readable microprogram and does much to help avoid syntax errors. In binary, only the programmer can catch a faulty 1 or 0; the microassembler can catch syntax errors or some faulty register specifications. No microassembler exists that can catch all logic errors in the implementation of a

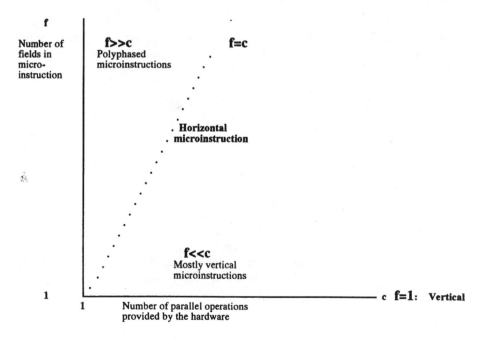

FIGURE 8.22   Microinstruction fields versus microcommands.

given instruction algorithm. It is still very easy to make mistakes. It should be noted that this level is a good compromise between convenience and cost.

The following is a typical example of a microprogram in the microassembler (it implements a 16-bit add on an 8-bit path and ALU):

```
CLC Clear        Carry
Lod   A          Get first part of first operand
Add   B          Add to first part of second operand
Sto   C          Give low byte of final result
Lod   a          Get second part of first operand
Adc   b          Add to second part of second operand and to carry bit
Sto   c          Give high byte of final result
JCS   Error      Jump to error routine if Result > 65536
Jmp   FetchNext
```

High-Level Languages for Microprogramming

Many higher-level languages have been designed and implemented: see a discussion of some design philosophies in Malik and Lewis [1978]. In principle a high-level language program is oriented toward the application it supports and is farther away from the hardware-detailed implementation of the machine it runs on. The applications supported are mostly other machine definitions (emulators) and implementations of some algorithms.

The objects defined and manipulated by high-level languages for microprogramming are therefore the virtual components of virtual machines. They are usually much the same as their real counterparts: registers, paths, ALUs, microoperations, etc. Furthermore, writing a microprogram is usually defining a machine architecture. It involves a lot of intricate and complicated details, but the algorithms implemented are mostly quite simple. The advantages offered by high-level languages to write better algorithms without getting lost in implementation details are therefore not exploited.

Firmware Implementation

Microprogramming usually requires the regular phases of design, coding, test, conformance acceptance, and documentation. It differs from other programming activities when it comes to the deliverable. The usual final product takes the form of hardware, namely a control memory, PROM, ROM, or other media containing the bit patterns equivalent to the firmware. These implementation steps require special hardware and software support. They include a linker, loader, PAL programmer, or ROM burner; a test program in a control memory test bench is also necessary.

Supporting Software

It is advisable to test the microprogram before its actual hard implantation, if the implantation process is irreversible or too costly to repeat. Software simulators have been implemented to allow thorough testing of the newly developed microprogram. Needless to say, these tools are very specialized to a given environment and therefore costly to develop, as their development cost cannot be distributed over many applications.

Emulation

Concept

In a microprogrammed environment a computer architecture is softly (or firmly) defined by the microprogram designed to implement its operations, its datapaths, and its machine-level instructions. It is easy to see that if one changes the microprogram for another one, then a new computer is defined. In this environment the desired operation is simulated by the execution of the firmware, instead of being the result of action on real hardwired components.

Since the word simulation was already in use for simulation of some system by software, the word emulation was chosen to mean simulation of an instruction set by firmware. Of course, "simulation" by hardware is not a simulation, but the real thing.

The general structure of an emulator consists of the following pseudocode algorithm:

```
BEGIN
Initialize Machine Components
Repeat

Fetch Instruction
Emulate Operation of the current instruction
Process interrupts
Update instruction counter

Until MachineIsOff

Perform shutdown procedure
END
```

Many variations exist, in particular to process interrupts within the emulation of a lengthy operation or to optimize throughput, but the general principle and structure are fairly constant.

Emulation of CPU Operation

One of the advantages of microprogramming is that the designer can implement his or her dream instructions simply by emulating its operation. We have seen already the code for a typical 16-bit adder, but it is not difficult to code a parity code generator, a cyclic redundancy check calculator, or an instruction that returns the eigenvalues of an $n * n$ matrix. This part is straight programming. One consideration is to make sure that the machine is still listening to the outside world (interruptions), or actively monitoring it (I/O flags) in order not to lose asynchronous data while looking for a particular pattern in a 1 megabyte string. Another consideration is to optimize memory usage by combining common processes for different operations. For example, emulating a 32-bit add instruction and emulating a 16-bit add instruction have common parts. This is, however, a programming concern not specific to emulation. RISC has made the point that there is not much to be gained by implementing complex instructions in microcode rather than in assembler.

I/O System and Interrupts

Programming support for I/Os and interrupts is more complicated than for straight machine instructions. This is due to the considerable speed differences between I/O devices and a CPU, the need for synchronization, the need for not losing any external event, and the concerns for optimizing processing time. Microprogramming offers considerable design flexibility, as these problems are more easily handled by programming than with hardware components.

Other Applications of Microprogramming

The main application of microprogramming is the emulation of a virtual machine architecture on a different host hardware machine. It is, however, easy to see that the concept of emulation can be broadened to other functions than the traditional hardware operation.

It is mainly a matter of point of view. Emulation and simulation are essentially the same process but viewed from different levels. Realizing a 64-bit addition and implementing a communication controller are qualitatively the same type of task. Once this is considered, there are theoretically no limits to the uses of microprogramming.

From the programmer's point of view, programming is the activity of producing, in some language, an implementation of some algorithm. If the language is at the very lowest level, as is the case with microprogramming, and at the same time the algorithm is filled with intricate data structures and complex decisions, the task might be enormous, but nothing says it cannot be done (except, maybe, experience). With this perspective of the field, we now look at some applications of microprogramming.

Design Optimization

Microprogramming helped design schemes aimed at improving parallel execution of operations, and thus optimizing performance—for example, very long instruction word (VLIW), or explicit parallel instruction computing (EPIC).

Operating System Support

One of the first applications, besides emulation, was to support some operating system functions. Since microprograms are closer to the hardware and programming directly in microcode removes the overhead of decoding machine-level instructions, it was thought that directly coding operating system (OS) functions would improve their performance. Success was achieved in some areas, such as virtual memory. In general, people write most OS functions in assembly language probably because the cost is not offset by the benefits, especially with rapidly changing OS versions. The problems raised by the human side of programming have changed the question "Should it be in microcode or in assembler?" to the question "Should it be in assembler or in C?" This parallels the CISC/RISC debate.

High-Level Languages Support

Early research was done also in the area of support for high-level languages. Support can be in the form of microprogrammed implementations of some language primitive (for example, the trigonometric functions) or support for the definition and processing of data structures (for example, trees and lists primitives). Many interesting research projects have led to esoteric laboratory machines. More common examples include the translate instructions, string searches and compares, or indexing multidimensional arrays.

The OO paradigm has also been incorporated into CPU design. See Van der Hoeven et al. [1993].

Paging, Virtual Memory

An early and typical application of microprogramming is the implementation of the paging algorithm for a virtual memory system. It is a typical application since it is a low-level function that must be time-optimized and is highly hardware dependent. Furthermore, the various maintenance functions which are required by the paging algorithms and the disk I/Os can be done during the idle time of the processing of other functions or during part of that processing in order to avoid I/O delays.

Diagnostics

Diagnostic functions have also been an early application of microprogramming. A firmware implementation is ideally suited to test the various components of a computer system, since the gates, paths, and units can be exercised in an isolated manner, therefore allowing one to precisely pinpoint the trouble area.

Controllers

Real-time controllers benefit from a microprogrammed implementation due to the speed gained by programming only the required functions, therefore avoiding the overhead of general-purpose instructions. Since the microprogrammer can better make use of the available parallelism in the machine, long processes can still support the asynchronous arrival of data by incorporating the interrupt polling at intervals in these processes.

High-Level Machines

Machines that directly implement the constructs of high-level languages can be easily implemented via microprogramming. For example, Prolog machines and Lisp machines have been tried. It is also possible to conceive of an application that is directly microcoded. Although this could provide a high-performance hardware, human errors and software engineering practice seem to make such a machine more of a curiosity than a maintainable system.

Defining Terms

Control memory: A memory containing a set of microinstructions (a microprogram) that defines the instruction set and operations of a CPU.

Emulator: The firmware that simulates a given machine architecture.

Firmware: Meant as an intermediate between software, which can be modified very easily, and hardware, which is practically unchangeable (once built); the word firmware was coined to represent the microprogram in control memory—i.e., the modifiable representation of the CPU instruction set.

High-level language for microprogramming: A high-level language more or less oriented toward the description of a machine. Emulators can more easily be written in a high-level language; the source code is compiled into the microinstructions for actual implementation.

Horizontal microinstruction: Theoretically, a completely horizontal microinstruction is made up of all the possible microcommands available in a given CPU. In practice, some encoding is provided to reduce the length of the instruction.

Microcommand: A small bit field indicating if a gate is open or closed, if a function is enabled or not, if a control path is active or not, etc. A microcommand is therefore the specification of some action within the control structure of a CPU.

Microinstruction: The set of microcommands to be executed or not, enabled or not. Each field of a microinstruction is a microcommand. The instruction specifies the new state of the CPU.

Vertical microinstruction: A completely vertical microinstruction would contain one field and therefore would specify one microcommand. An Op code is used to specify which microcommand is specified. In practice microinstructions that typically contain three or four fields are called vertical.

References

T. Agerwala, "Microprogram optimization: a survey," *IEEE Trans. Comput.*, vol. C25, no. 10, pp. 862–873, 1976.

J.D. Bagley, "Microprogrammable virtual machines," *Computer*, vol. 9, no. 3, pp. 38–42, 1976.

D.K. Banerji and J. Raymond, *Elements of Microprogramming*, Englewood Cliffs, NJ: Prentice-Hall, 1982.

J. Carter, *Microprocessor Architecture and Microprogramming, a State Machine Approach*, Englewood Cliffs, NJ: Prentice-Hall, 1996.

G.F. Casaglia, "Nanoprogramming vs. microprogramming," *Computer*, vol. 9, no. 1, pp. 54–58, 1976.

S.R. Das, D.K. Banerji, and A. Chattopadhyay, "On control memory minimization in microprogrammed digital computers," *IEEE Trans. Comput.*, vol. C22, no. 9, pp. 845–848, 1973.

S. Habib, Ed., *Microprogramming and Firmware Engineering Methods*, New York, NY: Van Nostrand Reinhold, 1988.

L.H. Jones, "An annotated bibliography on microprogramming," *SIGMICRO Newsletter*, Vol. 6, no. 2, pp. 8–31, 1975.

L.H. Jones, "Instruction sequencing in microprogrammed computers," *AFIPS Conf. Proc.*, Vol. 44, 1975, pp. 91–98.

K. Malik and T.J. Lewis, "Design objectives for high level microprogramming languages," in *Proc. 11th Ann. Microprogram. Workshop*, Englewood Cliffs, NJ: Prentice-Hall, 1978, pp. 154–160.

H. Okuno, N. Osato, and I. Takeuchi, "Firmware approach to fast lisp interpreter," *Twentieth Ann. Workshop on Microprogram., (MICRO-20)*, ACM, 1987.

J. Raymond and D.K. Banerji, "Using a microprocessor in an intelligent graphics terminal," *IEEE Computer*, vol. 9, no. 4, pp. 18–25, 1976.

M. Smotherman, *A Brief History of Microprogramming*, 1999. Online at http://www.cs.clemson.edu/~mark/uprog.html.

A.J. Van der Hoeven, P. Van Prooijen, E.F. Deprettere, and P.M. Dewilde, "A hardware design system based on object-oriented principles," *IEEE*, Proceedings of the Conference on European Design Automation, Amsterdam, The Netherlands, pp. 459–463, 1991.

M.V. Wilkes, W. Renwick, and D.J. Wheeler, "The design of the control unit of an electronic digital computer," *Proc. IEE*, 1958, pp. 121–128.

9

Programming

James M. Feldman
Northeastern University

Edward W. Czeck
Chrysatis Symbolic Design

Ted G. Lewis
Naval Postgraduate School

Johannes J. Martin
University of New Orleans

Michael D. Ciletti
University of Colorado

9.1 Assembly Language

James M. Feldman and Edward W. Czeck

The true language of computers is a stream of 1s and 0s—bits. Everything in the computer, be it numbers or text or program, spreadsheet or database or 3-D rendering, is nothing but an array of bits. The meaning of the bits is in the "eye of the beholder"; it is determined entirely by context. Bits are not a useful medium for human consumption. Instead, we insist that what *we* read be formatted spatially and presented in a modest range of visually distinguishable characters. 0 and 1 arranged in a dense, page-filling array do not fulfill these requirements *in any way*. The several languages that are presented in this handbook are all intended to make something readable to two quite different readers. On the one hand, they serve the human reader with his/her requirements on symbols and layout; on the other, they provide a grammatically regular language for interpretation by a **compiler**. A compiler, of course, is normally a program running on a computer, but human beings can and sometimes do play both sides of this game. They want to play with both the input and output. Such accessibility requires that not only the input but the output of the compilation process be comfortably readable by humans. The language of the input is called a **high-level language** (**HLL**). Examples are C, Pascal, Ada and Modula II. They are designed to express both regularly and concisely the kinds of operations and the kinds of constructs that programmers manipulate. The output end of the compiler generates **object code**—a generally unreadable, binary representation of machine language, lacking only the services of a **linker** to turn it into true machine language. The language that has been constructed to represent object code for human consumption is *assembly language*. That is the subject of this section.

Some might object to our statement of purpose for assembly language. While few will contest the concept of assembly language as the readable form of object code, some see writing assembly code as the way to "get their hands on the inner workings of the machine." They see it as a "control" issue. Since most HLLs today give the user reasonably direct ways to access hardware, where does the "control" issue arise? What assembly proponents see as the essential reason for having an assembly language is the option to optimize the "important" sections of a program by doing a better job of machine code generation than the compiler does. This perspective was valid enough when compilers were mediocre optimizers. It was not unlike the old days when a car came with a complete set of tools because you needed them. The same thing that has happened to cars has happened to compilers. They are engineered to be "fuel efficient" and perform their assigned functions with remarkable ability. When the cars or compilers get good enough and complex enough, the tinkerer may do more harm than good. IBM's superscalar RISC computer—the RS6000—comes with superb compilers *and no **assembler** at all*. The Pentagon took a long look at their costs of programming their immense array of computers. Contrary to popular legend, they decided to save money. The first amendment not withstanding, their conclusion was: "Thou shalt not assemble."

The four principal reasons for *not* writing assembly language are

- Any sizable programming job gets done at least four times faster in a HLL.
- Most modern compilers are good optimizers of code; some are superb.
- Almost all important code goes through revisions—maintenance. Reworking old assembly code is similar to breaking good encryption; it takes forever.
- Most important of all is portability. To move any program to a different computer, you must generate machine code for that new platform. With a program in a HLL, a new platform is almost free; all it requires is another pass through the compiler for the target platform. With assembly code, you are back to square one. Assembly code is unique to the platform.

Given all of that, the question naturally arises: Why have an article on assembly language? We respond with two reasons, both of which we employ in our work as teachers and programmers:

- An essential ingredient in understanding computer hardware and in designing new computer systems and compilers is a detailed appreciation of the operations of central processing units (CPUs). These are best expressed in assembly language. Our undergraduate Computer Engineering courses include a healthy dose of assembly language programming for this specific reason.
- If you are concerned about either CPU design or compiler effectiveness, you have to be able to look in great detail at the interface between them—*machine language*. As we have said, the easiest way to read machine language is by translating it to assembly language. This is one way to get assembly language, not by writing in it as a source of code but by running the object code itself through a backward translator called a **disassembler**. While many compilers will oblige you by providing an assembly listing if asked, often that listing does not include optimizations that occur only when the several modules are linked together, providing opportunities for truly global optimization. Some compilers "help" the reader by using **macros** (names for predefined blocks of code) in place of the real machine instructions and register assignments. The absence of the optimizations and the inclusion of unexpected macros can make the assembly listing almost useless for obtaining insight into the program's fine detail. The compilers that we have used on the DECstations and SPARC machines do macro inclusion. To see what is really going on in these machines, you must disassemble the machine code. That is precisely what the Think C® compiler on the Macintosh does when you ask for machine code. It disassembles what it just did in compiling and linking the whole program. What you see is what is really there. The code we present for the 68000 was obtained in that way.

These are important applications. Even if most or all other programming needs can be better met in HLLs, these applications are sufficient reason for many engineers to want to know something about assembly language.

There are other applications of assembly language, but they tend to be specific to rather specialized and infrequent tasks. For example, the back end of most HLL compilers is a machine code generator. To write one of those, you certainly must know something about assembly language. On rare occasions, you may find some necessary machine-specific transaction which is not supported by the HLL of choice or which requires some special micro optimization. A "patch" of assembly code is a way to fit this inexpressible thought into the program's vocabulary. These are rare events. The reason why we recommend to you this section on assembly code is that it improves your understanding of HLLs and of computer architecture.

We will take a single subroutine which we express in C and look at the machine code that is generated on two representative machines. The machines include two widely used *complex instruction set computers* (**CISCs**) and one *reduced instruction set computer* (**RISC**). These are the 68000®, the VAX®, and a SPARC®. We will have two objectives:

- To see how a variety of paradigms in HLLs are translated (or, in other words, to see what is really going on when you ask for a particular HLL operation)
- To compare the several architectures to see how they are the same and how they differ

The routine attempts to get a count of the number of numbers which occur in a block of text. Since we are seeking numbers and not *digits*, the task is more complex than you might first assume. This is why we say "attempts." The function that we present below handles all of the normal text forms:

- Integers, such as 123 or −17
- Numbers written in a fixed-point format, such as 12.3 or 0.1738
- Numbers written in a floating-point format, such as −12.7e+19 or 6.781E2

If our program were to scan the indented block of code above, it would report finding six numbers. The symbols that the program recognizes as potentially part of a number include the digits 0 to 9 and the symbols 'e', 'E', '.', '−' and '+'. Now it is certainly possible to include other symbols in legitimate numbers, such as HEX numbers or the like, but this little routine will not properly deal with them. Our purpose was not to handle all comers but to provide a routine with some variety of expression and possible application. Let us begin.

NumberCount()

We enter the program at the top with one pointer passed from the calling routine and a set of local variables comprising two integers and eight Boolean variables. Most of the Boolean variables will be used in pairs. The first element of a pair, for instance, *ees* of *ees* and *latche*, indicates that the current character is one of a particular class of non-numeric characters which might be found inside a number. If you consider that the number begins at the first digit, then these characters can occur legally only once within a given number. *ees* will be set true if the current character is the *first instance* of either 'e' or 'E'. The paired variable, *latche*, is set true if there has ever been one of those characters in the current number. The other pairs are *period* and *latchp* and *sign* and *latchs*.

There is also a pair of Booleans which indicate if the current character is a *digit* and if the scanner is currently *inside* a number. Were you to limit your numbers to integers, these two are the only Booleans which would be needed. At the top of the program, all Booleans are reset (made FALSE). Then we step through the block looking for numbers. The search stops when we encounter the first null [char(0)] marking the end of the block. Try running through the routine with text containing the three forms of number. You will quickly convince yourself that the routine works with all normal numbers. If someone writes "3..14" or "3.14ee6", the program will count 2 numbers. That is probably right in the first two cases. Who knows in the third?

Let us look at this short routine in C.

```
# define blk_length 20001
int NumberCount(char block[])
{ int count=0,inside=0,digit;
  int ees=0, latche=0, latchp=0, period=0, latchs=0, sign=0;
  char *source;

  source = block;
  do {
    digit = (*source >= '0') && (*source <= '9');
    period = (*source=='.') && inside && !latchp; && !latche;
    latchp = (latchp || period);
    ees = ((*source=='E') || (*source=='e')) && inside && !latche;
    latche = (latche || ees);
    sign = ((*source=='+') || (*source=='−')) && inside && latche && !latchs;
```

```
    latchs = (latchs || sign);
    if (inside) {
      if (!(digit || ees || period || sign)) inside=latchp=latche=latchs=0;
    }
    else if (digit) {
      count++;
      inside = 1;
    }
    source++;
    }
    while ((*source != '\0') && ((source-block)<blk_length+1));
    return count;
  }
```

To access values within the character array, the normal C paradigm is to step a pointer along the array. *Source* points at the current character in the array; *source* is the character ("what *source* points at"). *source* is initialized at the top of the program before the loop (source = block;) and incremented (source++;) at the bottom of the loop. Note the many repetitions of *source*. Each one means the same current character. If you read that expression as *the character which source is pointing to*, it looks like an invitation to fetch the same character from memory eight times. A compiler that optimizes by removing *common subexpressions* should eliminate all but the first such fetch. This optimization is one of the things that we want to look for.

For those less familiar with C, the meanings of the less familiar symbols are:

==	equal (in the logical sense)
!	not
!=	not equal
&&	and
||	or
count++	increment *count* by 1 unit (after using it)

C uses 0 as FALSE and anything else as TRUE.

Comparisons Down on the Factory Floor

Now let us see what we can learn by running this program through compilers on several quite different hosts. The items that we wish to examine include:

I. Subroutine operations comprising:
   A. Building the call block
   B. The call itself
   C. Obtaining memory space for local variables
   D. Accessing the call block
   E. Returning the function value
   F. Returning to the calling routine
II. Data operations
   A. Load and store
   B. Arithmetic
   C. Logical
   D. Text
III. Program control
   A. Looping
   B. if and the issue of multiple tests

Our objectives are to build three quite different pictures:
- An appreciation for the operations underlying the HLL statements
- An overview of the architectures of several important examples of CISC and RISC processors
- An appreciation for what a HLL optimizer should be doing for you

We will attempt to do all three all of the time.

Let us begin with the calling operations. Our first machine will be the MC68000, one of the classical and widely available CISC processors. It or one of its progeny is found in many machines and forms the heart of the Macintosh (not the PowerMac) and the early Sun workstations. Programmatically, the 68000 family shares a great deal with the very popular VAX family of processors. Both of these CISC designs derive in rather linear fashion from DEC's PDP-11 machines that were so widely used in the 1970s. Comparisons to that style of machine will be done with the SPARC, a RISC processor found in Sun, Solbourne, and other workstations.

Memory and Registers

All computers will have data stored in memory and some space in the CPU for manipulating data. Memory can be considered to be a long list of bytes (8-bit data blocks) with *addresses* (locations in the list) spanning some large range of numbers from 0 to typically 4 billion (4 GB). The memory is constructed physically by grouping chips so that they appear to form enormously deep columns of bytes, as shown in Figure 9.1. Since each column can deliver one byte on each request, the number of adjacent columns determines the number of bytes which may be obtained from a single request. Machines today have 1, 2, 4, or 8 such columns. (Some machines, the 68000 being our current example, have only 2 columns but arrange to have the CPU ask for two successive transfers to get a total of 4 bytes.) In general, the CPU may manipulate in a single step a datum as wide as the memory. For all of the machines which we will consider, that maximum datum size is 32 bits or 4 bytes. While convention would have us call this biggest datum a *word*, historical reason has led both the VAX and MC68000 to call it a *longword*. Then, 2 bytes is either a *halfword* or a *word*. We will use the VAX/68000 notation (longword, word, and byte) wherever possible to simplify the reading. To load data from memory, the CPU sends the address and the datum size to the memory and gets the datum as the reply. To store data, the address is sent and then the datum and datum size.

Some machines require that the datum be properly aligned with the stacking order of the memory columns in Figure 9.1. Thus, on the SPARC, a longword must have an address ending in 00 (xxx00 in Figure 9.1), and a word address must end in 0. The programmer who arranges to violate this rule will be greeted with an **address error**. Since the MC68000 has only two columns, it complains only if you ask for words or longwords with odd addresses. Successor models of that chip (68020, 30, and 40), like the VAX, accept any address and have the CPU read two longwords and do the proper repacking.

The instruction explicitly specifies the size and indicates how the CPU should calculate the address. An instruction to load a byte, for example, is `LB`, `MOVE.B`, or `MOVB` on the SPARC, MC68000, and VAX, respectively. These are followed immediately by an expression which specifies an address. We will discuss how to specify an address later. First, we must introduce the concept of a register.

The space for holding data and working on it in the CPU is the *register set*. Registers are a very important resource. Bringing data in from memory is quite separate from any operations on that data. Data in memory must first be fetched, then acted upon. Data in registers can be acted on immediately. Thus, the availability of registers to store very active

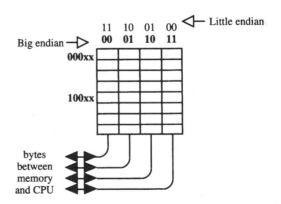

FIGURE 9.1 Memory arranged as 4 columns of bytes. The binary addressess are shown in the two formats widely used in computers. The illustration shows only 32 bytes in a 4 3 8 arrary, but a more realistic span would be 4 3 1,000,000 or 4 3 4,000,000 (4 MB to 16 MB).

variables and intermediate results makes a processor inherently faster. In some machines, most or all of the registers are tied to specific uses. The most prevalent example would be Intel's 80×86 processors, which power the ubiquitous PC. Such architectures, however, are considered quite old-fashioned. All of the machines that we are considering are of a type called *general register machines* in that they have a large group of registers which may be used for any purpose. The machines that we include have either 16 or 32 registers, with only a few tied to specific machine operations.

Table 9.1 shows the general register resources in the three machines. The SPARC is a little strange. The machine provides eight *global* registers and then a *window blind* of 128 registers which sits behind a frame which exposes 24 of the 128. A program can ask the machine to raise or lower the blind by 16 registers. That leaves an overlap of eight between successive yanks or rewinds. This arrangement is called a *multiple overlapping register set* (MORS). If you think of starting with register r8 at the bottom and r31 at the top, a yank of 16 on the blind will now have r49 at the top and r24 at the bottom. r24 to r31 are shared between the old set and the new. To avoid having to keep track of which registers are showing, the set of 24 are divided into what came *in* from the last set, those that are only *local*, and those that will go *out* to the next set. These names apply to going toward increasing numbers. In going the other direction, the *ins* of the current set will become the *outs* of the next set. Almost all other machines keep their registers screwed down to the local masonry, but you will see in a moment how useful a MORS can be. (Like other useful but expensive accessories, the debate is always on whether it is worth it [Patterson and Hennessy, 1989].)

Stack. Most subroutines define a number of local variables. NumberCount in C, for example, defines 10 local variables. While these local variables will often be created and kept in register, there is always some need for a bit of memory for each *invocation of* (call to) a subroutine. In the "good old days," this local storage was often tied to the block of code comprising the subroutine. However, such a fixed block means that a subroutine could never call itself or be called by something that it called. To avoid that problem (and for other purposes) a memory structure called a *stack* was invented which got its name because it behaved like the spring-loaded plate stack in a restaurant. Basically, it is a *last-in-first-out* (LIFO) structure whose top is defined by a pointer (address) which resides in a register commonly called the *stack pointer* or SP.

Heap. When a subroutine needs space to store local variables, it acquires that space on the stack. When the subroutine finishes, it returns that stack space for use by other routines. Thus, local variable allocations live and die with their subroutines. It is often necessary to create a data structure which is passed to other routines whose lives are independent of the creating routine. This kind of storage must be independent of the creator. To meet this need, the *heap* was invented. This is an expandable storage area managed by the system. You get an allocation by asking for it [*malloc (structure_size)* in C]. You get back a pointer to the allocation and the routine can pass that pointer to any other routine and then go away. When it comes time to dispose of the

TABLE 9.1   General Registers in the Three Machines

	Reg	Special	Names	Comments
MC68000	16	1	D0..D7 A0..A7	A(ddress) register operations are 32 bits wide. Address generation uses A registers as bases. D (data) registers allow byte, word, and longword operations. A7 is SP.
VAX	16	4	r0..r11 AP, FP, SP, PC	AP,FP,SP and PC hold the addresses of the argument block, the frame, the stack and the current place in the program, respectively. All data instructions can use any register.
SPARC	32 (136)	4	zero, g1..g7, i0..i5, FP, RA, l0..l7, o0..o5, SP, o7	The 4 groups of eight registers comprise: global (g), incoming parameters (i), local (l) and outgoing parameters (o). g0 is a hardwired 0 as a data source and a wastebasket as a destination. The registers are arranged as a window blind (see text) with the g's always visible and the others moveable in multiple overlapping frames of 24.

The special registers are within the set of general registers. Where a PC is not listed, it exists as a special register and can be used as an address when the program uses *program-relative* addressing.

allocation—that is, return the space for other uses—the program must do that actively by a deallocation call [*free*(*pointer*) in C]. Thus, one function can create a structure, several may use it, and another one can return the memory for other uses, all by passing the pointer to the structure from one to another.

Both heap and stack provide a mechanism to obtain large (or small) amounts of storage dynamically. Thus, large structures which are created only at run time need not have static space stored for them in programs that are stored on disk nor need they occupy great chunks of memory when the program does not need them. Dynamic allocation is very useful and all modern HLLs provide for it.

Since there are two types of dynamic storage, there must be some way to lay out memory so that unpredictable needs in either stack or heap can be met at all times. The mechanism is simplicity itself. The program is stuffed into low addresses in memory along with any static storage (e.g., globals) which are declared in the program. The entire remaining space is then devoted to dynamic storage. The heap starts right after the program and grows toward higher addresses; the stack goes in at the top of memory and grows down. The system is responsible to see that they never collide (a *stack crash*). When it all goes together, it looks like Figure 9.2 [Aho et al., 1986].

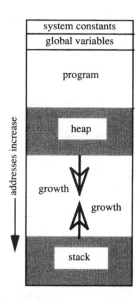

FIGURE 9.2 Layout of a program, static storage, and dynamic storage in memory.

There is one last tidbit that an assembly programmer must be aware of in looking at memory. Just as some human alphabets are written left to right and some right to left (not to mention top to bottom), computer manufacturers have chosen to disagree on how to arrange words in memory. The two schemes are called *big-endian* and *little-endian* (after which end of a number goes in the lowest-numbered byte and also after a marvelous episode in *Gulliver's Travels*). The easiest way to perceive how it is done in the two systems is to think of all numbers as being written in conventional order (left to right), but for big-endian you start counting on the upper left of the page and on little-endian you start counting on the upper right (see Figure 9.1). Since each character in a text block is *a number* of length 1 byte, this easy description makes big-endian text read in normal order (left to right) but little-endian text reads from right to left. Figure 9.3 shows the sentence "This is a sentence" followed by the two hexadecimal (HEX) numbers 01020304 and 0A0B0C0D

00			03
T	h	i	s
	i	s	
a		s	e
n	t	e	n
c	e	.	01
02	03	04	0A
0B	0C	0D	
18			1B

Big Endian
(SPARC, MC68000)

03			00
s	i	h	T
	s	i	
e	s		a
n	e	t	n
04	.	e	c
0D	01	02	03
	0A	0B	0C
1B			18

Little Endian
(VAX)

FIGURE 9.3 Byte numbering and number placement for big- and little-endian systems. Hexadecimal numbers are used for the memory addresses.

written to consecutive bytes in the two systems. Why must we bring this up? Because anyone working in assembly language must know how the bytes are arranged. Furthermore, two of the systems we are considering are big-endian and one (the VAX) is little-endian. Which is the better system? Either one. It is having both of them that is a nuisance.

As you look at Figure 9.3, undoubtedly you will prefer big-endian, but that is only because it appeals to your prejudices. In truth, either works well. What is important is that you be able to direct your program to go fetch the item of choice. In both systems, you use the lowest-numbered byte to indicate the item of choice. Thus, for the number 01020304, the address will be 13. For the big-endian system, 13 will point to the byte containing 04 and for the little-endian system, it will point at the byte containing 01.

Figure 9.3 contains a problem for some computers which we alluded to in the discussion of Figure 9.1. We have arranged the bytes to be four in a row as in Figure 9.1. That is the way that the memory is arranged in two of our three machines. (In the 68000, there are only two columns.) A good way to look at the fetch operation is that the memory always delivers a whole row and then the processor must acquire the parts that it wants and then properly arrange them. (This is the effect if not always the method.) Some processors—the VAX being a conspicuous example—are willing to undertake getting a longword by fetching two longwords and then piecing together the parts that it wants. Others (in our case, the 68000 and the SPARC) are not so accommodating. Those machines opt for simplicity and speed and require that the program keep its data aligned. To use one of those machines, you (or the compiler or assembler) must rearrange Figure 9.3 by inserting a null byte into Figure 9.2. This modification is shown in Figure 9.4. With this modification, all three machines could fetch the two numbers in one operation without rearrangement.

Look closely at the numbers 01020304 and 0A0B0C0D in Figure 9.4. Notice that for both configurations, the numbers read from left to right and that (visually) they appear to be in the same place. Furthermore, as pointed out in the discussion of Figure 9.3, the "beginning" or address of each of the numbers is identical. However, the byte that is pointed at by the address is not the same and the internal bytes do not have the same addresses. Getting big-endian and little-endian machines in a conversation is not easy. It proves to be even more muddled than these figures suggest. A delightful and cogent discussion of the whole issue is found in Cohen [1981].

The principal objective in this whole section has been accomplished if looking at Figure 9.4 and given the command to load a byte from location 0000 0019, you get the number 0B in the big-endian machine and 0C in the little-endian machine.

If you are not already familiar with storing structures in memory, look at the string (sentence) and ask how those letters get in memory. To begin with, every typeable symbol and all of the unprintable actions such as tabbing and carriage returns have been assigned a numerical value from the *ASCII code*. Each assignment is a byte-long number. What "This" really looks like (HEX, left to right) is 54 68 69 73. The spaces are HEX 20; the period 2E. With the alignment null byte at the end, this list of characters forms a proper C string. It is a structure of 20 bytes. A structure of any number of bytes can be stored, but from the assembly point of view, it is all just a list of bytes. You may access them two at a time, four at a time, or one

00			03
T	h	i	s
	i	s	
a		s	e
n	t	e	n
c	e	.	00
01	02	03	04
0A	0B	0C	0D

18   1B

Big Endian
(SPARC, MC68000)

03			00
s	i	h	T
	s	i	
e	s		a
n	e	t	n
00	.	e	c
01	02	03	04
0A	0B	0C	0D

1B   18

Little Endian
(VAX)

FIGURE 9.4   The same items as in Figure 9.3, but with justification of the long integers to begin on a longword boundary.

at a time. Any interpretation of those bytes is entirely up to the program. Unlike the HLL which requires that you tell it what each named variable is, assembly language knows only bytes and groups of bytes. In assembly language, the "T" can be thought of as a letter or the number 54 (HEX). Your choice. Or, more importantly, your program's choice.

Addressing

Now that we have both memory and addresses, we should next consider how these processors require that programmers specify the data that is to be acted upon by the instructions.

All of these machines have multiple modes of address. The VAX has the biggest vocabulary; the SPARC the smallest. Yet all can accomplish the same tasks. Four general types of address specification are quite common among assembly languages. These are shown in Table 9.2. They are spelled out in words in the table, but their usage is really developed in the examples which follow in this and the succeeding sections.

In Table 9.2, formats 1.4 and 1.5 and the entries in 4 require some expansion. The others will be clear in the examples we will present. Base-index addressing is the mechanism for dealing with subscripts. The base points at the starting point of a data structure, such as a string or a vector; the index measures the offset from the start of the structure to the element in question. For most machines, the index is simply a separate register which counts the bytes from the base to the item in question. If the items in the list are 4 bytes long, then to increment the index, you add 4. While that is not hard to remember, the VAX does its multiplication by the item length for you. Furthermore, it allows you to index any form of address that you can write. To show you what that means, consider expanding numbers stored in words into numbers stored in longwords. The extension is to preserve sign. The VAX provides specific instructions for conversions. If we were moving these words in one array to longwords in another array, we would write:

CVTWL (r4)[r5],(r6)[r5]   ;convert the words starting at (r4) to longwords starting at (r6)

Note that the same index, [r5], is used for both arrays. On the left, the contents of r5 are multiplied by 2 and added to r4 to get the address; on the right, the address is r5*4+r6. You would be saying: "Convert the 4th word to the 4th longword." This is undoubtedly compact and sometimes convenient. It is also unique to the VAX.

For the 68000, the designers folded both base-displacement and base-index into one mode and made room for word or longword indices. It looks like:

TABLE 9.2   Addressing Modes

1. Explicit addresses	**Example**	
1.1. Absolute addressing	765	The actual address written into the instruction.
1.2. Register indirect	(r3)	Meaning "the address is in register 3."
1.3. Base-displacement	−12(r3)	Meaning "12 bytes before the address in register 3."
1.4. Base-index	(r3,r4)	Meaning make an address by adding the contents of r3 and r4. This mode has many variations which are discussed below.
1.5. Double indirect	@5(r4)	Very uncommon! Means calculate an address as in 1.3, then fetch the longword there, and then use it as the address of what you really want.
2. Direct data specification		
2.1. Immediate/literal	#6 or 6	Meaning "use 6 as the datum." In machines which use #6, 6 without # means address 6. This is called "absolute addressing."
3. Program-relative		
3.1. Labels	loop:	The label (typically an alphanumeric ending in a colon) is a marker in the program which the assembler and linker keep track of. The common uses are to jump to a labeled spot or to load labeled constants stored with the program.
4. Address-modifying forms (CISC only)		
4.1. Postincrement	(sp)+	Same as 1.2 except that, after the address is used, it is incremented by the size of the datum in bytes and returned to the register from which it came.
4.2. Predecrement	−(sp)	The value in SP is decremented by the size of the datum in bytes, used as the address and returned to the register from which it came.

add.1 64(A3,D2.w),D3   ;address = (A3+64) +sign-extended(D2)

The 68000 limits the displacement to a signed byte, but other than that, it is indeed a rather general indexing format. If you do not want the displacement, set it to 0.

For the powerful but simple SPARC, the simple base-index form shown in 1.4 is all that you have (or need).

The double-indirect format, 1.5, is so rarely used that it has been left out of almost all designs but the VAX. What makes it occasionally useful is that subroutines get pointers to "pass by pointer" variables. Thus, if you want to get the variable, first you must load the address and then the variable. The VAX allows you to do this in one instruction. While that sounds compact, it is expensive in memory cycles. If you want to use that pointer again, it pays to have it in register.

The two items under heading 4 are strange at first. Their principal function is adding items to and removing them from a dynamic stack, or for C, to execute the operation *X++ or *(--X). The action may be viewed with the code below and the illustration of memory in Figure 9.2:

movl r4,-(sp)   ;make room on the stack (subtract 4 from SP) and put the
                                       contents of r4 in that spot
movl (sp)+, r4   ;take a longword off the stack, shorten the stack by 4 bytes,
                                       and put the longword in r4

RISCs abhor instructions which do two unrelated things. Instead of using a dynamic stack, they use a quasi-static stack. If a subroutine needs 12 bytes of stack space, it explicitly subtracts 12 from SP. Then it works from there with the base-displacement format (1.3) to reference any place in the block of bytes just defined. If you want to use a pointer and then increment the pointer, RISCs will do that as two independent instructions.

Let us consider one short section of MC68000 code from our sample program in C to see how these modes work and to sample some of the flavor of the language:

```
;ees = ((*source=='E') || (*source=='e')) && inside && !latche;
            CMPI.B     #$45,(A4)   ;'E'    "compare immediate"    literal hex 45,
                                                                                   what A4 points at
            BEQ        first        ;        "branch if equal"      to label first
            CMPI.B     #$65,(A4)   ;'e'    "compare immediate"    literal hex 65,
                                                                                   what A4 points at
            BNE        second       ;        "branch if not equal"  to label second
first:
            TST.W      D6           ;        "test word" (subtract 0)  D6 ('inside')
            BEQ        second       ;        "branch if equal"      to label second
            TST.W      D3           ;        "test word" (subtract 0)  D3 ('latche')
            BEQ        third        ;        "branch if equal"      to label third
second:
            MOVEQ      #00,D0       ;        "move quick"           literal 0 to D0
            BRA        fourth       ;        "branch always"        to label fourth
third:
            MOVEQ      #$01,D0      ;        "move quick"           literal 1 to D0
fourth:
            MOVE.W     D0,-6(A6)   ;        "move word"            from D0 to -6(FP)
```

There are all sorts of little details in this short example. For example, a common way to indicate a comment is to start with a ";". The assembler will ignore your comments. The "#" indicates a literal, and the "$" that the literal is written in hexadecimal notation. The VAX would use #^x to express the same idea. "Compare" means "subtract but save only the **condition codes** of the result" (*v* or *overflow*, *n* or *negative*, *z* or *zero*, and *c* or *carry*). Thus, the first two lines do a subtraction of whatever A4 is pointing at (*source) from the ASCII value for 'E' and then, if the two were equal (the result, zero), the program jumps to line 5. If *source is not 'E', then it simply goes to the next line, line 3. The instruction, TST.W D6, is quite equivalent to CMPI.W D6, #0, but the TST instruction is inherently shorter and faster. On a SPARC, where it would be neither shorter nor faster, TST does not exist.

Exactly what the assembler or linker does to replace the label references with proper addresses, while interesting, is not particularly germane to our current topic. Note that the range of the branch is somewhat limited. In the 68000, the maximum branch is ±32K and in the VAX a mere +127 to −128. If you need to go further, you must combine a branch with a jump. For example, if you were doing BEQ farlabel, you would instead do:

```
    BNE           nearlabel
    Jmp           farlabel      ;this instruction can go any distance
   nearlabel:
```

Follow through the example above until the short steps of logic and the addressing modes are clear. Then progress to the next section where we use the addressing modes to introduce the general topic of subroutine calling conventions.

Calling Conventions

Whenever you invoke a subroutine in a HLL, the calling routine (*caller*) must pass to the called routine (*callee*) the parameters that the subroutine requires. These parameters are defined at compile time to be either *pass-by-value* or *pass-by-pointer* (or *pass-by-reference*), and they are listed in some particular order. The convention for passing the parameters varies from architecture to architecture and HLL to HLL, but basically it always consists of building a *call block* which contains all of the parameters and which will be found where the recipient expects to find it.

Along with the passing of parameters, for each system, a convention is defined for register and stack use which establishes:

- Which registers must be returned from callee to caller with the same contents that the callee received (such registers are said to be *preserved across a call*)
- Which registers may be used without worrying about their contents (such registers are called *scratch registers*)
- Where the return address is to be found
- Where the value returned by a function will be found

The convention may be supported by hardware or simply a gentlemanly rule of the road. However the rules come into being, they define the steps which must be accomplished coming into and out of a subroutine. The whole collection of such rules forms the *calling convention* for that machine. In this section, we look at our three different machines to see how all accomplish the same tasks but by rather different mechanisms.

The two CISCs do almost all of their passing and saving on the stack. The call block will be built on the stack; the return address will be put on the stack; saved registers will be put on the stack. Only a few stack references are passed forward in register; the value returned by the function will be passed back in register.

How different is the SPARC! The parameters to be passed are placed in the *out* registers (six are available for this purpose). Only the overflow, if any, would go on the stack. In general, registers are saved by window-blinding rather than moving them to the stack. On return, data is returned in the *in* registers and the registers restored by reverse window-blinding.

MC68000 Call and Return. Let us look at the details for two of the machines. We start with the 68000, because that is the most open and "conventional." We continue with the function NumberCount. Only a single parameter must be passed—the pointer to the text block. The HLL callee sees NumberCount(block) as an integer (i.e., what will be returned), but the assembly program must do a call and then use the returned integer as instructed. A typical assembly routine would be:

```
MOVE.L    A2,-(SP)        ;  move pointer to block onto the stack
JSR       NumberCount     ;  save return address on the stack and start
                          ;  executing NumberCount
                          ;  do something with value returned in D0
```

The first instruction puts the pointer to the block, which is in A2, on the stack. It first must make room, so the "–" in –(A7) first subtracts 4 from A7 (making room for the longword) and then moves the longword into the space pointed to by the now-modified A7. The one instruction does two things: the decrementing of SP and the storing of the longword in memory.

```
MOVE.L    A2,-(A7)      A7 ⇐ A7-4    ;a7 = SP
                        M(A7) ⇐ A2   ;M(x) = memory(address x)
```

The next instruction, *jump subroutine* (JSR), does three things. It decrements SP (i.e., A7) by 4, stores the return address on the top of the stack, and puts the address of NumberCount in the *program counter*. We have just introduced two items which need specific definition:

Return address (RA):   This will always be the address of the instruction which the callee should return to. In the 68000 and the VAX (and all other CISCs), the RA points to the first instruction after the JSR. In the SPARC and almost any RISC, RA will point to the second instruction after JSR. That curious difference will be discussed later.

Program counter (PC):   This register (which is a *general register* on the VAX but a special register on the other machines) points to the place (memory location) in the machine language instruction stream where the program is currently operating. As each instruction is fetched, the PC is automatically incremented. The action of the JSR is to save the last version of the PC—the one for the next fetch—and replace it with the starting address of the routine to be jumped to.

Summing up these transactions in algebraic form:

Should you wonder how the address of NumberCount gets in there, the *linker*, which assigns each section of

```
JSR NumberCount      SP ⇐ SP-4              ;A7 = SP
                     M(SP) ⇐ PC            ;M(x) = memory(address x)
                     PC ⇐ address of NumberCount
```

code to its proper place in memory and therefore knows where all the labels are, will insert the proper address in place of the name.

This completes the call as far as building the call block, doing the call itself, and picking up the result. Had there been more parameters to pass, that first instruction would have been replicated enough times to *push* all of the parameters, one at a time, onto the stack. Now let us look at the conventions from the point of view of the callee. The callee has more work.

When the callee picks up the action, the stack and registers are as shown in Figure 9.5. With the exception of D0 and A7, the callee has no registers ... yet. The callee must make room for local variables in either register or memory. If it wants to use registers, it must save the user's data from the registers. The subroutine can get whatever space it needs on the stack. Only after the setup will it get down to work. The entire section of stack used for local variables and saving registers is called the callee's frame. It is useful to have a pointer (FP) to the bottom of the frame to provide a static reference to the return address, the passed parameters, and the subroutine's local variable area on the stack. In the 68000, the convention is to use A6 as FP. When our routine, NumberCount, begins, the address in A6 points to the start of the caller's frame. The first thing the callee must do is to establish a local frame. It does that with the instruction LINK.

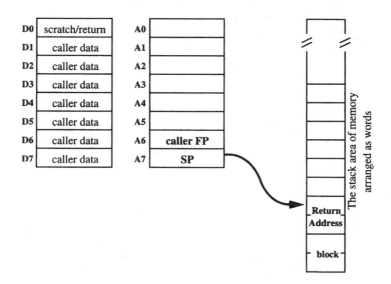

FIGURE 9.5 The stack area of the 68000's memory and the register assignments that the called subroutine sees as it is entered at the top. The registers all hold longwords, the size of an address. In typical PC/Macintosh compilers, integers are defined as 16-bit words. Accordingly, the stack area of memory is shown as words, or half the width of a register.

Typical of a CISC, each instruction does a large piece of the action. The whole entry operation for the 68000 is contained in two instructions:

```
LINK      A6,#$FFF8
MOVEM.L   D3-D7/A4,-(SP)
```

The first instruction does the frame making; the second does the saving of registers. There are multiple steps in each. Each double step of decrementing SP and moving a value onto the stack will be called a *push*. The steps are as follows:

```
LINK      A6,#$FFF8       ;push A6 (A7 < A7-4, M(A7) &z.lsquo; A6)
                          ;move A7 to A6 (SP to FP)
                          ;add FFF8 (-8) to SP (4 words for local variables)
MOVEM.L   D3-D7/A4,-(A7)  ;push 5 data registers (3..7) and 1 address
                          ;register (A4)
```

At this point, the stack looks like Figure 9.6.

The subroutine is prepared to proceed. How it uses those free registers and the working space set aside on the stack is the subject of the section on optimization in this chapter. For the moment, however, we simply assume that it will do its thing in exemplary fashion, get the count of the numbers, and return. We continue in this section by considering the rather simple transaction of getting back.

The callee is obliged to put the answer back where the caller expects to find it. Two paradigms for return are common. The one that our compiler uses is to put the answer in D0. The other common paradigm is to put the answer back on the stack. The user will have left enough room for that answer at FP+8, whether or not that space was also used for transferring parameters in. Using our paradigm, the return becomes:

```
MOVE.W $FFFC(A6),D0      ;answer from callee's stack frame [-4(FP)] to D0
MOVEM.L (A7)+,D3-D7/A4   ;registers restored to former values
UNLK A6                  ;SP < FP, FP < M(SP), SP < SP+4
RTS                      ;PC < M(SP), SP < SP+4
```

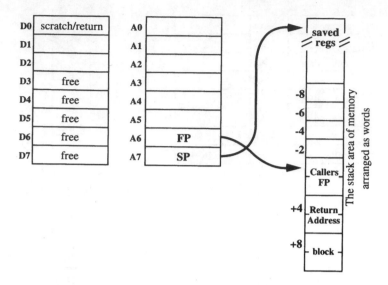

FIGURE 9.6   The stack area of the 68000's memory and the register situation just after MOVEM has been executed. The memory area between the two arrows is the subroutine's *frame*.

When all of this has transpired, the machine is back to the caller with SP pointing at *block*. The registers look like Figure 9.5 except for two important changes. SP is back where the caller left it and D0 contains the answer that the caller asked for.

Transactional Paradigms

The final topic in this section is the description of some of the translations of the simple and ordinary phrases of the HLLs into assembly language. We will show some in each of our three machines to show both the similarities and slightly different flavors that each machine architecture gives to the translation.

The paradigms that we will discuss comprise:

- Arithmetic
- Replacement
- Testing and branching, particularly multiple Boolean expressions
- Stepping through a structure

Many studies have shown that most computer arithmetic is concerned with addressing, testing, and indexing. In NumberCount there are several examples of each. For example, near the bottom of the program, there are statements such as:

```
count++;
```

For all three machines, the basic translation is the same: Add an *immediate* (a constant stored right in the instruction) to a number in register. However, for the CISCs, one may also ask that the number be brought in and even put back in memory. The three translations of this pair comprise:

MC68000	VAX	SPARC
ADDQ.W #$1,$FFFE(A6)	INCL R0	add %o2,1,%o2

Typical of the VAX, it makes a special case out of adding 1. There is no essential difference in asking it to add 1 by saying "1," but if one has a special instruction, it saves a byte of program length. With today's inexpensive memories, a byte is no longer a big deal, but when the VAX first emerged (1978), they were delivered with less memory than a PC or Mac would have today. The VAX, of course, can say ADDL #1, r0 just like the 68000, and for any number other than 1 or 0, it would. Note also that the VAX compiler chose to keep *count* in register, while in Think C® decided to put it on the stack [−2(SP)]. A RISC has no choice. If you want arithmetic, your numbers must be in register. However, once again, we are really talking about the length of the code, not the speed of the transaction. *All* transactions take place from registers. The only issues are whether the programmer can see the registers and whether a single instruction can include both moving the operands and doing the operand arithmetic. The RISC separates the address arithmetic [e.g., −2(SP)] from the operand arithmetic, putting each in its own instruction. Both get the job done.

The next items we listed were *replacement* and *testing and branching*. We have both within the statement:

```
digit = (*source >= '0') && (*source <= '9');
```

The translation requires several statements:

MC68000	VAX	SPARC
MOVE.B (A4),D3	clrb r1	add %g0,0,%o1
CMPI.B #$30,D3	cmpb @4(ap),#48	ldsb [%o2],%o0
BLT ZERO	blss ZERO	subcc %o0, 47,%g0
CMPI.B #$39,D3	cmpb @4(ap),#57	ble ZERO
BLE ONE	bgtr ZERO	nop
	incb r1	subcc %o0,57,%g0
		bg ZERO
		nop
		add %g0, 1,%o1
		add %o1,0,%l3
ZERO:	ZERO:	ZERO:

```
        MOVEQ #$00,D0
        BRA DONE
ONE:
        MOVEQ #$01,D0
DONE:
        MOVE.W D0,$FFF6(A6)
```

To begin with, all three do roughly the same thing. The only noticeable difference in concept is that the SPARC compiler chose to compare the incoming character (*source) to 47 (the character before '0') and then branch if the result showed the letter to be "less than or equal," while the other two compared it to '0' as asked and then branched if the result was "less than." No big deal. But let us walk down the several columns to see the specific details. Prior to beginning, note that all three must bring in the character, run one or two tests, and then set an integer to either *zero* (false) or *not zero* (true). Also, let it be said that each snatch of code is purportedly optimized, but at least with the small sample that we have, it looks as if each could be better. We begin with three parallel walkdowns. Notes as needed are provided below.

MC68000	VAX	SPARC
character from M→D#	Set (byte) DIGIT to 0	Set (byte) DIGIT to 0
Is (D3-'0') <=0?	Is (*source-'0') <=0?	character from Mfi out1
If <, branch to label ZERO	If <, branch to ZERO	Is (*source-'/') <=0?
Is (D3-'9') <=0?	Is (*source-'9') <=0?	If <=, branch to ZERO
If <=. branch to label ONE	If neither, branch to ZERO	Is (*source-'9') <=0?
	Add (byte) 1 to DIGIT	If neither, branch to
		Add 1 to DIGIT
ZERO:	ZERO:	ZERO:

MC68000:
Put a longword 0 in D0
Branch to label DONE
ONE:
Put a longword 1 in D0
DONE:
Put value in D0 into DIGIT

Notes:

1. Moving the character into register to compare it with '0' and '9':

   a. The first 68000 line moves the next character *as a byte* into register D3. The other 3 bytes will be ignored in the byte operations. Remember that the program had already moved the pointer to the string into A4.

   b. The SPARC does the same sort of thing with a pointer in %o2, except with the difference that it sign-extends the byte to a longword. Sign extension simply copies the *sign-bit* into the higher-order bits, effectively making 3E into 0000 003E or C2 into FFFF FFC2. That is what the mnemonic means: "LoaD Signed Byte."

   c. The VAX compiler takes a totally different approach—a rather poor one, actually. It leaves not only the byte in memory but even the pointer to the byte. Thus, every time it wants to refer to the byte—and it does so numerous times—it must first fetch the address from memory and then fetch the byte itself. This double memory reference is what @4(ap) means: "At the address you will find at address 4(ap)." The only thing that makes all this apparent coming and going even remotely acceptable is that the VAX will cache (place in a fast local memory) both the address and the character the first time that it gets them. Then, it can refer to them rapidly again. Cache references, however, are not as fast as register references.

2. Testing the character:

   The next line (3rd for the SPARC) does a comparison between 48 (or 47) and the character. *Compare* is an instruction which subtracts one operand from the other, but instead of putting the results somewhere, it stores only the facts on whether the operation delivered a negative number or zero or resulted in either an overflow or a carry. These are stored in **flags**, single bits associated with the arithmetic unit. The bits can contain only one result at a time. The 68000 and VAX must test immediately after the comparison or they will lose the bits. The SPARC changes the bits only when the instruction says so (the CC on the instruction — "change condition codes"). Thus, the subtraction can be remote from the test-and-branch.

   The SPARC is explicit about where to store the subtraction—in %g0. %g0 is a pseudo-register. It is always a 0 as a source and is a garbage can as a destination. The availability of the 0 and the garbage can serves all the same functions that the special instructions for zeros and comparisons do on the CISCs.

3. The differences in the algorithm to do the tests:

   There are two different paradigms expressed in these three examples. One says: "Figure out which thing you want to do and then do that one thing." The other says: "First set the result false and then figure out if you should set it true." While the second version would seem to do a lot of unnecessary settings to zero, the other algorithm will execute one less branch. That would make it roughly equivalent. However, the 68000 algorithm is definitely longer—uses more memory for code. That is not really much of an issue, but why put the result first into a temporary register and then where you really want it?

Compiler Optimization and Assembly Language

Compiler Operations

To understand the optimizing features of compilers and their relation to assembly language, it is best to understand some of the chores for the compiler. This section examines variable allocation and how it can be optimized, and the optimization task of constant expression removal. Examples of how compilers perform these operations are taken from the various systems used in the article.

Variable Allocation. Variables in high-level languages are an abstraction of memory cells or locations. One of the compiler's tasks is to assign the abstract variables into physical locations—either registers within the processor or locations within memory. Assignment strategies vary, but an easy and often-used strategy is to place *all* variables in memory. Easy, indeed, but wasteful of execution time in that it requires memory fetches for all HLL variables. Another assignment strategy is to assign as many variables to the registers as possible and then assign any remaining variables to memory; this method is typically sufficient, except when there is a limited number of registers, such as in the 68000. In these cases, the best assignment strategy is to assign registers to the variables which have the greatest use and then assign any remaining variables to memory. In examining the compilers and architecture used in this article, we find examples of all these methods.

In the unoptimized mode, VAX and Sparc compilers are among the many which take the easy approach and assign variables only to memory locations. In Figures 9.6 and 9.7, the variable assignments are presented for the unoptimized and optimized options. Note that only one or two registers are used, both as scratch pads, in the unoptimized option, whereas the optimization assigns registers to all variables. The expected execution time savings is approximately 42 of the 50 memory references per loop iteration. That does not include additional savings caused by compact code. Detailed comparisons are not presented since the interpretation of architectural comparisons is highly subjective.

Unlike the VAX and Sparc compilers, the 68000 compiler assigns variables to registers in both the unoptimized and unoptimized options; these assignments are depicted in Figures 9.7 and 9.8. Since there are only eight general-purpose data registers in the 68000 and two are assigned as scratch pads, only six of the program's ten variables can be assigned to registers. The question is how the 68000 compiler chose which variables to assign to registers and which to leave in memory. As might be expected, the compiler assigned registers based on their usage for the unoptimized option as well as the optimized. The exact counting strategy is unknown. However, a fair estimate, which yields the same assignment as the compiler, is to count only the variable's usage in the loop—the likely place for the program to spend most of its execution time. There are other ways to estimate the variable usage such as assigning weights to a reference based on its placement within a loop, its placement within a conditional (if-then-else) statement, etc. These estimates and how they are applied within a compiler can change the variable allocation as well as the efficiency of the code.

In the optimized case, a slightly different register assignment is used. This is because the optimizer created another character variable—*source—which it assigned to a register. The motivation for its creation and its assignment to a register is shown in the next section on constant expression removal.

Even though the assignment of variables to registers gives an improvement in performance, it is not always possible to assign a variable to a register. In C, one operation is to assign the address of a variable to a pointer-type variable (e.g., ip = & i). If i were assigned to a register, the operation would become invalid, because a register does not have a memory address. Although this example appears limited to the C language, the use of a variable's address is widespread when subroutine parameters are passed by reference. (Variables sent to a subroutine are either passed by reference, where the address of the variable is passed to the subroutine,

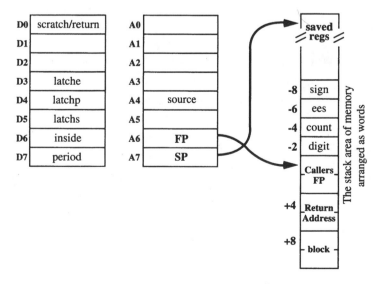

FIGURE 9.7 The stack area of the 68000's memory and the register assignments that the Think C® compiler made with global optimization *turned off*. The stack is shown just after the MOVEM instruction. The items in bold are as they would be after that instruction. While the registers all hold longwords, in typical PC/Macintosh compilers, integers are defined as words. This figure is the programmatic successor to Figure 9.6.

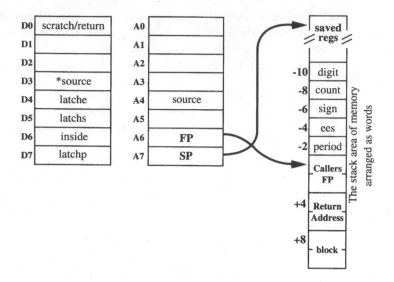

FIGURE 9.8   The stack area of the 68000's memory and the register assignments that the Think C® compiler made with global optimization *turned on*. The stack is shown just after the MOVEM instruction. The items in bold are as they would be after that instruction. This figure should be compared with Figure 9.7.

allowing modifications to the original variable, or they are passed by value, where a copy of the variable is passed to the subroutine.) When a parameter is passed by reference its address must be obtained and passed to the subroutine, an action commonly found in most languages. This action compounds the task of selecting candidate variables for register assignment.

Constant Expression Removal.   The task of allocating program variables to physical locations is accomplished by all compilers; we have shown that there are many ways to achieve this goal with varying ease or run-time performance. This section explores a compiler task which is done strictly for optimization—the removal of constant expressions. In exploring this task, we show strategies for the recognition of this optimization and also some caveats in their application.

Constant expressions are subexpressions within a statement which are unchanged during its execution. An obvious example is the expression *vector[x]* in the following conditional statement.

```
if (vector[x] < 100) && (vector[x] > 0) then ...
```

An astute coder who does not trust compilers would not allow two memory references for vector[x] in the same conditional statement and would rewrite the code by assigning vector[x] to a temporary variable and using that variable in the conditional. An astute compiler would also recognize the constant expression and assign vector[x] to a scratch pad register and use this register for both comparisons. This type of optimization, where small sections of code (typically one source line) are optimized, is called *peep-hole* optimization.

Within the example program, the assignment statement which checks if the character is a digit within the range from '0' to '9' is a statement which can benefit from this type of optimization. The C code lines, with the unoptimized and optimized SPARC assembly code, are listed below. Note that in addition to the constant expression removal the optimized code also assigns variables to registers.

```
digit = (*source >= '0') && (*source <= '9');
```

In translating this line on the SPARC, Figure 9.9 shows the 32 registers visible at any moment in the window-blinding SPARC. The top 24 shift by 16 in a normal call. The eight globals remain the same. The shift of the registers is accompanied by copying SP to o6 and the call instruction puts the return address into o7. Accordingly, a call wipes out the caller's o7. Register g0 serves as a 0 (as a source) and as a *wastebasket* (as a destination). *ld* loads a longword, and *ldsb* sign-extends a byte into a longword. The instruction after a branch is executed whether the

SPARC GCC no optimization

```
L2:   add 0, %g0, %o0        ; put 0 in o0 — i.e., assume a false result
      ld [%fp-92], %o1       ; get source from memory
      ldsb [%o1], %o1        ; get *source
      add 47, %g0, %o2       ; move 47 into o2
      subcc %o1, %o2, %g0    ; compare *source to '0'
      ble L5                 ; branch if less then, digit is false
      nop
      ld [%fp-92], %o1       ; get source again
      ldsb [%o1], %o1        ; get *source again
      add 57, %g0, %o2       ;
      subcc %o1, %o2, %g0    ; compare to '9'
      bg L5
      nop
      add 1, %g0, %o0        ; results is a true, change temporary
L5:                          ; jump target if result is false
      st %o0, [%fp-36]       ; move the result to variable.
```

SPARC GCC optimization

```
L2:   add 0, %g0, %o1        ; assume a false result from statement.
      ldsb [%o2], %o0        ; get *source save in register o0
      addcc -47, %o0, %g0    ; compare to '0'
      ble L5
      nop
      addcc -57, %o0, %g0    ; reuse *source and compare to '9'
      bg L5
      nop
      add 1, %g0, %o1        ; results is a true, so change temporary
L5:                          ; jump target if result is false
      add %o1, %g0, %l3      ; move the result to variable.
```

i7	return address	r31
i6	frame pointer	r30
i5	param #5	r29
i4	param #4	r28
i3	param #3	r27
i2	param #2	r26
i1	param #1	r25
i0	param #0	r24
l7	local	r23
l6	local	r22
l5	local	r21
l4	local	r20
l3	local	r19
l2	local	r18
l1	local	r17
l0	local	r16
o7	scratch	r15
o6	stack pointer	r14
o5	param out #5	r13
o4	param out #4	r12
o3	param out #3	r11
o2	param out #2	r10
o1	param out #1	r9
o1	param out #1	r8
g7	global	r7
g6	global	r6
g5	global	r5
g4	global	r4
g3	global	r3
g2	global	r2
g1	global	r1
g0	src=0,dst=WB	r0

FIGURE 9.9   SPARC register assignments.

branch is taken or not (delayed branching). An instruction such as add 47, %g0, %o2 adds a constant to 0 and puts it in the register. This is equivalent to move.l #47, d4 on the 68000. An *add* or *sub* with *cc* appended changes the condition codes. To do a *compare*, one uses *addcc* or *subcc* and puts the result in g0 (the wastebasket).

The same type of constant expression can be found and removed with a global perspective, typically from within loops. A simple example is the best way to describe how they can be removed and to offer some caveats when the compiler cannot see the global picture. The following example code updates each element in the vector by adding the first element scaled by a constant *y*. An observation shows that the subexpression, vector[0] * *y*, is constant throughout all executions of the loop. An obvious improvement in code is to calculate the product of vector[0] * *y* outside of the loop and store its value in a temporary variable. This is done in the second example.

Constant expression present
```
for(i= 0; i < size; i++)
{ vector[i] = vector[i] + (vector[0] * y); }
```

Constant expression removed
```
temp = vector[0] * y;
for(i= 0; i < size; i++)
{vector[i] = vector[i] + temp;}
```

Ideally, the compiler should find and remove these constant expressions from loops, but this is not as obvious as it may seem. Consider the above example if the following line were inserted in the loop:

```
y = vector[i-1] + vector[i+1]
```

If each source line is taken in isolation, y appears constant, but y is dependent on the loop index i. Hence before removing constant expressions, the compiler must map the dependencies of each variable on the other variables and the loop index. Additionally, other not-so-obvious dependencies—such as when two pointers modify the same structure—are difficult to map and can result in erroneous object code. This is one of the difficulties in optimizing compiler operation and why its extent is limited.

A subtle example for constant expression removal is found in our sample program in the reference to *source. In these statements, the character referenced (addressed) by source is obtained from memory. The pointer (address) *source* is changed only at the bottom of the loop and the memory contents addressed by *source* are static. A global optimization should obtain the character once at the top of each pass of the loop and save on subsequent memory references throughout. The 68000 C compiler with the optimization option determined *source to be constant throughout the loop and assigned register D3 to hold its contents. This saved seven of the eight memory accesses to *source in each loop pass. The unoptimized 68000 option, the SPARC, and the VAX compilers did not use global constant expression removal and must fetch the operand from memory before its use.

The Problems. With optimization yielding more efficient code resulting in improved system performance, why would you not use it? Our favorite, among the several reasons, is the following quote from compiler documentation: "Compiling with optimization may produce incorrect object code." Problems are caused by assumptions used by the compiler which are not held by the programmer. For example, an optimization which assumes that memory contents do not change with time is erroneous for multi-tasking systems which share memory structures and also for memory-mapped I/O devices, where memory contents are changed by external events. For these cases, the data in register may not match the newer data in memory.

Additionally, HLL debuggers do not always work well with the optimization option since the one-to-one correspondence between HLL code and the object code may have been removed by optimization. Consider the reassignment of *source to a data register which is performed by the 68000 C compiler. If a debugger were to modify the contents of *source, then it would have to know about the two locations where it is stored: the memory and the register. Other types of optimizations which may cause problems are when unneeded variables are removed or when code is resequenced. If a HLL debugger tries to single-step through HLL code, there may not be corresponding assembly code, and its execution will appear erroneous.

High-Level Language and Assembly Language Relations

In comparing the various assembly languages from the compiler-generated code, we have not presented a full vocabulary of assembly languages or the minutiae of the underlying machines. Exploring only the code generated by the compilers may lead one to believe that all assembly languages and processor architectures are pretty much the same. This is not really the case. What we have shown is that compilers typically use the same assembly language instructions regardless of the underlying machines. The compiler writer's motivation for this apparent similarity is not because all architectures are the same, but because it is difficult—arguably even nonproductive—for the compiler to take advantage of the complex features which some CPU architectures offer. An argument may be made that compilers generate the best code by developing code rather independently of the underlying architecture. Only in the final stages of code generation is the underlying platform's hardware architecture specifically considered [see Aho et al., 1986]. Differences in the architectures and assembly languages are plentiful; compilers typically do not and probably should not take advantage of such features.

The VAX is one of the best examples of an architecture having an almost extraordinary vocabulary, which is why it is often considered the prototypical *CISC* machine. What were the motivations for having this rich vocabulary if compilers simply ignore it? Early computer programming was accomplished through slightly alphabetized machine language—mnemonics for opcodes and sometimes for labels. Assembly language represented a vast improvement in readability, but even though FORTRAN, COBOL and Algol were extant at the same time as assembly language, their crude or absent optimization abilities led to the popular belief that *really good* programming was always done in assembly. This was accepted lore until early studies of optimization began to have an impact on commercial compilers. It is fair to say that this impact did not occur until the early 1980s. The VAX and the 68000 were products of the middle and late 1970s. It is no great surprise then to find that CISC computer architectures were designed to enable and assist the assembly language programmer. Such an objective promotes the inclusion of many complex features which the programmer might want to utilize. However, two facts emerged in the late 1970s which suggested that this rich vocabulary was provided at too high a cost for its benefit (for a more complete discussion of the emergence of the RISC concept, which really began with Seymour Cray in the 1960s, see [Feldman and Retter, 1994]):

- It was widely observed that the generation, testing, and maintenance of large programs in assembly code was extremely expensive when compared to doing the same task in a HLL. Furthermore, the compilers were improving to the point where they competed quite well with typical assembly code.
- Although initially not widely accepted, it was observed that the rich vocabulary made it very difficult to set up an efficient processing pipeline for the instruction stream. In essence, the assembly line was forced to handle too many special cases and slowed down under the burden. When the analysis of compiled programs showed that only a limited span of instructions was being used, these prescient designers decided to include only the heavily used instructions and to restrict even these instructions so that they would flow in unblemished, uniform streams through the production line. Because this focus resulted in noticeably fewer instructions— though that was not the essential objective—the machines were called *RISC*, a sobriquet that came out of a VLSI design project at Berkeley [Patterson and Hennessy, 1989].

Even though RISC hardware designs have increased performance in essence by reducing the complexity and richness of the assembly language, back at the ranch the unrepentant assembly language programmer still desired complex features. Some of these features were included in the assembly languages not as native machine instructions but essentially as a *high-level* extension to assembly language. A universal extension is the inclusion of *macros*. In some sense, a macro looks like a subroutine, but instead of a call to a distant block of code, the macro results in an inline insertion of a predefined block of code. Formally, a macro is a name which identifies a particular sequence of assembly instructions. Then, wherever the name of the macro appears in the text, it is replaced by the lines of code in the macro definition. In some sense, macros make assembly language a little more like a HLL. It makes code more readable, makes code maintenance a little faster and more reliable (fix the macro definition and you fix all of the invocations of the macro), and it speeds up the programmer's work.

Another extension to some assembly languages is extended mnemonics. Here the coder places a mnemonic in place of specific assembly language instructions; during code assembly the mnemonic is automatically translated to an optimal and correct instruction or instruction sequence. These free the coder from the management of low-level details, leaving the task to a program where it is better suited. Examples of extended mnemonics include *get* and *put,* which generate memory transfers by selecting the addressing mode as well as the specific instructions based on the operand locations. An increasingly common feature of assembly languages is the inclusion of structured control statements which emulate high-level language control-flow constructs such as: `if .. then .. else`, `for loops`, `while .. do loops`, `repeat .. until` loops, break, and next. These features remove the tedium from the programmer's task, allow for a more readable code, and reduce the cost of code development. An amusing set of examples are found in the assemblers that we have used on the SPARC. Architecture not withstanding, the assembly programmers wanted VAX assembly code! In spite of the absence of such constructs in the SPARC architecture, you find expressions such as CMP (compare) and MOV. Since these are easily translated to single lines of real SPARC code, their only raison d'etre is to keep the old assembly language programmers happy. Presumably, those who knew not the VAX are not writing SPARC assembly code.

Summary

After all this fuss over compilers and how they generate assembly code, the obvious question is "Why bother to write any assembly code at all?" Three reasons why "some assembly may be required" follow.

- A human writing directly in assembly language is probably better than an optimizing compiler in extracting the last measure of resources (e.g., performance, hardware usage) from the machine. That inner loop—the code where the processor spends most of its execution cycles—may need to be hand-optimized to achieve acceptable performance. Real-time systems, where the expedient delivery of the data is as critical as its correctness, are another area where the required optimization may be greater than that achievable by compilers. The disparity in performance between human optimizers and their machine competitors comes from two special capabilities of the human programmer. These are the ability to know what the program will be doing—forward vision based on the program's intent rather than its structure—and the ability to take advantage of special quirks or tricks that have no general applicability. If you really need to extract this last full measure of performance, assembly language is the route. The cost of doing such hand-optimization is much greater than the hours spent in doing it and getting it debugged. Special quirks and tricks expressible only in assembly language will not translate to another machine and may disappear even in an "upgrade" of the intended processor.

- There is overhead in using HLL conventions, some of which can be eliminated by directly coding in assembly language. A typical embedded processor does not need the full span of HLL conventions and support, such as parameter passing or memory and stack allocation. One can *get away with* such dangerous things as global variables which do not have to be passed at all. By eliminating these conventions, increased performance is obtained. It should be pointed out that code written without such standard conventions is likely to be *very* peculiar, bug-prone, and hard to maintain.

- HLLs provide only limited access to certain hardware features of the underlying machine. Assembly language may be required to access these features. Again, this makes the code unportable and hard to maintain, but small stubs of assembly code may be required to invoke hardware actions which have no representation in a HLL. For example, setting or clearing certain bits in a special register may not be expressible in C. While any address can be explicitly included in C code, how do you reference a register which has no address? An example of such usage is writing or reading into or out of the status register. Some machines map these transactions into special addresses so that C could be used to access them, but for the majority of machines which do not provide this route to the hardware, the only way to accomplish these actions is with assembly code. To this end, some C compilers provide an inline assembler. You can insert a few lines of assembly language right in the C code, get your datum into or out of the special register, and move right back to HLL. Those compilers which provide this nonstandard extension also provide a rational paradigm for using HLL variable names in the assembly statements. Where necessary, the name gets expanded to allow the variable to be fetched and then used.

These reasons are special; they are not valid for most applications. Using assembly language loses development speed, loses portability, and increases the maintenance costs. While this caveat is well taken and widely accepted, at least for the present, few would deny the existence of situations where assembly language programming provides the best or only solution.

Defining Terms

Address error: An exception (error interrupt) caused by a program's attempt to access unaligned words or longwords on a processor which does not accommodate such requests. The address error is detected within the CPU. This contrasts with problems which arise in accessing the memory itself, where a logic circuit external to the CPU itself must detect and signal the error to cause the CPU to process the exception. Such external problems are called *bus errors*.

Assembler: A computer program (*application*) for translating an assembly-code text file to an *object* file suitable for linking to become an executable image (*application*) in machine language. Some HLL compilers include an inline assembler, allowing the programmer to drop into and out of assembly language in the midst of a HLL program.

CISC: *Complex instruction set computer,* a name to mean "not a RISC," but generally one that offers a very rich vocabulary of computer operations at a cost of making the processor which must handle this variety of operations more complex, expensive, and often slower than a RISC designed for the same task. One of the benefits of a CISC is that the code tends to be very compact. When memory was an expensive commodity, this was a substantial benefit. Today, speed of execution rather than compactness of code is the dominant force.

Compiler: A computer program (*application*) for translating a HLL text file to an *object* file suitable for linking to become an executable image (*application*) in machine language. Some compilers do both compilation and linking, so their output is an application.

Condition codes: Many computers provide a mechanism for saving the characteristics of results of a particular calculation. Such characteristics as *sign, zero result, carry* or *borrow,* and *overflow* are typical of integer operations. The program may reference these flags to determine whether to branch or not.

Disassembler: A computer program which can take an executable image and convert it back into assembly code. Such a reconstruction will be true to the machine language but normally loses much of the convenience factors, such as *macros* and name equivalencies, that an original assembly language program may contain.

Executable image: A program in pure machine code and including all of the necessary header information that allows an operating system to load it and start running it. Since it can be run directly, it is *executable.* Since it represents the original HLL or assembly program it is an *image.*

Flags: See *Condition codes.*

High-level language (HLL): A computer programming language generally designed to be efficient and succinct in expressing human programming concepts and paradigms. To be contrasted with low-level programming languages such as *assembly language.*

Linker: A computer program which takes one or more object files, assembles them into blocks which are to fit in particular blocks in memory, and resolves all external (and possibly internal) references to other segments of a program and to libraries of precompiled subroutines. The output of the linker is a single file called an *executable image* which has all addresses and references resolved and which the operating system can load and run on request.

Macro: A single line of code-like text, defined by the programmer, which the assembler will then recognize and which will result in an inline insertion of a predefined block of code. In most cases, the assembler allows both hidden and visible local variables and local labels to be used within a macro. Macros also appear in some HLLs, such as C (the *define* paradigm).

Object code: A file comprising an intermediate description of a segment of a program. The object file contains binary data, machine language for program, tables of offsets with respect to the beginning of the segment for each label in the segment, and data that would be of use to debugger programs.

RISC: *Reduced instruction set computer,* a name coined by Patterson et al. at the University of California at Berkeley to describe a computer with an instruction set designed for maximum execution speed on a particular class of computer programs. Such designs are characterized by requiring separate instructions for load/store operations and arithmetic operations on data in registers. The earliest computers explicitly designed by these rules were designs by Seymour Cray at CDC in the 1960s. The earliest development of the RISC philosophy of design was given by John Cocke in the late 1970s at IBM. See *CISC* above for the contrasting approach.

References

A.V. Aho, R. Sethi, and J.D. Ullman, *Compiler Principles, Techniques and Tools,* Reading, Mass.: Addison-Wesley, 1986. A detailed text on the principles of compiler operations and tools to help you write a compiler. This text is good for those wishing to explore the intricacies of compiler operations.

D. Cohen, "On holy wars and a plea for peace," *Computer,* pp. 11–17, Sept. 1981. A delightful article on the comparisons and motivations of byte ordering in memory.

J. Feldman and C. Retter, *Computer Architecture: A Designer's Text Based on a Generic RISC,* New York: McGraw-Hill, 1994.

D. Patterson and J. Hennessy, *Computer Architecture, A Quantitative Approach,* San Mateo, Calif.: Morgan Kaufman, 1989. An excellent though rather sophisticated treatment of the subject. The appendices present a good summary of several seminal RISC designs.

9.2   High-Level Languages

Ted G. Lewis

High-level languages (**HLLs**), also known as higher-order languages (**HOLs**), have a rich history in the annals of computing. From their inception in the 1950s until advances in the 1970s, HLLs were thought of as simple mechanical levers for producing machine-level instructions (see Table 9.3). Removing the details of the underlying machine, and automatically converting from a HLL statement to an equivalent machine-level statement, releases the programmer from the drudgery of the computer, allowing one to concentrate on the solution to the problem at hand.

TABLE 9.3   Each statement of a HLL Translatges into More than One Statement in a Machine-Level Language Such as Assembler

Language	Typical Number of Machine Level Statements
FORTRAN	4–8
COBOL	3–6
Pascal	5–8
APL	12–15
C	3–5

Over the years, HLLs evolved into a field of study of their own, finding useful applications in all areas of computing. Some HLLs are designed strictly for solving numerical problems, and some for symbolic problems. Other HLLs are designed to control the operation of the computer itself, and yet even more novel languages have been devised to describe the construction of computer hardware. The number of human-crafted languages has multiplied into the hundreds, leading to highly special-purpose HLLs.

This evolution is best characterized as a shift away from the mechanical lever view of a HLL toward HLLs as notations for encoding **abstractions**. An abstraction is a model of the real world whose purpose is to de-emphasize mundane details and highlight the important parts of a problem, system, or idea. Modern HLLs are best suited to expressing such abstractions with little concern for the underlying computer hardware.

Abstraction releases the HLL designer from the bounds of a physical machine. A HLL can adopt a metaphor or arbitrary model of the world. Such unfettered languages provide a new interface between human and computer, allowing the human to use the machine in novel and powerful ways. Abstractions rooted in logic, symbolic manipulation, database processing, or operating systems, instead of the instruction set of a central processing unit (CPU), open the engineering world to new horizons. Thus, the power of computers depends on the expressiveness of HLLs.

To illustrate the paradigm shifts brought on by HLLs over the past 30 years, consider PROLOG, LISP, SQL, C++, and various operating system command languages. PROLOG is based on first-order logic. Instead of computing a numerical answer, PROLOG programs derive a conclusion. LISP is based on symbolic processing instead of numerical processing and is often used to symbolically solve problems in calculus, robotics, and artificial reasoning. SQL is a database language for manipulating large quantities of data without regard for whether it is numeric or symbolic. C++ is based on the **object-oriented paradigm**, a model of the world that is particularly powerful for engineering, scientific, and business problem solving.

None of these modern languages bear much resemblance to the machines they run on. The idea of a mechanical lever has been pushed aside by the more powerful idea of language as world builder. The kinds of worlds that can be constructed, manipulated, and studied are limited only by the HLL designer's formulation of the world as a paradigm.

In this section we answer some fundamental questions about HLLs: What are they? What do we mean by "high level"? What constitutes a paradigm? What are the advantages and disadvantages of HLLs? Who uses HLLs? What problems can be solved with these languages?

What Is a HLL?

At a rudimentary level, all languages, high and low, must obey a finite set of rules that specify both their syntax and semantics. **Syntax** specifies legal combinations of symbols that make up statements in the language. **Semantics** specifies the meanings attached to a syntactically correct statement in the language. To illustrate the difference between these two fundamental traits of all languages consider the statement, "The corners of the round table were sharp." This is syntactically correct according to the rules of English grammar, but what does it mean? Round tables do not have sharp corners, so this is a meaningless statement. We say it is semantically incorrect.

Statements of a language can be both syntactically and semantically correct and still be unsuitable for computer languages. For example, the phrase "... time flies ..." has two meanings: one as an expression of clock speed, and the other as a reference to a species of insects. Therefore, we add one other requirement for computer languages: there must be only one meaning attached to each syntactically correct statement of the language. That is, the language must be *unambiguous*.

This definition of a computer language does not separate a HLL from all other computer languages. To understand the features of HLLs that make them different from other computer languages, we must understand the concepts of mechanical translation and abstraction. Furthermore, to understand the differences among HLLs, we must know how abstractions are used to change the computing paradigm. But first, what is a HLL in terms of translation and abstraction?

Defining the syntax of a HLL is easy. We simply write rules that define all legal combinations of the symbols used by the language. Thus, in FORTRAN, we know that arithmetic statements obey the rules of algebra, with some concessions to accommodate keyboards. A **metalanguage** is sometimes used as a kind of shorthand for defining the syntax of other languages, thus reducing the number of cases to be listed.

Defining the semantics of a language is more difficult because there is no universally accepted metalanguage for expressing semantics. Instead, semantics is usually defined by another program that translates from the HLL into some machine-level language. In a way, the semantics of a certain HLL is defined by writing a program that unambiguously maps each statement of the HLL into an equivalent sequence of machine-level statements. For example, the FORTRAN statement below is converted into an equivalent machine-level sequence of statements as shown to the right:

X = (B**2 − 4*A*C)	PUSH B	
	PUSH #2	
	POWER	//B**2
	PUSH #4	
	PUSH A	
	PUSH C	
	MULT	//A*C
	MULT	//4*(A*C)
	SUB	//(B**2)−(4*(A*C))
	POP X	//X=

In this example, we assume the presence of a **pushdown stack** (see Figure 9.10). The PUSH and POP operations are machine-level instructions for loading/storing the top element of the stack. The POWER, MULT, and SUB instructions take their arguments from the top of the stack and return the results of exponentiation, multiplication, and subtraction to the top of the stack. The symbolic expression of calculation in fortran becomes a sequence of low-level machine instructions which often bear little resemblance to the HLL program.

The foregoing example illustrates the mechanical advantage provided by FORTRAN because one FORTRAN statement is implemented by many machine-level statements. Furthermore, it is much easier for a human programmer to read and write FORTRAN than to read and write machine-level

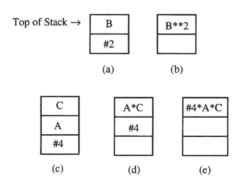

FIGURE 9.10 (a) The stack after PUSH #4 and PUSH B; (b) the stack after POWER; (c) the stack after PUSH #4, PUSH A, and PUSH C; (d) the stack after MULT; and (e) the stack after MULT a second time.

instructions. One major advantage of a HLL is the obvious improvement in program creation and, later on, its maintenance. As the size of the program increases, this advantage becomes larger as we consider the total cost to design, code, test, and enhance an application program.

The FORTRAN program containing the example statement is treated like input data by the translating program which produces a machine-level sequence as output. In general, the input data is called the **source program**, and the resulting translated output is called the **object program**. There are two ways to obtain an object program from a source program: compiling and interpreting.

In most cases, FORTRAN is translated by a compiler program. The idea behind a **compiler** is that the translator converts the source program in its entirety before any part of the resulting object program is actually run on the computer. That is, compiling is a two-step process. In some HLLs, however, it is impossible to entirely convert a source program into an object program until the program executes.

Suppose the storage for A, B, and C in the previous example is not known at the time the program is compiled. We might want to allocate storage on-the-fly while the program is running, because we do not know in advance that the storage is needed. This is an example of **delayed binding** of a variable to its storage location in memory.

Powerful languages such as Pascal and C permit a limited amount of delayed binding, as illustrated in the following example written in Pascal. This example also illustrates a limited amount of abstraction introduced by the HLL.

```
type      rnumber = real;                    {template}
          rptr = ^rnumber;                   {pointer}
var       Aptr, Bptr, Cptr: rptr;            {instance}
...
{later in the program...}
new(Aptr); read(Aptr^);                      {binding}
new(Bptr); read(Bptr^);
new(Cptr); read(Cptr^);
X := (Bptr^) * (Bptr^) − 4 * (Aptr^) * (Cptr^);
```

The **type** statement is an abstraction that defines a template and access mechanism for the variables A, B, and C that are to be created on-the-fly. The **var** statement is similar to the DIMENSION statement in that it tells the translator to allocate space for three pointers: Aptr, Bptr, and Cptr. Each of these allocations will point to the actual values of A, B, and C according to the previous **type** statement.

The actual allocation of storage is not known until the program executes the sequence of new() functions in the body of the program. Each new() function allocates space according to the **type** statement and returns a pointer to that space. To access the actual values in these storage spaces, the up arrow, ^, is written following the variable name. Thus, the read() function gets a number from the keyboard and puts it in the space pointed to by the pointer variable. Similarly, the value of X is computed by indirect reference to each value stored at the newly allocated memory location.

The purpose of this example is to illustrate the use of delayed binding in a HLL. Languages such as LISP and C++ require even greater degrees of delayed binding because of the abstractions they support. When the amount of delayed binding becomes so great that very little of the program can be compiled, we say that the HLL is an *interpreted language,* and the translator becomes an **interpreter** rather than a compiler. This crossover is often obscure, so some HLLs are translated by both a compiler and an interpreter. BASIC is a classic example of a HLL that is both interpreted and compiled.

The purpose of delayed binding is to increase the level of a HLL by introducing abstraction. Abstraction is the major differentiating feature between HLLs and other computer languages. Without abstraction and delayed binding, most HLLs would be no more powerful than a macro **assembler** language. However, with abstraction, HLLs permit a programmer to express ideas that transcend the boundaries of the physical machine.

We can now define HLL based on the concept of abstraction. *A HLL is a set of symbols which obey unambiguous syntactic and semantic rules: the syntactic rules specify legal combinations of symbols, and the semantic rules specify legal meanings of syntactically correct statements relative to a collection of abstractions.*

The notion of abstraction is very important to understanding what a HLL is. The example above illustrates a simple abstraction, e.g., that of data structure abstraction, but other HLLs employ much more powerful abstraction mechanisms. In fact, the *level of abstraction* of a HLL defines how high a HLL is. But, how do we measure the level of a HLL? What constitutes a HLL's height?

How High Is a HLL?

There have been many attempts to quantify the level of a programming language. The major obstacle has been to find a suitable measure of level. This is further complicated by the fact that nearly all computer languages contain some use of abstraction, and therefore nearly all languages have a "level." Perhaps the most interesting approach comes from information theory.

Suppose a certain HLL program uses P operators and Q operands to express a solution to some problem. For example, a four-function pocket calculator uses $P = 4$ operators for addition, subtraction, multiplication, and division. The same calculator might permit $Q = 2$ operands by saving one number in a temporary memory and the other in the display register. In a HLL the number of operators and operands might number in the hundreds or thousands.

We can think of the set of P operators as a grab bag of symbols that a working programmer selects one at a time and places in a program. Suppose each symbol is selected with probability $1/P$, so the information content of the entire set is

$$-\sum_1^P \frac{1}{P} \log\left(\frac{1}{P}\right) = \log(P)$$

Assuming the set is not depleted, the programmer repeats this process P times, until all of the operators have been selected and placed in the program. The information content contributed by the operators is $P \log(P)$, and if we repeat the process for selecting and placing all Q operands, we get $Q \log(Q)$ steps again. The sum of these two processes yields $P \log(P) + Q \log(Q)$ symbols. This is known as Halstead's **metric** for *program length* [Halstead, 1977].

Similar arguments can be made to derive the volume of a program, V, level of program abstraction, L, and level of the HLL, l, as follows.

P = Number of distinct operators appearing in the program
p = Total number of operators appearing in the program
Q = Number of distinct operands appearing in the program
q = Total number of operands appearing in the program
N = Number of operators and operands appearing in the program
V = Volume = $N \log_2(P+Q)$
L = Level of abstraction used to write the program $\approx (2/P)*(Q/q)$
l = Level of the HLL used to write the program = $L^2 V$
E = Mental effort to create the program = V/L

Halstead's theory has been applied to English (*Moby Dick*) and a number of programs written in both HLL and machine-level languages. A few results based on the values reported in Halstead [1977] are given in Table 9.4. This theory quantifies the level of a programming language: *PL/I* is higher level than Algol-58, but lower level than English.

In terms of the mental effort required to write the same program in different languages, Table 9.4 suggests that a HLL is about twice as high level as assembler language. That is, the level of abstraction of *PL/I* is more than double that of assembler. This abstraction is used to reduce mental effort and solve the problem faster.

TABLE 9.4 Comparison of Languages in Terms of Level, l, and Programming Effort E

Language	Level, l	Effort, E
English	2.16	1.00
PL/I	1.53	2.00
Algol-58	1.21	3.19
FORTRAN	1.14	3.59
Assembler	0.88	6.02

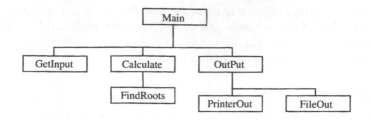

FIGURE 9.11 Hierarchical decomposition of procedural program.

HLLs and Paradigms

A programming **paradigm** is a way of viewing the world, e.g., an idealized model. HLLs depend on paradigms to guide their design and use. In fact, one might call HLL designers *paradigm engineers* because a good HLL starts with a strong model. Without such a model, the abstraction of a HLL is meaningless. In this section we examine the variety of paradigms embodied in a number of HLLs.

The **procedural paradigm** was the earliest programming paradigm. It is the basis of COBOL, FORTRAN, Pascal, C, BASIC, and most early languages. In this paradigm the world is modeled by an algorithm. Thus, an electrical circuit's behavior is modeled as a system of equations. The equations are solved for voltage, current, and so forth by writing an algorithmic procedure to numerically compute these quantities.

In the procedural paradigm a large system is composed of modules which encapsulate procedures which in turn implement algorithms. Hierarchical decomposition of a large problem into a collection of subordinate problems results in a hierarchical program structure. Hence, a large FORTRAN or C program is typically composed of a collection of procedures (subroutines in FORTRAN and functions in C) organized in layers, forming a tree structure, much like the organization chart of a large corporation (see Figure 9.11).

Hierarchy is used in the procedural paradigm to encapsulate low-level algorithms, thus abstracting them away. That is, algorithm abstraction is the major contributor to leveling in a procedural HLL. Figure 9.11 illustrates this layering as a tree where each box is a procedure and subordinate boxes represent procedures called by parent boxes, the top-most box is the most abstract, and the lowest boxes in the tree are the most concrete.

Intellectual leverage is limited to control flow encapsulation in most procedural languages. Only the execution paths through the program are hidden in lower levels. While this is an improvement over machine-level languages, it does not permit much flexibility. For example, algorithmic abstraction is not powerful enough to easily express non-numerical ideas. Thus, a C program is not able to easily model an electronic circuit as a diagram or object that can be reasoned about, symbolically.

One of the reasons procedural HLLs fail to fully hide all details of an abstraction is that they typically have weak models of data. Data is allowed to flow across many boundaries, which leads to problems with encapsulation. In FORTRAN, BASIC, Pascal, and C, for example, access to any data is given freely through globals, parameter passing, and files. This is called **coupling** and can have disastrous implications if not carefully controlled.

One way to reduce coupling in a procedural language is to eliminate side-effects caused by unruly access to data. Indeed, if procedures were prohibited from directly passing and accessing data altogether, many of the problems of procedural languages would go away. An alternative to the procedural paradigm is the **functional paradigm**. In this model of the world, everything is a function that returns a value. Data is totally abstracted away so that algorithms are totally encapsulated as a hierarchical collection of functions. LISP is the most popular example of a functional HLL [Winston and Horn, 1989].

A LISP statement that limits data access usually consists of a series of function calls; each function returns a single value which is used as an argument by another function and so on until the calculation is finished. For example, the FORTRAN statement $X = (B^{**}2 - 4^*A^*C)$ given earlier is written in functional form as follows:

ASSIGN(X, MINUS(SQUARE(B), TIMES(4, TIMES(A,C))))

This statement means to multiply A times C, then multiply the result returned by TIMES by 4, then subtract this from the result returned by SQUARE, and so forth. The final result is assigned to X.

One of the most difficult concepts to adjust to when using the procedural paradigm is the idea that all things are functions. The most significant implication of this kind of thinking is the replacement of loops with **recursion** and branches with guards. Recall that everything is a function that must return a value—even control structures. To illustrate, consider the functional (non-LISP) equivalent of the summation loop in FORTRAN, below.

```
S = 0
DO 20 I = 1,10      SUM(XList, N):
S = S + X(I)          N > 0 |
20 CONTINUE                 N is N − 1,
                            SUM is Head(XList) | SUM (TAIL(XList), N)
```

The functional form will seem strange to a procedural programmer because it is higher level, e.g., more abstract. It hides more details and uses functional operators HEAD (for returning the first element of XList), TAIL (for returning the N−1 tail elements of XList), and **is** for binding a value to a name. Also, notice the disappearance of the loop. Recursion on SUM is used to run through the entire list, one element at a time. Finally, the guard N > 0 prevents further recursion when N reaches zero.

In the functional program, N is decremented each time SUM is recursively called. Suppose N = 10, initially; then SUM is called 10 times. When N > 0 is false, the SUM routine does nothing, thus terminating the recursion. Interestingly, the additions are not performed until the final attempt to recurse fails. That is, when N = 0, the following sums are collected as the nested calls unwind:

```
SUM      : XList(10)
         : SUM + Xlist(9)
...        ...
         : SUM + XList(1)
```

Functional HLLs are higher level than procedural languages because they reduce the number of symbols needed to encode a solution as a program. The problem with functional programs, however, is their high execution overhead caused by the delayed binding of their interpreters. This makes LISP and PROLOG, for example, excellent **prototyping** languages but expensive production languages. LISP has been confined to predominantly research use; few commercial products based on LISP have been successfully delivered without first rewriting them in a lower-level language such as C. Other functional languages such as PROLOG and STRAND88 have had only limited success as commercial languages.

Another alternative is the **declarative paradigm.** Declarative languages such as Prolog and STRAND88 are both functional and declarative. In the declarative paradigm, solutions are obtained as a byproduct of meeting limitations imposed by constraints. Think of the solution to a problem as the only (first) solution that satisfies all constraints declared by the program.

An example of the declarative paradigm is given by the simplified PROLOG program below for declaring an electrical circuit as a list of constraints. All of the constraints must be true for the circuit() constraint to be true. Thus, this program eliminates all possible R, L, C, V circuits from consideration, except the one displayed in Figure 9.12. The declarations literally assert that Circuit(R, L, C, V) is a thing with "R connected to L, L connected to C, L connected to R, C connected to V, and V connected to R." This eliminates "V connected to L," for example, and leaves only the solution shown in Figure 9.12.

Circuit(R, L, C, V):
    Connected(R, L)
    Connected(L, C)
    Connected(L, R)
    Connected(C, V)
    Connected(V, R)

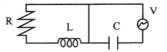

FIGURE 9.12 Solution to declaration for Circuit(R, L, C, V).

One interesting feature of declarative languages is their ability to represent infinite calculations. A declaration might constrain a solution to be in an infinite series of numbers, but the series may not need to be fully computed to arrive at an answer.

Another feature of such languages is their ability to compute an answer when in fact there may be many answers that meet all of the constraints. In many engineering problems, the first answer is as good as any other answer.

The declarative paradigm is a very useful abstraction for unbounded problems. Adding abstraction to the functional paradigm elevates declarative languages even higher. Solutions in these languages are arrived at in the most abstract manner, leading to comparatively short, powerful programs.

Perhaps the most common use of declarative languages is for construction of expert systems [Smith, 1988]. These kinds of applications are typically diagnostic. That is, they derive a conclusion based on assertions of fact. An electrical circuit board might be diagnosed with an expert system that takes symptoms of the ailing board as its input and derives a conclusion based on rules of electronic circuits—human rules of thumb given it by an experienced technician—and declarative reasoning. In this example, the rules are constraints expressed as declarations. The expert system program may derive more than one solution to the problem because many solutions may fit the constraints.

Declarative languages have the same inefficiencies as functional languages. For this reason, expert system applications are usually developed in a specialized declarative system called an *expert system shell*. A shell extracts the declarative or constraint-based capability from functional languages such LISP and PROLOG to improve efficiency. Often it is possible to simplify the shell so that early binding is achieved, thus leading to compiling translators rather than interpreters. Very large and efficient expert systems have been developed for commercial use using this approach.

Yet another paradigm used as the basis of modern languages such as C++ and Object Pascal is the *object-oriented programming (OOP) paradigm* [Budd, 1991]. OOP merges data and procedural abstractions into a single concept. In OOP, an object has both storage capacity and algorithmic functionality. These two abstractions are encapsulated in a construct called a **class**. One or more objects can be created by cloning the class. Thus, an **object** is defined as an instance of a class [Lewis, 1991].

OOP actually represents a culmination of ideas of procedural programming that have evolved over the past three to four decades. It is a gross oversimplification to say that OOP is procedural programming, because it is not, but consider the following evolution.

Procedure = Algorithm + Data Structures
Abstract Data Structure = Implementation Part + Interface Part
Class = Abstract Data Structure + Functions
Object = Class + Inheritance

The first "equation" states that a procedure treats algorithms and data separately, but the programmer must understand both the data structure and the algorithms for manipulating the data structures of an application. This separation between algorithms and data is a key feature of the procedural paradigm. During the 1970s structured programming was introduced to control the complexity of the procedural paradigm. While only partially successful, structured programming limited procedures to less powerful control structures by eliminating the GOTO and programs with labels. However, structured programming did not go far enough.

The next improvement in procedural programming came in the form of increased abstraction, called **ADT** (abstract data structures). An ADT separates the interface part of a procedure from its implementation part. Modula II and Ada™ were designed to support ADTs in the form of modules and packages. The interface part is an abstraction that hides the details of the algorithm. Programming in this form of the procedural paradigm reduces complexity by elevating a programmer's thoughts to a higher level of abstraction, but it still does not remove the problem of how procedures are related to one another.

Classes group data together into clusters that contain all of the functions that are allowed to access and manipulate the data. The class concept is a powerful structuring concept because it isolates the data portion of a program, thus reducing coupling and change propagation.

The class construct invented by the designers of Simula67 enforced the separation of interface and implementation parts of a module, and in addition introduced a new concept. Inheritance is the ability to do what

TABLE 9.5 Procedural versus Object-Oriented Thinking

Procedural	Object-Oriented
Instructions and data are separated.	Objects consist of both data and instructions.
Software design is linear, e.g., it progresses from design through coding and testing. This means change is difficult to accommodate.	Software design is interactive with coding and testing. This means change is easier to accommodate.
Programs are top-down decompositions of procedures, e.g., trees.	Programs are networks of objects that send messages to one another without concern for tree structure.
Program components are the real world, thus making programming more of a magic art.	Program components have abstractions of correspondence with the real world, thus making programming more of a discipline.
New programs are mostly custom built with little reuse from earlier programs. This leads to high construction costs and errors.	New programs are mostly specializations of earlier programs through reuse of their components. This leads to low construction costs and higher quality.

another module can do. Thus, inheritance relates modules by passing on the algorithmic portion of a module to other modules. Inheritance in a programming language like SmallTalk, Object Pascal, and C++ means even greater abstraction because code can be reused without being understood.

An object is an instance of a class. New objects inherit all of the functions defined on all of the classes used to define the parent of the object. This simple idea, copied from genetics, has a profound impact on both design and programming. It changes the way software is designed and constructed, i.e., it is a new paradigm.

Object-oriented thinking greatly differs from procedural thinking (see Table 9.5). In OOP a problem is decomposed into objects which mimic the real world. For example, the objects in Figure 9.12 are resistor, inductor, capacitor, and voltage source. These objects have real-world features (state) such as resistance, inductance, capacitance, and voltage. They also have behaviors defined by sinusoidal curves or phase shifts. In short, the real-world objects have both state and function. The state is represented in a program by storage and the function is represented by an algorithm. A resistor is a program module containing a variable to store the resistance and a function to model the behavior of the resistor when it is subjected to an input signal.

The objects in an object-oriented world use inheritance to relate to one another. That is, objects of the same class all inherit the same functions. These functions are called **methods** in SmallTalk and **member functions** in C++ [Ellis and Stroustrup, 1990]. However, the state or storage attributes of objects cloned from the same class differ. The storage components of an object are called **instance variables** in SmallTalk and **member fields** in C++.

The wholism of combining data with instructions is known as *ADTs*; the concept of sending messages instead of calling procedures is the *message-passing paradigm*; the concept of interactive, nonlinear, and iterative development of a program is a consequence of an object's **interface specification** being separated from its **implementation part**; the notion of modeling the real world as a network of interacting objects is called *OOD* (object-oriented design); the concept of specialization and **reuse** is known as *inheritance*; and OOP is the act of writing a program in an object-oriented language while adhering to an object-oriented view of the world.

Perhaps an analogy will add a touch of concreteness to these vague concepts. Suppose automobiles were constructed using both technologies. Table 9.6 repeats the comparison of Table 9.5 using an automobile design and manufacturing analogy.

We illustrate these ideas with a simple C++ example. The following code declares a class and two subclasses which inherit some properties of the class. The code also shows how interface and implementation parts are separated and how to override unwanted methods. Figure 9.13 depicts the inheritance and class hierarchy for this example.

```
class Node{
    public:                     //The interface part...
      Node() {}                 //Constructor function
      virtual~Node() {}         //Destructor function
      virtual int eval() { error(); return 0;}      //Override this function
    }
```

TABLE 9.6   Analogy with an Automobile Manufacturer

Procedural	Object-Oriented
Vendors and Assemblers work from their own plans. There is little coordination between the two.	Vendors and Assemblers follow the same blueprints; thus the resulting parts are guaranteed to fit together.
Manufacturing and design are sequential processes. A change in the design causes everyone to wait while the change propagates from designers to workers on the production line.	Design interacts with production. Prototypes are made and discarded. Production workers are asked to give suggestions for improvement or on how to reduce costs.
Changes on the manufacturing floor are not easily reflected as improvements to the design on the drafting board.	Changes in implementation rarely affect the design as interfaces are separated from implementation. Thus, the materials may change, but not the need for the parts themselves.
New cars are designed and constructed from the ground up, much like the first car ever produced.	New cars are evolutionary improvement to existing base technology, plus specializations that improve over last year's model.

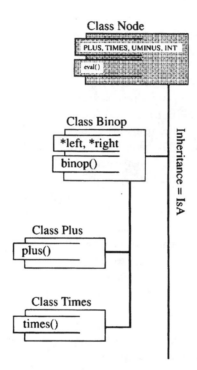

FIGURE 9.13   Partial class hierarchy for a C++ program that simulates a pocket calculator. The shaded Node class is an abstract class that all other classes use as a template.

The Node class consists of public functions that are to be overridden by descendants of the class. We know this because the functions are virtual, which in C++ means we expect to replace them later. Therefore we call this an **abstract class**. Also, Node() is the name of both the constructor and destructor member functions. A constructor is executed when a new object is created from the class, and the destructor is executed when the object is destroyed. These two functions take care of initialization and garbage collection which must be performed before and after dynamic binding of objects to memory space. Figure 9.13 shows this Node as an abstract class from which all other subclasses of this application are derived.

Now, we create a subclass that inherits the properties (interface) of Node() and adds a new property, e.g., Binop. Binop is an abstraction of the binary operators of a pocket calculator, which is the real-world object being simulated by this example program. The expression to be calculated is stored in a binary tree, and Binop sprouts a new left and right subtree when it is created and deletes this subtree when it is disposed.

```
class Binop : public Node {        //Derive Binop from Node
   public:
       Node *left, *right;         //Pointers to left and right subtrees
       ~Binop() { delete left; delete right;} //Collect garbage
       Binop( Node *lptr, Node *rptr) {left = lptr; right = rptr;}
}
```

Next, we define further specializations of Binop: one for addition of two numbers, Plus(), and the other for multiplication, Times(). The reader can see how to extend this to other operators.

```
class Plus: public Binop {
    public:                             //Add member functions to Binop
        Plus( Node *lptr, Node *rptr): Binop(lptr, rptr) {} //Use Binop
        int eval() { return left->eval()+right->eval();}       //Do Addition
};
class Times: public Binop {
    public:                             //Add member functions to Binop
        Times( Node *lptr, Node *rptr): Binop( lptr, rptr) {}  //Use Binop
        int eval() { return left->eval()*right->eval();}       //Do Multiply
};
```

In each case, the special-purpose operations defined in Plus() and Times() reuse Binop's code to perform the pointer operations. Then they add a member function eval() to carry out the operation. This illustrates reuse and the value of inheritance.

At this point, we have a collection of abstractions in the form of C++ classes. An object, however, is an instance of a class. Where are the objects in this example? We must dynamically create the required objects using the new function of C++.

```
Node *ptr = new Plus ( lptr, rptr);      //Create an object and point to it
int result = ptr->eval();                //Add
delete ptr;
```

The foregoing code instantiates an object that ptr points to, sends a message to the object telling it to perform the eval() function, and then disposes of the object. This example assumes that lptr and rptr have already been defined elsewhere.

Clearly, the level of abstraction is greatly raised by OOP. Once a class hierarchy is established, the actual data processing is hidden or abstracted away. This is the power of the OOP paradigm.

The pure object-oriented languages such as SmallTalk80 and CLOS have achieved only modest success due to their unique syntax and heavy demands on computing power. Hybrid HLLs such as Object Pascal and C++ have become widely accepted because they retain the familiar syntax of procedural languages, and they place fewer demands on hardware.

Although OOP is an old technology (circa 1970), it began to gain widespread acceptance in the 1990s because of the growing power of workstations, the increased use of graphical user interfaces, and the invention of hybrid object-oriented languages such as C++. Typically, C++ adds 15–20% overhead to an application program due to delayed binding of objects to their methods. Given that the power of the hardware increases more than 20% per annum, this is an acceptable performance penalty. In addition, OOP is much more suitable for the design of graphical user-interface-intensive applications because the display objects correspond with programming objects, thus simplifying design and coding. Finally, if you know C, it is a small step to learn C++.

Summary and Conclusions

HLLs: What are they? What do we mean by "high level"? What constitutes a paradigm? What are the advantages and disadvantages of HLLs? Who uses HLLs? What problems can be solved with these languages?

HLLs are human inventions that allow humans to control and communicate with machines. They obey rules of syntax and unambiguous semantics which are combined to express abstract ideas. HLLs are called "high level" because they express abstractions.

We have chosen to define the level of a HLL in terms of the information content of its syntax and semantics. The Halstead measure of language level essentially says that the higher a HLL is, the fewer symbols are needed to express an idea. Thus, if language A is higher than language B, a certain program can be expressed more succinctly in A than B. This is clearly the case when comparing HLLs with various machine-level languages, where a single statement in the HLL requires many statements in the machine-level language.

HLLs differ from one another in the abstractions they support. Abstract views of the world are called paradigms, and the guiding principle of any HLL is its programming paradigm.

We have compared the following programming paradigms: procedural, functional, declarative, and object-oriented. Procedural programming has the longest history because the first HLLs were based on low-level abstractions that are procedural. FORTRAN, COBOL, C, and Pascal are classical examples of the procedural languages.

Functional and declarative languages employ higher levels of abstraction by restricting the world view to simple mechanisms: mathematical functions and constraints. It may seem odd that such restrictions increase the level of abstraction, but languages like LISP and PROLOG hide much of the detail found to be necessary in the procedural paradigm. This increases the measure of level defined in this section.

Object-oriented programming embraces a novel abstraction that seems to fit the world of computing: objects. In this paradigm, the world is modeled as a collection of objects that communicate by sending messages to one another. The objects are related to each other through an inheritance mechanism that passes on the algorithmic behavior from one class of objects to another class. Inheritance permits reuse and thus raises the programming abstraction to a level above previous paradigms.

The future of HLLs is uncertain and unpredictable. It is unlikely that anyone in 1970 would have predicted the acceptance of functional, declarative, or object-oriented paradigms in the 1990s. Therefore, it is unlikely that the following predictions bear much relationship to computing in the year 2000. However, it is instructive to project a few scenarios and explain their power.

Functional and declarative programming result in software that can be mathematically analyzed, thus leading to greater assurances that the software actually works. Currently these paradigms consume too much memory and machine cycles. However, in the year 2000, very high-speed machines will be commonplace. What will we use these powerful machines for? One answer is that we will no longer be concerned with the execution efficiency of a HLL. The drawbacks of functional and declarative languages will fade, to be replaced by concern for the correctness and expressiveness of the HLL. If this occurs, functional and declarative languages will be the preferred HLLs because of the elevated abstractions supported by the functional and declarative paradigms. Applications constructed from these HLLs will exhibit more sophisticated logic, communicate in non-numeric languages such as speech and graphics, and solve problems that are beyond the reach of current HLLs.

Object-orientation is an appealing idea whose time has come. OOP will be to the 1990s what structured programming was to the 1970s. Computer hardware is becoming increasingly distributed and remote. Networks of workstations routinely solve problems in concert rather than as stand-alone systems. This places greater demands on flexibility and functionality of applications. Consider the next step beyond objects—servers:

Server = Object + Process

A server is an object that is instantiated as an operating system process or task. The server sends messages to other servers to get work done. The servers "live" on any processor located anywhere on the network. Software is distributed and so is the work. OOP offers the greatest hope for distributing applications in this fashion without loss of control. Should this scenario come true, the OOP paradigm will not only be appropriate, but contribute to greater HLL leverage through reusable objects, distributed servers, and delayed binding of methods to these servers.

Object-oriented languages, databases, and operating systems are on the immediate horizon. Graphical user-interface servers such as X-Windows already exist, lending credibility to this scenario. At least for the near future, HLLs are most likely to become identical with the object-oriented paradigm.

Defining Terms

Abstract class: A class consisting of only an interface specification. The implementation part is unspecified, because the purpose of an abstract class is to establish an interface.

Abstraction: Abstraction in computer languages is a measure of the amount of separation between the hardware and an expression of a programming idea. The level of abstraction of a high-level language defines the level of that language.

ADT: An abstract data type (ADT) is a software module that encapsulates data and functions allowed to be performed on that data. ADTs also separate the interface specification of a module from the implementation part to minimize coupling among modules.

Assembler: A computer program for translating symbolic machine instructions into numerical machine instructions. Assemblers are considered low-level languages for programming a computer.

Class: A specification for one or more objects that defines state (data) and functions (algorithms) that all objects may inherit when created from the class. A class is a template for implementing objects.

Compiler: A computer program that translates the source program statements of a high-level language into lower-leveled object program statements. Compilers differ from interpreters in that they do not immediately perform the operations specified in the source program. Instead, a compiler produces an object program that in turn performs the intended operations when it is run.

Coupling: A measure of the amount of interaction between modules in a computer program. High coupling means that a change in one module is likely to affect another module. Low coupling means there is little impact on other modules whenever a change is made in one module.

Declarative paradigm: A programming paradigm in which the world is modeled as a collection of rules and constraints.

Delayed binding: The process of postponing the meaning of a programming object until the object is manipulated by the computer. Delayed binding is used by interpreters and compilers, but more often it is employed by interpreters.

Functional paradigm: A programming paradigm in which the world is modeled as a collection of mathematical functions.

HLL (also HOL): A HLL is a set of symbols which obey unambiguous syntactic and semantic rules: the syntactic rules specify legal combinations of symbols, and the semantic rules specify legal meanings of syntactically correct statements relative to a collection of abstractions.

Implementation part: The definition or algorithm for a programming module which gives the details of how the module works.

Instance variables: Data encapsulated by an object.

Interface specification: The definition of a programming module without any indication of how the module works.

Interpreter: A computer program that translates and performs the intended operations of the source statements of a high-level language program. Interpreters differ from compilers in that they immediately perform the intended operations specified in the source program, and they do not produce an object program.

Member fields: Instance variables of a C++ object.

Member functions: Methods defined on a C++ object.

Metalanguage: A formal language for defining other languages. A metalanguage is typically used to define the syntax of a high-level language.

Methods: Functions allowed to be performed on the data of an object.

Metric: A measure of a computer program's complexity, clarity, length, difficulty, etc.

Object: An instance of a class. Objects have state (data) and function (algorithms) that are allowed to manipulate the data.

Object-oriented paradigm: A programming paradigm in which the world is modeled as a collection of self-contained objects that interact by sending messages. Objects are modules that contain data and all functions that are allowed to be performed on the encapsulated data. In addition, objects are related to one another through an inheritance hierarchy.

Object program: Machine form of a computer program, which is the output from a translator.

Paradigm: An idealized model, typically used as a conceptual basis for software design. Programming paradigms dictate the approach taken by a programmer to organize, and then write, a computer program.

Procedural paradigm: A programming paradigm in which the world is modeled as a collection of procedures which in turn encapsulate algorithms.

Prototyping: A simplified version of a software system is a prototype. Prototyping is the process of designing a computer program through a series of versions; each version becomes a closer approximation to the final one.

Pushdown stack: A data structure containing a list of elements which are restricted to insertions and deletions at one end of the list, only. Insertion is called a push operation and deletion is called a pull operation.

Recursion: A procedure is called recursive if it calls itself.

Reuse: Programming modules are reused when they are copied from one application program and used in another. Reusability is a property of module design that permits reuse.

Semantics: The part of a formal definition of a language that specifies the meanings attached to a syntactically correct statement in the language.

Source program: Symbolic form of a computer program, which is the input to a translator.

Syntax: The part of a formal definition of a language that specifies legal combinations of symbols that make up statements in the language.

References

T. Budd, *Object-Oriented Programming*, Reading, Mass.: Addison-Wesley, 1991.

M.A. Ellis and B. Stroustrup, *The Annotated C++ Reference Manual*, Reading, Mass.: Addison-Wesley, 1990.

M.H. Halstead, *Elements of Software Science*, New York: Elsevier North-Holland, 1977.

T.G. Lewis, *CASE: Computer-Aided Software Engineering*, New York: Van Nostrand Reinhold, 1991.

P. Smith, *Expert System Development in Prolog and Turbo-Prolog*, New York: Halsted Press, 1988.

P.H. Winston and B.K.P. Horn, *LISP*, Reading, Mass.: Addison-Wesley, 1989.

9.3   Data Types and Data Structures

Johannes J. Martin

The study of *data types* and *data structures* is a part of the discipline of computer programming. The terms refer to the two aspects of data objects: their usage and their implementation, respectively. The study of data *types* deals with the *identification* of (abstract) data objects in the context of a programming project and with methods of their more or less formal *specification*; the study of data *structures*, on the other hand, is concerned with the *implementation* of such objects using already existing data objects as raw material.

Concretely, the area addresses a basic problem of programming: the reduction of complex objects, such as vectors, tensors, text, graphic images, sound, functions, directories, maps, corporate organizations, models of ecosystems or machinery, or anything else that a program may have to deal with, to the only native objects of digital computers: arrays of binary digits (bits). The fundamental problem of this reduction is managing program complexity. Two organizational tools are essential to its solution: abstraction and hierarchical structuring. Abstraction refers to the separation of *what* computational objects are used for from *how* they are reduced to (i.e., implemented by means of) simpler ones. Hierarchical structuring refers to breaking this reduction up into small manageable steps. Through several steps of abstraction more and more complex objects are constructed, each one reduced to the previous, simpler generation of objects. This process ends when the desired objects have been composed.

Abstract Data Types

An **abstract data type** is one or more *sets* of computational objects together with some basic operations on those objects. One of these sets is defined by the type either by enumeration or by generating operations and is called the *carrier set* of the type. Customarily it is given the same name as the type. All other sets are called auxiliary sets of the type. In exceptional cases a type may have more than one carrier set.

The heart of the specification of an abstract data type is the definition of its functions, their syntax and semantics. Their syntax is specified by their **functionalities** and their semantics by algebraic axioms. For more details see Martin [1986].

With sets A and B, the expression $A \rightarrow B$ denotes the set of all functions that have the domain A and the codomain B. Functions $f \in A \rightarrow B$ (traditionally denoted by $f: A \rightarrow B$) are said to have the functionality $A \rightarrow B$.

The collection of basic operations does not need to be minimal. It should, however, be rich enough so that all other operations that one might wish to perform on the objects of the carrier set can be expressed exclusively by these basic operations. The type *Boolean*, for example, consists of the set of Boolean values, Boolean = {true, false}, with the operations *not*, *and*, and *or*.

In general, things are not quite this simple. To be useful for programming purposes, even the type Boolean requires at least one additional function. This function, called a *conditional expression*, provides a choice of one of two given values depending on a given Boolean value. It has the form:

f: Boolean 3 SomeType 3 SomeType $\rightarrow$ SomeType

and, with a, b $\in$ SomeType, is defined by:

f (true, a, b) = a   and
f (false, a, b) = b

The syntactical form of conditional expressions varies for different programming languages that provide this construct. For example, in the language C it has the form:

Boolean ? SomeType : SomeType.   /* with the result type of SomeType */

The set SomeType is an auxiliary set of the type Boolean.

Fundamental Data Types

The fundamental types listed next are supported by almost all modern high-level programming languages (reference books on Pascal, Modula II, C, and Ada are listed among the references at the end of this section):

Integer, Real (sometimes called Float), Character, and Boolean

Since their carrier sets are ordered (one of the operations of these types is $\leq$), these types are also called scalar types. All provide operations for comparing values; in addition, Integer and Real come equipped with the usual arithmetic operations $(+,-,^*,/)$ and Boolean with the basic logical operations (not, and, or). Most computers support these operations by hardware instructions. Thus, while bit arrays are the original native objects of a digital computer, the fundamental scalar types may be viewed as the given elementary building blocks for the construction of all other types.

Type Constructors

Enumerated Types

Beginning with Pascal, modern languages provide a rather useful constructor for scalar types, called enumerated types. Enumerated types have finite (small) carrier sets that the programmer defines by enumerating the constants of the type (specified as identifiers). For example, if the type Boolean were not available in Pascal, its carrier set could simply be defined by:

type Boolean = (false, true)

In Pascal, enumerated types are automatically equipped with operations for comparison as well as with the functions succ and pred; i.e., successor and predecessor. In the above example, succ(false) = true, pred(true) = false; succ(true) and pred(false) are not allowed.

Records

Values of scalar types can be arranged into tuples by a construct called a *record* (also called a *structure* in some languages). Into programming languages, records introduce the equivalent of the Cartesian product. Tuples can be viewed as abstract data types. Consider pairs as an example:

The type *Pairs:*

Carrier set:	*Pairs*,
Auxiliary sets:	A, B;
Operations:	pair $\in A \; 3 \; B \longrightarrow Pairs$;
	first $\in Pairs \longrightarrow A$;
	scnd $\in Pairs \longrightarrow B$;

where $\forall \; a \in A, b \in B$
first (pair (a, b)) = a; scnd (pair(a, b)) = b;

Using Pascal notation this type is defined by:

type Pairs = **record** first: A; scnd: B **end**

By providing the so-called field names, first and scnd, the programmer implicitly chooses the names for the selector functions. Pascal does not provide the function "pair." Instead, it permits the declaration of variables of type Pairs whose component values may be set (by assignment) and retrieved:

p: Pairs;	{declaration of p as a variable of type Pairs}
p.first := a; p.scnd := b;	{p now has the value of "pair(a,b)" above}
if p.first = x then... else...	{p.first is Pascals notation for "first(p)"}

The sets A and B can be of any type including records. Furthermore, since there is no restriction on the number of fields records may have, they can represent arbitrary tuples.

Arrays

Arrays permit the construction of simple function spaces, I $\longrightarrow$ A. The domain I, a scalar type — in some languages restricted to a subset $\{0, 1,..., n\}$ of the integers — is called the index set of the array; A is an arbitrary type. In Pascal, the mathematical notation f $\in$ I $\longrightarrow$ A assumes the form:

f: **array**[I] **of** A

I has to be a finite scalar type (e.g., 0 ... 40). The function *f* can now be defined by associating values of type A with the values of the domain I using the assignment operation:

f[i] := a; where i $\in$ I and a $\in$ A

Application of *f* to a value j $\in$ I is expressed by f[j]. This expression has a value of type A and can be used as such.

As with records, Pascal allows the naming of the function space I $\longrightarrow$ A by the definition:

type myFunctions = **array**[I] **of** A

and the subsequent declaration of a specific function (array) by:

f: myFunctions

Functions of several arguments are represented by so-called multidimensional arrays:

$f \in I_1 \; 3 \ldots 3 \; I_n \longrightarrow A$

is defined by

f: **array**[$I_1, \ldots, I_n$] **of** A

Variant Records

Variant records model *disjoint* (also called *tagged*) *unions*. In contrast to an ordinary union $C = A \cup B$, a disjoint union $D = A + B$ is formed by tagging the elements of A and B before forming D such that elements of D can be recognized as elements of A or of B. In programming, this amounts to creating variables that can be used to house values of both type A and type B. A tag field, (usually) part of the variable, is used to keep track of the type of the value currently stored in the variable. In Pascal, $D = A + B$ is expressed by:

```
type          tagType  =  (inA, inB) {an enumerated type};
              D        =  record
                            case kind: tagType of
                              inA: (aValue: A);
                              inB: (bValue:B);
                            end.
```

Variables of type D are now used as follows:

```
                    mix:      D;                    {mix is declared to be of type D}
              mix.kind              :=       inA;
              mix.aValue            :=       a;.
              . . .
              if mix.kind = inA
                  then {do something with mix.aValue, which is of type A}
                  else {do something with mix.bValue, which is of type B}
```

Conceptually, only one of the two fields, mix.aValue or mix.bValue, exists at any one time. The proper use of the tag is policed in some languages (e.g., Ada) and left to the programmer in others (e.g., Pascal).

An Example of a User-Defined Abstract Data Type

Most carrier sets are assumed to contain a distinguished value: *error*. *Error* is not a proper computational object: a function is considered to compute *error* if it does not return to the point of its invocation due to some error condition. Functions are called **strict** if they compute the value *error* whenever one or more of their arguments have the value *error*.

The following example models a cafeteria tray stack with the following operations:

1. Create a new stack with *n* trays.
2. Remove a tray from a stack.
3. Add a tray to the stack.
4. Check if the stack is empty.

Specification:

Cts (cafeteria tray stacks)   is the carrier set of the type;
Boolean and *Integer*   are auxiliary sets;
newStack, remove, add, isEmpty   are the operations of the type
where

```
              newStack       ∈ Integer      → Cts; {create a stack of n trays}
              remove, add    ∈ Cts          → Cts;
              isEmpty        ∈ Cts          → Boolean;
```

Axioms (logical expressions that describe the semantics of the operations): All functions are strict and, for all non-negative values of n,

1. remove (newStack(n))   = if n = 0 then *error* else newStack(n − 1);
2. add(newStack(n))   = newStack(n + 1);
3. isEmpty (newStack(n))   = (n = 0).

These axioms suffice to describe the desired behavior of Cts exactly; i.e., using these axioms, arbitrary expressions built with the above operations can be reduced to *error* or newStack(m) for some *m*.

Implementation:
For the representation of Cts (i.e., its data structure) we will choose the type Integer.

type   Cts = integer;

function newStack(n: Integer):Integer;
    begin if n < 0   **then** error ('n must be >= 0')   **else**   newStack:= n **end**;

function remove (s: Cts):Cts;
    begin if s = 0   **then** error ('stack empty')   **else**   remove:= s − 1 **end**;

function add (s: Cts):Cts;   **begin** add:= s + 1 **end**;

function isEmpty (s: Cts):Boolean;   **begin** isEmpty:= (s = 0) **end**;

Above, "error" is a function that prints an error message and stops the program.

Dynamic Types

The carrier sets of dynamic types contain objects of vastly different size. For these types, variables (memory space) must be allocated dynamically; i.e., at run time, when the actual sizes of objects are known. Examples for dynamic types are character strings, lists, tree structures, sets and graphs. A classical example of a dynamic type is a special type of a list: a queue. As the name suggests, a queue object is a sequence of other objects with the particular restrictive property that objects are inspected at and removed from its front and added to its rear.

Specification of Queues

Carrier set: *Queues*
Auxiliary sets: *Boolean*, *A*   (*A* contains the items to be queued)
Operations: newQueue, isEmpty, queue, pop, front

1. newQueue $\in$ *Queues*;   {a new, empty queue}
2. isEmpty $\in$ *Queues* $\rightarrow$ *Boolean*;   {check if a queue is empty}
3. queue $\in$ *A×Queues* $\rightarrow$ *Queues*;   {add an object to the rear of a queue}
4. front $\in$ *Queues* $\rightarrow$ *A*;   {return front element for inspection}
5. pop $\in$ *Queues* $\rightarrow$ *Queues*;   {remove front element from a queue}

Axioms: All functions are strict and, for a $\in$ A and s $\in$ Queues,

```
      isEmpty (newQueue);       (i.e., isEmpty (newQueue)    is true)
not   isEmpty (queue(a, s));    (i.e., isEmpty (queue(a, s))    is false)
      pop (newQueue)      = error;
      pop (queue(a, s))   = if s = newQueue
                                 then newQueue else queue(a, pop(s));
      front(newQueue)     = error;
      front(queue(a, s))  = if s = newQueue
                                 then a else front(s).
```

Implementation of Queues

The following implementation represents queues of the form queue(a, s) by the ordered pair (a, s) and a new, empty queue by the null pair *nil*. For the moment we assume that a data type, *Pairs*, which provides pairs on demand at run time, already exists and is defined as follows:

Specification of Pairs

Carrier Set: *Pairs*
Auxiliary sets: *Boolean, A*

Operations: nil, isNil, pair, first, scnd

1. nil $\vdash$ Pairs,   {a distinguished pair, the null pair}
2. isNil $\in$ Pairs $\rightarrow$ Boolean,   {test for the null pair}
3. pair $\in$ A 3 Pairs $\rightarrow$ Pairs,   {combine an item and a pair to a new pair}
4. first $\in$ Pairs $\rightarrow$ A,   {the first component; i.e., the item}
5. scnd $\in$ Pairs $\rightarrow$ Pairs,   {the second component; i.e., the pair}

Axioms: All functions are strict and, for $a \in A$ and $p \in$ Pairs,

isNil(nil);   (i.e. isNil(nil)   is true)
not isNil(pair(a,p));   (i.e. isNil(pair(a,p))   is false)
first (nil)   = error;
first(pair(a,p)) = a;
scnd (nil)   = error;
scnd(pair(a,p)) = p;

With pairs, queues may now be implemented as follows:

```
type    Queues = Pairs;
function newQueue    :Queues;         begin newQueue := nil      end;
function isEmpty (s: Queues)   :Boolean;   begin isEmpty  := (s = nil)   end;
function queue (x: A; s: Queues)   :Queues;  begin queue    := pair(x, s)   end;
function pop (s: Queues) : Queues;
begin
   if isNil(s)
      then error ('cannot pop empty queue')
      else if scnd(s) = nil
            then pop := nil
            else pop := pair (first(s), pop (scnd(s)))
   end;
function front (s: Queues): A;
begin
   if isNil(s)
      then error ('an empty queue does not have a front')
      else if scnd(s) = nil
            then front := first(s)
            else front := front(scnd(s))
   end;
```

The logic of these programs echoes the axioms. Such implementations are sometimes not very efficient but are useful for prototype programs, since the probability of their correctness is very high. The queues behave as *values*; i.e., the functions queue(a,s) and pop(s) do not modify the queues s but compute new queues; after the execution of, e.g., s1 := pop(s) there are two independent queue values, s and s1. This is exactly the behavior postulated by the axioms. However, practical applications frequently deal with *mutable objects*, objects that can be modified. With mutable objects memory may often be used more efficiently, since it is easier to decide when a memory cell, used, e.g., for storing an ordered pair, is no longer needed and thus may be recycled. If queues are viewed as mutable objects, the operations *queue* and *pop* are implemented as procedures that modify a queue. In order to apply the style of axioms introduced above for the description of mutable objects, these objects are best viewed as containers of values. The *procedure* queue(a, qobj), for example, takes the queue value, e.g., s, out of the container qobj, applies the *function* queue(a, s) (described by the axioms), and puts the result back into qobj.

If more than one place in a program needs to maintain a reference to such an object, the object must be implemented using a *head cell:* a storage location that represents the object and that is not released. The following implementation uses a head cell with two fields, one pointing to the front and one to the rear of the queue. We assume that the type *Pairs* has two additional functions:

    6. pairH $\in$ Pairs 3 Pairs $\to$ Pairs;   {create a pair with 2 pair fields}
    7. firstH $\in$ Pairs $\to$ Pairs;   {retrieve the first field of such a 2-pair cell}

and three procedures:

    8. setfirstH (p: Pairs; q: Pairs);   {change the firstH field of q to p}
    9. setscnd (p: Pairs; q: Pairs);   {change the scnd field of q to p}
  10. delete (s: Pairs)   {free the storage space occupied by s}

```
type    Queues = Pairs;
procedure newQueue(var q : Queues);      begin q          := pairH(nil, nil)      end;
function isEmpty (s : Queues) : Boolean;  begin isEmpty  := (firstH(s) = nil)      end;
procedure queue (x : A; s : Queues);
  var temp: Pairs;
  begin temp := pair(x, nil);
      if isNil(firstH(s)) then setfirstH(temp, s) else setscnd(temp, scnd(s));
      setscnd (temp, s);
      end;
function pop (s : Queues) : Queues;
  var temp: Pairs;
  begin
      if isNil(firstH(s))
          then error ('cannot pop empty queue')
          else begin temp := firstH(s); setfirstH(scnd(temp), s); delete(temp) end
      end;
function front (s : Queues) : A;
  begin
      if isNil(firstH(s))
          then error ('an empty queue does not have a front')
          else front:= first(firstH(s))
      end;
```

Compared to the value implementation given earlier, this implementation improves the performance of *front* and *pop* from $O(n)$ to $O(1)$.

An algorithm has $O(f(n))$ (pronounced: order $f(n)$ or proportional to $f(n)$) time performance if there exists a constant c, such that, for arbitrary values of the input size n, the time that the algorithm needs for its computation is $t \le c \cdot f(n)$.

Most modern programming languages support the implementation of the type *Pairs* (*n-tuples*) whose instances can be created dynamically. It requires two operations, called *new* and *dispose* in Pascal, that dynamically allocate and deallocate variables, respectively. These operations depend on the concept of a *reference* (or *pointer*), which serves as a name for a variable. References always occupy the same storage space independently of the type of variable they refer to. The following implementation of *Pairs* explains the concept.

```
type CellKind = (headCell, bodyCell);
     Pairs     = ^PairCell; {Pairs are references to PairCells}
     PairCell = record    tail: Pairs;
                     case  kind: CellKind of
                           headCell: (frnt: Pairs);
                           bodyCell: (val: A)
                     end;
function pair(item: A; next:Pairs):Pairs;
  var p: Pairs;
  begin
     new(p, bodyCell); {a new "bodyCell" is created and accessible through p}
     p^.kind := bodyCell; p^.val := item; p^.tail := next; pair := p
  end;
function first(p: Pairs):A;
  begin
     if p = nil then error('...')
          else if p^.kind = bodyCell then first:= p^.val else error('...')
  end;
procedure setfirstH(p, q:Pairs);
  begin
     if q - nil then error('...')
          else if q^.kind = headCell then q^.frnt:= p else error('...')
  end;
```

(Note: The Pascal constant **nil** denotes the null pointer, a reference to nothing.)

function isNil(p: Pairs): Boolean; **begin** isNil := (p = **nil**) **end**;

The reader should have no difficulty filling in the rest of the implementation of Pairs. Most of the algorithms on dynamic data structures were developed in the 1960s; still, an excellent reference is Knuth [1973].

More Dynamic Data Types

Stacks and Lists with a Point of Interest

A queue is the type of (linear) list used to realize first-come-first-served behavior. In contrast, another linear type, the *stack*, realizes last-come-first-served behavior. Sometimes it is necessary to scan a list object without dismantling it. This is accomplished by giving the list a point of interest (see Figure 9.14). This requires four additional operations:

```
restart(l: List);              {moves point of interest to beginning of list l}
current(l:List):A;             {returns object at point of interest}
advance(l:List);               {advances point of interest by one toward end of list l}
endOfList(l: List): Boolean;   {true, if end of list has been reached}
```

The type can be extended further by allowing insertions and deletions at the point of interest.

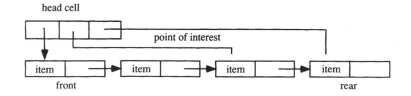

FIGURE 9.14   A list implementation with a point of interest and access to front and rear.

N-ary, Binary, and General Trees

An ***n*-ary tree** is the smallest set containing the *empty tree* and all ordered $n+1$-tuples $t = (a, t_1, \ldots, t_n)$ where a is member of some auxiliary set and the t_i are n-ary trees. The element a is called the *root element* or simply the root of t and the t_i are called *subtrees* of t.

Note that in this sense, a list is a unary tree. Binary trees used as searchtrees access finite ordered sets.

A **binary searchtree** is a tree that accesses a set. A tree t *accesses* a set s if the root of t is some element a of s, and s_1, called the left subtree of t, accesses the subset $\{x | x \in s$ and $x < a\}$ *and* s_2, *called the right subtree of t, accesses the subset* $\{x | x \in s$ and $x > a\}$.

If the left and right subtrees of the above definition are of similar size, then the time for finding an element in the set is proportional to $\log(n)$ where n is the cardinality of the set.

Quaternary trees (usually called quad trees) and octonary trees (called oct trees) are used to access two-dimensionally and three-dimensionally organized data, respectively. As with lists, the implementation of n-ary trees is based on $n+1$-tuples. A minimal set of operations for binary trees includes:

nilTree	$\in$ Trees;		{the empty tree, represented by nil}
isNil	$\in$ Trees	$\rightarrow$ Boolean;	{test if tree is empty}
tree	$\in$ A 3 Trees 3 Trees	$\rightarrow$ Trees;	{build tree from an item and subtrees}
root	$\in$ Trees	$\rightarrow$ A;	{retrieve root item of tree}
left	$\in$ Trees	$\rightarrow$ Trees;	{retrieve left subtree}
right	$\in$ Trees	$\rightarrow$ Trees;	{retrieve right subtree}

A **general tree** is the smallest set containing all order pairs $t = (a,s)$ where a is a member of some auxiliary set and s is a possibly empty list of general trees. The element a is called the root element or simply the root of t and the trees in s are called subtrees of t.

Note that there is no empty general tree; the simplest tree has a root and an empty list of subtrees. General trees are useful for the representation of hierarchical organizations such as the table of contents of a book or the organization of a corporation.

Functions, Sets, Relations, Graphs

Functions with reasonably small domains can be represented by arrays, as described earlier. Similarly, sets formed from a reasonably small universal set, relations on small domains, and graphs with not too many vertices can be represented by their characteristic functions implemented as bit arrays. In fact, Pascal provides a type constructor for sets that are derived from small universal sets.

Frequently domains are far too large for this approach. For example, the symbol table that a compiler of a programming language maintains is a function from the set of valid identifiers to some set of attributes. The set of valid identifiers is infinite, or, if some length limitation is imposed, finite but exceedingly large (there are nearly 300 billion identifiers of 8 characters or less). For most of its domain a symbol table returns the default value *new* or *not found*. It is therefore economical to store the mapping only for those domain values that map to a value different from the default value. A function of this sort is usually specified as follows:

Specification of Functions:

Carrier Set:	*Functions*	
Auxiliary sets:	*Dom, Cod*	(domain and codomain)
Operations:	newFun, apply, update	

1. newFun	$\in$ Functions,	(returns default everywhere)
2. apply	$\in$ Functions 3 Dom	$\rightarrow$ Cod,
3. update	$\in$ Functions 3 Dom 3 Cod	$\rightarrow$ Functions;

Axioms: All functions are strict and, for $x, z \in$ Dom, $y \in$ Cod and $f \in$ Functions,

apply(newFun, x)   = default;
apply (update(f,x,y), z) $=$ if $x = z$ then y else apply (f, z);

An implementation based on these axioms amounts to representing a function as a list of those of its individual mappings that differ from *default* and leads to an **O(1) performance** for *update* and an O(n) performance of *apply*. Better is an implementation by binary searchtrees with a performance of O(log(n)) for both *apply* and *update*.

Hashing

The fastest method for the implementation of functions is *hash coding* or *hashing*. By means of a *hash function*, h ∈ Dom → 0 .. k − 1, the domain of the function is partitioned into k sections and each section is associated with an index. For each partition a simple list implementation is used and the lists are stored in an array A: array[1 .. k − 1]. In order to compute apply(f,x) or update(f,x,y), the list at A[hash(x)] is searched or updated. If the hash function has been properly chosen and if k and the number of function values different from default are of similar size, then the individual lists can be expected to be very short and independent of the number of nondefault entries of the function; thus performance for apply and update is O(1).

The above discussion applies also to sets, relations, and graphs, since these objects can be represented by their characteristic functions.

Object-Oriented Programming

In languages that support object-oriented programming, *classes* (i.e., types) of objects are defined by specifying (1) the variables that each object will own as *instance variables* and (2) operations, called *methods,* applicable to the objects of the class. As a difference in style, these methods are not invoked like functions or procedures, but are *sent* to an object as a *message.* The expression [*window moveTo: x: y*] is an example of a message in the programming language *Objective C*, a dialect of C. Here the object *window*, which may represent a window on the screen, is instructed to apply to itself the method *moveTo* using the parameters x and y.

New objects of a class are created—usually dynamically—by *factory methods* addressed to the class itself. These methods allocate the equivalent of a record whose fields are the instance variables of the object and return a reference to this record, which represents the new object. After its creation an object can receive messages from other objects.

To data abstraction, object-oriented programming adds the concept of inheritance: from an existing class new (sub)classes can be derived by adding additional instance variables and/or methods. Each subclass inherits the instance variables and methods of its superclass. This encourages the use of existing code for new purposes. As an example, consider a class of a *list* objects. Each object has two instance variables pointing to the front and the rear of the list. In Objective C, the specification of the interface for this list class, i.e., the declaration of the instance variables and headers (functionalities) of the methods, has the following form:

```
@interface MyLists: Object   /* Object is the universal (system) class from which
                                 all classes are derived directly or indirectly */
{                            /* declaration of the instance variables;
    listRef front;             listRef is the type of a pointer to a list assumed
    listRef rear;              to be defined elsewhere */
}
– initList;                  /* initializes instance variables with null pointers */
– (BOOL) isEmpty;            /* test for empty list; note: parameter list is implied */
– add: (item) theThing;      /* item is the type of things on the list */
– pop;
– (item) front;
@end
```

As a companion of the *interface* file there is also an *implementation* file that contains the executable code for the methods of the class. A list with a point of interest can be defined as a subclass of MyList as follows:

@interface   <u>ScanList</u> : **MyList** /* ScanList is made a subclass of MyList */

{   listRef pointOfInterest; }
– restart;
– (BOOL)endOfList;
– advance;
– (Item)current;
@end

If we also need a list that can add and delete at the point of interest, we define:

@interface   <u>InsertionList</u> : **ScanList**
{ }   /* there are no new instance variables */
– insert: (item) theThing;
– shrink;   /* removes item at the point of interest */
@end

If a subclass defines a method already defined in the superclass, the new definition overrides the old one. Suppose we need a list where items are kept in ascending order:

@interface   <u>SortedList</u> : **MyList**
{ }
– add : (item) theThing;   /* this version of add inserts *theThing* at the proper
                                       place to keep the list sorted */

@end

Defining Terms

Abstract data type: One or more sets of computational objects together with some basic operations on those objects. One of these sets is defined by the type either by enumeration or by generating operations and is called the *carrier set* of the type. Customarily it is given the same name as the type. All other sets are called auxiliary sets of the type. In exceptional cases a type may have more than one carrier set.

Binary searchtree: A tree that accesses a set. A tree t accesses a set s if the root of t is some element a of s, and s_1, called the left subtree of t, accesses the subset $\{x \mid x \in s \text{ and } x < a\}$, and s_2, called the right subtree of t, accesses the subset $\{x \mid x \in s \text{ and } x > a\}$.

Functionality: With sets A and B, the expression $A \rightarrow B$ denotes the set of all functions that have the domain A and the codomain B. Functions $f \in A \rightarrow B$ (traditionally denoted by f: $A \rightarrow B$) are said to have the functionality $A \rightarrow B$.

General tree: The smallest set containing all ordered pairs $t = (a, s)$ where a is member of some auxiliary set and s is a possibly empty list of general trees. The element a is called the root element or simply the root of t and the trees in s are called subtrees of t.

N-ary tree: The smallest set containing the *empty tree* and all ordered $n+1$-tuples $t = (a, t_1, \ldots, t_n)$ where a is member of some auxiliary set and the t_i are n-ary trees. The element a is called the root element or simply the root of t and the t_i are called subtrees of t.

O(f(n)) performance: An algorithm has $O(f(n))$ (pronounced: order f(n) or proportional to f(n)) time performance if there exists a constant c, such that, for arbitrary values of the input size n, the time that the algorithm needs for its computation is $t \leq c \cdot f(n)$.

Strictness: Most carrier sets are assumed to contain a distinguished value: *error*. *Error* is not a proper computational object: a function is considered to compute *error* if it does not return to the point of its invocation due to some error condition. Functions are called *strict* if they compute the value *error* whenever one or more of their arguments have the value *error*.

References

A. Drozdek, *Data Structures and Algorithms in C++ Course Technology,* 2nd ed., Pacific Grove, CA: Brooks/ Cole 2000.

T.H. Cormen, *Introduction to Algorithms,* New York, NY: McGraw-Hill, 1995.

K. Jensen and N. Wirth, *Pascal: User Manual and Report,* Berlin, Germany: Springer-Verlag, 1974.

D.E. Knuth, *The Art of Computer Programming,* vol. 1, Reading, MA.: Addison-Wesley, 1973, Chap. 2.

J.J. Martin, *Data Types and Data Structures,* C.A.R. Hoare, Series Ed., Englewood Cliffs, NJ: Prentice-Hall International, 1986.

B.R. Preiss, *Data Structures and Algorithms,* New York, NY: Wiley, 1998.

S. Sahni, *Fundamentals of Data Structures in C++,* Rockville, MD: Computer Science Press, 1995.

R. Sedgewick, *Algorithms in C++,* Saddle River, NJ: Pearson, 2001.

B. Stroustrup, *The C++ Programming Language,* 3rd ed., Reading, MA.: Addison-Wesley, 2000.

N. Wirth, *Programming in Modula-2,* Berlin, Germany: Springer-Verlag, 1983.

Further Information

There is a wealth of textbooks on data structures. Papers on special aspects of data types and their relation to programming languages are found regularly in periodicals such as *ACM Transactions on Programming Languages and Systems,* the *Journal of the Association for Computing Machinery, IEEE Transactions on Computers, IEEE Transactions on Software Engineering,* or *Acta Informatica.*

9.4   The Use of Hardware Description Languages in Computer Design

Michael D. Ciletti

Introduction

Engineers use hardware description languages (HDLs) and electronic design automation (EDA) tools to design the integrated circuits of high-performance computers, signal/image processors, and other complex digital systems. Software tools execute major design steps while describing, analyzing, and managing a vast amount of data. At the heart of the design flow for an integrated circuit is a computer language that lets the designer model the circuit's structure and functionality. Classical design methodology relied heavily on schematics as the vehicle for describing a design, but today language-based descriptions dominate the design flow. Modern design methodology has made possible designs of enormous size and complexity accompanied by significant gains in the designer's productivity and efficiency.

Productivity Gains

No engineer alone can correctly manage the details of state-of-the-art integrated circuits (ICs) containing millions of transistors, but an EDA tool using a language-based description of the design easily handle the complexity and size of their databases. Even small designs rely on HDL-based descriptions because designers have to quickly produce correct designs targeted for an ever-shrinking market and window-of-opportunity for new products. Language-based designs are portable and independent of technology, allowing design teams to migrate designs to keep pace with improvements in technology. HDLs are a convenient medium for integrating third-party intellectual property with a proprietary design.

Although these benefits are important, the most significant gain is had by using HDLs in a design paradigm that relies on synthesis tools to automatically create a circuit realization from an HDL description of functionality. Today designers build a software prototype/model of the design, verify its functionality, and then use a synthesis tool to automatically optimize the circuit and create a netlist in a physical technology. HDLs and synthesis tools let an engineer focus attention on functionality, rather than individual transistors or gates. HDLs support the engineer by automatically synthesizing a circuit that will realize the desired functionality while meeting constraints. Moreover, alternative architectures can be generated from a single HDL model and evaluated quickly to perform design tradeoffs.

A hardware description language can serve as a platform for several tools: design entry, design verification, fault analysis/simulation, timing analysis/verification, and synthesis. This scope of use eliminates the need to translate descriptions between languages used by different tools, or to translate databases.

Two hardware description languages are widely used and supported by several EDA tool vendors: Verilog[TM][1] and VHDL[2], the VHSIC hardware description language. Both languages are IEEE standards. Analog design languages, such as Spice [3], are used in verifying critical timing paths of a design, but are impractical when used to verify the functionality of a large scale circuit. Mixed analog and digital languages are emerging from EDA vendors [4], and system-level design languages (e.g., SystemVerilog) are in development [5]. This article will focus attention on the Verilog[TM] HDL [6] for computer design, and introduce its important features.

Design Methodology

Before presenting some details of hardware description languages, we will consider where they fit in the design process. A simplified version of a language-based design flow for an application specific integrated circuit (ASIC) is shown in Figure 9.15. The functionality, timing requirements, operational considerations, and other relevant attributes of a design must be specified first. The design's speed, silicon area, power, and other constraints may be part of the specification. After the features and performance of a computer have been specified, designers partition the design along functional lines to create manageable design tasks then enter the design flow by describing the design in a computer-based hardware description language, usually either Verilog or VHDL.

Hardware description languages provide the design team with a spectrum of design styles ranging from explicit structural descriptions to implicit, abstract, or algorithmic descriptions. Structural descriptions compose a design by interconnecting simple structural objects, such as gates, or more complex functional units, e.g., a bit-slice microcontroller. Structural composition is similar to placing components on a schematic, but the description is a readable text instead of a schematic. Abstract models are also written in a readable text format, but without structural detail (examples will clarify these distinctions).

The functionality of a design is verified (step 4) either by simulation or by formal methods. The design flow iterates back to step 3 until the design has been verified. The design is integrated and verified to meet overall functional specifications (step 5) before being signed off for synthesis (step 6). Then the language-based description is passed off to a synthesis tool, which optimizes the logic and maps the abstract circuit into specific hardware, such as an FPGA or a cell library of gates and other parts (step 7). As an additional precaution, the functionality of the synthesized circuit can be compared to that of the original behavioral description to confirm their equivalence (step 8).

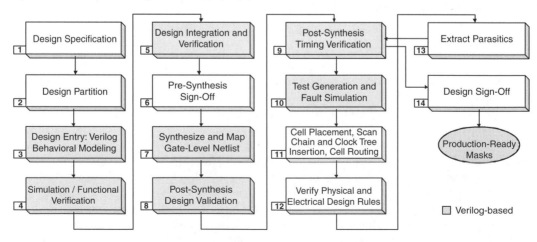

FIGURE 9.15 Design flow for an ASIC methodology.

Although the synthesis process is intended to meet timing specifications, the synthesized circuit is checked to verify that speeds are adequate on critical paths (step 9). Failure at this step requires that the design be resynthesized until timing margins are met. This may lead to (1) transistor resizing, (2) architectural modifications/substitutions, (3) consumption of more silicon area, and (4) consumption of more power. The synthesized gate-level design is analyzed to confirm that it has no undetectable faults; test patterns are also generated at this point in the flow (step 10). The issue of testability questions whether the chips that come off the fabrication line can in fact be tested to verify that they operate correctly. Testing considers process-induced faults, not design errors. The problem is daunting, for an ASIC chip might have millions of transistors, but only a few hundred package pins that can be used to probe the internal circuits. Here the designer might embed additional special circuits that guarantee that an ASIC can be tested, either alone or on a printed circuit board. This step in the design flow determines the strategy that will be implemented to ensure testability.

In cell-based technology the individual cells are integrated to form a global mask set by a computer-based tool that places and interconnects (routes) the individual gates on a template of the silicon chip (step 11). Physical design rules are checked (step 12) to ensure that the geometrical constraints imposed by the fabrication technology are satisfied. Rules governing electrical issues, such as crosstalk, are dealt with at this step. Finally, parasitic capacitance induced by the layout are extracted by a software tool and then used to produce a more accurate verification of the electrical characteristics and timing performance of the design (step 13). Finally, sign-off is achieved (step 14) and approval is given to produce the production-ready mask set. The shaded steps in the design flow in Figure 9.15 are supported by computer-based tools coupled with a hardware description language. For steps 11 to 13 the description of the design consists of the geometric data used in the photomasking steps of the fabrication process.

Design Entry

Design entry creates a language-based description of a digital circuit. The description may be structural, behavioral, or a mixture of these basic styles. A structural description consists of interconnected functional objects, such as gates. A behavioral description consists of an abstract representation of signal relationships with little or no apparent correspondence to physical hardware.

Example 9.1. The Verilog-based description given below as *Shift_Register_1* corresponds to the 4-bit shift register shown in Figure 9.16. The flip-flops (*Dflop*) themselves are previously designed functional units, used here as "instantiated" design objects.

```
module Shift_Register_1 (Data_out, Data_in, clock, reset);
    input        clock, reset;
    input        [3:0] Data_in;
    output       Data_out;
    wire         w1, w2, w3;

    Dflop M3 (w3, Data_in, clock, reset);
    Dflop M2 (w2, w3, clock, reset);
    Dflop M1 (w1, w2, clock, reset);
    Dflop M0 (Data_out, w1, clock, reset);
endmodule
```

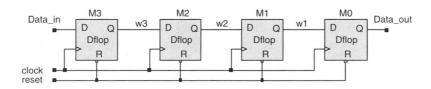

FIGURE 9.16 Shift register example.

Instantiation is analogous to placing and interconnecting components on a PC board. The instantiations in *Shift_Register_1* list four distinctly named and interconnected copies of *Dflop* to form the structure of a shift register. Also note that each instance of *Dflop* bears a name, *M0 ... M3*.

The description of *Shift_Register_1* in Example 9.1 is said to be *structural* because it consists of functional objects and their structural connections, i.e., the individual flip-flops and the structural relationships between their ports (terminals). (Note: the bold-faced items are reserved/keywords of the language). In the Verilog HDL a description is encapsulated by the keyword pair **module ... endmodule**. The code entered between these keywords consists of semicolon-terminated statements which, as a whole, describe the relationship between signals entering and leaving the module. The description of *Shift_Register_1* also reveals the language's ability to represent hierarchical decomposition of a design, for modules *M0... M3* are considered to be "child" modules of a "parent" module, *Shift_Register_1*. Modules that are instantiated within other modules are said to be "nested" in the parent module. The mechanism of nesting modules within other modules creates a natural hierarchy of the design. Nesting can be done to any depth supported by a simulator.

It is important to note that the statements in a language description of a module do not specify a sequence of execution like those of other programming languages, such as C. Instead, a Verilog mode describes a relationship between signals. In this example the relationship is imposed by the declared structure of the design. Verilog descriptions represent a circuit, and can be used with a simulator to reveal the signals that would evolve under the influence of stimulus.

The interface between a module and its environment is called a *port*. The Verilog language supports three port modes: *input*, *output* and *inout* (bi-directional). An input port is driven by its environment; the signal value at the port is determined external to the module. The value of a signal at an output port is determined by the module itself, and may be referenced by the environment. In *Shift_Register_1* the ports for *Data_out* and *Data_in* are declared as 4-bit vectors, and those of *clock* and *reset* are declared as (default) scalars.

The declaration of a module must identify the ports and their modes. In this example the order in which signals are listed in the port determined how external signals are bound to the description within the module (name association is an option too, but won't be discussed here).

The variables used in a design can correspond to signals in a circuit; variables may have one of four logic values: 0, 1 x, or z, where x is an ambiguous value (0 or 1) and z denotes high impedance (unconnected). Note that the declaration in *Shift_Register_1* includes wires ($w1$, $w2$, $w3$) used to connect the flip-flops. A complete description of the design requires a declaration of *Dflop*.

Example 9.2. For illustration an abstract description of a D-type flip-flop having asynchronous reset is given below.

```
module Dflop (q_out, data, clock, reset);
  output      q_out;
  input       data, clock, reset;
  reg         q_out;
always @ (posedge clock or negedge reset) begin
  if (reset == 0) q_out <= 0; else q_out <= data;
end
endmodule
```

The ports of *Dflop* are scalars, and the description consists of a cyclic behavior declared by the keyword **always** and conditioned by the active edges of *clock* and *reset*. Edge semantics are built into the language for rising (**posedge**) and falling (**negedge**) edges. Cyclic behaviors (e.g., **always @ (...) begin ... end**) execute when their event control expression (e.g., **posedge** clock **or negedge** reset) changes. If *clock* has a rising edge or if *reset* has a falling edge, the statement (**if** (reset == 0)...) executes. If *reset* is low, then the output is set to 0; otherwise, the operator (<=) assigns to *q_out* the value of data.

The statements associated with a cyclic behavior execute in sequence. After they execute, the activity flow (under simulation) returns to the event control expression and activity suspends until the expression has an event (i.e., changes). Then the cycle repeats. This type of simulation activity is said to be "event-driven," because computations occur only when signals change. Event-driven simulation is relatively efficient and capable of simulating circuits whose size precludes analog simulation with SPICE or similar tools.

Note that the style used here declares the functionality of a flip-flop but does not describe a gate-level (structural) realization. In this style, variables that are assigned value by a cyclic behavior must be declared as a "register" variable. One type of a register variable is a *reg*. A register variable retains its value until it is reassigned a value when a statement executes. They are required by a cyclic description because, in the absence of gate-level models, variables that abstractly model signals must retain a logic value as time evolves in simulation.

Verilog contains a family of primitives having predefined functionally corresponding to common combinational logic gates, pass transistors, pull-up/down loads, and transmission gates. The language also supports user-defined combinational and sequential primitives in a truth-table format. Primitives can be instantiated with or without propagation delay. In general, primitives can be used to build structural models having a desired functionality.

Example 9.3. A structural model of a transparent latch formed by a cross-coupled pair of nand primitive gates will be presented next. The gates, G1 and G2, are instantiated with unit propagation delay (#1), and the corresponding schematic is shown in Figure 9.17.

> **module** Nand_Latch (q_out, q_out_bar, preset_bar, clear_bar);
> **output**   q_out, q_out_bar;
> **input**   preset_bar, clear_bar;
>
> **nand** #1 G1 (q_out, preset_bar, q_out_bar);
> **nand** #1 G2 (q_out_bar, clear_bar, q_out);
> **endmodule**

The Verilog HDL offers alternative behavioral descriptions of the same functionality. Designers using synthesis tools rely primarily on (abstract) behavioral descriptions and let the tools determine the structure and physical implementation that meets the performance specifications. One style of behavioral modeling of combinational logic relies on language operators and "continuous assignment" statements.

Example 9.4. Consider a circuit that compares two 2-bit words, A and B, and asserts three signals indicating whether A is less than B, equal to B, or greater than B. A designer could develop Boolean algebraic equations describing the circuit, and then use language operators to write the corresponding Verilog description given below, where "~" denotes bitwise complement, "|" denotes bitwise-or, and "&" denotes bitwise-and.

> **module** Comparator_1 (A_lt_B, A_gt_B, A_eq_B, A1, A0, B1, B0);
> **output**   A_lt_B, A_gt_B, A_eq_B;
> **input**   A1, A0, B1, B0;
> **assign**   A_lt_B = (~A1) & B1 | (~A1) & (~A0) & B0 | (~A0) & B1 & B0;
> **assign**   A_gt_B = A1 & (~B1) | A0 & (~B1) & (~B0) | A1 & A0 & (~B0);

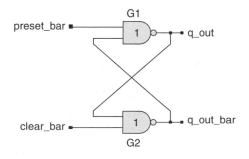

FIGURE 9.17 Schematic for a NAND latch.

    assign A_eq_B = (~A1) & (~A0) & (~B1) & (~B0) | (~A1) & A0 & (~B1) & B0 | A1 & A0 & B1 & B0 | A1 & (~A0) & B1 & (~B0);
endmodule

The model in *Comparator_1* is equivalent to a gate-level description, and is referred to as "implicit combinational logic." Continuous assignments use abstract expressions to describe a relationship between the value of a target variable, e.g., *A_lt_B* and other variables. A simulator will set up a monitoring mechanism so that whenever the right-hand side expression of a continuous assignment changes during simulation, the left-hand side is automatically updated.

Example 9.5. The continuous assignments in the previous example have an alternative and simpler form that are more useful when the datapath is large. In *Comparator_2* the expressions on the right-hand side of the assignments are treated as Boolean values that are either true (1) or false (0). Note that, in contrast to the description of *Comparator_1*, the size of the description of *Comparator_2* is independent of the wordlength, which is declared with the keyword ***parameter***. The relational operators ($<$, $>$, $==$) are built into the language and automatically accommodate arbitrary word size.

```
module Comparator_2 (A_lt_B, A_gt_B, A_eq_B, A, B);
    parameter                  word_length = 16;
    output                     A_lt_B, A_gt_B, A_eq_B;
    input [word_length-1:0]     A, B;

    assign A_lt_B = (A < B);
    assign A_gt_B = (A > B);
    assign A_eq_B = (A == B);
endmodule
```

An alternative description of the comparator consists of an algorithm written within an abstract cyclic behavior. *Comparator_3* has the same functionality as the previous examples.

```
module Comparator_3 (A_lt_B, A_gt_B, A_eq_B, A, B);
    parameter                  word_length = 16;
    output                     A_lt_B, A_gt_B, A_eq_B;
    input [word_length-1:0]     A, B;
    reg                        A_lt_B, A_gt_B, A_eq_B;
always @ (A, B) // Cyclic behavior
 begin
 A_lt_B = 0;
 A_gt_B = 0;
 A_eq_B = 0;
 if (A==B) A_eq_B = 1;
   else if (A > B) A_gt_B = 1;
     else A_lt_B = 1;
 end
endmodule
```

The model in *Comparator_3* does not have an obvious gate-level counterpart. However, the algorithm is readable, and clearly expresses the functionality intended by the designer. A synthesis tool will create an implementation that is identical or equivalent to that synthesized for *Comparator_2*.

Structural descriptions are useful in partitioning a large design. Continuous assignments are useful for modeling combinational logic, but this style can be cumbersome when the logic involves many variables. Consequently, the contemporary emphasis is on abstract behavioral descriptions of functionality or algorithmic descriptions, using language operators and procedural constructs without structural details.

Example 9.6. An alternative behavioral model of the shift register presented in Example 9.1 is given below to illustrate the compact representation that is supported by a language description. If the shift register has a 32-bit word length the model parameter, *word_length*, can be edited to accommodate a particular datapath. The concatenation operator, {...}, forms a vector (bus) by aggregating objects. In this example the assignment to *Data_Reg* clearly represents the register transfers that must occur at each clock cycle.

```
module Shift_Register_2 (Data_in, Data_out, clock, reset);
    input           Data_in, clock, reset;
    output          Data_out;
    parameter       word_length = 32;
    reg             [word_length-1: 0]        Data_reg;
    assign          Data_out = Data_reg[0];
  always @ (negedge reset or posedge clock)
    begin
    if (reset == 0) Data_reg <= 0;
    else Data_reg <= {Data_in, Data_reg[word_length-1:1]};
    end
endmodule
```

Note: All of the above examples are written in a style that can be synthesized automatically by a synthesis tool for a given word size.

Functional Verification

The functionality of an integrated circuit can be verified using simulators or formal verification tools, with the latter being essential for large circuits. We will illustrate only simulation. One attractive feature of a hardware description language is that the same language can be used to describe a design and to write a testbench that verifies its functionality. A Verilog testbench is itself a module, and it contains an instantiation of the model that is to be verified along with statements that stimulate the model under the control of a simulator.

Example 9.7. A Verilog model of a three-bit, up-down counter with controls to load data and disable the count action is given below, along with a simple testbench. The results of simulating *up_down_counter* with this testbench are shown in Figure 9.18.

```
module up_down_counter (Count, Data_in, count_up, load, counter_on, clock, reset);
    input               count_up, load, counter_on, clock, reset;
    input       [2: 0]  Data_in;
    output      [2: 0]  Count;
    reg         [2: 0]  Count;
  always @ (posedge reset or posedge clock)
    if (reset == 1) Count <= 0; else
    if (load == 1) Count <= Data_in; else
    if (counter_on == 1) begin
      if (count_up == 1) Count <= Count +1;
      else Count <= Count -1;
    end
endmodule

module test_up_down_counter ();
  wire  [2:0]   Count;
  reg   [2:0]   Data_in;
  reg           count_up, load, counter_on, clock, reset;

  up_down_counter M1 (Count, Data_in, count_up, load, counter_on, clock, reset);
```

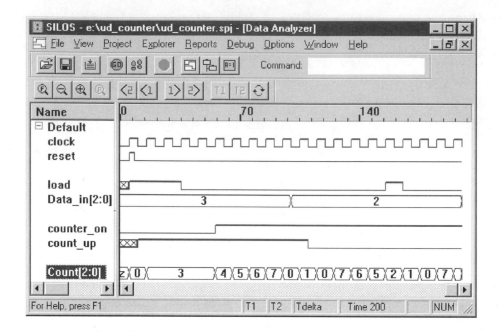

FIGURE 9.18   Simulation results for a Verilog model of an up-down counter.

initial #200 **$finish**;

initial begin clock = 0; **forever** #5 clock = ~ clock; **end**
initial fork
  begin Data_in = 3; #100 Data_in = 2; **end**
  begin reset = 0; #5 reset = 1; #3 reset = 0; **end**
  begin #5 load = 1; #30 load = 0; # 120 load = 1; #10 load = 0; **end**
  begin #10 count_up = 1; #100 count_up = 0; **end**
  begin counter_on = 0; #55 counter_on = 1; **end**
join
endmodule

The testbench applies the prescribed input signals to *up_down_counter*, and selected output signals are displayed by the Silos III[TM] simulator [7]. The testbench demonstrates the action of the *load, counter_on,* and *count_up* signals. An indefinite loop statement (**forever**) is used to describe a clock signal. The testbench includes a single-pass behavior (keyword: **initial**) to describe the input waveforms. A single-pass behavior is similar to a cyclic (**always**) behavior. Its statements execute in sequence, subject to timing controls (e.g., #5), but the behavior executes only once. It expires after the last statement has executed. The Verilog **fork ... join** construct describes parallel activity threads having a list of assignment statements, one for each input control signal. In a given thread an assignment of value to a control signal has to wait until the previous statement in the thread has executed. Thus, the delay shown with a statement is relative to the completion of the previous statement. Because the behavior generating the clock signal is an indefinite loop, a built-in systems task, **$finish**, is used to terminate the simulation after 200 time steps and return control to the operating system. The testbench exercises only a few signal patterns. In general, testbenches require careful, systematic development to provide a high level of confidence in the design.

Design Synthesis

Verilog descriptions of combinational logic, state machines, and other sequential circuits can be synthesized routinely by synthesis tools offered by several vendors.

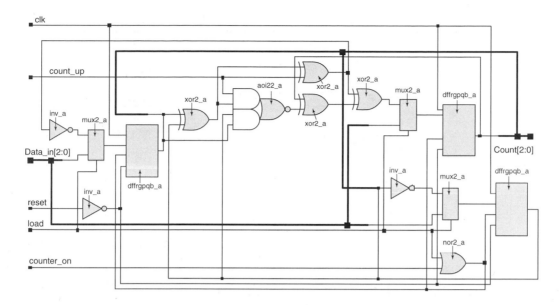

FIGURE 9.19   Circuit synthesized from *up_down_counter.*

Example 9.8. The results of synthesizing *up_down_counter* are shown in Figure 9.19. The synthesis engine (Synopsys Design Compiler[TM][8]) used a target technology of a standard cell library. The particular flip-flop selected by the tool consists of a gated D-type flip-flop with active low reset, and an internal datapath that feeds the output back to the input if the gate input (G) is low. Otherwise, the flip-flop operates like an ordinary flip-flop.

Design Example

RISC SPM.   As a final comprehensive example, key portions of a reduced instruction set computer (RISC) will be modeled with the Verilog HDL to illustrate how the language is used in computer design. We will discuss the overall architecture of the machine shown in Figure 9.20, *RISC_SPM*, a RISC-architecture stored program machine. We will describe the functional units, and then model the machine's ALU and controller.

The machine's architecture consists of a processor, a controller, and memory that stores instructions and data. The machine operates by fetching, decoding, and executing instructions to: (1) control the arithmetic and logic unit (*ALU*), (2) change the contents of storage registers, (3) change the contents of the program counter (*PC*), instruction register (*IR*) and the address register (*ADD_R*), (4) change the contents of the memory, (5) fetch data and instructions from memory, and (6) control the data busses. The instruction register holds the instruction that is currently being executed; the program counter contains the address of the next instruction to be executed; the address register holds the address of the memory location that will be addressed next by a read or write operation.

RISC SPM

Processor.   The processor consists of registers, datapaths, control lines, and an ALU that performs arithmetic and logic operations on its operands, as specified by the opcode held in the instruction register. A multiplexer, *Mux_1*, determines the source of data that is bound for *Bus_1*; another datapath mux, *Mux_2*, determines the source of data bound for *Bus_2*. Four general-purpose registers (*R0, R1, R2, R3*) and the program counter (*PC*) drive *Mux_1*. The contents of *Bus_1* can be steered to the ALU, the memory, or to *Bus_2* (via *Mux_2*). *Mux_2* is driven by the *ALU*, *Mux_1*, and the *memory*. Thus, an instruction can be fetched from the memory, placed on *Bus_2*, and loaded into the instruction register. A word of data can be fetched from memory and steered to a general-purpose register, or to the operand register (*Reg_Y*) prior to an

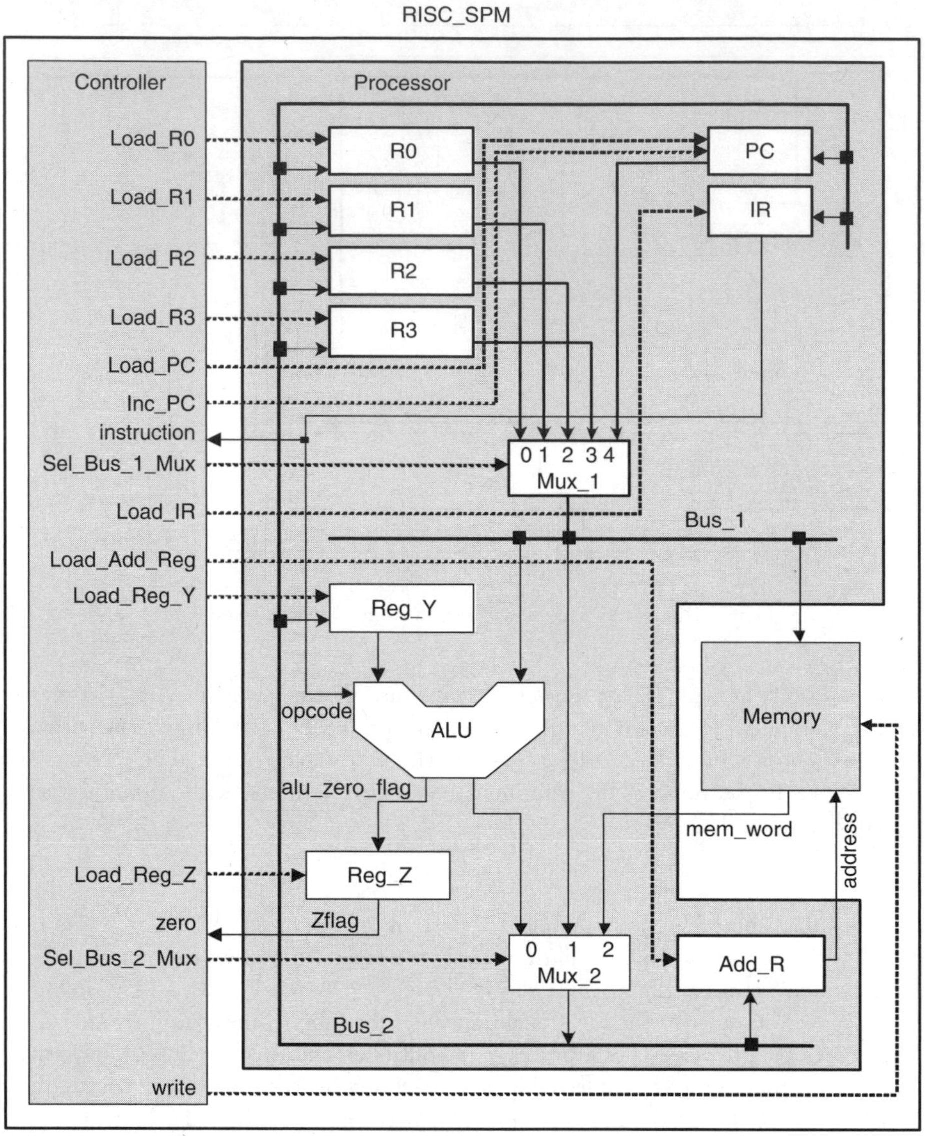

RISC_SPM

FIGURE 9.20   Architecture of a RISC stored-program machine.

operation of the *ALU*. The result of an *ALU* operation can be placed on *Bus_2*, loaded into a register, and subsequently transferred to memory. The register (*Reg_Z*) holds a flag indicating that the result of an *ALU* operation is zero.

ALU.   The ALU has two operand datapaths, *data_1* and *data_2*; its simple instruction set is shown below:

Instruction	Action
ADD	Form *data_1* + *data_2*
SUB	Form *data_1* − *data_2*
AND	Form the bitwise-and: *data_1* & *data_2*
NOT	Form the bitwise Boolean complement of *data_1*

The code below describes the *ALU* of *RISC_SPM*. The description includes comments (single-line: //, multiple line: /* ... */) for clarity.

```
    module Alu_RISC (alu_zero_flag, alu_out, data_1, data_2, sel);
     parameter word_size = 8;
     parameter op_size = 4;
    // Opcodes
     parameter NOP = 0, ADD = 1, SUB = 2, AND = 3;
     parameter NOT = 4, RD = 5, WR = 6, BR = 7, BRZ = 8;
     output                    alu_zero_flag;
     output [word_size-1: 0]    alu_out;
     input [word_size-1: 0]     data_1, data_2;
     input [op_size-1: 0]      sel;
     reg                       alu_out;

    assign alu_zero_flag = ~|alu_out;
    always @ (sel, data_1, data_2)
     case (sel)
       NOP:      alu_out = 0;
       ADD:      alu_out = data_1 + data_2;        // Reg_Y + Bus_1
       SUB:      alu_out = data_2 - data_1;
       AND:      alu_out = data_1 & data_2;
       NOT:      alu_out = ~ data_2;               // Gets data from Bus_1
       default:  alu_out = 0;
     endcase
    endmodule
```

The ALU is modeled as combinational logic described by a cyclic behavior (**always** block) that is activated whenever the datapaths or the select bus change. Notice that parameters make the description more readable.

Controller. The machine's timing and operations are governed by the controller. It must steer data to the destination specified by the instruction being executed. The design of the controller is strongly dependent on the specification of the machine's ALU, datapath resources, and the clocking scheme. For simplicity a single clock will be used here, and instruction commences on a single (rising) edge of the clock. The controller generates control signals by monitoring the state of the processing unit (i.e., the zero flag of the ALU) and the instruction register. The signals produced by the controller are listed below:

Control Signal	Action
Load_Add_Reg	Loads the address register
Load _PC	Loads *Bus_2* to the program counter
Load_IR	Loads *Bus_2* to the instruction register
Inc_PC	Increments the program counter
Sel_Bus_1_Mux	Selects among the *Program_Counter*, *R0*, *R1*, *R2*, and *R3* to drive *Bus_1*
Sel_Bus_2_Mux	Selects among *Alu_out*, *Bus_1*, and memory to drive *Bus_2*
Load_R0	Loads general-purpose register *R0*
Load_R1	Loads general-purpose register *R1*
Load_R2	Loads general-purpose register *R2*
Load_R3	Loads general-purpose register *R3*
Load_Reg_Y	Loads *Bus_2* to the register *Reg_Y*
Load_Reg_Z	Stores output of *ALU* in register *Reg_Z*
write	Loads *Bus_1* into the *SRAM* memory at the location specified by the address register

The control unit must (1) determine when to load registers, (2) select the path of data through the multiplexers, (3) determine when data should be written to memory and (4) control the three-state busses in the architecture.

opcode				source		destination	
0	0	1	0	0	1	1	0

(a)

opcode				source		destination	
0	1	1	0	1	0	don't care	don't care
address							
0	0	0	1	1	1	0	1

(b)

FIGURE 9.21   Format of a (a) short instruction, and (b) a long instruction.

The instructions that control the machine are stored in memory as a "machine language" program. The design of the controller depends on the processor's instruction set, i.e., the instructions that are available to a program that is to be executed, and on the machine's architecture. A machine-language program consists of a stored sequence of 8-bit words (bytes). An instruction can have a short or a long format, depending on the operation. Short instructions require one byte of memory and have the format shown in Figure 9.21(a). Each short instruction has a 4-bit opcode, a 2-bit source register address, and a 2-bit destination register address. Each long instruction requires 2 bytes of memory. The first word contains a 4-bit opcode, and its remaining 4 bits can be used to specify a pair of 2-bit source and destination registers depending on the instruction. The second word of a long instruction holds the address to the memory word that holds an operand required by the instruction [Figure 9.21(b)]. The machine's instruction mnemonics are described below.

Single-Byte Instruction	Action
NOP	No operation is performed; all registers retain their values. The addresses of the source and destination register are don't-cares. They have no effect.
ADD	Adds the contents of the source and destination registers and loads the result into the destination register.
AND	Forms the bitwise-and of the contents of the source and destination registers and loads the result into the destination register.
NOT	Forms the bitwise complement of the content of the source register and loads the result into the destination register.
SUB	Subtracts the content of the source register from the destination register and loads the result into the destination register.

Two-Byte Instruction	Action
RD	Fetches a memory word from the location specified by the second byte and loads the result into the destination register. The source register bits are don't-cares.
WR	Writes the contents of the source register to the word in memory specified by the address held in the second byte. The destination register bits are don't-cares.

BR	Branches the activity flow by loading the program counter with the word at the location (address) specified by the second byte of the instruction. The source and destination bits are don't-cares.
BRZ	Branches the activity flow by loading the program counter with the word at the location (address) specified by the second byte of the instruction if the zero flag register is asserted.

The machine's program counter holds the address of the next instruction to be executed. An assertion of the external reset is asserted and loads the program counter with 0, indicating that the bottom of memory holds the next instruction that will be fetched. For single-cycle instructions the instruction at the address in the program counter is loaded into the instruction register, and the program counter is incremented under the action of the clock. The output of the instruction decoder determines the resulting action on the datapaths and the ALU. A long instruction requires an additional clock cycle during which the second byte of the instruction is fetched from memory at the address held in the program counter to complete the instruction. While two-cycle operations are being executed intermediate contents of the ALU may be meaningless.

Controller Design

The controller of a digital computer is implemented as a finite state machine (FSM), and various tools and tables can be used to describe their logic [9]. A synchronous state machine makes its state transitions between states at the active edges of the clock. The machine's inputs and its current state determine its outputs and its next state. The machine's states must be specified for a given architecture, instruction set, and clocking scheme used in the design. The specification can be accomplished by identifying what steps must occur to effect the results of the instruction. In this example we will use an algorithmic state machine (ASM) chart [8] to describe the machine's activity. ASM charts present a clear picture of how machines operate under the influence of their instructions. ASM charts use square boxes to denote machine states, diamonds to denote decisions, i.e., input-dependent activity flow, and boxes with rounded corners to denote mealy-type asserted output signals. (Signals not explicitly asserted are considered to be deasserted.)

The RISC SPM has three phases of operation: fetch, decode, execute. Fetching gets an instruction from memory; decoding decodes the instruction, manipulates datapaths, and loads registers; execution carries out the instruction's operation. The fetch phase requires two clock cycles: one to load the address register and one to retrieve the addressed word from memory. The decode phase takes one cycle. Depending on the instruction, the execution phase may require zero, one or two more cycles. A *NOT* instruction can execute in the same cycle that the instruction is decoded. Single-byte instructions, such as *ADD*, take one cycle to execute during which the results of the operation are loaded into the destination register. The source register can be loaded during the decode phase. A two-byte instruction will take two cycles to execute. For example, in executing the *RD* instruction, one cycle is used to load the address register with the second byte, and one to retrieve the word from the memory location addressed by the second byte and load it into the destination register. The eleven states of *RISC_SPM* are listed below with the control actions that must occur in each state.

S_idle	State entered after reset is asserted. No action.
S_fet1	Load the address register with the content of the program counter (Note: PC is initialized to the starting address by reset action). The state is entered at the first active clock after reset is de-asserted, and is revisited after a *NOP* instruction is decoded.
S_fet2	Load the instruction register with the word addressed by the address register, and increment the program counter to point to the next location in memory, in anticipation of the next instruction or data fetch.
S_dec	Decode the instruction register and assert signals to control datapaths and register transfers.

S_ex1   Execute the *ALU* operation for a single-byte instruction, conditionally assert the
             zero flag, and load the destination register.

S_rd1   Load the address register with the second byte of an *RD* instruction, and
             increment the *PC*.

S_rd2   Load the destination register with the memory word addressed by the byte
             loaded in *S_rd1*.

S_wr1   Load the address register with the second byte of a *WR* instruction, and
             increment the *PC*.

S_wr2   Load the destination register with the memory word addressed by the byte
             loaded in *S_wr1*.

S_br1   Load the address register with the second byte of a *BR* instruction,
             and increment the *PC*.

S_br2   Load the program counter with the memory word addressed by
             the byte loaded in *S_br1*.

S_halt   Default state to trap failure to decode a valid instruction.

Figure 9.22 shows a portion of the overall ASM chart [10] for the *NOP, ADD, SUB* and *AND* instructions (all single-byte instructions), and Figure 9.23 shows the ASM chart for the *RD* operation. Once the ASM chart has been built, the designer can write the Verilog description of the controller to match the behavior implied by the chart. This process unfolds in stages. First, the functional units are declared according to the partition of the machine. Then their ports and variables are declared and checked for syntax. Then the individual units are described and debugged. The last step is to integrate the design and verify that it has correct functionality.

The design of the control unit is straightforward. First, the model's ports and variables are declared. Then the datapath multiplexers are described with continuous assignments using the conditional (?...:) operator. This conditional operator acts like a software switch to evaluate a right hand expression on the basis of whether the expression immediately preceding the "?" symbol is true or false. If true, the first expression after the "?" is evaluated; otherwise, the expression after the ":" is evaluated. In this example, the conditional operator is nested repeatedly within the continuous assignment. The controller is modeled by two cyclic behaviors. One governs the synchronous state transitions, and the other describes the combinational logic for the "next state" and the outputs (the control signals generated by the controller). The combinational logic is described by a case statement that parses the state of the machine, then assigns the next state and outputs. States for which the next state and outputs depend on the opcode as well as the state, are described by a second case statement evaluating the opcode. For illustration we show the decoding of the NOP, ADD, SUB, and AND instructions. The reader is encouraged to compare the Verilog description to the ASM chart in Figure 9.22, develop the remaining ASM charts, and complete the design of the controller.

```
module Control_Unit (Load_R0, Load_R1, Load_R2, Load_R3, Load_PC, Inc_PC,
Sel_Bus_1_Mux, Load_IR, Load_Add_R, Load_Reg_Y, Load_Reg_Z, Sel_Bus_2_Mux,
write, instruction, zero, clk, rst);
parameter word_size = 8, op_size = 4, state_size = 4;
parameter src_size = 2, dest_size = 2, Sel1_size = 3, Sel2_size = 2;

// State Codes
parameter S_idle = 0, S_fet1 = 1, S_fet2 = 2, S_dec = 3, S_ex1 = 4, S_rd1 = 5, S_rd2 = 6;
parameter S_wr1 = 7, S_wr2 = 8, S_br1 = 9, S_br2 = 10, S_halt = 11;
// Opcodes
parameter NOP = 0, ADD = 1, SUB = 2, AND = 3, NOT = 4;
parameter RD = 5, WR = 6, BR = 7, BRZ = 8;
// Source and Destination Codes
parameter R0 = 0, R1 = 1, R2 = 2, R3 = 3;
```

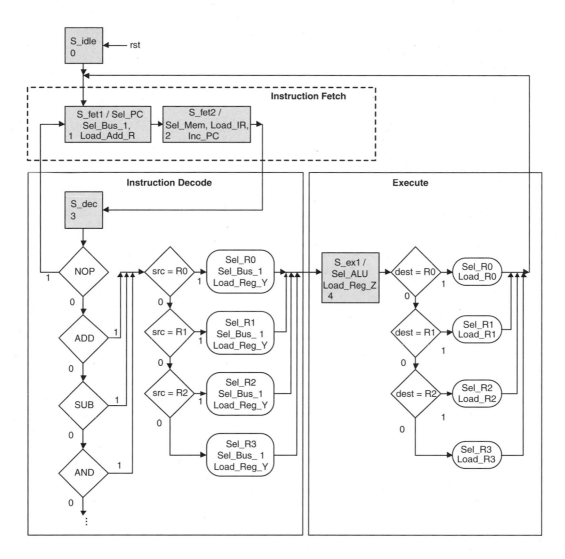

FIGURE 9.22 ASM chart for the *NOP*, *ADD*, *SUB* and *AND* instructions.

output Load_R0, Load_R1, Load_R2, Load_R3, Load_PC, Inc_PC;
output [Sel1_size-1: 0] Sel_Bus_1_Mux;
output Load_IR, Load_Add_R, Load_Reg_Y;
output Load_Reg_Z;
output [Sel2_size-1: 0] Sel_Bus_2_Mux;
output write;
input [word_size-1: 0] instruction;
input zero;
input clk, rst;
reg [state_size-1: 0] state, next_state;
reg Load_R0, Load_R1, Load_R2, Load_R3, Load_PC, Inc_PC;
reg Load_IR, Load_Add_R, Load_Reg_Y;
reg Sel_ALU, Sel_Bus_1, Sel_Mem;
reg Sel_R0, Sel_R1, Sel_R2, Sel_R3, Sel_PC;

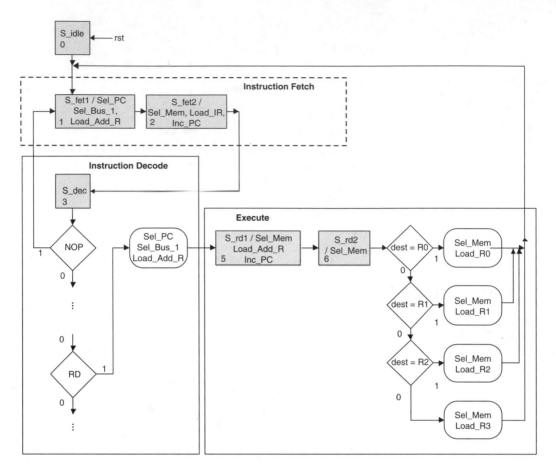

FIGURE 9.23 ASM chart for the *RD* instruction.

reg Load_Reg_Z, write;

reg err_flag;

wire [op_size-1: 0] opcode = instruction [word_size-1: word_size - op_size];

wire [src_size-1: 0] src = instruction [src_size + dest_size -1: dest_size];

wire [dest_size-1: 0] dest = instruction [dest_size -1: 0];

assign Sel_Bus_1_Mux[Sel1_size-1: 0] = Sel_R0 ? 0:

                        Sel_R1 ? 1:
                        Sel_R2 ? 2:
                        Sel_R3 ? 3:
                        Sel_PC ? 4: 3'bx; // 3-bits, sized number

assign Sel_Bus_2_Mux[Sel2_size-1: 0] = Sel_ALU ? 0:

                        Sel_Bus_1 ? 1:
                        Sel_Mem ? 2: 2'bx;

always @ (**posedge** clk **or negedge** rst) **begin**   // State Transitions
      if (rst == 0) state <= S_idle; **else** state <= next_state; **end**

```
always @ (state or opcode or zero) begin // Next state and output logic
    Sel_R0 = 0;
    Sel_R1 = 0;
    Sel_R2 = 0;
    Sel_R3 = 0;
    Sel_PC = 0;
    Load_R0 = 0;
    Load_R1 = 0;
    Load_R2 = 0;
    Load_R3 = 0;
    Load_PC = 0;
    Inc_PC = 0;
    Load_IR = 0;
    Load_Add_R = 0;
    Load_Reg_Y = 0;
    Load_Reg_Z = 0;
    Sel_Bus_1 = 0;
    Sel_ALU = 0;
    Sel_Mem = 0;
    write = 0;
    err_flag = 0;
case (state)
S_idle:     next_state = S_fet1;
S_fet1:     begin
                next_state = S_fet2;
                Sel_PC = 1;
                Sel_Bus_1 = 1;
                Load_Add_R = 1;
            end
S_fet2:     begin
                next_state = S_dec;
                Sel_Mem = 1;
                Load_IR = 1;
                Inc_PC = 1;
            end
S_dec:      case (opcode)
                NOP: next_state = S_fet1;
                ADD, SUB, AND: begin
                next_state = S_ex1;
                case (src)
                  R0: begin Sel_R0 = 1; Sel_Bus_1 = 1; Load_Reg_Y = 1; end
                  R1: begin Sel_R1 = 1; Sel_Bus_1 = 1; Load_Reg_Y = 1; end
                  R2: begin Sel_R2 = 1; Sel_Bus_1 = 1; Load_Reg_Y = 1; end
                  R3: begin Sel_R3 = 1; Sel_Bus_1 = 1; Load_Reg_Y = 1; end
                  default: err_flag = 1;
                endcase end
                ... // Decode additional opcodes.
             .. // Decode additional states
            endcase
        end
endmodule
```

Notice that the controller is described by two cyclic behaviors. The first makes state transitions at the rising edge of the clock. The second describes the combinational next-state logic of the machine's output signals and next state as a function of the decoded present state and the instruction. This style is widely used by designers, and readily synthesizes the controller.

Summary

The previous discussion and examples have presented some of the main constructs and features of the Verilog HDL and demonstrated its use in computer design. The reader is advised to consult the references for an in-depth treatment of additional language features and for the latest syntax [11].

References

1. *IEEE Standard Hardware Description Language Based on the Verilog Hardware Description Language*, Language Reference Manual (LRM), IEEE Std. 1364–1995, Piscataway, NJ: The Institute of Electrical and Electronic Engineers, 1996.

2. *IEEE Standard VHDL Language Reference Manual* (LRM), IEEE Std. 1076–1987, Piscataway, NJ: The Institute of Electrical and Electronic Engineers, 1988.

3. L.W. Negel, *SPICE2: A Computer Program to Simulate Semiconductor Circuits*, Memo ERL-M520, Department of Electrical Engineering and Computer Science, University of California at Berkeley, 9 May 1975.

4. D. Fitzpatrick and I. Miller, *Analog Behavioral Modeling with the Verilog-A Language*, Boston, MA: Kluwer Academic Publishers, 1998.

5. See www.Acceellera.org.

6. M.D. Ciletti, *Modeling, Synthesis, and Rapid Prototyping with the Verilog HDL*, Upper Saddle River, NJ: Prentice-Hall, 1999.

7. Simucad, Inc. Vendor supplied documentation, Milpitas, California.

8. Synopsys, Inc. Vendor supplied documentation, Mountain View, California.

9. R.H. Katz, *Contemporary Logic Design*, Redwood City, CA: Benjamin/Cummings Publishing Co., 1994.

10. M.D. Ciletti, *Advanced Digital Design with the Verilog HDL, 2003*, Upper Saddle River, NJ: Prentice-Hall, 2003.

11. M.D. Ciletti, *Starter's Guide to Verilog 2001*, Upper Saddle River, NJ: Prentice-Hall, 2004.

10

Input and Output

Solomon Sherr

Westland Electronics

10.1 Input Devices*

Input devices are those portions of computer, data processing, and information systems that perform the essential function of providing some means for entering commands and data into the system. Therefore, input devices are found in all such systems, but are treated here as a separate equipment group, independent of the total system configuration. However, the place of input devices in a representative computer system may be clarified by reference to Figure 10.1, which shows the interface of the main input device categories in relation to the portions of the generalized system that accept the inputs. The categories and the devices listed in Table 10.1 are the subject of this section.

Keyboards

Keyboards are essentially electromechanical devices, and are still ubiquitous, in spite of the inroads of other input devices. The primary type of keyboard in use as an input device is the alphanumeric (A/N) form, well known in its typewriter application, but with various additions and expansions consisting of numeric and special function keys. This type of keyboard is with a standard QWERTY format, so named because of the layout of the top left alpha keys, for the A/N portion, a separate numeric set to the right, and a group of function keys at the top. Other layouts for the A/N portion have been proposed and at least one (Dvorak) accepted by the American National Standards Institute (ANSI), but it has not received much use in spite of its advantages in increased efficiency. At present, the overwhelming majority of system keyboards still use the QWERTY layout, and it is the only one considered here.

A keyboard consists of a number of keyswitches whose exact structure is of prime importance in keyboard design. The relevant characteristics of keyswitch operation are life, actuation force, travel distance, and feedback. Accepted values are shown in Table 10.2 for different keyswitch designs. The elastomer type is preferred to a limited extent over the other two when the electronic audio feedback is included. This indicates that some type of audio feedback is desirable. One form of keyswitch design using an elastomer or "molded boot" is shown in Figure 10.2(a), in which the boot consists of two collapsible domes. In this design, the internal movement of the keyswitch is completely silent so that some source of sound must be added to achieve the desired audible feedback. The snap switch design shown in Figure 10.2(b) has built-in sound and achieves a small reduction in insertion errors over the elastomer design with audio feedback.

---

*The material contained in this section is a shortened version of that which appears in *Electronic Displays*, 2nd ed., by Sol Sherr, Chapter 6, Section 6.1, 1993, published by John Wiley & Sons, Inc., and is reprinted here by permission.

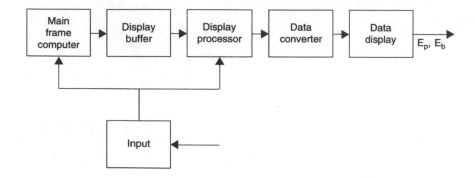

FIGURE 10.1   Generalized display-system block diagram. (*Source:* After S. Sherr, *Electronic Displays*, New York: John Wiley & Sons, 1979. With permission.)

TABLE 10.1   List of Input Devices

Category	Designation	Operation Mode
Keyboard	Alphanumeric	Electromechanical
Keyboard	Function	Electromechanical
Pointing	Light pen	Screen pointing
Pointing	Touchscreen	Screen pointing
Pointing	Pen tablet	Tablet pointing
Coordinates	Digitizer	X-Y conversion
Coordinates	Data tablet	X-Y location
Cursor	Mouse	Movement
Cursor	Trackball	Movement
Cursor	Joystick	Movement
Image	Scanner	Conversion
Verbal	Voice	Conversion

TABLE 10.2   Keyboard Parameter Values

Parameter	Snap Switch	Elastomer	Foam Pad
Key travel	3.8 mm	3.2 mm	3.8 mm
Force	>60 gm	>50 gm	>30 gm
Life	10 million cycles	10 million cycles	10 million cycles
Feedback	Audio mechanical	Audio electric	Tactile

The life requirement is estimated on the basis of workstation users operating at approximately half the accepted rate of 20 million actuations per key used for electronic typewriters. The actual layout and content of the keyboard may vary greatly, ranging from the standard typewriter arrangement, through different combinations of alphanumerics and symbols, to the special-function keyboards that contain legends and symbols specific to the particular application. However, the outputs of each type are the same in that they must contain coded signals that relate the action to be performed by the information system to that defined by the key being operated, in terms of the input code of the system. Thus, many of the keyboards output the ASCII code, and the system is usually designed so that it can accept this type of standard code. Incidentally, ASCII, the acronym for American Standard Code for Information Interchange, is the standard means for encoding alphanumerics and a group of selected symbols for transmission to a display system, among others. It is the standard code used in the United States and most other English-speaking countries and corresponds to the ISO seven-bit code. The seven-bit ASCII is usually used, and it should be noted that for serial data

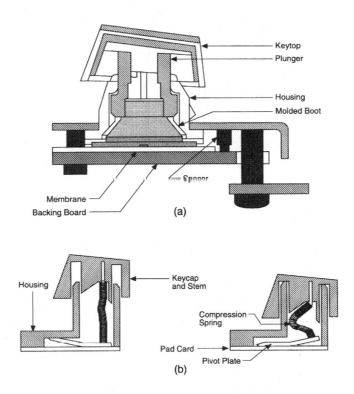

FIGURE 10.2 (a) Elastomer-type keyswitch. (b) Snap switch. (*Source:* After H. Brunner et al., "Effects of key action design on keyboard preference and throughput performance," Micro Switch. With permission.)

transmission an eighth bit is added for parity. Various keyboard arrangements are possible, and many variants are found in particular applications. The means for coding the key operation may be through magnetic reed relays, solid-state circuits, or more exotic devices such as Hall effect sensors. These device characteristics are only incidental to the operation and beyond the scope of this chapter. Similarly, we do not discuss the human-factors aspects of keyboard design, not because they are not important, but because, apart from the visual considerations, the other factors have to do with tactile and physical features best left to others.

Light Pen

The light pen initially was a very popular means for accomplishing manual input to the random deflection information display systems, but fell out of favor when raster systems became more popular due to its being somewhat difficult to use with raster systems. This device goes by a misleading name, as it does not emit light and is not a pen other than being somewhat similar to one in its physical appearance. However, when we consider its functional characteristics, the validity of the term becomes apparent, as it is used to cause the electron beam to "write" patterns on the cathode ray tube (CRT) that are defined by the motion of the light pen on the CRT faceplate.

The light pen operates by sensing the existence or nonexistence of a pulse of light at the point on the screen of the CRT or surface of any other light-emitting device where the point of the pen is placed. This is accomplished by means of the circuit shown in Figure 10.3, where the light pulse is collected and transmitted through the fiber optics to a light-sensitive device that converts the light pulse into an electrical pulse which is shaped by some form of electronics (of which a Schmitt trigger is one example). We need not concern ourselves with the exact form of the electronics except to note that this pulse is then sent to the

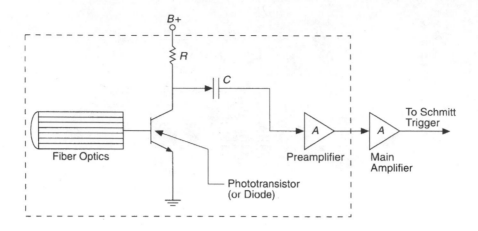

FIGURE 10.3   Light pen schematic. (*Source:* After S. Sherr, *Electronic Displays*, New York: John Wiley & Sons, 1979, p. 388. With permission.)

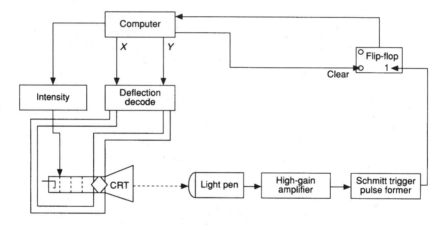

FIGURE 10.4   Block diagram of light pen computer system. (*Source:* S. Sherr, *Electronic Displays*, New York: John Wiley & Sons, 1979, p. 389. With permission.)

computer, as shown in Figure 10.4, and provides a complete, closed-loop system. As the electronic pulse occurs at the time when the light pulse passes under the light pen, the computer is informed of the location at which the designated operation is to be performed and may proceed accordingly. Thus, the light pen is a pointing device that designates a point on the display screen and can be used as an input device. Various light pen programs have been written to expand the capabilities of the original one, and it should be noted that the light pen is coming back into favor as improvements in accuracy, ease of operation, and reliability occur.

There are two characteristics of light pen operation that affect the capabilities of this input device. The first is the sensitivity, given by

$$S = E_L \mu_p A_p A_m \mu_s \mu_f t_L \tag{10.1}$$

where E_L = illuminance at photodetector, μ_p = photodetector sensitivity, A_p = preamplifier gain, A_m = main amplifier gain, μ_s = Schmitt trigger sensitivity, μ_f = flip-flop sensitivity, and t_L = optical loss.

Equation (10.1) may be used to calculate the light output required from the display surface, which may be a CRT or other light-emitting device, but with the limitation that most of the flat panel units are matrix driven and must track the drive sequence in order to know the location of the light pen from the drive pulse timing.

When phosphors are involved as for the CRT, vacuum fluorescent displays (VFDs), thin-film electroluminescent (TFEL) units, and color liquid crystal displays (LCDs), the phosphor delays must be entered into the timing, and the total delay is given by

$$E_0 = E_i(1 - e - t/\tau) \qquad (10.2)$$

where E_0 = voltage at triggering element, E_i = voltage equivalent of phosphor light output, t = time, and τ = sum of all delays.

These delays set limits to the positional accuracy, as the computer tracking the signal will be in error by this amount. Other inaccuracies are due to the dimensions of the optical pickup surface, all of which somewhat negate the simplicity of operation. The result is the parameter values shown in Table 10.3.

TABLE 10.3 Light Pen Data

Field of View	Response Time	Sensitivity
0.02–0.08 in.	120–150 ns	0.02–0.04 ft.L

Data Tablet (Graphics, Digitizer)

A very convenient means for data entry, retaining some of the ease of operation of the light pen but with much better accuracy, are the various forms of data tablets available. These tablets differ from the light pen in another significant way in that they do not require a moving spot of light to detect the location of the beam or direct it to a new location. This need for a moving light spot made the light pen difficult to use with the data tablets initially designed to overcome this limitation while still using a device with a pen-like input. The first successful example was the Rand tablet, a digital device that used an X–Y assembly from which a wand placed above some point on the X–Y wire matrix could pick up pulse generator output that fed X and Y electrical pulses into the matrix. By determining the number of pulses in a time period, the location of the wand is established. Another similar device used magnetostrictive rather than electrical signals to accomplish the same result, and this location is converted into display coordinates used to position a cursor on the CRT screen. The cursor may then be used as a visual feedback element so that the operator can correct the position of the wand until the cursor is properly placed. At this time the information from the tablet may also be transferred to either the host computer or the resident desktop or portable computer, as desired. Since the cursor is not used to signal its position to a pickup device, as is the case with the light pen, it may be used with any type of display system, including the non-light-emitting flat panel displays. Another advantage of the tablet is that it may be used to position cursors in the blank areas of the display, where no light pulses are available unless they are specially generated by the light pen.

There have been numerous improvements and new developments using a variety of technologies that include magnetostrictive, electromagnetic, electrostatic or capacitive, scanned X–Y grid, resistive, and sonic. Of these, electromagnetic tablets dominate the digitizer market, and sonic is of interest because it does not require a tablet, but most of the other technologies are essentially restricted to touch input devices covered later. As noted previously, electromagnetic is the most popular technology for high-performance digitizer tablets. Operation is based on transformer principles, whereby a conductor carrying ac creates a magnetic field around it that induces a current in a second conductor. The digitizer tablet uses the amplitude and phase of the induced current to determine digitizing data. The tablet contains an X–Y pattern of conductors beneath its surface, in a manner similar to the Rand Tablet, but instead of counting pulses in a time period a circular conductor is used as the pick-up element for the induced current. This coil is placed on the tablet surface, and its position is determined by measuring the phase and amplitude of the current in the coil. Its center is interpolated by sweeping through the X–Y grid lines and demodulating the signal in the coil to determine the phase reversal point, or by calculating this point using digitized data fed into a microprocessor. The X–Y coordinates may be resolved to better than 0.025 mm using either of these two techniques.

Another digitizer technology is the one that uses the measurement of the time required for sound waves to travel from a source to movable microphone pickups. This sonic technology has the advantage that no special digitizing board is required, and either a stylus or a cursor can be used as the digitizer. Two sonic sources are contained in an *L* frame so that both *X* and *Y* coordinates can be determined by calculating the time it takes for the sound wave to reach the microphones contained in the pickup device. This calculation is made on the basis of sound traveling at 345 m/s at 20°C, and the accuracy is dependent on stable ambient conditions. This tends to limit the resolution to about 300 lpi, and the accuracy to ±0.1%. The device may also be implemented with a single sonic source as the digitizing means and a pair of microphones located outside the digitizing area. In this case the location of the transducer is calculated by triangulation and converted into Cartesian coordinates.

Digitizers are used primarily for inputting accurate coordinate data from maps and engineering drawings. Their high accuracy requirements have led to relatively high prices. Alternative means for inputting data are the data and graphics tablets that meet most input requirements at a lower cost and accuracy. The main technology is still electromagnetic, and the units are essentially the same as the digitizers, but with lower accuracies. However, several of the other technologies have also been used to achieve lower costs. Most successful among them are the capacitive and resistive versions, which may also be used as digitizers. The capacitive units, also termed electrostatic, use capacitive coupling where the coupling between the tablet and the cursor or stylus is determined by the capacitance made up of the tablet surface as one plate and the pickup element as the other. In this case, the capacitance is given by

$$C = f(^{\text{TM}}A/d) \tag{10.3}$$

where C = capacitance, ^{TM} = permittivity of dielectric, A = relative area of two plates, d = distance between plates, and f = proportionality factor.

A scanned grid approach is used to determine the location of the cursor. As in the electromagnetic tablet, an *X–Y* grid of conductors is embedded in the tablet, with semiconductor switches on each line providing contact on

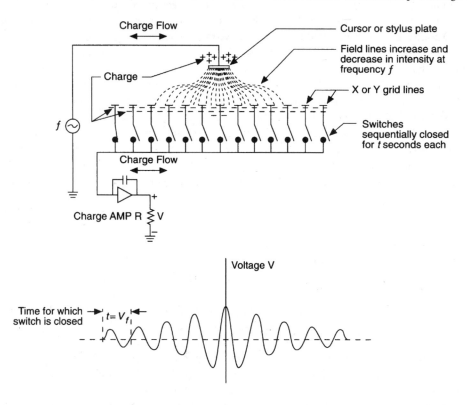

FIGURE 10.5   Capacitive technology. (*Source:* After T. E. Davies et al., "Digitizers and input tablets," in *Input Devices*, S. Sherr, Ed., New York: Academic Press, 1988, p. 186. With permission.)

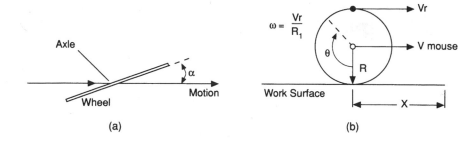

(a)   (b)

FIGURE 10.6 Wheel showing velocities and slip angle. (*Source:* After C. Goy, "Mice," in *Input Devices*, S. Sherr, Ed., New York: Academic Press, 1988, p. 225. With permission.)

a scanned basis. The charge flowing from each capacitance is summed through a summing amplifier as shown in Figure 10.5. The resultant voltage peaks twice, once for the X and once for the Y lines, as they are scanned. The peak positions are digitized by means of a counter that starts at the beginning of the scan, and runs at some multiple of the scan rate. The digital values represent the coordinates of the cursor location.

Mouse

The mouse has gone a long way from its original invention by Engelbart in 1965, through its redesign at Xerox and introduction by Apple as a main input device, and its general acceptance by computer users as an important addition to the group of input devices. It should be noted, in passing, that the mouse is essentially an upside-down trackball, although the latter is now being referred to as an upside-down mouse. However, the trackball came first and is described further in the next section.

Mice contain motion-sensing elements and are operated by moving mechanical or optical elements. One form uses wheels and shafts to drive the sensing elements, as shown schematically in Figure 10.6. The angular velocity (ω) of the wheel and shaft is given by

$$\omega = V_r/R \quad \text{rad/s} \tag{10.4}$$

where V_r = velocity of wheel and R = wheel radius.

The rotation angle (θ) is given by

$$(\theta) = X/R \quad \text{rad} \tag{10.5}$$

where X = distance moved.

This type of mouse has two sets of wheels and shafts, one for horizontal and the other for vertical motion.

A more popular type of mechanical mouse is the one that uses a ball for the motion sensing device, as shown in Figure 10.7. Again, the velocity of the ball circumference equals the velocity of the mouse, and the angular velocity is given by

$$\omega = V/R_1 \quad \text{rad/s} \tag{10.6}$$

where R_1 = shaft radius.

The smaller the shaft the more rapid its rotation for a given mouse velocity. Another form of the ball-and-shaft mouse is the one that uses an optical interrupter, as shown in Figure 10.8. In this form, the light from the light-emitting diodes (LEDs) is interrupted by the coded disks that are rotated by the shafts, and is then picked up by the phototransistors and converted into the digital signal that

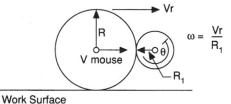

FIGURE 10.7 Ball and shaft. (*Source:* C. Goy, "Mice", in *Input Devices*, S. Sherr, Ed., New York: Academic Press, 1988. With permission.)

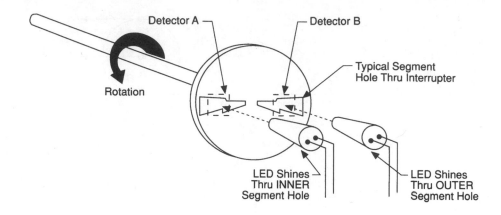

FIGURE 10.8   Optical interrupter. (*Source:* C. Goy, "Mice," in *Input Devices*, S. Sherr, Ed., New York: Academic Press, 1988, p. 229. With permission.)

represents the disk rotation. An optical interrupter is also used for the optomechanical mouse, and here the interrupter contains a set of slots; as the interrupter rotates quadrature signals are created that correspond to the shaft rotation.

In addition to the shaft and optomechanical mice, an early form of mouse used multiturn potentiometers connected to the wheels, and the output voltage that represented the motion varied in direct proportion to the mouse motion. The voltage was then converted by means of an analog-to-digital converter into digital form for input to the computer.

Finally, there are the true optical mice that use a special surface that is printed with a set of geometric shapes, usually a grid of lines or dots, that are illuminated and focused on a light detector. The most common form uses a grid made up of orthogonal lines, with the vertical and horizontal lines printed in different colors. These colors absorb light at different frequencies so that the optical detectors can differentiate between horizontal and vertical movement of the mouse. If such a structure is used as the mouse, then the photodetector will pick up a series of light-dark impulses consisting of the reflections from the mirror surface and the grid lines and convert them into square waves. A second LED and photodetector that is mounted orthogonally to the first is used to detect motion in the orthogonal direction, and the combination of the two inks avoids confusion between the two directions of motion. The system then counts the number of impulses created by the mouse motion and converts the result into motion information for the cursor. This type of mouse has the advantage that no mechanical elements are required.

Trackball

As noted previously, the trackball uses technology similar to the mouse, but preceded it as an input device. Thus, the comment that it is an upside-down mouse should be reversed. The movable element is housed in an assembly, and the assembly remains stationary so that much less desk space is required than for the mouse. In addition, the trackball may be mounted on a keyboard so that very little additional desk space is needed. The movable element can be the same as used in the mouse, and the output can be a set of bits corresponding to the coordinates to which the cursor should be driven, or where the command should be carried out. The output format is essentially equivalent to that used for the mouse, and the same protocols are used.

The typical trackball has an X and Y optical encoder that generates a pulse for each 0.76 mm of incremental motion of the ball. This means that the pulse train may range from 10 to 2500 pulses per second (pps), depending on how fast the ball is rotated. This is much more rapid than required for satisfactory updates, which need not be greater than about 100 times per second. This can easily be accomodated by the RS-232 protocol using an eight-bit word. Thus, the trackball is an excellent alternative for the mouse, and is rapidly returning to a preferred position as an input device.

Joystick

The joystick has not achieved much acceptance as an input device for electronic display systems, except for video games, although it has been the preferred control for many types of aircraft. However, it can be used to some extent in display systems other than those used in video games, and therefore warrants inclusion in this section. There are two basic types of joysticks, termed "displacement" and "force-operated". A typical displacement joystick may have two or three degrees of freedom. The activating means may vary from as few as four switches mounted 90 degrees apart, to full potentiometers for analog output, and optical encoders for digital output. A third axis may be added by allowing the handle to rotate and drive a third potentiometer. Spring forces of 5 to 10 lbs. are usual for the other two axes, and displacements go from 6 to 30 degrees.

The force joystick operates by responding to pressure on the handle to generate the *X–Y* coordinates. It may be either a two- or three-dimensional version, with the same types of handles as for the displacement joysticks. However, it is difficult to use a rotating handle for the third dimension because some force is usually transmitted to the other dimensions causing crosstalk. Therefore, a separate lever is preferred. The force is detected by means of piezoelectric sensors that are bonded to the handle rod, and a voltage source is applied across the network, as shown in Figure 10.9. The output is taken from the strain gauge and the analog voltage will be proportional to the amount of force. The same type of protocol and output circuitry may be used as for the displacement unit, and both can generate either position or rate data. An exponential curve with a dead zone threshhold is preferred for pulse rates in order to avoid starting pulse rate uncertainties, with the first pulse starting as soon as the threshhold is exceeded.

Touch Input

Touch input devices come in two basic forms, either placed directly on the display surface, or as a separate panel attached to the computer system. In its second form it is essentially a data tablet differing mainly in that it acts as another display unit with some form of a touch-sensitive surface. In this implementation it is the same as the *Touchscreen* input device, and this discussion concentrates on the technologies used for Touch-screens. There are five different technologies used for touch input devices, which are capacitive or resistive overlays, piezoelectric, light beam interruption, and surface acoustic wave. The system may be divided into the sensor unit, which senses the location of the pointing element, and the controller that interfaces with the sensor and communicates the location information to the system computer. Since the controller is an electronic device that does not use technology different from the computer it is not covered here. The main differences among the different touch input devices are due to the choice of sensor technology, and the discussion concentrates on these technologies.

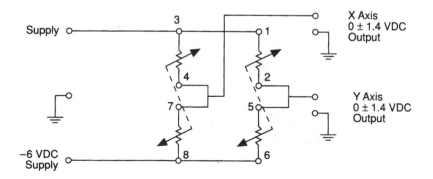

FIGURE 10.9 Schematic connections in a force joystick. (*Source:* After D. Doran, "Trackballs and joysticks," in *Input Devices*, S. Sherr, Ed., New York: Academic Press, 1988, p. 260. With permission.)

Capacitive. Capacitive overlay technology is illustrated in Figure 10.10 where a transparent metallic coating is placed over the display screen and the finger or stylus capacitance is sensed to determine the touch location. The overlay may consist of a group of separate sections etched into the surface with each separate section connected to the controller, or a continuous surface connected at the four corners. The first form is termed discrete capacitive, and touch location is determined by having each section sequentially connected to an oscillator circuit where the frequency of oscillations is affected by the pointing device. The oscillation frequency is measured and compared to a stored reference frequency. If the frequency difference is large enough then it is recognized as a touch at that location. It is a simple system, but suffers from low resolution and slow response so that it is only practical for menu selection.

The analog capacitive system uses the same metallic overlay, but the metallic surface is continuous rather than etched. The connections at the four ends are each connected to a separate oscillator, and the frequency of each is measured and stored. Then when the overlay is touched the change in capacitance will have a different effect on the frequency of each oscillator. These are measured and the differences are used to determine the coordinates of the touch by means of an algorithm. This technique is capable of much higher resolution (250×250) than the digital approach and is preferred for graphics or other high-density displays.

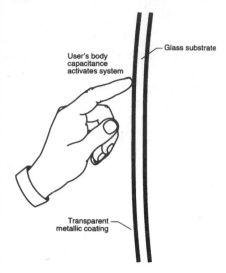

FIGURE 10.10 Capacitive overlay technology. (*Source:* After A.B. Carrell and J. Carstedt, "Touch input technology," *SID Sem. Lecture Notes*, p. 15.30, 1987. With permission. Courtesy Society for Information Display.)

Resistive. Resistive overlay technology requires a more complex assembly consisting of two layers, as illustrated in Figure 10.11. The layers both contain transparent metallic surfaces and are separated by spacers so that an air gap exists between the layers in the absence of any pressure on the touch panel. The metallic layers face each other and when the outer panel is pressed the metallic layers make contact and form a conductive path at the point of contact. When a voltage is applied between the top of the outer layer and the bottom of the inner layer, the two layers act as a voltage divider, and the voltage at the point of contact may be measured in the *X* and *Y* directions by applying the voltage in first one and then the other direction. The measured voltages are then transmitted to the controller where they are converted into coordinates which are then sent to the computer.

The panel may be discrete, in which the conductive coating on the top layer is etched in one direction and that on the bottom layer in the other direction, or analog, where the conductive coatings in both layers are continuous. In the discrete case, the panel then acts as an *X–Y* matrix, and the resolution is determined by the

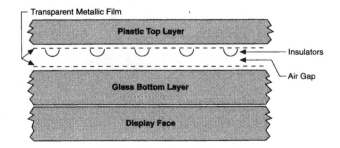

FIGURE 10.11 Resistive overlay technology. (*Source:* After A.B. Carrell and J. Carstedt, "Touch input technology," *SID Sem. Lecture Notes*, p. 15.31, 1987. With permission. Courtesy Society for Information Display.)

number of etched lines. The analog configuration requires the addition of linearization networks on each edge of the panel so that a large-area resistor is created with a voltage drop in one direction. Other linearization techniques are also possible, but only the four-element system is described here as shown in Figure 10.12. In this arrangement, one of the layers acts as the large-area resistor and the other as a voltage probe where either can function in either role. For the Y coordinate value the top layer is the voltage probe, and the voltage is applied by the controller to the bottom layer. Similarly, the X coordinate is found by connecting the voltage to the top layer and making the bottom layer into the voltage probe. In either type of system, the resolution can be very high, but the transmissivity is reduced to under 80% due to the multiple layers.

Piezoelectric. The piezoelectric technology uses pressure-sensitive transducers as the means for determining the location of the touch, as shown in Figure 10.13. The sensor is a glass plate with transducers connected to the four corners. Pressure on the plate causes readings to occur at each of the transducers, which depend on the location of the pressure. Thus, the controller can measure the readings and obtain the coordinates by means of a proper algorithm. This technique allows a high-transmissivity plate to be used that can be curved to follow the CRT face plate curvature, but it allows only a limited number of touch points to be used.

Light Beam Interruption. This is a fairly straightforward technology that requires a matrix of light sources and detectors facing each other in the X and Y directions. When the beams from the X and Y light sources are

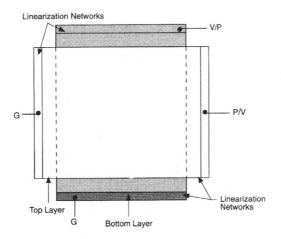

FIGURE 10.12 Four-wire analog resistive. (*Source:* A.B. Carrell and J. Carstedt, "Touch input technology," *SID Sem. Lecture Notes*, p. 15.32, 1987. With permission. Courtesy Society for Information Display.)

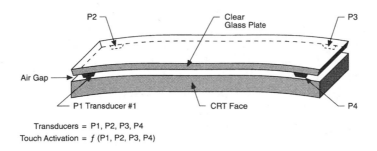

FIGURE 10.13 Piezoelectric technology. (*Source:* A.B. Carrell and J. Carstedt, "Touch input technology," *SID Sem. Lecture Notes*, p. 15.34, 1987. With permission. Courtesy Society for Information Display.)

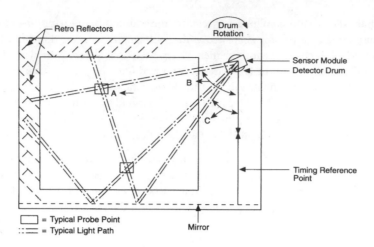

FIGURE 10.14   Rotating infrared beam technology. (*Source:* A.B. Carrell and J. Carstedt, "Touch input technology," *SID Sem. Lecture Notes*, p. 15.34, 1987. With permission. Courtesy Society for Information Display.)

interrupted, this is sensed by the facing light detectors and the signals are sent to the controller. The light beams are turned on sequentially by pulsing the LEDs and thus create a full matrix of light beams without requiring each of them to be on continuously. This system does not reduce the screen transmissivity as there is no obstruction of the screen output, but it is limited in resolution to the number of LED detector pairs that can be placed on the periphery of the screen.

Another approach to light interruption is to use a rotating beam of light, which has the advantage that only one light source and detector pair is required. This technology is depicted in Figure 10.14 and consists of a LED and a light detector placed inside a rotating drum which has a slit that allows light to be transmitted outside the drum. The light is swept across the surface and strikes the retroreflectors that sends it back directly to the detector. The beam scan is sampled 256 times on each scan, and Figure 10.14 shows how two angles of interruption are created, angle B by direct interruption, and angle C by mirror reflection interruption. The result is that the location of the interruption can be calculated by comparing the two angles. Again, there is no obstruction of the screen but a moving element must be added, and parallax errors may occur.

Pen-Based Computing.   This is an application for touch input devices that is growing at a rapid rate. The input device comes in several forms, each of which can recognize hand printing with the special operating system and software recognizing this type of input. The pen-based input device comes in several forms, of which the one termed TouchPen™ can function both as a digitizer with a touch tablet, and as the touch input device with a touch input pen-based computer system. A second one is that developed by Wacom, Inc., primarily for the GO Systems computer, but used by other pen-based systems as well. Finally, a third unit is that made by Scriptel Corp. and used by Wang Laboratories in its system.

TouchPen™ was developed by Microtouch Systems, Inc., initially for use in GridPad made by the Grid Systems Corp. It is essentially a high-resolution digitizer consisting of an all-glass tablet that can be used with a number of stylus input operating systems to digitize handwriting. It is basically a touch input device using resistive techniques to digitize the handwriting appearing on the display surface of pen-based computer systems. The glass tablet is placed on the display surface and the system pen is used to transmit the digitized data to the computer. As noted previously, the tablet may also be used as a standard touch input device.

The second form of pen-based input device is one that uses electromagnetic technology and consists of a grid of wires that transmit radio waves that are picked up by a tuned circuit in the stylus. This circuit resonates at its own frequency and transmits that signal back to the wires at the grid location it is touching. The pen also transmits its signal to the computer, which turns off the grid transmission, and locates the position of the pen by determining which of the grid wires pick up the pen signal. The pen does not need to actually touch the display surface and does not require any power, which is an advantage somewhat counteracted by the higher cost.

Finally, the Scriptel unit is similar to that made by Microtouch, but differs in that it uses electrostatic technology and is also similar to the capacitive touch panel.

Surface Acoustic Wave (SAW). This technology is more recent than the others and has not received wide acceptance as yet. It is based on the transmission through the glass of SAWs generated by transducers mounted on the glass overlay. These waves are detected by receivers also mounted on the glass, and the time of arrival of the waves at the receivers is known because the wave velocity is known. The placing of a finger on the glass weakens the signal and the location of the finger can be determined by the difference in its effect on the SAW.

There are two types of SAW systems in use, namely those using reflective techniques and those using attenuation as the source of position information. The reflective systems are similar to sonar where the time from the source to the pointing finger and then from the finger to the receiver is measured to arrive at finger location. The attenuation technology is illustrated in Figure 10.15 and consists of two transducers, two receivers, and four reflector strips, all mounted on a glass substrate. One transducer-receiver pair is used for X and the other for Y location. Figure 10.15 shows the X axis pair, and the transducer transmits a burst of acoustic energy in a horizontal wave. The wave is partially reflected by the top reflector strips and travels down to the bottom strip where the reflectors are at an angle such that it is reflected to the lower left corner receiver. The wave now has a long rectangular shape, and each point in time corresponds to a specific vertical path across the substrate. The Y axis is scanned in the same fashion after the X wave dies out. Then, when the finger touches the substrate, its water content absorbs some of the energy in the wave, and the wave is attenuated. The dip in the wave amplitude corresponds to the amount of absorbed energy, and the time of the lowest point can be determined, allowing the location of the finger to be calculated. Finally, in addition to the X and Y coordinates, a Z coordinate can be determined, depending on how hard the user presses. This depends on surface contact, which affects the amount of attenuation. The advantages of this system are high resolution, speed of transmission, and the

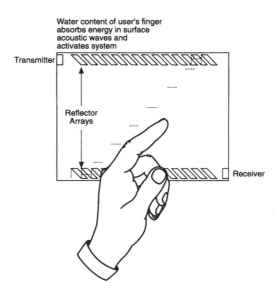

FIGURE 10.15 Attenuation SAW technology. (*Source:* A.B. Carrell and J. Carstedt, "Touch input technology," *SID Sem. Lecture Notes,* p. 15.35, 1987. With permission. Courtesy Society for Information Display.)

availability of a Z axis component. Its main disadvantages are the variation in moisture content in fingers and sensitivity to local moisture on the substrate. However, it is being used in developmental units and should be considered as another input device technology.

Scanners

Scanners are a means for inputting text and/or images directly into the computer system, thus avoiding the need for retyping and redrawing information contained in other sources. It is a relatively convenient way to avoid repetition if the data to be entered already exist in readable form. This is done by special image-recognition software that accompanies the scanning hardware, and can transfer an entire image containing both text and illustrations, but without the capability to modify the image. However, the addition of optical character recognition (OCR) software allows the entered text to be modified as if it were entered by typewriter. This can greatly simplify entering and editing text from some preexistent source and has resulted in a proliferation of devices that can perform this function.

These devices come in two main forms, *hand-held* and *page* scanners, with or without OCR software in addition to the standard image-recognition software. A typical hand-held scanner consists of a light source, a light-sensitive device such as a charge-coupled device (CCD) array, and the electronics to actuate the elements of the array sequentially under software control. The scanner window is placed over the page, and is moved down or across the page so that the window covers as much of the page as falls within the capability of the software. The light source is reflected from the page to the CCD and the charge in the CCD is modified by the reflectivity of the printed material.

The window area ranges from 4 to 5 in. in width by 0.5 in. in height and may be moved through 14 to 20 in., so that a fairly large area may be covered in a single manual scan. Images wider than the maximum window may be scanned in two passes, and the OCR software can stitch the two scans together into a single image, although this procedure requires considerable care in scanning so that the scans line up properly. Therefore, when images wider than the window of the hand-held scanner are to be scanned, it is advisable to use a *flatbed scanner* which can handle a full 8.5 in. by 11 in. page, or some of the larger scanners than can accept large drawings and input them into the computer system. Resolutions of 400 dpi and higher, with up to 250 levels of gray and 24 bits of color resolution are available. Thus, scanners offer a wide variety of choice and performance capabilities, and are powerful input devices when prepared data in visual form is to be entered into the computer system.

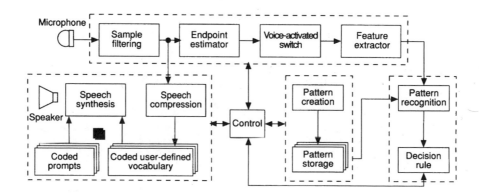

FIGURE 10.16   Block diagram of speech recognition and synthesis chip. (*Source:* After M. Leonard, "Speech poised to join man-machine interface," *Electronic Design*, pp. 43–48, September 26, 1991. With permission.)

Voice

Voice input is an intriguing approach to data input, with particular attractiveness to managers who want a simple and direct means for inputting data and commands. For many years, this technology tended to promise more than it could achieve, but recent developments have brought it to the point where it can be considered as

TABLE 10.4 Input Device Functional Evaluation

Input Device	Control	Function Data/Text	Data/Graphics	Total
Keyboard	E	E	P	9
Light Pen	G	G	E	10
Tablet	E	G	E	11
Mouse	E	F	E	11
Trackball	E	G	E	11
Joystick	F	F	G	5
Touchscreen	G	F	G	8
Scanner	F	E	G	9
Voice	G	F	P	6
Total	29	23	28	80

E = Excellent = 5; G = Good = 4; F = Fair = 3; P = Poor = 2

TABLE 10.5 Representative Performance Parameters

Input Device	Parameter	Value
Light pen	Response time	150–500 ns
	Spectral response	4200–9500 A
	Luminous sensitivity	0.03–0.7 nts
	Field of view	0.02–0.1 in.
	Ambient rejection	350 nts
Data tablet (digitizers)	Resolution (l/in.)	100–2000
	Accuracy (in.)	0.0005–0.02
	Active area (in.)	12 × 12–60 × 120
	Active height (in.)	0.02–2.5
	Digitizing rate (pps)	100–350
	Transducers	Stylus, puck, cursor
Mouse	Resolution	10–1000 dpi
	Speed	1–20 in./s
	Accuracy	25–1000 dpi
Trackball	Resolution	100–1000 cpi
	Speed	1200–9600 BPS
	Accuracy	100–1000 dpi
	Ball diameter	1.5–2.5 in.
Joystick	Travel	25–30°
	Accuracy	5–10%
	Repeatability	1%
Touchscreen	Resolution	256 × 256–4096 × 4096
	Transmissivity	60–100%
	Viewing area (in.)	3 × 4.5–15 × 20
	Speed	80–200 touch pts./s
Scanner	Resolution (dpi)	75–1600
	Scan rate (in./s)	0.5–2.0
	Scanning width (in.)	4.1–36 gray shades 32–256
	Scan time (s/page)	1–30
Voice	Active vocabulary	13–5000 words

a viable input means. This has been due to new developments in software that make it possible to minimize the amount of training required and increase the success rate to close to 100%.

One basic approach to speech recognition is represented by the block diagram shown in Figure 10.16. This is a system that is built around a special chip developed by Texas Instruments. This system uses templates and special algorithms for recognizing the input speech patterns. The system is speaker dependent, with the capability of storing up to 32 word templates and user-defined phrases. The output portion may be superfluous when the system is used only for inputting data and commands, but can be a useful adjunct to the visual response. Other techniques such as speaker-independent and phoneme-recognition systems are also available. Vocabularies range from 50 to 5000 active words, and both isolated and connected words can be recognized, although the larger numbers tend to be associated with isolated word systems. In general, it seems feasible that a combination of speech input and pen-based computing may find a viable market.

Summary

The multiplicity of input devices that are available makes it difficult to determine which is most suitable for any specific set of requirements. However, the limited functional comparison of the input devices covered in

TABLE 10.6   Input Devices—Advantages and Disadvantages

Device	Advantages	Disadvantages
Keyboard	Simple operation	Requires many keys
	Well known	Requires training
	Standard interface	No graphics
Light pen	Eye-hand coordination	Arm fatigue
	Low cost models	Limited resolution
	No desk space required	May block display
Graphic tablet	Natural hand movements	Eye-hand conflict
	Screen not blocked	Requires desk space
	No parallax	Breakable stylus
	Good for graphics	Poor for A/N entry
Mouse	Small space needed	Some space needed
	Low cost	Slow transmission
	Screen viewing	Low resolution
	Any surface may be used	Grid for optical
	(Optical) noiseless	Mechanical noise
Trackball	High resolution	Poor for A/N input
	Fixed desk space	Slow transmission
	Screen viewing	Mechanical noise
	Tactile feedback	3-D difficult
Joystick	Fixed desk space	Low accuracy
	Low fatigue	Low resolution
	Low cost	No A/N input
Touchscreen	Eye-hand coordination	Arm fatigue
	Minimal training	May block display
	Minimum input errors	Varied resolution
	User acceptance	Parallax
	No special commands	Slow data entry
Scanner	Full A/N page input	Hand scanner width
	Color scan input	High cost for color
	High resolution	Slow input
	OCR software	Compatibility
Voice	Ease of use	Limited words
	Minimal training	Machine training
	No special devices	Graphics difficult

this section shown in Table 10.4 may be of some use, and in any event is a starting point in this evaluation. It should be noted that what appears best at one time may become unpopular or obsolete at a later time, as occurred for light pens and trackballs, both of which have come back into favor.

In addition to the generalized evaluation shown in Table 10.4, it is also of interest to examine representative performance parameters. These are shown in Table 10.5 and while representative do not necessarily cover the range of performance parameters offered. More data may be obtained from the vendors of these devices.

Advantages and Disadvantages

Input devices make up one of the functional groups of the display systems, and their technical characteristics are covered in some detail at the beginning of this chapter, with performance information provided in Table 10.5 containing characteristic parameter values for each type, as available. The following material expands somewhat on that information by placing these devices in the context of a full graphics display system and evaluating the functions that the various types of input devices perform in that type of system in terms of their advantages and disadvantages. It is of some interest to compare the advantages and disadvantages of each type at this point, as listed in Table 10.6. This is an imposing list and may be used to aid in choosing the best input devices for specific applications. It also concludes this section on input devices.

Defining Terms

Data tablet/digitizer: A device consisting of a surface, usually flat, and incorporating means for selecting a specific location on the surface of the device and transmitting the coordinates of this location to a computer or other data processing unit that can use this information for moving a cursor on the screen of the display unit.

Joystick: An input device somewhat in the form of the navigation control device found in early aircraft and operating in a somewhat similar manner by generating series of pulses whose frequency or number depend on how far, with what force, and in what direction the control stick is moved from the central position.

Keyboards: Electromechanical devices consisting of sets of keys labeled with alphanumeric, numeric, and functional designations that enable the user to describe and define the operation to be performed.

Light pen: Neither a pen or a light source but rather an input device in the shape of a pen that operates by sensing the existence or nonexistence of light pulses at specific locations on the surface of a display device and uses this information to signal the computer as to the location of the pen.

Mouse: An input device based on a much older type known as a trackball and fancifully named because it bears only a casual resemblance to a mouse. It consists of a roller ball that is moved on a flat surface and causes orthogonal potentiometers or other types of X–Y-position signal generators to move and produce electrical signals defining the desired coordinates of the cursor on the screen so that the cursor can be moved to that position.

Scanners: Means for converting hard copy into electrical signals that can be entered into a computer or data processing system. The usual means for accomplishing such conversion is to move a light beam over the surface containing the data either by hand or automatically and using arrays of light-sensitive devices to convert the reflected light into electrical pulses.

Touch input: A means for selecting a location on the surface of the display unit using a variety of technologies that can respond to the placing of a finger or other pointing device on the surface. These are essentially data panels placed either on the display surface or between the user and the display surface.

Trackball: The earliest version of an input device using a roller ball, differing from the mouse in that the ball is contained in a unit that can remain in a fixed position while the ball is rotated. It is sometimes referred to as an upside-down mouse, but the reverse is more appropriate as the trackball came first.

Voice: Means for enabling a computer or data processing system to recognize spoken commands and input data and convert them into electrical signals that can be used to cause the system to carry out these commands or accept the data. Various types of algorithms and stored templates are used to achieve this recognition.

References

H. Brunner et al., "Effects of key action design on keyboard preference and throughput performance," Micro Switch.

A.B. Carrell, and J. Carstedt, "Touch input technology," *SID Sem. Lec. Notes*, pp. 15.30–15.35, 1987.

T.E. Davies et al., "Digitizers and input tablets," in *Input Devices*, S. Sherr, Ed., New York: Academic Press, 1988, p. 186.

D. Doran, "Trackballs and joysticks," in *Input Devices*, S. Sherr, Ed., New York: Academic Press, 1988, pp. 251–262.

C. Goy, "Mice," in *Input Devices*, S. Sherr, Ed., New York: Academic Press, 1988, pp. 225–232.

M. Leonard, "Speech poised to join man-machine interface," *Elec. Des.*, pp. 43–48, Sept. 26, 1991.

S. Sherr, *Electronic Displays*, New York: Wiley, 1979, pp. 323, 388–389.

Further Information

Electronic Displays, 2nd ed., by Sol Sherr and published by John Wiley & Sons, Inc., contains an extensive and detailed discussion of other aspects of display systems and technology, as well as a somewhat expanded version of this section. In addition, *Input Devices,* edited by Sol Sherr, and *Output Hardcopy Devices,* edited by Robert C. Durbeck and Sol Sherr, both published by Academic Press, include extensive discussions of a wide variety of devices.

The Society for Information Display (SID) sponsors a yearly symposium at which a large amount of information on new developments in information display as well as tutorials and seminars on basic information display topics are presented and made available in published form. In addition, it publishes two journals, namely, *Proceedings of the Society for Information Display* and *Information Display.* Other relevant meetings and publications are those sponsored by the Computer Society and Electron Devices groups of the IEEE, the SIGGRAPH group of the Association for Computing Machines (ACM), and the National Computer Graphics Association (NCGA).

11

Secure Electronic Commerce

Mostafa Hashem Sherif

AT&T

Electronic commerce is the set of totally dematerialized relations that economic agents establish with each other. Partial or complete dematerialization of commercial transactions allows remote operation using a telecommunications network. Payments can be made with several forms of electronic money such as the monetary value stored in electronic purses within programmable integrated circuit (smart) cards, in wallets residing on the user's computer, or in a network account.

Security of commerce has leaped to the forefront because of changes in the business environment, in the legal framework, and in the technology. The separation of functions that were once within the province of a vertically integrated enterprise and the outsourcing of responsibilities have multiplied the number of participants in the end-to-end information processing of a transaction, thus increasing the security risks. Also, the use of the IP protocol in remote commercial applications, even though designers of that protocol did not have security in mind, requires a careful analysis of the potential threats and ways to circumvent them. Legislation has been promulgated to specify the controls and the audits for electronic monetary transactions, and to specify the measures to protect the records collected. This is particularly needed in countries, like Singapore, where three fourths of the volume of non-cash transactions are done with electronic purses. Starting 2008, Singapore plans to accord the money stored in electronic purses the same legal status as cash.

The degree of security offered in electronic commerce must be commensurate with the amount of the transaction and the value of the goods and services being bought. For example, the protection afforded to micropayments (payments less than $10) is not at the same level as that given to the transfer of large amounts among financial institutions. Ways to reduce the cost of security include lighter cryptography, offline authorization of payments, and the grouping of transactions before requesting financial compensation.

Depending on the nature of the economic agents, electronic commerce applications fall within one of four main categories of business relations:

1. Business-to-business
2. Business-to-consumer
3. Point-of-sale (face-to-face) operations
4. Peer-to-peer commerce (without any payment intermediary)

For each category, the management of security has to find the appropriate method to secure the telecommunication infrastructure, the exchanges associated with the purchasing transaction, the payment and financial compensation, and the back office operations of the merchant including supply chain management. This is includes the software used for the various applications.

The telecommunication network can be packet-switched (e.g., the Internet) or circuit-switched (traditional telephone network) with wireline or wireless access. The large scale use of smart cards as a means for payment requires a new secure infrastructure. This infrastructure includes a network of secure terminals including recharging points to add monetary value to the rechargeable cards, secure gateways to the financial networks, and low-cost but secure card readers that are tamper-resistant and equipped with a security module to carry out the necessary cryptographic functions. The availability and reliability of the telecommunication service depend on the physical infrastructure and the network design, as well as on the operating procedures of the network operator.

Once this generic requirement has been met, solutions specific to electronic commerce address the following aspects:

1. Protection of the exchanges between the merchant and the buyer on the one side and the merchant and its financial institution on the other. This includes protection of the various servers involved.
2. Protection of the payment and the associated financial messages.
3. In consumer applications, verification of the person's credit worthiness, establishment of payment threshold for a given person, and fraud detection and management. The latter requires surveillance of activities at points of sale and observation of short-term events and long-term trends.
4. Protection of the merchandise, particularly merchandise in the virtual domain.

From a transactional viewpoint attacks can be either passive or active. Passive attacks rely on the observation of the traffic to deduce some information on the exchange (e.g., bank account or the nature of the purchase). Active attacks consist of intentional manipulations to effect some of the parameters of the transaction (e.g., shipping address, account information, etc.). The granularity of the level of protection determines at what layer of the open system interconnection (OSI) model the security services are implemented. In general, the finer the granularity, the higher the layer.

The main security services in a generic electronic transaction are confidentiality, integrity, authentication, access control and non-repudiation. Anonymity and nontraceability are two services is that are specific to payment mechanisms. The basic principle of all these services is the use of mathematical functions to reshuffle the original message into an unreadable form before transmission. After the message is received only the authenticated recipient is allowed to restore the original text.

Confidentiality means that the exchanged messages are not divulged to an unauthorized third party. This is achieved by cryptography, whether symmetric or asymmetric. Symmetric cryptography uses a shared key between the sender and the receiver. It is speedier than asymmetric cryptography for the same amount of protection provided that secure distribution channels are used to share the keys. Asymmetric or public key cryptography obviates the need for a secure key distribution channel. In on-line systems, public key cryptography can be combined with symmetric cryptography. In this way, public key encryption is used at the beginning of an electronic commerce session to exchange the shared secret that will be used throughout the session for symmetric encryption.

Integrity means that the message cannot be changed without leaving a trace. A one-way hash function is used to produce a signature of the original message to be compared at the receiving end with the signature computed upon arrival of the message. (A hash function has the property that it allows an easy computation of the signature from the original message, but that the reverse computation is very difficult. In addition, the probability that two different messages would produce the same signature is very low, while small differences between two messages produce signatures that are widely apart). The integrity of the message is verified if both signatures are identical. This means that the contents of the messages or of their sequence cannot be manipulated without detection: this impedes the falsification of payment instructions or the generation of spurious instructions. A blind signature is a special procedure for verifying the integrity of a message (e.g., a purchase order) without disclosing its content.

Identification of the participants in a transaction is the verification of a pre-existing relation between an entity and a characteristic such as a password, a cryptographic key, or some biometric property. One entity may possess several distinct identifiers, one for each relationship.

Authentication of the participants is the corroboration of the identity that an entity claims with the guarantee of a trusted third party. The authentication credentials may be an X.509 certificate, a Kerberos ticket, or on an identity password pair. Authentication is necessary to ensure the service of nonrepudiation. Banking organizations are now issuing and distributing digital certificates to their members and their clients. The main organizations are Identrus and the Global Trust Authority.

Access control is the process of ensuring that only the authorized entities whose identities have been duly authenticated can gain access to the protected resources. There are two types of access control mechanisms: identity-based and role-based. Identity-based access control uses the authenticated identity to determine and enforce the access rights. For role-based access control, the access privileges depend on the job function and its context such as the type of operation request or the time of day. *Nonrepudiation* is a legal construct based on the accumulation and preservation of proofs that can be verified by a third party (tax authority, judiciary, etc.).

The services that are more specific for payment transactions are *anonymity* and *nontraceability*. Anonymity means that the identity of the payer is not used explicitly to settle the transaction. This feature contradicts the aim of tailoring the service to individual preferences because such a personalization establishes some relationship between the payer's identity and the transaction. In the case of remote payments, anonymity of the communication is a necessary condition for payment anonymity, because once the source has been identified, it is more difficult to hide the identity of the originator. However, some degree of anonymity can be established in "mix networks" with public key cryptography, as proposed by David Chaum. In point-of-sale applications, a smart card with offline verification is a reasonable way to provide anonymity. Nontreaceability means that, in addition to the payer being anonymous, two payments made by the same person could not be linked to each other.

IPSec is a protocol suite used to secure the communications between two peers at the network layer. It offers authentication, confidentiality, and encryption, as well as key management. It is used to secure virtual private networks in one of two modes: the transport mode and the tunnel mode. The transport mode secures the communication between two hosts, while the tunnel mode is useful when one or both ends of the connection is a trusted entity such as a firewall. This trusted entity then provides the security services to the endpoint that it is connected to. The tunnel mode is also employed when a router provides the security services to the traffic that it is forwarding.

The secure sockets layer (SSL) and the transport layer security (TLS) are two protocols widely used to secure the connection between a client and server. They supply a simple mechanism to protect the exchanges between two points over transport control protocol (TCP); thus, they operate between the transport layer and the application layer of the OSI reference model. Their modular architecture allows them to evolve without disturbing the whole structure. TLS is derived from SSL and has been standardized by the Internet Engineering Task Force (IETF). Wireless transport layer security (WTLS) is the result of a complete revision of TLS to meet the constraints of wireless environment. It was designed to be part of the Wireless Access Protocol (WAP) environment to provide mobile terminals with Internet access. However, WTLS and TLS are not compatible, so that a gateway must ensure interoperability between TLS and WTLS, thereby raising additional security concerns. This has prevented widespread use of WTLS.

The main computational load of SSL/TSL comes from the cryptographic operation during session establishment. To alleviate this load a session can be resumed without a new cryptographic exchange. Another way is to use an accelerator between the client and the server. The use of accelerators, however, opens the door to additional security threats because there may be now two back-to-back SSL/TLS sessions given that SSL cannot handle multiparty transactions.

There are several protocols for the centralized management of access control of a large number of clients or users. The most common are the *Remote Authentication Dial-in User Service* (RADIUS) and the *Terminal Access Controller Access System* (TACACS). Both require the establishment of secret keys between each network element and the server controlling the access. Note that both server-to-client authentication and user-to-client authentication are outside the scope of RADIUS. Commercial systems add one of two basic systems for end-user authentication: one-time passwords or challenge-response identification. RADIUS does not include provisions for congestion control so that, unless some protections are made, large networks may suffer degraded performance and data loss when congestion happens.

Without adequate authentication and access control, electronic commerce sites are vulnerable to denial of service attacks that can prevent normal access by legitimate users. These attacks overwhelm the resources (routers, servers, switches, etc.) with spurious or superfluous tasks. Nevertheless, these attacks are inevitably associated with IP networks for two reasons: network control data and user data intermingle on the same physical and logical bandwidth, and the IP has no policy for admission control because it is a connectionless protocol.

Secure Electronic Transaction (SET) was designed with bank card security in mind. It operates at the application layer independently of the lower layers, even though most of its intended use was over TCP/IP networks. In SET each message exchange is encrypted and authenticated. SET allows multiparty transactions through a mechanism called dual signature. This is a procedure to protect the integrity of message that carries two distinct components that the cardholder sends, one to the merchant and the other to the payment gateway. With this procedure the cardholder can send a purchase order to the merchant and a payment instruction to the bank within the same message, each encrypted with a different algorithm and/or parameters. The merchant then extracts the purchase order, reads it, and verifies the integrity of the payment instructions without reading them. Once the merchant is satisfied, he forwards the whole message to the payment gateway with an indication that he has accepted the order. In turn, the payment gateway can verify the integrity of the merchant acceptance and read the totality of the payment data. In this way, the bank knows the financial details of the transaction without knowing the subject of the transaction. This mechanism avoids unnecessary exchanges because each recipient can read the part of the message that it has to receive and verify the integrity of the other part without knowing its content. The price is the complexity of the implementation, a factor that has impeded the commercial success of SET.

Hybrid SSL/SET architectures for payment protection combine the security advantages of SET with the simplicity of SSL. The solution relies on a payment intermediary to act as a proxy for its client with respect to the SET infrastructure. The intermediary acts a SET/SSL gateway and reduces the cryptographic load on both the buyer and the merchant.

Another approach to secure the multiparty association for bank card transactions is the 3-D Secure program used in the "Verified by Visa" program. 3-D Secure establishes four point-to-point SSL/TLS connections to link the buyer, the merchant, and the payment gateway. However, management of four links for each transaction poses salability problems.

Measures for logical security in card transactions include authentication of the holder, of the card, and of the card reader, and securing all the communication channels with the host system. A new generation of integrated circuit cards takes advantage of the microprocessor computational capabilities to add more functions to the payment applications. Although the security of previous generations of bank cards was proprietary, the Europay, MasterCard, Visa (EMV) specifications have standardized many aspects of security management. Visa and American Express have selected the JavaCard architecture to share the resources in multi-application cards among several applications. In addition, the security of the integration of smart cards and local computers acting as access terminals has been addressed in several initiatives such as OpenCard for Java and PC/SC for Windows©.

In summary, achieving security is complex and building scale to this security is extremely difficult. Because electronic commerce is a scale business, the security of electronic commerce forces a complete revision of the whole value chain. Some of the important issues that must be addressed relate to the protection of intellectual property, civil rights, privacy rights, as well as taxation, fraud prevention, etc. These issues are not only technological but challenge the fundamental assumptions of society.

12

Software Engineering

Carl A. Argila
Software Engineering Consultant

Paul C. Jorgensen
Grand Valley State University

12.1 Tools and Techniques*

Carl A. Argila

The last decade has seen a revolution in software engineering tools and techniques. This revolution has been fueled by the ever-increasing complexity of the software component of delivered systems. Although the software component of delivered systems may not be the most expensive component, it is usually, however, "in series" with the hardware component; if the software doesn't work, the hardware is useless.

Traditionally, software engineering has focused primarily on computer programming with ad hoc analysis and design techniques. Each software system was a unique piece of intellectual work; little emphasis was placed on architecture, interchangeability of parts, reusability, etc. These ad hoc software engineering methods resulted in the production of software systems which did not meet user requirements, were usually delivered over budget and beyond schedule, and were extraordinarily difficult to maintain and enhance.

In an attempt to find some solutions to the "software crisis," large governmental and private organizations motivated the development of so-called "waterfall" methods. These methods defined formal requirement definition and analysis phases, which had to be completed before commencing a formal design stage, which in turn had to be completed before beginning a formal implementation phase, etc. Although waterfall methods were usually superior to ad hoc methods, large and complex software systems were still being delivered, over budget and beyond schedule, which did not meet user requirements. There were several reasons for this. First, waterfall methods focus on the generation of *work products* rather than "engineering." Simply put, writing documents is not the same as doing good engineering. Second, the waterfall methods do not support the *evolution* of system requirements throughout the development life cycle. Also, the prose English specifications produced within the waterfall methods are not well suited to describing the complex behaviors of software systems.

The basic, underlying philosophy of how software systems should be developed changed dramatically in 1978 when Tom DeMarco published his truly seminal book, *Structured Analysis and System Specification* [DeMarco, 1979]. DeMarco proposed that software systems should be developed like any large, complex engineering systems — by first building scale models of proposed systems so as to investigate their behavior. This *model-based software engineering* approach is analogous to that used by architects to specify and design

---

*The material in this article was originally published in *The Electrical Engineering Handbook*, Richard C. Dorf, Ed., Boca Raton, FL: CRC Press, 1993.

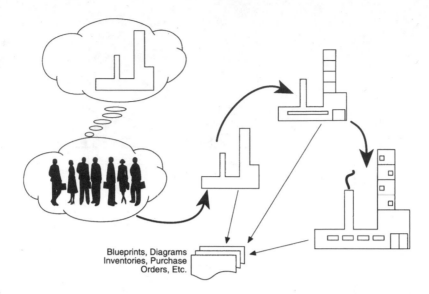

FIGURE 12.1   Model-based software engineering.

large complex buildings (see Figure 12.1). We build scale models of software systems for the same reason that architects build scale models of houses, so that users can visualize living with the systems of the future. These models serve as vehicles for communication and negotiation between users, developers, sponsors, builders, etc. Model-based software engineering holds considerable promise for enabling large, complex software systems to be developed on budget, within schedule, while meeting user requirements [see Harel, 1992].

As shown in Figure 12.2, a number of specific software development models may be built as part of the software development process. These models may be built by different communities of users, developers, customers, etc. Most importantly, however, these models are built in an *iterative* fashion. Although work products (documents, milestone reviews, code releases, etc.) may be delivered chronologically, models are built iteratively throughout the software system's development life cycle.

In Figure 12.3 we illustrate the distinction between *methodology*, *tool*, and *work product*. A number of differing software development methods have evolved, all based on the underlying model-based philosophy. Different methods may in fact be used for the requirements and analysis phases of project development than for design and implementation. These differing methods may or may not integrate well. Tools such as computer aided software engineering (**CASE**) may support all, or only a part, of a given method. Work products, such as document production or code generation, may be generated manually or by means of CASE tools.

This article will present a synopsis of various practical software engineering techniques which can be used to construct software development models; these techniques are illustrated within the context of a simple case study system.

Approach

One of the most widely accepted approaches in the software engineering industry is to build two software development models. An **essential model** captures the behavior of a proposed software system, independent of implementation specifics. An essential model of a software system is analogous to the scale model of a house built by an architect; this model is used to negotiate the *essential* requirements of a system between customers and developers. A second model, an **implementation model**, of a software system describes the technical aspects of a proposed system within a particular implementation environment. This model is analogous to the detailed blueprints created by an architect; it specifies the *implementation* aspects of a system to those who will do the construction. These models [described in Argila, 1992] are shown in Figure 12.4. The essential and implementation models of a proposed software system are built in an iterative fashion.

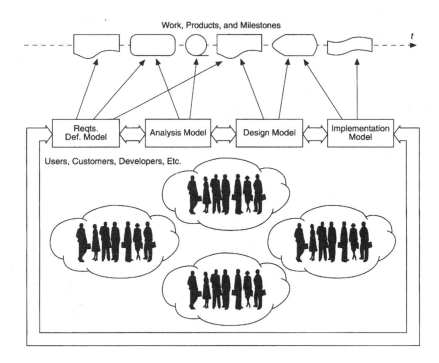

FIGURE 12.2   Modeling life cycle.

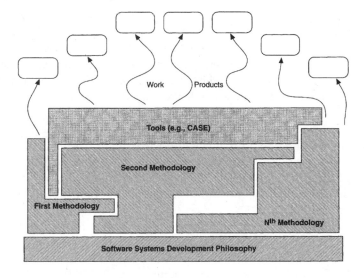

FIGURE 12.3   Methods, tools and work products.

Methods

The techniques used to build the essential and implementation models of a proposed software system are illustrated by means of a simple case study. The radio button system (RBS) is a component of a fully automated, digital automobile sound system. The RBS monitors a set of front-panel *station selection buttons* and performs station selection functions.

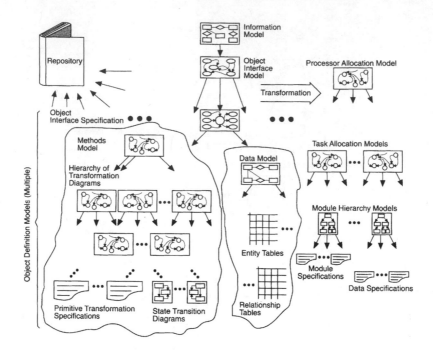

FIGURE 12.4   Software engineering methods overview.

When a station selection button is momentarily depressed, the RBS causes a new station to be selected. This selection is made on the basis of station-setting information stored within the RBS. The RBS can "memorize" new station selections in the following manner: When a given station selection button is depressed longer than momentarily (say, for more than 2 seconds), the currently selected station will be "memorized." Future momentary depressions of this button will result in this "memorized" station being selected.

The RBS also performs a muting function. While a station is being selected, the RBS will cause the *audio system* to mute the audio output signal. The RBS will also cause the audio output signal to be muted until a new station selection has been successfully memorized.

The RBS interfaces with the front-panel station selection buttons by "reading" a single-byte memory location. Each bit position of this memory location is associated with a particular front-panel station selection button. The value of 0 in a given bit position indicates that the corresponding button is *not* depressed. The value of 1 in that bit position indicates that the corresponding button *is* depressed. (For example, 0000 0000 indicates no station selection buttons are currently depressed; 0000 0010 indicates that the second button is currently depressed, etc.)

The RBS interfaces with the *tuning system* by means of a common memory location. This single-byte memory location contains a non-negative integer value which represents a station selection. (For example, 0000 0000 might represent 87.9 MHz, 0000 0001 might represent 88.1 MHz, etc.) The RBS may "read" this memory location to "memorize" a current station selection. The RBS may also "write" to this memory location to cause the tuning system to select another station.

Finally, the RBS interfaces with the audio system by sending two signals. The RBS may send a MUTE-ON signal to the audio system causing the audio system to disable the audio output. A MUTE-OFF signal would cause the audio system to enable the audio output.

Information Modeling

The construction of an **information model** is fundamental to so-called object-oriented approaches. An information model captures a "view" of an application domain within which a software system will be built.

Information models are based on entity-relationship diagrams and underlying textual information. A sample information model for the RBS is shown in Figure 12.5. Entities (shown as rectangles) represent "things" or **objects** in the application domain. Entities may be established by considering principal nouns or noun phrases in the application domain. Entities have *attributes* associated with them which express the qualities of the entity. Entities participate in *relationships;* these are shown as diamonds in the entity-relationship diagram. Relationships may be determined by considering principal verbs or verb phrases in the application domain. Relationships have *cardinality* associated with them and entities may participate *conditionally* in relationships. Finally, there are special kinds of relationships which show *hierarchical relationships* between objects.

Essential Modeling

The essential model consists of a number of graphical components with integrated textual information. Figure 12.6 shows the **object collaboration model** for the RBS. This model depicts how a collection of objects

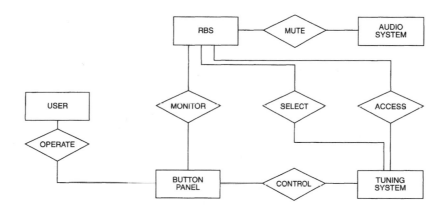

FIGURE 12.5 RBS information model.

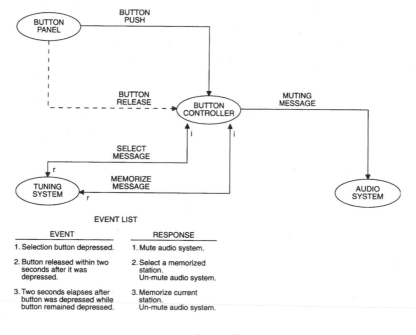

FIGURE 12.6 RBS object collaboration model.

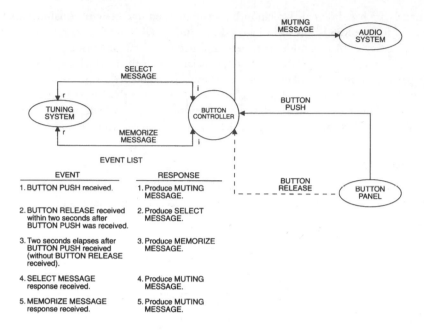

FIGURE 12.7   RBS object interface specification.

or entities can communicate (by exchanging messages) to perform the proposed system functions. An *event list* is part of this model; it shows what responses must be produced for a given external stimulus.

For each object there is an **object interface specification** (as shown in Figure 12.7) which shows the public and private interfaces to an object. An event list is also associated with this specification; it shows how the object will respond to external stimuli. A hierarchy of **transformation diagrams** is associated with each object specification (as shown in Figure 12.8 for the RBS). This diagram defines all of the functions or "methods" which the object performs. Some behavior may be expressed by means of a **state transition diagram** (Figure 12.9).

Implementation Modeling

Two principal activities must be accomplished in transitioning from the essential to the implementation model. First, all of the methods and data encapsulated by each object must be mapped to the implementation environment. This process is illustrated in Figure 12.10. Second, all of the details which were ignored in the essential model (such as user interfaces, communication protocols, hardware limitations, etc.) must now be accounted for.

Each component of the essential model must be allocated to hardware processors. Within each hardware processor, allocation must be continued to the *task* level. Within each task, the computer program controlling that task must be described. This latter description is accomplished by means of a **module structure chart.** As illustrated in Figure 12.11, for one component of the RBS, the module structure chart is a formal description of each of the computer program units and their interfaces.

CASE Tools

The term *computer-aided software engineering* (CASE) is used to describe a collection of tools which automate all or some of various of the software engineering life cycle phases. These tools may facilitate the capturing, tracking and tracing of requirements, the construction and verification of essential and implementation models and the automatic generation of computer programs. Most CASE tools have an underlying

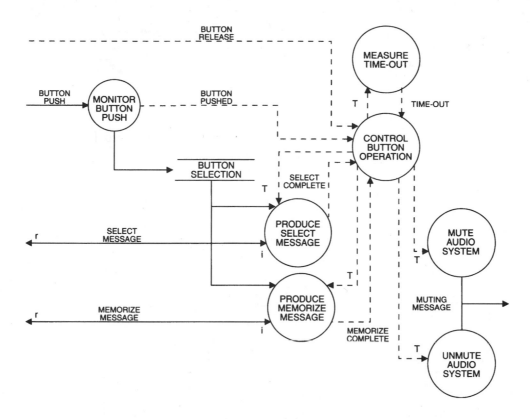

FIGURE 12.8 RBS transformation diagram.

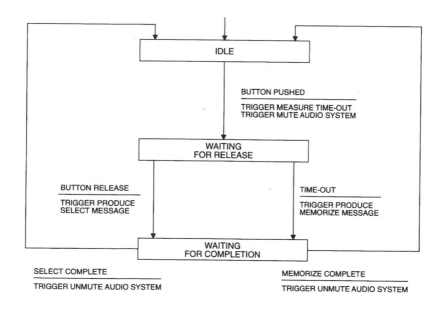

FIGURE 12.9 RBS state transition diagram.

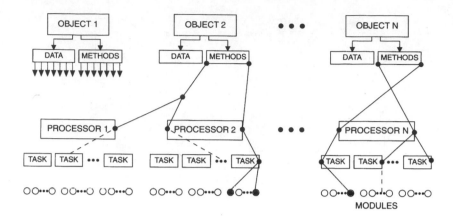

FIGURE 12.10   Implementation modeling.

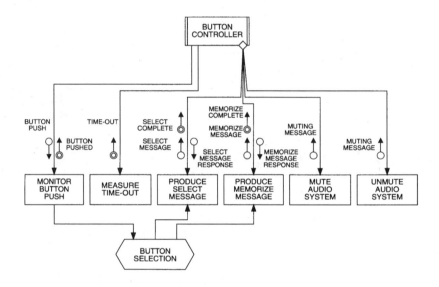

FIGURE 12.11   RBS module structure chart.

project repository which stores project-related information, both textual and graphical, and uses this information for producing reports and work products.

CASE tool features may include:

- Requirements for capture, tracing and tracking
- Maintenance of all project-related information
- Model verification
- Facilitation of model validation
- Document production
- Configuration management
- Collection and reporting of project management data
- CASE data exchange

Defining Terms

CASE: Computer-aided software engineering. A general term for tools which automate various of the software engineering life cycle phases.

Essential model: A software engineering model which describes the behavior of a proposed software system independent of implementation aspects.

Implementation model: A software engineering model which describes the technical aspects of a proposed system within a particular implementation environment.

Information model: A software engineering model which describes an application domain as a collection of objects and relationships between those objects.

Module structure chart: A component of the implementation model; it describes the architecture of a single computer program.

Object: An "entity" or "thing" within the application domain of a proposed software system.

Object collaboration model: A component of the essential model; it describes how objects exchange messages in order to perform the work specified for a proposed system.

Object interface specification: A component of the essential model; it describes all of the public and private interfaces to an object.

State transition diagram: A component of the essential model; it describes event-response behaviors.

Transformation diagram: A component of the essential model; it describes system functions or "methods."

References

C. Argila, "Object-oriented real-time systems development" (video course notes), Los Angeles: University of Southern California IITV, June 11, 1992.

G. Booch, *Object-Oriented Design with Applications*, Redwood City, CA.: Benjamin/Cummings, 1991.

P. Coad and E. Yourdon, *Object-Oriented Analysis*, 2nd ed., New York, NY: Prentice-Hall, 1991.

P. Coad and E. Yourdon, *Object-Oriented Design*, New York, NY: Prentice-Hall, 1991.

T. DeMarco, *Structured Analysis and System Specification*, New York, NY: Prentice-Hall, 1979.

D. Harel, "Biting the silver bullet," *Computer*, January 1992.

J. Rumbaugh, M. Blaha, W. Lorenson, F. Eddy, and W. Premerlani, *Object-Oriented Modeling and Design*, New York, NY: Prentice-Hall, 1991.

S. Shlaer and S. Mellor, *Object-Oriented Systems Analysis: Modeling the World in Data*, New York, NY: Prentice-Hall, 1988.

S. Shlaer and S. Mellor, *Object Life-Cycles: Modeling the World in States*, New York, NY: Prentice-Hall, 1992.

P. Ward and S. Mellor, *Structured Development for Real-Time Systems*, New York, NY: Prentice-Hall, vol. 1, 1985; vol. 2, 1985; vol. 3, 1986.

E. Yourdon and L. Constantine, *Structured Design*, 2nd ed., New York, NY: Prentice-Hall, 1975, 1978.

Further Information

A video course presenting the software engineering techniques described here is available [see Argila, 1992]. The author may be contacted for additional information and comments at (800) 347–6903.

12.2   Software Testing

Paul C. Jorgensen

Introduction

From the 1960s through today, many attempts have been made to improve software development: they include the use of higher-level languages, structured programming, an emphasis on design, requirements modeling, object-oriented programming, agile development and most recently, test-driven development. Despite this 40-year history of improvements, the need for software testing persists. IEEE has defined a series of useful terms for software testing. An *error* is a mistake made by a person. When an error is recorded in a description of the eventual software (such as a requirements model, a design document or source code), the error becomes a *fault*. A *failure* occurs when the code corresponding to a fault executes. When a failure is recognized, it becomes an *incident* [IEEE, 1983]. In this context, the goal of software testing is to devise test cases so that, when executed by the software being tested, an incident will reveal the presence of a fault. At a minimum, a software test case contains:

- Preconditions describing the state of the software before executing the test case
- An interleaved sequence of inputs and expected outputs
- A place to record observed outputs
- A pass/fail assessment

When expected and observed outputs are consistent, the test case passes; otherwise, an incident has occurred, demonstrating the presence of a fault. Software testing can never demonstrate the absence of faults.

Software testing has three distinct levels: unit, integration and system. Each has unique problems and goals. Unit testing is restricted to a single procedural module (or an object-oriented class), and is concerned with faults within the unit's scope. Integration testing assumes that units have been rigorously tested and is concerned with faults outside the scope of individual units. System testing is conducted at the level of eventual customer use.

Fundamental Testing Approaches

When software executes a test case, an execution-time behavior occurs. This makes test cases highly dynamic. Given a particular program, imagine a universe of discourse of software behaviors and consider two particular sets: specified behaviors, S, and implemented behaviors, P (see Figure 12.12).

The goal of software testing at any level is to determine the extent to which the specified and implemented behaviors (the intersection of sets S and P in Figure 12.12) coincide. The relative complements (S–P and P–S) are both problematic.

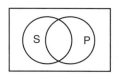 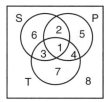

FIGURE 12.12 Specified and implemented behaviors and test cases.

The former signifies specified behaviors that have not been implemented, and the latter denotes unspecified behaviors that have been implemented. The set, T, of test cases completes the Venn diagram of software behaviors and testing. The significance of the eight regions is summarized in Table 12.1.

A *test method* is a systematic way to identify test cases. There are two fundamental test methods: specification-based (also called functional testing or black-box testing) and code-based (also called structural

TABLE 12.1

Region	Description
1	Specified and implemented behaviors with corresponding test cases.
2	Specified and implemented behaviors that are not tested.
3	Specified behaviors that are not implemented and are revealed by test cases.
4	Unspecified behaviors that are implemented and correspond to test cases. Failure of these cannot be determined because they are not specified.
5	Implemented behaviors that are neither specified nor tested. Very problematic.
6	Specified behaviors that are neither implemented nor tested. Also very problematic.
7	Spurious test cases of no utility.
8	Behaviors of no interest.

testing, or white- or clear-box testing). Specification-based testing pertains to regions 1, 2 and 3 in Figure 12.12, and code-based testing pertains to regions 1, 2 and 4. In this context, the goal of software testing is to reduce regions 5 and 6. Test coverage metrics (discussed later) provide some answers for region 5. The main hope for region 6 is to conduct effective software technical inspections, which are beyond the scope of this chapter.

Specification-Based Test Methods

For ease of description, consider a program that is a function of two input variables, *x* and *y*. (Extending this to more realistic programs will be obvious.) The input space is the shaded region shown in Figure 12.13. The simplest specification-based test method focuses on the variables' boundary values and makes two important assumptions: faults occur near the extreme values (recognized in practice) and are confined to a single variable.

Boundary value testing identifies five values of interest: the minimum, a value slightly greater than the minimum, a nominal value, a value slightly less than the maximum, and the maximum. Test cases are generated by holding all variables at their nominal values except one, and that variable assumes the other four values. This is repeated for each variable and is illustrated in graph A of Figure 12.14. Robust boundary-value testing is a simple extension obtained by adding invalid values of variables below the minimum and above the maximum (see graph B of Figure 12.14).

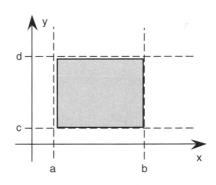

FIGURE 12.13 Input space of a function of two variables.

Murphy's Law suggests that the single fault assumption may not be appropriate. If this is a concern, the Cartesian product of the test cases is formed to identify *worst-case boundary-value* test cases (graph C) and *robust worst-case boundary-value* test cases (graph D). These four forms are elementary, and commercially available test support products can generate partial test cases for these methods (No product can generate the expected output portion of the test cases). All forms of boundary-value testing are vulnerable to twin deficiencies: gaps of untested functionality and redundant test cases. Worse, specification-based testing can never recognize if these deficiencies occur [Jorgensen, 2002].

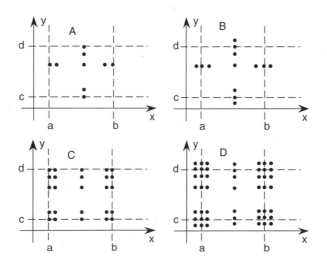

FIGURE 12.14 Variations of boundary-value testing.

The equivalence relation from discrete mathematics provides a response to the twin deficiencies of gaps and redundancies in boundary-value testing. An equivalence relation defined on a set induces a partition of the set, where a partition is a set of subsets of the original set such that the subsets are disjoint (have no redundancies) and their union is in the original set (with no gaps). Program variables are in an equivalence class if the program treats all class members similarly (the similar treatment assumption). Equivalence class testing identifies useful equivalence relations on the input variable space and then selects one test case from each equivalence class. The boundary value example is extended here for variables x and y:

Valid equivalence classes for x: $[a, b)$, $[b, c)$, $[c, d)$, $[d, e]$

Valid equivalence classes for y: $[m, n)$, $[n, p)$, $[p, q]$

Invalid equivalence classes for x: $x < a$, $x > e$

Invalid equivalence classes for y: $y < m$, $y > q$

Equivalence class test methods are best used when the program being tested has input data that can be separated into "similar treatment" classes. A good example is the calculation of automobile insurance premiums in cases where driver age ranges are important. When the similar treatment assumption is warranted, the *weak normal equivalence class* test method (WN) uses one input value from each equivalence class of valid values. An early focus of equivalence class testing was on invalid data values. Traditional programs frequently had 80% of their source code dedicated to detecting invalid data, a response to the "Garbage-In, Garbage-Out" mantra. As with boundary-value testing, the robust form adds invalid values. *Weak robust equivalence class* test cases (a seemingly inconsistent name) are illustrated in Figure 12.15 as WR. Note that weak normal test cases (WN) are a subset of weak robust test cases.

The adjective "weak" refers to the fact that each test case is a member of two equivalence classes. The WR test case in the lower left of Figure 12.15 is in both the invalid $x < a$ class and the invalid $y < m$ class. The weak forms make the single fault assumption. If this is unwarranted, the Cartesian product of the equivalence classes results in *strong normal equivalence class* and *strong robust equivalence class* test cases, shown as SN and SR in Figure 12.16.

The four forms of boundary-value testing and the four forms of equivalence class testing share the two assumptions of validity and fault multiplicity summarized in Table 12.2. There is one remaining assumption common to all forms of boundary-value and equivalence class testing summarized in Table 12.2: the independent variable assumption. This assumption is inherent in the worst-case and strong methods. Any time a Cartesian product is performed, it is assumed that the operands are truly independent. If there are dependencies among the variables, elements of the Cartesian product can be infeasible or impossible. The final

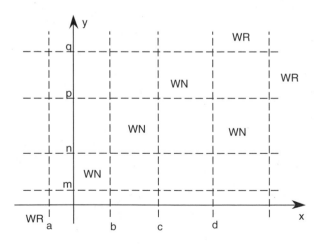

FIGURE 12.15   Weak normal and weak robust equivalence class test cases.

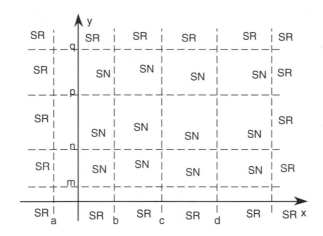

FIGURE 12.16   Strong normal and strong robust equivalence class test cases.

TABLE 12.2   Assumptions of Boundary Value and Equivalence Class Testing

	Normal Values	Robust Values
Single fault	Boundary value testing	Robust boundary value testing
	Weak normal equivalence class	Weak robust equivalence class
Multiple fault	Worst case boundary value testing	Robust worst case boundary value testing
	Strong normal equivalence class	Strong robust equivalence class

TABLE 12.3   A Sample Decision Table

	R1	R2	R3	R4	R5	R6	R7	R8
C1	T	T	T	T	F	F	F	F
C2	T	T	F	F	T	T	F	F
C3	T	F	T	F	T	F	T	F
A1			X	X	X		X	X
A2					X	X	X	
A3			X	X		X		X
A4					X		X	
Impossible	X	X						

refinement to specification-based testing is to use decision tables to sort dependencies among the variables. Decision tables consist of conditions, actions and rules. In a *limited entry decision table* (LEDT), conditions are binary, so an LEDT with n conditions will have rules of 2^n. The action portion of a decision table indicates the actions to be performed under each combination of condition values. Table 12.3 is a sample limited entry decision table with three conditions, four actions and eight rules.

Decision tables can be manipulated algebraically, and analyzed to determine completeness, consistency and redundancy. They also can be algebraically simplified. The sample in Table 12.3 is complete (three conditions with eight unique rules) and there is no inconsistency. The "impossible" entry denotes impossible combinations of conditions. In this example, C1 and C2 cannot both be true. Note that C3 is irrelevant, so rules R1 and R2 could be combined. Similarly, rules R3 and R4 can be combined, as can rules R5 and R7. The resulting simplified table is in Table 12.4. No additional simplifications are possible.

TABLE 12.4   The Simplified Decision Table

	R1, R2	R3, R4	R5, R7	R6	R8
C1	T	T	F	F	F
C2	T	F	–	T	F
C3	–	–	T	F	F
A1		X	X		X
A2			X	X	
A3		X		X	X
A4			X		
Impossible	X				

If there are dependent variables in a program to be tested, these should be identified with appropriate conditions and use of the "impossible" action. Remaining rules will correspond to legitimate test cases, and the algebraic simplification further reduces any (logical) redundancy.

Code-Based Test Methods

The starting point of most code-based test methods is the *program graph*, a directed graph where nodes correspond to source statement fragments and edges indicate possible sequential execution. Figure 12.17 shows the program graph of a sample program expressed in (language neutral) pseudo-code.

The precepts of structured programming require a single entry and a single exit. Therefore, a program graph of a structured program will have a single source node and a single sink node. When a program executes a test case, some path of nodes is traversed. Code-based testing begins with the idea of path testing, where sets of test cases correspond to different paths through the program graph. Here, the meaning of the "similar treatment" assumption is absolutely clear: test cases with "similar treatment" traverse the same path in the program graph. The literature includes simplistic objections to path-based testing. A notable, persistent example is found in Schach [2005], which presents a simple program graph with a loop that can be executed up to 18 times. It yields 4.77 trillion distinct execution paths. The author asserts that this

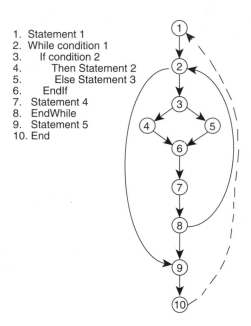

1. Statement 1
2. While condition 1
3.   If condition 2
4.   Then Statement 2
5.   Else Statement 3
6.   EndIf
7.   Statement 4
8. EndWhile
9. Statement 5
10. End

FIGURE 12.17   Program graph of structured code.

large number of program paths dooms the idea of path-based testing. Objections such as this miss the point of code-based testing: using information derived from the program graph to provide insight into the utility of a set of test cases.

The cyclomatic complexity of a strongly connected directed graph is shown by the formula $V(G) = e - n + p$, where e is the number of edges, n is the number of nodes and p is the number of connected regions. Graphs of structured programs need an extra edge (the dotted edge in Figure 12.17) that extends from the unique sink node back to the unique source node to be strongly connected and directed graph. The cyclomatic complexity formula for graphs of structured programs simplifies to $V(G) = e - n + 2$ because p is always equal to 1, and the added edge from sink to source node is ignored. Cyclomatic complexity describes the number of linearly independent cycles in a strongly connected directed graph; this translates to the number of linearly independent paths from the source to the sink node in the graph of a structured program. The cyclomatic complexity of the example in Figure 12.17 is $V(G) = 12 - 10 + 2$, leaving four essentially distinct paths to be tested. (The cyclomatic complexity of the 4.77 trillion example is just 6.)

A *test coverage metric* is a criterion that gives an indication of the extent that a set of test cases exercises a program. Program graphs support three obvious test-coverage metrics:

M1: every node
M2: every edge
M3: every path

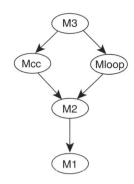

FIGURE 12.18 Lattice of test coverage metrics.

Given a program, its program graph and a set of test cases, the test cases satisfy the M1 coverage metric if, when they are executed, every node corresponding to a statement fragment is traversed. M1 coverage is minimal. Most experts prefer M2 coverage, and M3 is sensible only for loop-free programs.

Just as specification-based test methods are vulnerable to gaps and redundancies, path-based testing has its deficiency: infeasible paths. Logical dependencies among variables can cause paths that never can be executed. In Figure 12.17, for example, suppose condition 2 is always true when condition 1 is true. In that case, the edge from node 3 to node 5 can never be traversed, and any path containing this edge is infeasible. Path-based testing is vulnerable because infeasible paths cannot be recognized from the program graph. The only way to identify infeasible paths is to go back to the nodes' meaning and analyze dependencies in the source code. In the general case, this cannot be programmed and is equivalent to the Halting Problem.

Closer examination of source codes yields coverage metrics "between" M2 and M3. Compound conditions require special test considerations. If condition 2 in Figure 12.17 is a compound condition comprised of three simple conditions, the truth table of the simple conditions can identify additional test cases. Loops present another coverage problem. Most experts are satisfied with coverage that traverses both the normal path of a loop and its exit from the loop. Loops therefore require two test cases and nested loops multiply accordingly. If these coverage metrics are denoted as Mcc and Mloop, respectively, a lattice of test coverage metrics is created (Figure 12.18) where the directed edge means "stronger than." Test coverage metrics now include:

M1: every node
M2: every edge
Mcc: M2 plus truth table coverage of compound
conditions
Mloop: M2 plus normal and exit loop traversal
M3: every path

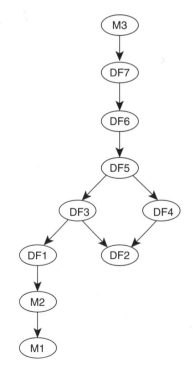

FIGURE 12.19 Combined lattice of test coverage metrics.

There are two useful refinements of path-based testing. Dataflow testing [Rapps, 1985] annotates the nodes of a program graph with the information about variables that occur in the statement fragment corresponding to the node. A node *n* is a *defining node of variable V*, written Def (V, n), if the variable *V* receives a value at node *n*. Similarly, node *n* is a *usage node of a variable V*, written Use(V, n) if the value of V is used at node n. A *define/use path for variable V*, written DU-Path, is a path in the program graph that begins with a defining node of V and ends with a usage node. If there is only one defining node in a DU-Path, it is a *definition-clear path* (written DC-Path). There are two types of usage nodes, predicate use (written P-use) and computation use (C-use). Dataflow testing adds this information to the program graph and then defines several test coverage metrics:

DF1: All predicate uses
DF2: All Defs
DF3: All P-uses, some C-uses
DF4: All C-uses, some P-uses
DF5: All uses
DF6: All DC-Paths
DF7: All DU-Paths

Because the dataflow test coverage metrics are based on important characteristics of source code, they provide a refined framework of test coverage that can be customized to individual programs.

The second refinement to path-based testing is the use of program slices. Given a program, P, and its program graph, the *slice on a variable V at statement fragment n*, written [S (V, n)], is the set of all statement fragments in P contributing to the value of V at statement fragment n. Program slices are distinct from DU-Paths. Generally, slices form their own lattice, and a slice lattice does not conform well to the Rapps–Weyuker lattice in Figure 12.19. Slices are close to the meaning of a program, and they replicate how developers think when trying to isolate a fault [Gallagher, 1991].

References

IEEE Computer Society, *IEEE Standard Glossary of Software Engineering Terminology*, ANSI/IEEE Std 729-1983.

K.B. Gallagher and J.R. Lyle, "Using program slicing in software maintenance," *IEEE Trans. Software Eng.*, vol. SE-17, no. 8, pp. 751–761, 1991.

P.C. Jorgensen, *Software Testing—A Craftsman's Approach*, 2nd ed., Boca Raton, FL: CRC Press, 2002.

S. Rapps and E.J. Weyuker, "Selecting software test data using data flow information," *IEEE Trans. Software Eng.*, vol. SE-11, no. 4, pp. 367–375, 1985.

S.R. Schach, *Object-Oriented & Classical Software Engineering*, 6th ed., New York, NY: McGraw-Hill, 2005.

13

Computer Graphics

Nan C. Schaller
Rochester Institute of Technology

Evelyn P. Rozanski
Rochester Institute of Technology

13.1 Introduction

Computer graphics is everywhere: on the web, in our homes (high-end appliances, computers, game consoles and television — weather, news and broadcast sports), in our cars (GPS systems), in the movies and arcades and on our persons (PDAs, mobile phones and wearable computers). The term computer graphics refers to the generation, representation, manipulation, processing and visual display of data using a computer. Computer-generated images may be two-dimensional (2D) or three-dimensional (3D); they may be animated or still, and portray real scenes, imagined scenes, or things not normally visible such as forces. Image processing is closely related to computer graphics but omits the initial generation phase and starts instead with an image captured by some other device.

Image synthesis in computer graphics involves four steps that closely mimic photography. The first is modeling, creating, and placing objects including light sources in the scene. The second is the modeling and placement of a synthetic camera. This is often modeled as a simple pinhole camera, although much more complex models have been used, such as the one discussed by Kolb [1995]. The third is rendering the scene, which utilizes the material properties of the objects (their reflectivity, refraction, transmissiveness and textures) as well as the properties of the light sources (spot, point, ambient, color, attenuation, etc.) to model the transport of light in the scene and its projection onto the *viewplane*, which is analogous to film. Lastly, tone reproduction may be used to take the results of the lighting calculation and produce an image suitable for display on a particular device [Geigel, 2004].

Today's computer graphics are produced using computer systems that have a graphics-processing unit (GPU), one or more graphical display devices, and one or more input devices such as a keyboard, mouse, digitizer or data glove. Graphical display devices include monitors, printers, plotters, video and film as well as 3D displays.

Computer graphics is an integral component of a wide variety of applications. It is used in cartography and in the business world, where applications range from presentation graphics to desktop publishing.

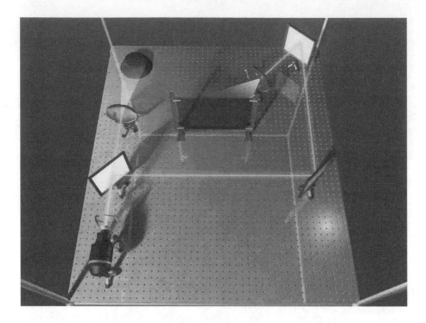

FIGURE 13.1   Image from an animated visualization about how holograms are created. Software: Maya; Hardware: PC. Artists: Orde Stevanoski and Hardeep Kharbanda, MFA students at Rochester Institute of Technology.

Highly interactive real-time systems are used in games and in flight simulators. In engineering, computer-aided design/computer-aided manufacturing (CAD/CAM) systems allow users to create, store, manipulate and test objects and designs. Fully integrated systems allow standard component parts libraries to be incorporated into a product's design. Product design and drafting information is fed directly into manufacturing operations. Another engineering graphics application is very large scale integration (VLSI) design. Collaborative computer-aided engineering systems enable engineering teams to work currently rather than serially [Kasik, 2000].

Graphics are useful for visualizing physical phenomena and data as well as the volume visualization of complex datasets. An example of volume visualization is the medical modeling of anatomy using MRI data [Kaufman, 1998]. Similarly, data visualization tools help people explore and explain data. Interacting with a carefully designed visual representation of data can help an engineer form mental models that let him or her perform specific tasks more effectively [Munzner, 2002]. It is often easier to formulate such mental models while examining a visual representation of such data rather than examining the underlying numbers, especially as there may be gigabytes of data. Figure 13.1 is an image from an animated visualization of how holographs are created, while Figure 13.2 is an image from an educational tool used to teach optics.

In art and animation, computer graphics has taken the drudgery out of transforming and redrawing objects. It has been used to enhance cell animation as well as to produce glitzy Hollywood special effects. Synthetic actors have become so realistic that they are almost indistinguishable from live actors.

Virtual reality (VR) is one of the most spectacular uses of graphics. Its goal is to immerse users in effective, real-time, synthesized, 3D environments [Feiner, 2000]. VR uses high-resolution graphics terminals, head-mounted displays, CAVEs®* (projection-based VR systems that surround the viewer with four screens), and *Responsive Workbench*-type [RWB, 2004] environments to provide the user with a stereo view of a virtual world and the ability to navigate through it. These systems use tracking devices to determine the current position of the user and input devices such as data gloves and head-mounted displays. VR applications abound and include data and scientific visualization, simulation, games and architecture. Utilizing perception techniques and period illumination enables the accurate reconstruction of VR archeological sites. VR is used in education and training

---

*Registered trademark of the Electronic Visualization Laboratory at the University of Illinois at Chicago.

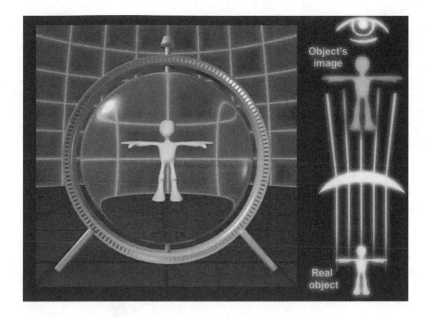

FIGURE 13.2 Image from a tool for teaching optics. Software: Maya; Hardware: PC. Artists: Orde Stevanoski and Hardeep Kharbanda, MFA students at Rochester Institute of Technology.

along with less sophisticated graphics. Collaborative VR allows users in a variety of locations to participate in a common virtual environment. Augmented reality, an emerging application area, takes virtual reality a step further and fits somewhere between reality and virtual reality. At its simplest, it is the overlaying of computer-generated imagery on top of the real world using see-through displays. At its best, it adds graphical objects as well as sounds, *haptics* (touch), and smell into reality. Augmented reality has applications in many areas, including medicine, games and the military. The most recent (November and December) issues of *IEEE Computer Graphics and Applications* have contained summary reports about the current state of VR.

Computer graphics owes its ever-improving performance to continuing research in three interrelated areas: hardware, software and algorithm development. Each of these areas is discussed below.

13.2 Graphics Hardware

Graphics Processing Units (GPUs)

GPUs are components in graphics systems that power all levels of graphics visualization systems. They use a fast bus and act as coprocessors to the CPU. Today's GPUs are very fast, faster often than their coprocessor CPU. This speed is made possibly by parallelism as well as stream processing. GPUs treat computer graphics primitives such as vertices and pixels as streams of data. Multiple programmable processing units act in concert with data flowing between them. In a GPU, a vertex processor transforms and processes points, and a fragment processor computes pixel color. As a stream processor, a GPU performs simple operations and exploits parallelism by running the same program for each pixel. In recent years, users have been given the ability to program GPUs directly, creating small programs, known as *shaders*, to perform real-time rendering effects such as bump mapping or shadows using relatively low-level software such as Cg [Shader Tech, 2004] or directX [Microsoft, 2004; Macedonia, 2003].

Graphical Output Devices

Computer graphics systems use a variety of output components for displaying computer-generated images. These can be classified into two groups: hard copy technologies and display technologies.

Hard Copy Technologies

Hard copy technologies include printers, pen plotters, electrostatic plotters, laser printers, ink-jet plotters, thermal transfer plotters, and film recorders [Foley, 1996]. These devices use either a raster or vector drawing style. Raster drawing uses discrete dots whereas vector drawing uses a continuous drawing motion to create the image. Raster display devices are compared to one another with respect to dot size and the number of dots per inch, known as *addressability*: the closer the dots, the smoother the image; the smaller the dot, the finer the detail. *Resolution* is related to dot size and is the number of distinguishable lines per inch. High-resolution devices produce fine detail, smooth lines, and crisp images.

Color is achieved using a variety of methods. Some devices use multicolored ribbons with single print heads, multiple print heads with different ribbons, or over-striking to combine colors. Other devices use color pens, spray (e.g., ink jet), toner (e.g., laser printer and electrostatic plotters), or pigment from colored wax paper (e.g., thermal transfer).

Hard copy devices vary in the number of colors and intensity levels, addressability, dot size, cost, image quality and speed. The laser printer is perhaps the most common, high-quality output device in this category [Foley, 1996].

One of the most interesting hard copy devices is the 3D printer or 3D rapid prototyper, which can use *fused deposition modeling* (applying material in layers) to allow engineers to create a range of physical 3D models and functional test parts directly from their CAD programs. These models can be made with a variety of materials and treated to enhance strength and durability as well as temperature resistance.

Display Technologies

Displays are, for the most part, characterized by their responsiveness to a moving image. As with hard copy technologies, display technologies vary in performance and cost. Comparisons are generally based on power consumption, screen size, depth, weight, ruggedness, brightness, addressability, contrast, intensity levels or number of colors possible per picture element (*pixel*), viewing angle and relative cost. It should be noted that several of the technologies mentioned below are covered in more detail elsewhere in this handbook.

1. **CRT**: Perhaps the most common graphics display is the cathode ray tube (CRT), used in televisions. A CRT is composed of five parts: (1) the electron gun, which when heated emits electrons at an appropriate rate; (2) the control grid, which regulates the flow of electrons; (3) the focusing system, which concentrates the beam into a fine point; (4) the deflection system, which directs the beam to the appropriate location; and (5) the phosphor screen, which glows when bombarded with the electron beam. The *persistence* of the phosphor is defined as the time from the removal of the excitation source to when the phosphorescence has decayed to 10 percent of the initial light output [Foley, 1996]. The persistence of the phosphor used determines how frequently the screen will need to be *refreshed* or redrawn. Color is produced by laying triads of red-green-blue (RGB) phosphors on the screen and using three electron guns, one for each color, to excite the phosphor for each pixel. The CRT scans the image, one row at a time, from a matrix whose elements correspond to the individual pixels or points on the screen. This matrix is referred to as the *frame buffer* and allows for a constant refresh rate, usually at least 60 times per second. Systems may also have more than one frame buffer (*double buffering*) to facilitate faster image generation. CRT displays are generally high resolution (1024 × 1280), SVGA (768 ×1024), NTSC (~350 × 480), and HDTV (720 × 1280 and 1080 × 1920) [Bailey, 2003].
2. **Liquid crystal displays (LCDs)** use two sheets of polarizing material sandwiching a liquid-crystal solution. This type of display is refreshed one row at a time using matrix addressing. When an electric current passes though the solution, crystals are polarized in such a way as to block out part of the backlighting to form the desired image [Ortiz, 2004]. These devices are light in weight, rugged, and have low power consumption and fair intensity.
3. **Plasma panels** consist of arrays of neon bulbs between glass plates. The displays may be monochrome or color depending on whether there is a single bulb or three (red, green, and blue) bulbs for each pixel. These devices excel in screen size, weight, ruggedness and brightness characteristics.

4. **Electroluminescent displays** also use grid-like structures for addressing elements and consist of a thin film of phosphorescent substance, a zinc sulfide doped with manganese, between two plates, one with horizontal wires embedded in it and the other with vertical. The light-emitting material can be made to glow at the intersections, creating a point of light. These displays are available in color and have excellent brightness characteristics.

5. **Head-mounted displays, stereoscopic displays, heads up displays, and all *Workbench*-type displays** are 3D displays often used in VR and augmented reality applications. The technology used is varied but combine graphics from each eye's viewpoint to provide a 3D image. Special glasses are required in some cases to get the stereoscopic effect. For example, in a *Responsive Workbench*-type display, mirrors and projectors are used to create computer-generated stereoscopic views on a horizontal tabletop display surface. These images are viewed through shutter glasses to generate the 3D effect and a six degrees-of-freedom tracking system tracks the user's head to maintain the correct point of view type [RWB, 2004].

Some current display technology research areas include the following:

1. **Organic light-emitting Diode (OLED)** displays use organic light-emitting polymers to eliminate the need for the backlighting used in LCDs. The color of the emitted light is determined by the structure of the polymers used [Ortiz, 2003].

2. **Smart displays** are portable, battery-operated monitors that have integrated wireless support to communicate with a nearby base PC. These displays come with a stylus and onscreen keyboard for input [Ortiz, 2003].

3. **Flexible displays** are built from thin plastic film or other material that bends, unlike than traditional glass, which aids portability. They may be rolled up and carried in a pocket [Ortiz, 2003].

4. **High dynamic range displays** can display a wide luminance range, for example, from 0.1 cd/m^2 to 10,000 cd/m^2 [Seetzen, 2003].

5. Research continues in all types of **3D immersive displays**, including holographic.

It should be noted that the field of *color science* is used in computer graphics and by hardware vendors to produce predictable color results on individual output systems, and color fidelity when moving color images between the ever-increasing variety of imaging devices and media.

13.3   Graphics Software

Graphics software has changed dramatically since its inception. In the 1970s and early 1980s, there were few graphics software tools. The first available packages were CAD/CAM packages designed for engineering. Most other engineering applications required users to develop their own programs to solve their graphics problems. These programs were written using low-level graphical commands or calls to some standard or quasi-standard graphical routines. Most of these systems were developed for mainframe computer environments. A trend, begun in the late 1980s, resulted in a change in computing hardware environments as well as in software approaches. Today, the predominant hardware platforms are personal computers and powerful UNIX workstations. Customized, stand-alone software tools are often used to create graphics instead of writing programs. Software development uses standard languages and graphical user interfaces. The technical community relies on the ever-increasing power of computers to support new software packages to manipulate and visually display complex data in real-time.

Engineering Software Packages

Many commercial scientific and engineering software packages have graphics functionality. Some of these allow the engineer to extend the capabilities of the system by programming their own application specific add-on graphics modules. The *IEEE Spectrum Focus Report: Software* in November 1991 suggested that these packages fall into the following categories:

1. **Logic simulation systems for application-specific integrated circuits (ASICs)** are used to design and display schematics of multigate ASICs constructed from large functional building blocks. Each block can represent a finite-state machine with several states and gates, specified in a relevant hardware description language.

2. **Electromagnetic design and simulation systems** are used to simulate the electromagnetic fields of operating printed-circuit boards. Multilayers of a board may be displayed, with colors indicating field densities in lines.

3. **Data acquisition, analysis, display and technical reporting systems** have compute-intensive analysis routines and enhanced data visualization capability. These packages might be used to produce plots and graphs based on acquired data that are displayed in several windows at once; changes made in one window result in the recalculation and update of the information displayed in all corresponding windows.

4. **Packages that do mathematical calculations and graphics for visualization** are used for operations such as curve fitting, evaluation of integrals, statistical analysis, signal processing and numerical analysis.

5. **Digital signal processors for embedded systems** use customized graphics software to analyze, manipulate and display the signals generated by the operating equipment.

There are other types of graphics systems used by engineers as well. For example, lighting engineers use rendering and visualization tools such as Radiance [Radiance, 2004] or Lightscape [Lightscape, 2004] to visualize their designs.

Other Graphics Software Packages

A similar plethora of graphics packages is available for most fields. For example, Windows Office users have charting and drawing tools at their fingertips. Animators use packages that simplify the modeling, rendering and animating of their scenes.

General Purpose Libraries and Packages

An engineer could, if need be, build his or her own graphics tools. Traditionally, graphical software systems are programmed in high-level languages with interfaces to standard or quasi-standard software packages. These packages often provide device independence by allowing systems to drive a wide variety of display devices, and application portability by isolating the programmer from machine-specific graphics commands. Such portability allows the programmer to move an application from one system to another without modifying his or her code.

The first quasi-standard graphics library was ACM/SIGGRAPH's Core system developed in 1977 and revised in 1979. While it was not a formally recognized standard, it did fulfill a role as a baseline specification for graphics libraries [Foley, 1996]. Two official standard libraries are GKS-3D (the 3D Graphical Kernel System), and PHIGS/PHIGS+ (the Programmer's Hierarchical Interactive Graphics System). Both libraries support graphics primitives, such as lines, polygons and character strings, as well as their attributes. GKS allows primitives to be grouped into segments. PHIGS/PHIGS+ allows segments to be nested and uses a database structure that allows for selective editing and manipulation of the model. Both support geometric transformations, i.e., scaling, translation, and rotation. These packages have influenced the shape of today's quasi-standard libraries which include the cross-platform OpenGL, a low-level graphics rendering and imaging library that sends commands to the GPU [Shreiner, 2004], as well as packages that are used to program GPUs directly such as Cg [Shader Tech, 2004] and DirectX [Microsoft, 2004].

Plotting and Page Description Languages

In most cases, it is unlikely that an engineer would have to work directly with plotting packages or page description languages, but engineers should be aware that graphics can be programmed directly in either when necessary. Generally, plotting packages consist of graphics routines, much like those in the packages mentioned above, that are callable from a high-level language program and handle both 2D and 3D images.

Page description languages are desktop publishing formats that produce graphical output on a printer, display or other output devices. They are used in application programs such as composition systems and illustrators where text, graphical shapes and sampled images are combined into a single document. The dominant language in this category is PostScript, which is a simple interpretive programming language with powerful graphics capabilities. PostScript communicates the description of a document to a printing system in a high level, device-independent manner. It features the construction of arbitrary shapes, which may self-intersect and which may be painted, transformed, cropped or rendered. Postscript commands are embedded in a general purpose programming language. PostScript programs can be created, transmitted and interpreted using an ASCII source, which allows for easy document interchange [Adobe, 1990].

Web Graphics

No current discussion of graphics software would be complete without mention of graphics support for the web. On a high level, this comes in packages designed to facilitate web page development, i.e., "desktop publishing" for the web. On a low level, engineers can utilize Web3D packages such as VRML, Java3D, and X3D as well as Java and scripting languages such as JavaScript and Micromedia Flash's ActionScript, to add and manage graphics on web pages. Transmission speed of the graphics and scene descriptions for web pages and applications is a major concern.

13.4   Graphics Algorithms

Software packages would not be possible without the continued development and improvement of graphics algorithms. A brief overview of graphics algorithms and techniques is included as these have a tendency to migrate from software to hardware, and as an engineer, it is possible to be involved with this migration.

The lowest level algorithms, scan conversion algorithms, are used to determine which pixels are involved with the drawing and filling of graphics primitives, such as lines, triangles, circles and polygons. These are often implemented in hardware.

Graphics Modeling

There is a wide variety of ways in which geometric data may be represented in computer graphics systems, from vertices and vectors to precise canonical definitions (circle, sphere, cone) to general parametric forms [Bezier, nonuniform rational b-splines (NURBS), multiresolutions] [Kasik, 2000].

In traditional graphics systems, image data are modeled and stored as Cartesian coordinates or as vectors. These data are manipulated using geometric transformations, such as scaling, translation and rotation, in a reference system known as the world coordinate system (WCS). The units of the WCS can be whatever the user needs: inches, millimeters or miles. Each physical device has its own device coordinate system (DCS). A viewing transformation is used to take the WCS image data to its corresponding device-specific coordinates: A *window*, a portion of the world model, is selected to be shown in an area of the display known as the *viewport*. *Clipping* is used to eliminate any data outside the selected area. These values are sometimes converted to intermediate coordinates known as normalized device coordinates (NDC), in which all values range from 0 to 1. These are then easily adjusted to any DCS needed. In 3D, a view volume is used instead of a window to limit what is displayed. Perspective or parallel projection is used to convert the model's 3D coordinates to the appropriate DCS. This view volume is dependent on the selection of the location of the viewpoint (eye) in WCS, the 2D window, and the type of projection as well as selectable front and back clipping planes.

Other approaches to modeling use feature-based systems such as solid or geometric modeling. Solid modeling systems use constructive solid geometry to build complicated objects. These systems have a descriptive language that uses a database of 3-D primitive objects such as blocks, cylinders, spheres, wedges, cones and tori. These solid objects are combined to form other solids using operators such as union, intersection and difference. The resultant object can then be named, saved and positioned anywhere in a model. Attributes stored with the objects allow them to be displayed in wire-frame format or as a completely rendered image [Teicholz, 1985].

Several procedural modeling techniques have been introduced to ease the difficulty of creating complex scenes directly. In general, these enable the creation of complex objects with fairly small pieces of code. Examples include fractals and particle systems. Fractals, geometrical self-similar objects with a fractional dimension that were introduced by Mandelbrot [1982], form a powerful tool for generating objects that resemble natural phenomena such as mountains, trees and coastlines. Prusinkiewicz and others use grammar-based fractal techniques to generate realistic plants [Prusinkiewicz, 1990]. Reeves developed particle systems to facilitate the creation of the Genesis scene in Star Trek's *Wrath of Khan* [Reeves, 1983]. Particle systems have been used to generate a variety of natural phenomena such as fire, waterfalls, grass and plants as well as fireworks. These systems consist of many small objects (particles) and use stochastic processes to determine particle characteristics such as lifetime, color, size and motion. Physical forces such as gravity are often incorporated to create realistic behavior. Figure 13.3 illustrates the creation of fog and three very different balls using procedural modeling techniques. A good overview of such algorithms and their current applications is available [Ebert, 2003].

Other algorithms model physical or psychological forces. For example, physically based techniques use the laws of physics to create realistic movement of items such as hair, cloth, snow and rain, and to demonstrate material characteristics such as elasticity, bouncing and breaking [Baraff, 2003]. Behavioral animation is used to model the behavior of groups such as flocks of birds and schools of fish so that animators need not be concerned with the behavior of each individual in the flock [Reynolds, 1999].

Volume visualization algorithms might be of greatest interest to the engineer. In these image-based algorithms, the line between computer graphics and image processing is blurred. Here several 2D images, such as a series of CAT scans or MRIs, are used to develop a 3D model. Generally, the same structures are identified in successive images, 3D coordinates are assigned, and connections are made from one image to one another to form a 3D model [Kaufman, 1998].

FIGURE 13.3 Image created using procedural modeling, noise and fog. Software: Renderman, Hardware: PC. Artist: Michael J. Murdoch, M.S., computer science student at Rochester Institute of Technology.

Other image-based techniques include modeling a 3D scene based on a series of photographs or video. This technique presents interesting issues when trying to model areas that are occluded in the images and thus never seen [Debevec, 2002].

Modeling research in graphics today has gone beyond solely visible representation to include acoustic and haptic modeling.

Rendering

A long-term goal of computer graphics has been the production of photorealistic images, resulting in much research devoted to the area of shading and illumination. These algorithms are used to model the transport of light through the scene. Shading algorithms range from simple flat shading of individual polygons to the successively more complex and more realistic Gouraud, Phong and Cook–Torrance models [Foley, 1996], among others.

On the illumination end, there are local illumination models such as Phong as well as global illumination algorithms, such as ray tracing, which does a good job with shiny and transparent objects, radiosity, which does a good job with perfectly diffuse surfaces, and hybrid algorithms that combine the best of both techniques [Foley, 1996]. Recent advances in the illumination area include photon mapping, which adds accurate rendering of caustics [Jensen, 2001], and image-based lighting, which uses light values captured by photographing a real scene and applies them to synthetic scenes [Debevec, 2002]. The latter often involves high-dynamic range images where the range of illuminance values is much broader than most output devices can handle. Tone reproduction algorithms [Devlin, 2002] can be used to map the image data into the appropriate range for a particular device. Figure 13.4 illustrates image quality that may be created using ray tracing augmented with photon mapping.

Rendering, or producing the finished image of a model, not only involves the application of shading and illumination models but also includes algorithms for visible surface determination and texture or bump mapping. Visible surface determination algorithms determine which surfaces are visible from the current viewpoint and which are occluded. Z-buffering [Foley, 1996], a brute force visible surface determination algorithm, is commonly implemented in hardware. Texture and bump mapping are algorithms used to add detail to a model without adding geometry [Foley, 1996]. For example, texture mapping might be used to map

FIGURE 13.4 Ray traced ring with caustics done with self-programmed software using a PC. The checkerboard pattern is produced using procedural texture mapping. Artist: Kevin Pazirandeh, software engineering student at Rochester Institute of Technology.

FIGURE 13.5   The wood background in this image is created using texture mapping. Software: Modeling — Discreet 3DS Max, Rendering — RenderMan; Hardware: PC. Artist: Kevin Pazirandeh, software engineering student at Rochester Institute of Technology.

an image of wood grain or marble to a polygon in the model, or bump mapping might be used to create an orange-peel-like surface on a sphere. The latter is accomplished by altering surface normals before applying illumination algorithms. Image-based rendering is the high end of this type of algorithm, capturing texture details from several photos to produce output image textures [Debevec, 2002]. Both Figure 13.4 and Figure 13.5 were created utilizing texture mapping to enhance detail without adding geometry.

Computer graphics researchers are also studying acoustic and haptic rendering primarily to make VR environments more immersive. The more immersive the environment, the stronger the measure of *presence*, a term used in the study of perception to indicate how intensely the viewer feels that he or she is actually immersed or present in the scene.

As mentioned previously, a long-term goal of computer graphics has been to produce photorealistic graphics. As a sign of just how far computer graphics has come, researchers are now studying nonphotorealistic rendering (NPR). There are a myriad of reasons why NPR might be desirable, such as the production of images that look like oil or watercolor paintings, or when photorealism is so detailed that it hides what is of importance [Finkelstein, 2003].

Interaction Algorithms

The ability to create applications that react appropriately and in a timely fashion to input commands or data is one of the reasons that computer graphics is so powerful. Such interactions may involve selecting menu items or choosing parts of an image displayed on the screen. The immediacy of the feedback provided to the user is of utmost importance.

Graphics input devices include, among others, the mouse, special purpose keyboards using buttons or dials, data gloves and other VR devices, touch panels and screens, light pens, graphics tablets, joysticks, 3D digitizers, trackballs and voice systems. Each of these devices is capable of sending appropriate values to the graphics program for action [Hearn, 2004]. Graphics algorithms respond to such input based on logical device categories or by using callback functions.

The fields of human–computer interaction (HCI) and perception play a role in creating packages and images that respond and behave in a reasonable fashion. Perception also plays a role in deciding when and where to apply higher cost rendering algorithms.

Parallel Graphics

Many of the algorithms described above are computationally intensive. For example, it is possible for the rendering of a single image of a complex model to take many hours. Parallel computing, applying multiple CPUs to perform a single task, can play a role in improving the turnaround time for such images. For example, this could be done by sending each cell to a different processor when producing an animated film, or by having individual processors render different sets of pixels in a single image [Chalmers, 2002].

A more complete description of many of these algorithms and techniques can be found in most computer graphics textbooks [Foley 1996; Hearn 2004].

13.5   The Future

Current algorithmic advances are available in the proceedings of conferences such as the Association for Computing Machinery's (ACM's) SIGGRAPH, Eurographics, AFRIGRAPH, GRAPHITE, Nicograph, and the International Game Developers Conference. Some of these organizations jointly sponsor workshops for specialized graphics areas, such as the SIGGRAPH/Eurographics Workshop on Graphics Hardware.

In addition, two professional computing organizations publish periodicals that are specifically devoted to advances in computer graphics: *ACM Transactions on Graphics* and *IEEE Computer Graphics and Applications*.

Continuing research into the many facets of computer graphics ensures its advancement. The *Vision 2000* (January/February 2000) issue of *IEEE Computer Graphics and Applications* provides a particularly enticing glimpse into what the future may hold for computer graphics.

References

Adobe Systems Incorporated, *PostScript Language Reference Manual*, 2nd ed., Reading, MA: Addison-Wesley, 1990.

M. Bailey and A. Glassner, "Introduction to computer graphics," Course Notes, ACM SIGGRAPH Computer Graphics, *30th International Conference on Computer Graphics and Interactive Techniques*, San Diego, CA, 2003.

D. Baraff and A. Witkin, "Physically based modeling," Course Notes, ACM SIGGRAPH Computer Graphics, *30th International Conference on Computer Graphics and Interactive Techniques*, San Diego, CA, 2003.

M. Brown, *Understanding PHIGS, TEMPLATE*, San Diego, CA: Megatek Corporation, 1985.

P. Debevec and L. McMillan, Eds., Image-Based Modeling, Rendering, and Lighting Issue of *IEEE Comput. Graphics Appl.*, vol. 22, no. 2, 2002.

K. Devlin, A. Chalmers, A. Wilkie, and W. Purgathofer, "Tone reproduction and physically based spectral rendering," in *State of the Art Reports, Eurographics 2002*, D. Fellner, and R. Scopignio, Eds., The Eurographics Association, 2002, p. 101.

A. Chalmers, T. Davis and E. Reinhard, Eds., *Practical Parallel Rendering*, Wellesley, MA: A.K. Peters, 2002.

D.S. Ebert, F.K. Musgrave, D. Peachey, K. Perlin, and S. Worley, *Texturing and Modeling, A Procedural Approach*, 3rd ed., Los Altos, CA: Morgan Kaufmann, 2003.

N. England, "Graphics hardware," *IEEE Comput. Graphics Appl. (Vision 2000)*, vol. 20, no. 1, p. 46, 2000.

A. Finkelstein and L. Markosian, Eds., NonPhotorealistic Rendering Issue of *IEEE Comput. Graphics Appl.*, 23, 4, 2003.

J.D. Foley, A. Van Dam, S.K. Feiner, and J.F. Hughes, *Computer Graphics: Principles and Practice in C*, 2nd ed., Reading, MA: Addison-Wesley, 1996.

S. Feiner and D. Thalmann, Eds., Virtual Reality Issue of *IEEE Comput. Graphics Appl.*, vol. 20, no. 6, p. 24, 2000.

J.M. Geigel and N.C. Schaller, "Virtual photography — a framework for teaching image synthesis," Educators Program, ACM SIGGRAPH Computer Graphics, *31st International Conference on Computer Graphics and Interactive Techniques*, SIGGRAPH 2004, Los Angeles, CA, 2004.

D. Hearn and M.P. Baker, *Computer Graphics with OpenGL*, Englewood Cliffs, NJ: Prentice-Hall, 2004.

H.W. Jensen, *Realistic Image Synthesis Using Photon Mapping*, Wellesley, MA: A.K. Peters, 2001.

D.J. Kasik, "Viewing the future of CAD," *IEEE Comput. Graphics Appl. (Vision 2000)*, vol. 20, no. 6, p. 34, 2000.

A. Kaufman, "Advances in volume visualization," Course Notes, *25th International Conference on Computer Graphics and Interactive Techniques*, SIGGRAPH '98, Orlando, FL, 1998.

C. Kolb, D. Mitchell and P. Hanrahan, "A realistic camera model for computer graphics," ACM SIGGRAPH Computer Graphics, *Proc. 22nd Annu. Conf. Comput. Graphics Interact. Tech.*, Los Angeles, CA, 1995, p. 317.

Lightscape, http://www.lightscape.com/, June 2004.

M. Macedonia, "The GPU enters computing's mainstream," *IEEE Comput.*, vol. 36, no. 10, p. 106, 2003.

B.B. Mandelbrot, *The Fractal Geometry of Nature*, San Francisco, CA: W.H. Freeman, 1982.

Microsoft, http://msdn.microsoft.com/directx/, June 2004.

T. Munzner, "Information visualization," *IEEE Comput. Graphics Appl.*, vol. 22, no. 1, p. 20, 2002.

S. Ortiz, "New monitor technologies are on display," *IEEE Comput.*, vol. 36, no. 2, p. 13, 2003.

P. Prusinkiewicz and A. Lindenmayer, *The Algorithmic Beauty of Plants (The Virtual Laboratory)*, Berlin, Germany: Springer-Verlag, 1990.

Radiance, http://radsite.lbl.gov/radiance/HOME.html, June 2004.

W.T. Reeves, "Particle systems — a technique for modeling a class of fuzzy objects," *ACM Trans. Graphics*, vol. 2, no. 2, p. 91, 1983.

Responsive Workbench (RWB), http://graphics.stanford.edu/projects/RWB/, June 2004.

C. Reynolds, "Steering behaviors for autonomous characters," in *Proc. Game Developers Conf.*, San Jose, CA, p. 763, 1999.

H. Seetzen, L.A. Whitehead, and G. Ward, "High dynamic range display using low and high resolution modulators," in *Proc. Soc. Inf. Display Int. Symp.*, Baltimore, MD, 2003, p. 1450.

Shader Tech, http://www.cgshaders.org/, June 2004.

D. Shreiner, M. Woo, J. Neider, and T. Davis, *OpenGL Programming Guide*: The Official Guide to Learning OpenGL, Version 1.4, 4th ed., Reading, MA: Addison Wesley, 2003.

E. Teicholz, Ed., *CAD/CAM Handbook*, New York, NY: McGraw-Hill, 1985.

14
Computer Networks

Matthew N.O. Sadiku
Prairie View A&M University

Cajetan M. Akujuobi
Prairie View A&M University

The coming of the information age has brought about unprecedented growth in telecommunications-based services driven primarily by the Internet, the information superhighway. Within a short period of time, the volume of data traffic transported across communications networks has grown rapidly and now exceeds the volume of voice traffic. While voice networks, such the ubiquitous telephone network, have been in use for over a century, computer data networks are a recent phenomenon.

A computer communications network is interconnection of different computing devices to enable them communicate among themselves. As shown in Figure 14.1, computer networks are generally classified into three groups on the basis of their geographical scope:

- Local area networks (LANs) spanning a building, a campus, or an enterprise with a total span of 2 km.
- Metropolitan area networks (MANs) spanning a city, with a total span of 100 km.
- Wide area networks (WANs) spanning a nation or globe.

These networks differ in geographic scope, types of organization using them, types of services provided and transmission techniques. On the one hand, the local area network (LAN) is used in connecting equipments owned by the same organization over relatively short distances. Its performance degrades as the area of coverage becomes large. Thus LANs have limitations of geography, speed, traffic capacity and the number of stations they are able to connect. On the other hand, the wide area network (WAN) provides long-haul communication services to various points within a large geographical area, e.g., a nation or continent. With some of the characteristics of LANs and some reflecting WANs, the metropolitan area network (MAN) embraces the best features of both. MAN is designed to extend over a city or metro area, and it may be owned by a private company or by the public.

We begin this chapter by looking at the Open Systems Interconnection (OSI) reference model, which is commonly used to describe the functions involved in data communication networks. We then examine different LANs, MANs and WANs including the Internet.

14.1 OSI Reference Model

There are at least two reasons for needing a standard protocol architecture such as the OSI reference model. First, the uphill task of understanding, designing and constructing a computer network is made more

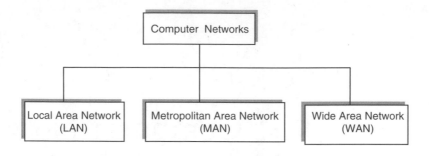

FIGURE 14.1   Classification of computer communication networks.

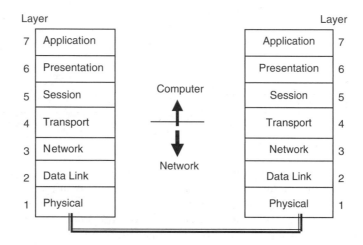

FIGURE 14.2   The OSI reference model.

manageable by dividing it into structured smaller subtasks. Second, the proliferation of computer systems has created heterogeneous networks: different vendors, different models from the same vendor, different data formats, different network management protocols, different operating systems. A way to resolve this heterogeneity is for vendors to abide by the same set of rules. Attempts to formulate these rules have preoccupied standards bodies such as the International Standards Organization (ISO), International Telecommunication Union (ITU), Institute of Electrical and Electronics Engineers (IEEE), American National Standards Institute (ANSI), British Standards Institution (BSI), and European Computer Manufacturers Association (ECMA). Here we consider the more universal standard protocol architecture developed by ISO.

The International Standards Organization (ISO) divides the task of networking computers into seven layers so that manufacturers can develop their own applications and implementations within the guidelines of each layer. In 1978 the ISO set up a committee to develop a seven-layer model of network architecture (initially for WANs), known as the OSI. The model serves as a means of comparing different layers of communication networks. Also, the open model is standard-based rather than proprietary-based; one system can communicate with another system using interfaces and protocols that both systems understand. Network users and vendors have "open systems" in which any standard computer device would be able to interoperate with others.

The OSI reference model is a seven-layer hierarchy that groups the functional requirements for moving information across a network. The seven layers of the OSI model are shown in Figure 14.2 and briefly explained as follows. We begin with the application layer (layer 7) and work our way down.

- *Application Layer*: This layer (layer 7) allows transferring information between application processes. It is implemented with host software. It is composed of specific application programs and its content varies with individual users. By application we mean a set of information processing desired by the user. Typical applications (or user programs) include login, password check, word processing, spreadsheet, graphics program, document transfer, electronic mailing system, virtual terminal emulation, remote database access, network management, bank balance, stock prices, credit check, inventory check and airline reservation. Examples of application layer protocols are Telnet (remote terminal protocol), file transfer protocol (FTP), simple mail transfer protocol (SMTP), remote login service (rlogin), and remote copy protocol (RCP).

- *Presentation Layer*: This layer (layer 6) presents information in a way that is meaningful to the network user. It performs functions such as translation of character sets, interpretation of graphic commands, data compression/decompression, data reformating and data encryption/decryption. Popular character sets include American Standard Code for Information Interchange (ASCII), Extended Binary Coded Decimal Interchange Code (EBCDIC), and Alphabet 5.

- *Session Layer*: A session is a connection between users. The session layer (layer 5) establishes the appropriate connection between users and manages dialog between them, i.e., controlling starting, stopping and synchronization of the dialog. It decides the type of communication such as two-way simultaneous (full duplex), two-way alternate (half-duplex), one-way, or broadcast. It is also responsible for checking for user authenticity and providing billing. For example, login and logout are the responsibility of this layer. IBM's NetBIOS (Network Basic Input/Output System), NetWare's SPX (Sequenced Packet Exchange), Manufacturing Automation Protocol (MAP), and Technical and Office Protocol (TOP) operate at this layer.

- *Transport Layer*: This layer (layer 4) uses the lower layers to establish reliable end-to-end transport connections for the higher layers. Its other function is to provide the necessary functions and protocols to satisfy a quality of service (QoS) (expressed in terms of time delay, throughput, priority, cost and security) required by the session layer. It creates several logical connections over the same network by multiplexing end-to-end user addresses on to the network. It fragments messages from the session layer into smaller units (packets or frames) and reassembles the packets into messages at the receiving end. It also controls the end-to-end flow of packets, performs error control and sequence checking, acknowledges successful transmission of packets, and requests retransmission of corrupted packets. For example, the Transmission Control Protocol (TCP) of TCP/IP and Internet Transport Protocol (ITP) of Xerox operate at this level.

- *Network Layer*: This layer (layer 3) handles routing procedure and flow control. It establishes routes (virtual circuits) for packets to travel and routes the packets from their source to destination and controls congestion. (Routing is of greater importance on MANs and WANs than on LANs.) It carries addressing information that identifies the source and ultimate destination. It also counts transmitted bits for billing information. It ensures that packets arrive at their destination in a reasonable amount of time. Examples of protocols designed for layer 3 are X.25 packet switching protocol and X.75 gateway protocol, both by ITU. Also, the Internet protocol (IP) of TCP/IP and NetWare's Internetwork Packet Exchange (IPX) operate at this layer.

- *Data Link Layer*: This layer (layer 2) specifies how a device gains access to the medium specified in the physical layer. It converts the bit pipe provided by the physical layer into a packet link, which is a facility for transmitting packets. It deals with procedures and services related to the node-to-node data transfer. A major difference between the data link layer and the transport layer is that the domain for the data link layer is between adjacent nodes whereas that of the transport layer is end-to-end. In addition, the data link layer ensures error-free delivery of data; hence it is concerned with error-detection, error correction, and retransmission. The error control is usually implemented by performing checksums on all bits of a packet after a cyclic redundancy check (CRC) process. This way any transmission errors can be detected. The layer is implemented in hardware and is highly dependent of the physical medium. Typical examples of data link protocols are Binary Synchronous Communications (BSC), Synchronous

TABLE 14.1   Summary of the Functions of OSI Layers

Layer	Name	Function
7	Application Layer	Transfers information between application processes
6	Presentation Layer	Syntax conversion, data compression and encryption
5	Session Layer	Establishes connection and manages a dialog
4	Transport Layer	Provides end-to-end transfer of data
3	Network Layer	End-to-end routing and flow control
2	Data Link Layer	Medium access, framing and error control
1	Physical Layer	Electrical/mechanical interface

Data Link Control (SDLC), and High-level Data Link Control (HDLC). For LANs and MANs, the data link layer is decomposed by IEEE into the media-access control (MAC) and the logical link control (LLC) sublayers.

- *Physical Layer*: This layer (layer 1) consists of a set of rules that specifies the electrical and physical connection between devices. It is implemented in hardware. It is responsible for converting raw bits into electrical signals and physically transmitting them over a physical medium such as coaxial cable or an optical fiber between adjacent nodes. It provides standards for electrical, mechanical and procedural characteristics required to transmit the bit stream properly. It handles frequency specifications, encoding the data, defining voltage or current levels, defining cable requirements, defining the connector size, shapes, and pin number, etc. RS-232, RS-449, X.21, X.25, V.24, IEEE 802.3, IEEE 802.4 and IEEE 802.5 are examples of physical-layer standards.

A summary of the functions of the seven layers is presented in Table 14.1. The seven layers are often subdivided into two. The first consists of the lower three layers (physical, data link and network layers) and is known as the *communications subnetwork*. The upper three layers (session, presentation and application layers) are termed the *host process*. The upper layers are usually implemented by networking software on the node. The transport layer is the middle layer, separating the data-communication functions of the lower three layers and the data-processing functions of the upper layers. It is sometimes grouped with the upper layers as part of the host process or grouped with the lower layers as part of data transport.

14.2   Local Area Networks

A local area network (LAN) is a computer network that spans a geographically small area. It consists of two or more computers that are connected together to share expensive resources such as printers, exchange files, or allow electronic communications. Most LANs are confined to a single building or campus. They connect workstations, personal computers, printers and other computer peripherals. Users connected to the LAN can use it to communicate with each other. LANs are capable of transmitting data at very fast rates, much faster than data can be transmitted over a telephone line, but the distances are limited. Also, since all the devices are located within a single establishment, LANs are usually owned and maintained by an organization. A key motivation for using LANs is to increase the productivity and efficiency of workers.

LANs differ from MANs and WANs by geographic coverage, data transmission and error rates, topology and data routing techniques, ownership, and sometimes by the type of traffic. Unique characteristics that differentiate LANs include:

- LANs generally operate within a few kilometers, spanning only a small geographical area.
- LANs usually have very high bit rates, ranging from 1 Mbps to 10 Gbps.
- LANs have a very low error rate, say $1:10^8$.
- A LAN is often owned and maintained by a single private company, institution or organization using the facility.

OSI		SNA	
7	Applications	7	Transaction services
6	Presentation	6	Presentation services
5	Session	5	Data flow control
4	Transport	4	Transmission control
3	Network	3	Path control
2	Data link	2	Data link control
1	Physical	1	Physical control

FIGURE 14.3 A comparison of SNA and OSI models.

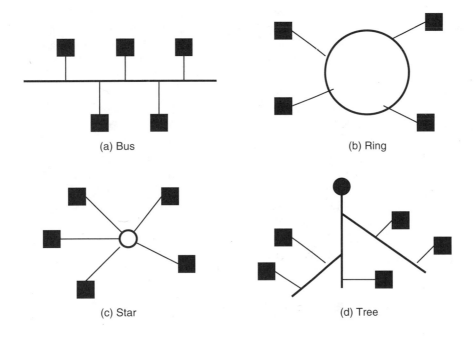

(a) Bus

(b) Ring

(c) Star

(d) Tree

FIGURE 14.4 Typical LAN topologies.

There are different kinds of LANs. The following features differentiate one LAN from another:

- *Topology*: The geometric arrangement of devices on the LAN. As shown in Figure 14.4, this can be bus, ring, star, or tree.
- *Protocols*: These are procedures or rules that govern the transfer of information between devices connected to a LAN. Protocols are to computer networks what languages are to humans.

- *Media*: The transmission medium connecting the devices can be twisted-pair wire, coaxial cables, or fiber optic cables. Wireless LANs use radio waves as media. Of all these media, optic fiber is the fastest but the most expensive.

Common LANs include Ethernet, token ring, token bus, and star LAN. For bus or tree LANs, the most common transmission medium is coaxial cable. The two common transmission methods used on coaxial cable are baseband and broadband. A baseband LAN is characterized by the use of digital technology; binary data are inserted into the cable as a sequence of pulses using a Manchester or Differential encoding scheme. A broadband LAN employs analog signaling and a modem. The frequency spectrum of the cable can be divided into channels using frequency division multiplexing (FDM). One of the most well-known applications of broadband transmission is the community antenna television (CATV). However, baseband LANs are more prevalent.

The IEEE has established the following eight committees to provide standards for LANs:

- IEEE 802.1 — standard for LAN/MAN bridging and management
- IEEE 802.2 — standard for logical link control protocol
- IEEE 802.3 — standard for CSMA/CD protocol
- IEEE 802.4 — standard for token bus MAC protocol
- IEEE 802.5 — standard for token ring MAC protocol
- IEEE 802.7 — standard for broadband LAN
- IEEE 802.10 — standard for LAN/MAN security
- IEEE 802.11 — standard for wireless LAN

Token ring is a network architecture which uses token-passing technology and a ring-type network structure. Although token ring is standardized in IEEE 802.5 standard, its use has faded to only a few organizations. Ethernet [IEEE 802.3] is the most popular and the least expensive high-speed LAN.

Ethernet is a LAN architecture developed by the Xerox Corporation in cooperation with DEC and Intel in 1976. The IEEE 802.3 standard refined the Ethernet and made it globally accepted. Ethernet has since become the most popular and most widely deployed LAN in the world.

Conventional Ethernet uses a bus or star topology and supports data transfer rates of 10 Mbps. It uses a protocol known as carrier sense multiple access with collision detection (CSMA/CD) as an access method to handle simultaneous demands. Each station or node attached to the Ethernet must sense the medium before transmitting data to see if any other station is already transmitting. If the medium appears to be idle, then the station can begin to send data. If two stations sense the medium idle and transmit at the same time, collision may take place. When such a collision occurs, the two stations stop transmitting, wait, and try again later after a randomly chosen delay period. The delay period is determined using binary exponential backoff.

Ethernet is one of the most widely implemented LAN standards. It has been estimated that there are more than 600 million existing Ethernet nodes today. A newer version of Ethernet, called Fast Ethernet (or 100Base-T) supports data transfer rates of 100 Mbps. Gigabit Ethernet (or 1000Base-T) delivers at 1 Gbps. The 10 Gbps format of Ethernet has been available since 2002.

14.3   Metropolitan Area Networks

Metropolitan area networks are basically an outgrowth of LANs. As the demand for information increases, first-generation network standards such as Ethernet and token ring cannot handle the enormous volume of data. A variety of users and applications drive the requirements for metropolitan area networks (MANs). These requirements include volume of traffic, cost, scalability, security, reliability, compatibility with existing and future networks, and management issues. To meet these requirements, several proposals have been made for MAN protocols and architectures. Of these proposed MANs, fiber distributed data interface (FDDI) and distributed queue dual-bus (DQDB) have emerged as standards. Since FDDI is the only standard that has survived the test of time, only FDDI will be considered here.

The FDDI is a dual token ring that supports data rates of 100 Mbps and uses optical fiber media. The FDDI specification recommends an optical fiber with a core diameter of 62.5 microns and a cladding diameter of 125 microns. The advantages of fiber optics over electrical media and the inherent advantages of a ring design contribute to the widespread acceptance of FDDI as a standard.

FDDI is a collection of standards formed by the ANSI X3T9.5 task group and set by the American National Standard Institute (ANSI) over a period of ten years. The standards produced by the task group cover physical hardware, physical and data link protocol layers, and a conformance testing standard. The original standard, known as FDDI-I, provides the basic data-only operation. An extended standard, FDDI-II, supports hybrid data and real-time applications.

FDDI is a follow-on to IEEE 802.5 (token ring) in that FDDI is based on token-ring mechanics. Although the FDDI MAC protocol is similar (but not identical) to token ring, there are some differences. Unlike in token ring, FDDI performs all networking monitoring and control algorithms in a distributed way among active stations and does not need an active monitor (hence the term "distributed" in FDDI.). Whenever any device is down, other devices reorganize and continue to function including token initialization, fault recovery, clock synchronization and topology control.

The key highlights of FDDI are summarized as follows:

- ANSI standard through the X3T9.5 committee
- Dual counter-rotating ring topology for fault tolerance
- Data rate of 100 Mbps
- Total ring loop of size 100 km
- Maximum of 500 directly attached stations or devices
- 2 km maximum distance between stations
- Variable packet size (4,500 bytes, maximum)
- 4B/5B data encoding scheme to ensure data integrity
- Shared medium using a timed-token protocol
- Variety of physical media, including fiber and twisted pair
- 62.5/125-μm multimode fiber-optic based network
- Low bit error rate of 10^{-9} (one in one billion)
- Compatibility with IEEE 802 LANs by use of IEEE 802.2 LLC
- Distributed clocking to support large number of stations
- Support for both synchronous and asynchronous services

FDDI has two types of nodes: stations and concentrators. The stations transmit information to other stations on the ring and receive from them. Concentrators are nodes which provide additional ports for attachments of stations to the network. A concentrator receives data from the ring and forwards it to each of the connected ports sequentially at 100 Mbps. While a station may have one or more MAC, a concentrator may or may not have a MAC. As shown in Figure 14.5, each FDDI station is connected to two rings, a primary and secondary, simultaneously. Stations have active taps on the ring and operate as repeaters. This allows FDDI network to be very large without signal degradation. The network uses its primary ring for data transmission, while the secondary ring can be used either to ensure fault tolerance, or for data. When a station or link fails, the primary and secondary rings form a single one-way ring, isolating the fault while maintaining a logical path among users. Thus, FDDI's dual-ring topology and connection management functions establish a fault-tolerance mechanism.

FDDI was developed to conform to the OSI reference model. FDDI divides the physical layer of the OSI reference model into two sublayers: physical layer medium dependent (PMD) and physical layer (PHY), while the data link layer is split into two sublayers: media access control (MAC) and IEEE 802.2 logical link control (LLC). A comparison of FDDI architectural model to the lower two layers of the OSI model along with the summary of the functions of the FDDI standards is illustrated in Figure 14.6. The FDDI MAC uses a timed-token rotation (TTR) protocol for controlling access to the medium. With this protocol the MAC in each station measures the time that has elapsed since the station last received a token. Each station on the FDDI ring uses three timers to regulate its operation. During the network initialization process, all stations connected to the ring negotiate for a target token rotation time (TTRT). The value of TTRT is determined small enough to

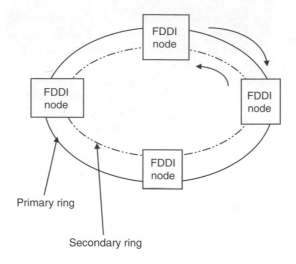

FIGURE 14.5 Dual self-healing counter-rotating rings of FDDI.

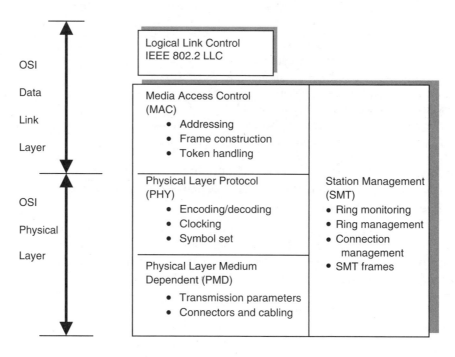

FIGURE 14.6 Summary of the functions of the FDDI standards.

satisfy the real-time constraints of synchronous traffic. This TTRT value (typically 8 ms) is set equal to the value of the token-rotation time (TRT), which is used to monitor the amount of elapsed time between subsequent arrivals of token at a station. Each station keeps track of the time it was last visited by a token.

The station management (SMT) controls the other three layers (PMD, PHY and MAC) and ensures proper operation of the station. It handles such functions as initial FDDI ring initialization, station insertion and removal, ring stability, activation, connection management, address administration, scheduling policies, collection of statistics, bandwidth allocation, performance and reliability monitoring, bit error monitoring, fault detection and isolation, and ring reconfiguration.

14.4   Wide Area Networks

A WAN is an interconnected network of LANs and MANs. A WAN connects remote LANs and ties remote computers together over long distances. Computers connected to a WAN are often connected through public networks such as the telephone system. They can also be connected through leased lines or satellites. WANs are, by default, heterogeneous networks that consist of a variety of computers, operating systems, topologies and protocols. The largest WAN in existence is the Internet.

Because of the long distance involved, WANs are usually developed and maintained by a nation's public telecommunication companies (such as AT&T in the U.S.), which offer various communication services to the people. Today's WANs are designed in the most cost-effective way using optical fiber. Fiber-based WANs are capable of transporting voice, video and data with no known restriction to bandwidth. Such WANs will remain cutting edge for years to come. There is also the possibility of connecting networks using wireless technologies.

Circuit and Packet Switching

For a WAN communication is achieved by transmitting data from the source node to the destination node through a network of intermediate switching nodes. Thus, unlike a LAN, a WAN is a switched network. There are many types of switched networks, but the most common methods of communications are circuit switching and packet switching. Circuit switching is a much older technology than packet switching. Circuit switching systems are ideal for communications that require data to be transmitted in real time. Packet-switching networks are more efficient if some amount of delay is acceptable.

Circuit switching is a communication method in which a dedicated path (channel or circuit) is established for the duration of a transmission. This is a type of point-to-point network connection. A switched circuit is maintained while the sender and recipient are communicating, as opposed to a dedicated circuit which is held open regardless of whether data is being sent or not. The most common circuit-switching network is the telephone system.

Packet switching is a technique whereby the network routes individual packets of data between different destinations based on addressing within each packet. A packet is a segment of information sent over a network. Any message exceeding a network-defined maximum length (a set size) is broken up into shorter units, known as packets. Packet-switching is the process by which a carrier breaks up messages (or data) into these segments, bundles, or packets by the source data terminal equipment (DTE) before they are sent. Each packet is switched and transmitted individually through the network and can even follow different routes to its destination and may arrive out of order. Most modern WAN protocols such as TCP/IP, X.25, and frame relay are based on packet switching technologies. Besides data networks such as the Internet, wireless services like Cellular Digital Packet Data (CDPD) employ packet switching.

X.25

For roughly 20 years X.25 was the dominant player in WAN packet switching technology until frame relay, SMDS and ATM appeared. X.25 has been around since the mid 1970s and so is pretty well debugged and stable. It was originally approved in 1976, and subsequently was revised in 1977, 1980, 1984, 1988, 1992 and 1996. It is currently one of the most widely used interfaces for data communication networks. There are literally no data errors on modern X.25 networks.

X.25 is a communications packet switching protocol designed for the exchange of data over a WAN. It represents a standard, a network, or an interface protocol. It is a popular standard for packet-switching networks approved in 1976 by the International Telecommunication Union–Telecommunication Standardization Sector (ITU-T) for WAN communications. It defines how connections between user devices and network devices are established and maintained. X.25 utilizes a connection-oriented service that insures that packets are transmitted in order. Through statistical multiplexing, X.25 enables multiple users to share bandwidth as it becomes available, therefore ensuring flexible use of network resources among all users. X.25 is also an interface protocol in that it spells the required interface protocols that enable a data terminal equipment (DTE) to

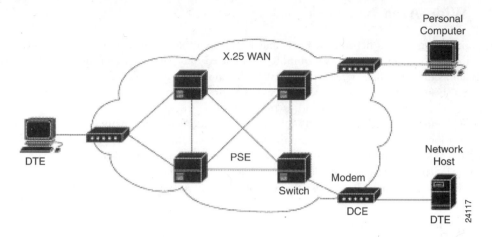

FIGURE 14.7   DTEs, DCEs and PSEs make up an X.25 network.

communicate with data circuit-terminating equipment (DCE), which provides access to the network. The DTE-DCE link provides full-duplex multiplexing, allowing a virtual circuit to transmit in either direction.

X.25 network devices fall into three general categories: data terminal equipment (DTE), data circuit-terminating equipment (DCE), and packet switching exchange (PSE). DTE devices are user end systems that communicate across the X.25 network. They are usually terminals, personal computers or network hosts, and are located on the premises of individual subscribers. DCE devices are the carrier's equipment such as modems and packet switches that provide the interface between DTE devices and a PSE, and are generally located in the carrier's facilities. PSEs are switches that compose the bulk of the carrier's network. They transfer data from one DTE device to another. Figure 14.7 illustrates the relationships between the three types of X.25 network devices.

The packet assembler/disassembler (PAD) is a device commonly found in X.25 networks. PADs are used when a DTE device is too simple to implement the full X.25 functionality. The PAD is located between a DTE device and a DCE device, and it performs three primary functions: buffering, packet assembly and packet disassembly. The PAD buffers data sent to or from the DTE device. It also assembles outgoing data into packets and forwards them to the DCE device (this includes adding an X.25 header). Finally, the PAD disassembles incoming packets before forwarding the data to the DTE (this includes removing the X.25 header).

A virtual circuit is a logical connection created to ensure reliable communication between two network devices. Two types of X.25 virtual circuits exist:

- *Switched virtual circuits (SVCs)* — SVCs are very much like telephone lines; a connection is established, data are transferred and then the connection is released. They are temporary connections used for sporadic data transfers.
- *Permanent virtual circuits (PVCs)* — a PVC is similar to a leased line in that the connection is always present. Permanent virtual circuits (PVCs) are permanently established connections used for frequent and consistent data transfers. Therefore, data may always be sent without any call setup.

Maximum packet sizes vary from 64 bytes to 4096 bytes, with 128 bytes being a default on most networks.

X.25 users are typically large organizations with widely dispersed and communications-intensive operations in sectors such as finance, insurance, transportation, utilities and retail. For example, X.25 is often chosen for zero-error tolerance applications by banks involved in large-scale transfers of funds, or by government utilities that manage electrical power networks.

Frame Relay

Frame relay is a simplified form of packet switching (similar in principle to X.25) in which synchronous frames of data are routed to different destinations depending on header information. It is basically an interface

Physical Layout of a Sample Frame Relay Network

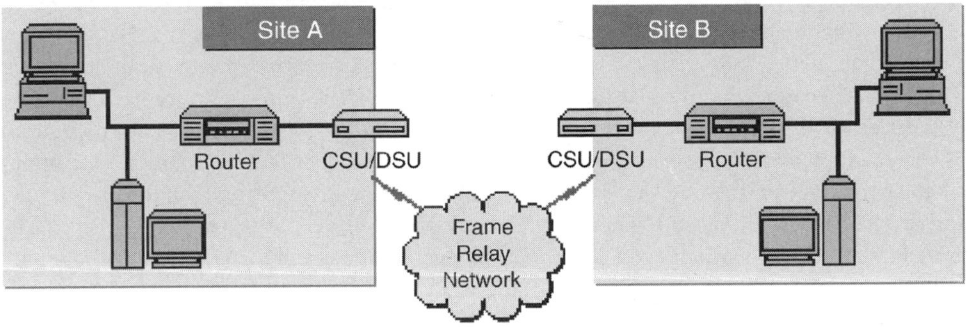

FIGURE 14.8 Physical layout of a typical frame relay network.

used for wide-area networking. It is used to reduce the cost of connecting remote sites in any application that would typically use expensive leased circuits.

Frame relay is an interface, a method of multiplexing traffic to be submitted to a WAN. Carriers build frame relay networks using switches. The physical layout of a sample frame relay network is depicted in Figure 14.8. The CSU/DSU is the channel service unit/data service unit. This unit provides a "translation" between the telephone company's equipment and the router. The router actually delivers information to the CSU/DSU over a serial connection much like the computer uses a modem, only at a much higher speed.

All major carrier networks implement *permanent virtual circuits* (PVCs). These circuits are established via contract with the carrier and typically are built on a flat-rate basis. Although *switched virtual circuits* (SVCs) have standards support and are provided by the major frame relay backbone switch vendors, they have not been widely implemented in customer equipment or carrier networks.

Two major frame relay devices are the FRAD and routers. Standalone frame relay access devices (FRADs) typically connect small remote sites to a limited number of locations. FRAD is also known as frame relay assembler/disassembler. Frame relay routers offer more sophisticated protocol handling than most FRADs. They may be packaged specifically for frame relay use, or they may be general purpose routers with frame relay software.

Frame relay is the fastest growing WAN technology in the U.S. In North America it is fast taking on the role that X.25 has had in Europe. It is used by large corporations, government agencies, small businesses, and even Internet service providers (ISPs). The demand for frame relay services is exploding and for two very good reasons — speed and economics. frame relay is consistently less expensive than equivalent leased services and provides the bandwidth needed for other services like LAN routing, voice and fax.

14.5   ISDN and ATM Networks

ISDN stands for Integrated Services Digital Network. ISDN is a high speed communication network that allows voice, data, text, graphics, music, video and other source material to be transmitted simultaneously across the world using end-to-end digital connectivity. "Digital network" means that the user is given access to a telecom network that ensures high quality transmission via digital circuits, while "integrated services" refers to the simultaneous transmission of voice, video and data services over the same wires. This way, computers can connect directly to the telephone network without first converting their signals to an analog form using modems. This integration brings with it a host of new capabilities combining voice, data, fax and sophisticated switching. Because ISDN uses the existing local telephone wiring, it is equally available to home and business customers. ISDN was intended to eventually replace the traditional plain old telephone service (POTS) phone lines with a digital network that would carry voice, data and video.

ISDN service is available today in most major metropolitan areas and probably will be completely deployed throughout the U.S. very soon. Many ISPs now sell ISDN access. However, the idea of using existing copper wiring to provide this network decreased ISDN capabilities, in reality. When the digital video systems started to develop in the 1980s, it was soon noticed that the maximum bandwidth (2.048 Mbps) of the ISDN was not enough. That is why broadband ISDN (BISDN) was born.

BISDN is a digital network operating at data rates in excess of 2.048 Mbps (the maximum rate of standard ISDN). BISDN is a second generation of ISDN. Broadband ISDN is not only an improved ISDN, but also a complete redesign of the "old" ISDN, now called narrowband ISDN. It consists of ITU-T communication standards designed to handle high-bandwidth applications such as video. The key characteristic of broadband ISDN is that it provides transmission channels capable of supporting rates greater than the primary ISDN rate. Broadband services are aimed at both business applications and residential subscribers.

BISDN's foundation is cell switching, and the international standard supporting it is *Asynchronous Transfer Mode* (ATM). Because BISDN is a blueprint for ubiquitous worldwide connectivity, standards are of the utmost importance. Major strides have been made in this area by the ITU-T during the past decade. More recently the ATM Forum has advanced that agenda.

ATM is a fast packet-oriented transfer mode based on asynchronous time division multiplexing. The words *transfer mode* say that this technology is a specific way of transmitting and switching through the network. The term *asynchronous* refers to the fact that the packets are transmitted using asynchronous techniques (e.g., on demand), and the two end-points need not have synchronized clocks. ATM will support both circuit switched and packet switched services. ATM can handle any kind of information (i.e., voice, data, image, text and video) in an integrated manner.

An ATM network is made up of an ATM switch and ATM endpoints. An ATM switch is responsible for cell transit through an ATM network. An ATM endpoint (or end system) contains an ATM network interface adapter. Examples of ATM endpoints are workstations, routers, digital service units (DSUs), LAN switches, and video coder/decoders (CODECs). An ATM network consists of a set of ATM switches interconnected by point-to-point ATM links or interfaces. ATM switches support two primary types of interfaces: user–network interface (UNI) and network–network interface (NNI). The UNI connects ATM end systems (such as hosts and routers) to an ATM switch. The NNI connects two ATM switches.

In ATM the information to be transmitted is divided into short 53-byte packets or cells. There are reasons for such a short cell length. First, ATM must deliver real time service at low bit rates. Thus, the size allows ATM to carry multiple forms of traffic. Both time-sensitive traffic (voice) and time-insensitive traffic (data) can be carried with the best possible balance between efficiency and minimal packetization delay. Second, using short, fixed-length cells allows for time-efficient and cost-effective hardware such switches and multiplexers.

Each ATM cell consists of 48 bytes for the information field and 5 bytes for the header. The header is used to identify cells belonging to the same virtual channel, and thus is used in appropriate routing. The ATM cell structure is shown in Figure 14.9. The cell header comes in two forms: the UNI header and the NNI header. The UNI is described as the point where the user enters the network. The NNI is the interface between networks. The typical header therefore looks like that shown in Figure 14.10 for the UNI. The header is slightly different for NNI, as shown in Figure 14.11.

ATM is connection-oriented and connections are identified by the virtual channel identifier (VCI). A virtual channel (VC) represents a given path between the user and the destination. A virtual path (VP) is created by multiple virtual channels heading to the same destination. The relationship between virtual channels and virtual paths is illustrated in Figure 14.12. A virtual channel is established at connection time and torn

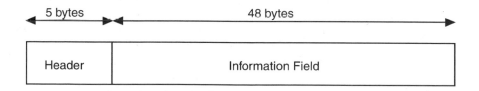

FIGURE 14.9   ATM cell structure.

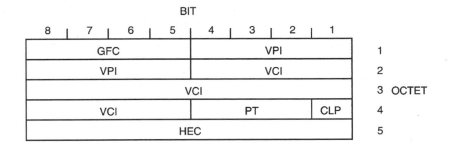

VPI Virtual path identifier   PT   Payload type
VCI Virtual channel identifier   CLP Call loss priority
HEC Header error control   GFC Generic flow control

FIGURE 14.10   ATM cell header for UNI.

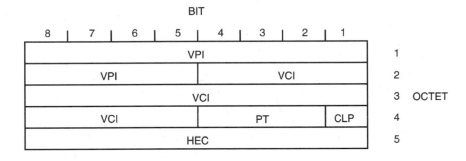

VPI Virtual path identifier   PT   Payload type
VCI Virtual channel identifier   CLP Call loss priority
HEC Header error control

FIGURE 14.11   ATM cell header for NNI.

FIGURE 14.12   Relationship between virtual channel, virtual path and transmission path.

down at termination time. The establishment of the connections includes the allocation of a virtual channel identifier (VCI) and/or virtual path identifier (VPI), and also includes the allocation of the required resources on the user access and inside the network. These resources, expressed in terms of throughput and quality of service (QoS), can be negotiated between user and network either before the call setup or during the call. Having both virtual paths and channels make it easy for the switch to handle many connections with the same origin and destination.

ATM can be used in existing twisted pair, fiber-optic, coaxial, hybrid fiber/coax (HFC), SONET/SDH, T1, E1, T3, E3, E4, etc., for LAN and WAN communications. ATM is also compatible with wireless and satellite communications.

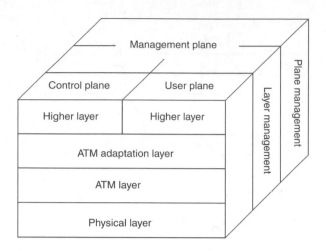

FIGURE 14.13   BISDN protocol reference model.

Figure 14.13 depicts the architecture for the BISDN protocol. It is evident that the BISDN protocol uses a three-plane approach. The user plane (U-plane) is responsible for user information transfer including flow control and error control. The U-plane contains all of the ATM layers. The control plane (C-plane) manages the call-control and connection-control functions. The C-plane shares the physical and ATM layers with the U-plane, and contains ATM adaptation layer (AAL) functions dealing with signaling. The management plane (M-plane) includes plane management and layer management. This plane provides the management functions and the capability to transfer information between the C- and U-planes. The layer management performs layer-specific management functions, while the plane management deals with the complete system. Figure 14.13 also shows how ATM fits into BISDN. The ATM system is divided into three functional layers, namely the physical layer, the ATM layer, and the ATM adaptation layer. The physical layer defines the transmission medium, encoding, clocking and any necessary electrical-to-optical transformation. The ATM layer provides functions such as cell-loss priority, cell construction, generic flow control, and connection assignment and removal. The AAL allows various information transfer protocols that are not based on ATM.

BISDN access can be based on a single optical fiber per customer site. A variety of interactive and distribution broadband services is contemplated for BISDN: high-speed data transmission, broadband video telephony, corporate videoconferencing, video surveillance, high-speed file transfer, TV distribution (with existing TV and/or high-definition television), video on demand, LAN interconnection and hi-fi audio distribution.

14.6   Internet

The Internet is a global network of computer networks (or WAN) that exchange information via telephone, cable television, wireless networks and satellite communication technologies. It is used by an increasing number of people worldwide. As a result, the Internet has been growing exponentially with the number of machines connected to the network and the amount of network traffic virtually doubling each year. The Internet today is fundamentally changing our social, political and economic structures, and in many ways circumventing geographic boundaries.

Internet Protocol Suite

The Internet is a combination of networks, including the Arpanet, NSFnet, regional networks such as NYsernet, local networks at a number of universities and research institutions, and a number of military networks. Each network on the Internet contains anywhere from two to thousands of addressable devices or nodes (computers) connected by communication channels. All computers do not speak the same language,

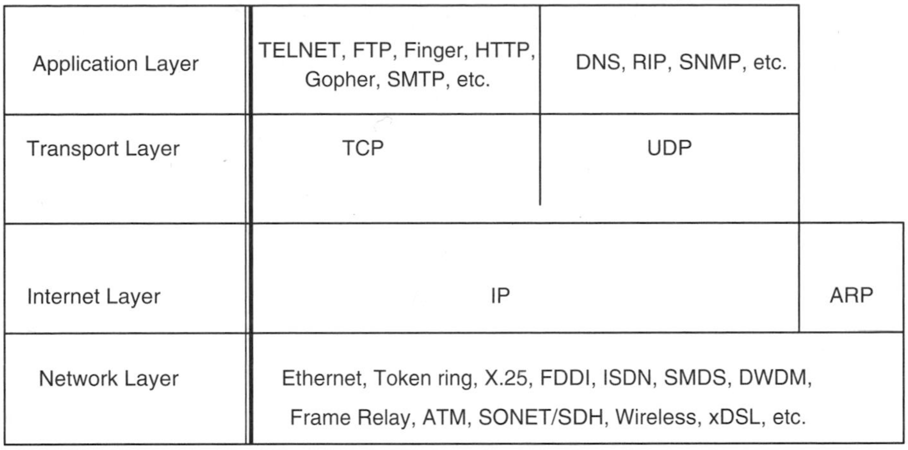

Application Layer	TELNET, FTP, Finger, HTTP, Gopher, SMTP, etc.	DNS, RIP, SNMP, etc.	
Transport Layer	TCP	UDP	
Internet Layer	IP		ARP
Network Layer	Ethernet, Token ring, X.25, FDDI, ISDN, SMDS, DWDM, Frame Relay, ATM, SONET/SDH, Wireless, xDSL, etc.		

FIGURE 14.14   Abbreviated Internet protocol suite.

but if they are going to be networked they must share a common set of rules known as *protocols*. That is where the two most critical protocols, Transmission Control Protocol/Internetworking Protocol (TCP/IP), come in. Perhaps the most accurate name for the set of protocols is the *Internet protocol suite*. (TCP and IP are two of the protocols in this suite). TCP/IP is an agreed upon standard for computer communication over Internet. The protocols are implemented in software that runs on each node.

The TCP/IP is a layered set of protocols developed to allow computers to share resources across a network. Figure 14.14 shows the Internet protocol architecture. Figure 14.14 is by no means exhaustive, but shows the major protocols and application components common to most commercial TCP/IP software packages and their relationship.

As a layered set of protocols, Internet applications generally use four layers:

- *Application Layer*: This is where application programs that use the Internet reside. It is the layer with which end users normally interact. Some application-level protocols in most TCP/IP implementations include FTP, TELNET and SMTP. For example, FTP (File Transfer Protocol) allows a user to transfer files to and from computers that are connected to the Internet.
- *Transport Layer:* Controls the movement of data between nodes. TCP (Transmission Control Protocol) is a connection-based service that provides services need by many applications. UDP (User Datagram Protocol) provides connectionless services.
- *Internet Layer:* Handles addressing and routing of the data. It is also responsible for breaking up large messages and reassembling them at the destination. IP (Internet Protocol) provides the basic service of getting datagrams to their destination. ARP (Address Resolution Protocol) figures out the unique address of devices on the network from their IP addresses.
- *Network Layer*: Supervises addressing, routing and congestion control. Protocols at this layer are needed to manage a specific physical medium such as Ethernet or a point-to-point line.

TCP/IP is built on connectionless technology. IP provides a *connectionless, unreliable, best-effort* packet delivery service. Information is transferred as a sequence of datagrams. Those datagrams are treated by the network as completely separate.

TCP sends datagrams to IP with the Internet address of the computer at the other end. The job of IP is simply to find a route for the datagram and get it to the other end. In order to allow gateways or other intermediate systems to forward the datagram, it adds its own header, as shown in Figure 14.15. The main things in this header are the source and destination Internet address (32-bit addresses, like 128.6.4.194), the protocol number, and another checksum. The source Internet address is simply the address of your

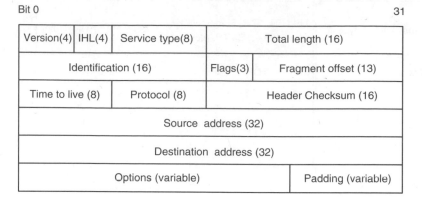

FIGURE 14.15   IP header format (20 bytes).

machine. The destination Internet address is the address of the other machine. The protocol number tells IP at the other end to send the datagram to TCP. Although most IP traffic uses TCP, there are other protocols that can use IP, so one has to tell IP which protocol to send the datagram to. Finally, the checksum allows IP at the other end to verify that the header was not damaged in transit. IP needs to be able to verify that the header did not get damaged in transit, or it could send a message to the wrong place. After IP has tacked on its header, the message looks like what is in Figure 14.15.

Addresses and Addressing Scheme

For IP to work every computer must have its own number to identify itself. This number is called the IP address. You can think of an IP address as similar to your telephone number or postal address. All IP addresses on a particular LAN must start with the same numbers. In addition, every host or router on the Internet has an address that uniquely identifies it and also denotes the network on which it resides. No two machines can have the same IP address. To avoid addressing conflicts, the network numbers have been assigned by the InterNIC (formerly known simply as NIC).

Blocks of IP addresses are assigned to individuals or organizations. The *network* part of the address is common for all machines on a local network. It similar to a postal zip code that is used by a post office to route letters to a general area. The rest of the address on the letter (i.e., the street and house number) is relevant only within that area. It is only used by the local post office to deliver the letter to its final destination. The *host* part of the IP address performs this same function. There are five types of IP addresses:

- *Class A* format: 126 networks with 16 million hosts each; an IP address in this class starts with a number between 0 and 127.
- *Class B* format: 16,382 networks with up to 64K hosts each; an IP address in this class starts with a number between 128 and 191.
- *Class C* format: 2 million networks with 254 hosts each; an IP address in this class starts with a number between 192 and 223.
- *Class D* format: Used for multicasting, in which a datagram is directed to multiple hosts.
- *Class E* format: Reserved for future use.

The IP address formats for classes A, B and C are shown in Figure 14.16.

IPv6

Most of today's Internet uses Internet Protocol Version 4 (IPv4), which is now nearly 25 years old. Due to the phenomenal growth of the Internet, the rapid increase in palmtop computers, and the profusion of smart

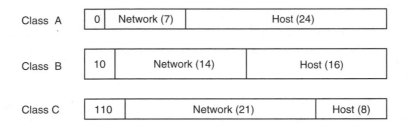

FIGURE 14.16   IP address formats.

cellular phones and PDAs, the demand for IP addresses has outnumbered the limited supply provided by IPv4. In response to the shortcomings of IPv4, the Internet Engineering Task Force (IETF) approved IPv6 in 1997. IPv4 will be replaced by Internet Protocol Version 6 (IPv6), which is sometimes called the Next Generation Internet Protocol (or IPng). IPv6 adds many improvements and fixes a number of problems in IPv4, such as the limited number of available IPv4 addresses.

With only a 32-bit address field, IPv4 can assign only 2^{32} different addresses, i.e., 4.29 billion IP addresses, which are inadequate in view of the rapid proliferation of networks and the two-level structure of the IP addresses (network number and host number). To solve the problem of severe IP address shortage, IPv6 uses 128-bit addresses instead of the 32-bit addresses of IPv4. That means, IPv6 can have as many as 2^{128} IP addresses, which is roughly 3.4×10^{38} or about 340 billion billion billion billion unique addresses.

The IPv6 packet consists of the IPv6 header, the routing header, the fragment header, the authentication header, the TCP header, and application data. The IPv6 packet header is of fixed length, whereas the IPv4 header is of variable length. The IPv6 header consists of 40 bytes as shown in Figure 14.17. It consists of the following fields:

- *Version* (4 bits): This is the IP version number, which is 6.
- *Priority* (4 bits): This field enables a source to identify the priority of each packet relative to other packets from the same source.
- *Flow Label* (24 bits): The source assigns the flow label to all packets that are part of the same flow. A flow may be a single TCP connection or a multiple of TCP connections.
- *Payload Length* (16 bits): This field specifies the length of the remaining part of the packet following the header.
- *Next Header* (8 bits): This identifies the type of header immediately following the header.
- *Hop Limit* (8 bits): This is to set some desired maximum value at the source and the field denotes the remaining number of hops allowed for the packet. It is decremented by 1 at each node the packet passes, and the packet is discarded when the hop limit becomes zero.
- *Source Address* (128): The address of the source of the packet.
- *Destination Address* (128 bits): The address of the recipient of the packet.

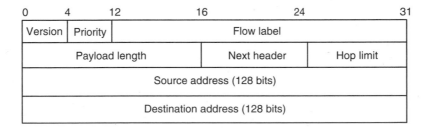

FIGURE 14.17   IPv6 header format.

There are three types of IPv6 addresses:

1. Unicast is used to identify a single interface.
2. Anycast identifies a set of interfaces. A source may use an anycast address to contact any node from a group of nodes.
3. Multicast identifies a set of interfaces. A packet with multicast address is delivered to all members of the group.

IPv6 is expected to gradually replace IPv4, with the two coexisting for a number of years during a transition period. IPv6 may be most widely deployed in mobile phones, PDAs and other wireless terminals in the future.

References

U. Black, *Data Networks: Concepts, Theory and Practice,* Englewood Cliffs, NJ: Prentice-Hall, 1989.

M.N.O. Sadiku, *Metropolitan Area Networks*, Boca Raton, FL: CRC Press, 1995.

W. Stallings, *ISDN and Broadband ISDN with Frame Relay and ATM*, 4th ed., Upper Saddle River, NJ: Prentice-Hall, 1999.

15

Fault Tolerance

Barry W. Johnson
University of Virginia

15.1 Introduction

Fault tolerance is the ability of a system to continue correct performance of its tasks after the occurrence of hardware or software faults. A **fault** is simply any physical defect, imperfection, or flaw that occurs in hardware or software. Applications of fault-tolerant computing can be categorized broadly into four primary areas: long-life, critical computations, maintenance postponement, and high availability. The most common examples of long-life applications are unmanned space flight and satellites. Examples of critical-computation applications include aircraft flight control systems, military systems, and certain types of industrial controllers. Maintenance postponement applications appear most frequently when maintenance operations are extremely costly, inconvenient, or difficult to perform. Remote processing stations and certain space applications are good examples. Banking and other time-shared systems are good examples of high-availability applications. Fault tolerance can be achieved in systems by incorporating various forms of redundancy, including hardware, information, time, and software redundancy [Johnson, 1989].

15.2 Hardware Redundancy

The physical replication of hardware is perhaps the most common form of fault tolerance used in systems. As semiconductor components have become smaller and less expensive, the concept of hardware redundancy has become more common and more practical. There are three basic forms of hardware redundancy. First, *passive* techniques use the concept of fault masking to hide the occurrence of faults and prevent the faults from resulting in **errors**. Passive approaches are designed to achieve fault tolerance without requiring any action on the part of the system or an operator. Passive techniques, in their most basic form, do not provide for the detection of faults but simply mask the faults. An example of a passive approach is triple modular redundancy (TMR), which is illustrated in Figure 15.1. In the TMR system three identical units perform identical functions, and a majority vote is performed on the output.

The second form of hardware redundancy is the *active* approach, which is sometimes called the *dynamic* method. Active methods achieve fault tolerance by detecting the existence of faults and performing some action to remove the faulty hardware from the system. In other words, active techniques require that the system perform reconfiguration to tolerate faults. Active hardware redundancy uses fault detection, fault location, and fault recovery in an attempt to achieve fault tolerance. An example of an active approach

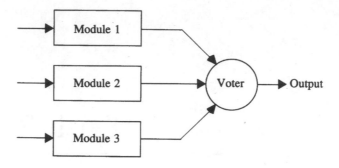

FIGURE 15.1   Fault masking using triple modular redundancy (TMR). (*Source:* B.W. Johnson, *Design and Analysis of Fault-Tolerant Digital Systems,* Reading, Mass.: Addison-Wesley, 1989, p. 52. With permission.)

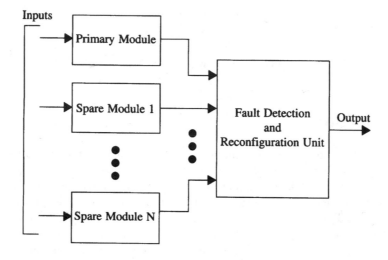

FIGURE 15.2   General concept of standby sparing.

to hardware redundancy is standby sparing, which is illustrated in Figure 15.2. In standby sparing one or more units operate as spares and replace the primary unit when it fails.

The final form of hardware redundancy is the *hybrid* approach. Hybrid techniques combine the attractive features of both the passive and active approaches. Fault masking is used in hybrid systems to prevent erroneous results from being generated. Fault detection, fault location, and fault recovery are also used in the hybrid approaches to improve fault tolerance by removing faulty hardware and replacing it with spares. Providing spares is one form of providing redundancy in a system. Hybrid methods are most often used in the critical-computation applications where fault masking is required to prevent momentary errors, and high reliability must be achieved. The basic concept of the hybrid approach is illustrated in Figure 15.3.

15.3   Information Redundancy

Another approach to fault tolerance is to employ redundancy of information. Information redundancy is simply the addition of redundant information to data to allow fault detection, fault masking, or possibly fault tolerance. Good examples of information redundancy are error detecting and error correcting codes, formed by the addition of redundant information to data words or by the mapping of data words into new representations containing redundant information [Lin and Costello, 1983].

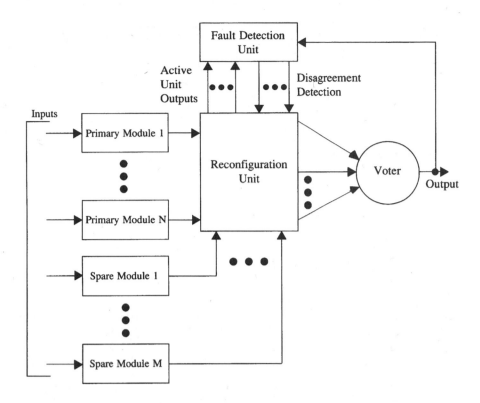

FIGURE 15.3 Hybrid redundancy approach. (*Source:* B.W. Johnson, *Design and Analysis of Fault-Tolerant Digital Systems*, Reading, Mass.: Addison Wesley, 1989, p. 70. With permission.)

In general, a *code* is a means of representing information, or data, using a well-defined set of rules. A *code word* is a collection of symbols, often called digits if the symbols are numbers, used to represent a particular piece of data based upon a specified code. A *binary code* is one in which the symbols forming each code word consist of only the digits 0 and 1. A code word is said to be *valid* if the code word adheres to all of the rules that define the code; otherwise, the code word is said to be *invalid*.

The *encoding operation* is the process of determining the corresponding code word for a particular data item. In other words, the encoding process takes an original data item and represents it as a code word using the rules of the code. The *decoding operation* is the process of recovering the original data from the code word. In other words, the decoding process takes a code word and determines the data that it represents.

It is possible to create a binary code for which the valid code words are a subset of the total number of possible combinations of 1s and 0s. If the code words are formed correctly, errors introduced into a code word will force it to lie in the range of illegal, or invalid, code words, and the error can be detected. This is the basic concept of the *error detecting codes*. The basic concept of the *error correcting code* is that the code word is structured such that it is possible to determine the correct code word from the corrupted, or erroneous, code word.

A fundamental concept in the characterization of codes is the *Hamming distance* [Hamming, 1950]. The *Hamming distance* between any two binary words is the number of bit positions in which the two words differ. For example, the binary words 0000 and 0001 differ in only one position and therefore have a Hamming distance of 1. The binary words 0000 and 0101, however, differ in two positions; consequently, their Hamming distance is 2. Clearly, if two words have a Hamming distance of 1, it is possible to change one word into the other simply by modifying one bit in one of the words. If, however, two words differ in two bit positions, it is impossible to transform one word into the other by changing one bit in one of the words.

The Hamming distance gives insight into the requirements of error detecting codes and error correcting codes. We define the *distance* of a code as the minimum Hamming distance between any two valid code words. If a binary code has a distance of two, then any single-bit error introduced into a code word will result in the erroneous word

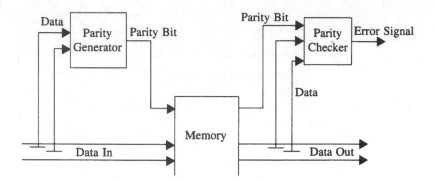

FIGURE 15.4   Use of parity coding in a memory application. (*Source:* B.W. Johnson, *Design and Analysis of Fault-Tolerant Digital Systems,* Reading, Mass.: Addison-Wesley, 1989, p. 85. With permission.)

being an invalid code word because all valid code words differ in at least two bit positions. If a code has a distance of 3, then any single-bit error or any double-bit error will result in the erroneous word being an invalid code word because all valid code words differ in at least three positions. However, a code distance of 3 allows any single-bit error to be corrected, if it is desired to do so, because the erroneous word with a single-bit error will be a Hamming distance of 1 from the correct code word and at least a Hamming distance of 2 from all others. Consequently, the correct code word can be identified from the corrupted code word.

In general, a binary code can correct up to c bit errors and detect an additional d bit errors if and only if

$$2c + d + 1 \leq H_d$$

where H_d is the distance of the code [Nelson and Carroll, 1986]. For example, a code with a distance of 2 cannot provide any error correction but can detect single-bit errors. Similarly, a code with a distance of 3 can correct single-bit errors or detect a double-bit error.

A second fundamental concept of codes is *separability*. A *separable code* is one in which the original information is appended with new information to form the code word, thus allowing the decoding process to consist of simply removing the additional information and keeping the original data. In other words, the original data is obtained from the code word by stripping away extra bits, called the code bits or check bits, and retaining only those associated with the original information. A *nonseparable code* does not possess the property of separability and, consequently, requires more complicated decoding procedures.

Perhaps the simplest form of a code is the parity code. The basic concept of parity is very straightforward, but there are variations on the fundamental idea. Single-bit parity codes require the addition of an extra bit to a binary word such that the resulting code word has either an even number of 1s or an odd number of 1s. If the extra bit results in the total number of 1s in the code word being odd, the code is referred to as *odd parity*. If the resulting number of 1s in the code word is even, the code is called *even parity*. If a code word with odd parity experiences a change in one of its bits, the parity will become even. Likewise, if a code word with even parity encounters a single-bit change, the parity will become odd. Consequently, a single-bit error can be detected by checking the number of 1s in the code words. The single-bit parity code (either odd or even) has a distance of 2, therefore allowing any single-bit error to be detected but not corrected. Figure 15.4 illustrates the use of parity coding in a simple memory application.

Arithmetic codes are very useful when it is desired to check arithmetic operations such as addition, multiplication, and division [Avizienis, 1971]. The basic concept is the same as all coding techniques. The data presented to the arithmetic operation is encoded before the operations are performed. After completing the arithmetic operations, the resulting code words are checked to make sure that they are valid code words. If the resulting code words are not valid, an error condition is signaled. An arithmetic code must be invariant to a set of arithmetic operations. An arithmetic code, A, has the property that $A(b*c) = A(b)*A(c)$, where b and c are operands, * is some arithmetic operation, and $A(b)$ and $A(c)$ are the arithmetic code words for the operands b

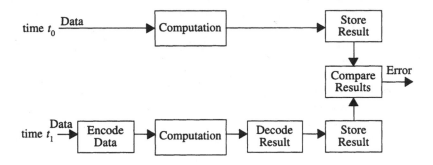

FIGURE 15.5 Time redundancy concept. (*Source:* B.W. Johnson, *Design and Analysis of Fault-Tolerant Digital Systems,* Reading, Mass.: Addison-Wesley, 1989, p. 137. With permission.)

and c, respectively. Stated verbally, the performance of the arithmetic operation on two arithmetic code words will produce the arithmetic code word of the result of the arithmetic operation. To completely define an arithmetic code, the method of encoding and the arithmetic operations for which the code is invariant must be specified. The most common examples of arithmetic codes are the *AN* codes, residue codes, and the inverse residue codes.

15.4   Time Redundancy

Time redundancy methods attempt to reduce the amount of extra hardware at the expense of using additional time. In many applications, the time is of much less importance than the hardware because hardware is a physical entity that impacts weight, size, power consumption, and cost. Time, on the other hand, may be readily available in some applications. The basic concept of time redundancy is the repetition of computations in ways that allow faults to be detected. Time redundancy can function in a system in several ways. The fundamental concept is to perform the same computation two or more times and compare the results to determine if a discrepancy exists. If an error is detected, the computations can be performed again to see if the disagreement remains or disappears. Such approaches are often good for detecting errors resulting from transient faults but cannot provide protection against errors resulting from permanent faults.

The main problem with many time redundancy techniques is assuring that the system has the same data to manipulate each time it redundantly performs a computation. If a transient fault has occurred, a system's data may be completely corrupted, making it difficult to repeat a given computation. Time redundancy has been used primarily to detect transients in systems. One of the biggest potentials of time redundancy, however, now appears to be the ability to detect permanent faults while using a minimum of extra hardware. The fundamental concept is illustrated in Figure 15.5. During the first computation or transmission, the operands are used as presented and the results are stored in a register. Prior to the second computation or transmission, the operands are encoded in some fashion using an encoding function. After the operations have been performed on the encoded data, the results are then decoded and compared to those obtained during the first operation. The selection of the encoding function is made so as to allow faults in the hardware to be detected. Example encoding functions might include the complementation operator and an arithmetic shift.

15.5   Software Redundancy

Software faults are unusual entities. Software does not break as hardware does, but instead software faults are the result of incorrect software designs or coding mistakes. Therefore, any technique that detects faults in software must detect design flaws. A simple duplication and comparison procedure will not detect software faults if the duplicated software modules are identical, because the design mistakes will appear in both modules.

The concept of N self-checking programming is to first write N unique versions of the program and to develop a set of acceptance tests for each version. The acceptance tests are essentially checks performed on the results produced by the program and may be created using consistency checks and capability checks, for example. Selection logic, which may be a program itself, chooses the results from one of the programs that passes the acceptance tests. This approach is analogous to the hardware technique known as hot standby sparing. Since each program is running simultaneously, the reconfiguration process can be very fast. Provided that the software faults in each version of the program are independent and the faults are detected as they occur by the acceptance tests, then this approach can tolerate $N - 1$ faults. It is important to note that the assumptions of fault independence and perfect fault coverage are very big assumptions to make in almost all applications.

The concept of N-version programming was developed to allow certain design flaws in software modules to be tolerated [Chen and Avizienis, 1978]. The basic concept of N-version programming is to design and code the software module N times and to vote on the N results produced by these modules. Each of the N modules is designed and coded by a separate group of programmers. Each group designs the software from the same set of specifications such that each of the N modules performs the same function. However, it is hoped that by performing the N designs independently, the same mistakes will not be made by the different groups. Therefore, when a fault occurs, the fault will either not occur in all modules or it will occur differently in each module, so that the results generated by the modules will differ. Assuming that the faults are independent the approach can tolerate $(N - 1)/2$ faults where N is odd.

The recovery block approach to software fault tolerance is analogous to the active approaches to hardware fault tolerance, specifically the cold standby sparing approach. N versions of a program are provided, and a single set of acceptance tests is used. One version of the program is designated as the primary version, and the remaining $N - 1$ versions are designated as spares, or secondary versions. The primary version of the software is always used unless it fails to pass the acceptance tests. If the acceptance tests are failed by the primary version, then the first secondary version is tried. This process continues until one version passes the acceptance tests or the system fails because none of the versions can pass the tests.

15.6   Dependability Evaluation

Dependability is defined as the quality of service provided by a system [Laprie, 1985]. Perhaps the most important measures of dependability are reliability and availability. Fundamental to reliability calculations is the concept of failure rate. Intuitively, the *failure rate is* the expected number of **failures** of a type of device or system per a given time period [Shooman, 1968]. The failure rate is typically denoted as λ when it is assumed to have a constant value. To more clearly understand the mathematical basis for the concept of a failure rate, first consider the definition of the reliability function. The **reliability** $R(t)$ of a component, or a system, is the conditional probability that the component operates correctly throughout the interval $[t_0, t]$ given that it was operating correctly at the time t_0.

There are a number of different ways in which the failure rate function can be expressed. For example, the failure rate function $z(t)$ can be written strictly in terms of the reliability function $R(t)$ as

$$z(t) = \left(\frac{-\mathrm{d}R(t)/\mathrm{d}t}{R(t)} \right)$$

Similarly, $z(t)$ can be written in terms of the unreliability $Q(t)$ as

$$z(t) = -\frac{\mathrm{d}R(t)/\mathrm{d}t}{R(t)} = \frac{\mathrm{d}Q(t)/\mathrm{d}t}{1 - Q(t)}$$

where $Q(t) = 1 - R(t)$. The derivative of the unreliability, $\mathrm{d}Q(t)/\mathrm{d}t$, is called the *failure density function.*

The failure rate function is clearly dependent upon time; however, experience has shown that the failure rate function for electronic components does have a period where the value of $z(t)$ is approximately constant.

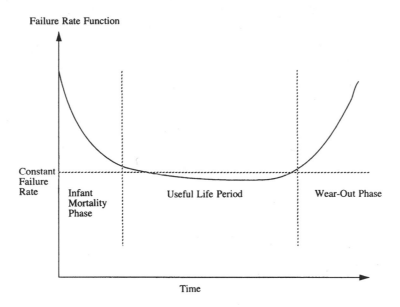

Time

FIGURE 15.6 Bathtub curve relationship between the failure rate function and time. (*Source:* B.W. Johnson, *Design and Analysis of Fault-Tolerant Digital Systems*, Reading, Mass.: Addison-Wesley, 1989, p. 173. With permission.)

The commonly accepted relationship between the failure rate function and time for electronic components is called the bathtub curve and is illustrated in Figure 15.6. The bathtub curve assumes that during the early life of systems, failures occur frequently due to substandard or weak components. The decreasing part of the bathtub curve is called the early-life or infant mortality region. At the opposite end of the curve is the wear-out region where systems have been functional for a long period of time and are beginning to experience failures due to the physical wearing of electronic or mechanical components. During the intermediate region, the failure rate function is assumed to be a constant. The constant portion of the bathtub curve is called the useful-life phase of the system, and the failure rate function is assumed to have a value of λ during that period. λ is referred to as the failure rate and is normally expressed in units of failures per hour.

The reliability can be expressed in terms of the failure rate function as a differential equation of the form

$$\frac{dR(t)}{dt} = -z(t)R(t)$$

The general solution of this differential equation is given by

$$R(t) = e^{-\int z(t)dt}$$

If we assume that the system is in the useful-life stage where the failure rate function has a constant value of λ, the solution to the differential equation is an exponential function of the parameter λ given by

$$R(t) = e^{-\lambda t}$$

where λ is the constant failure rate. The exponential relationship between the reliability and time is known as the *exponential failure law*, which states that for a constant failure rate function, the reliability varies exponentially as a function of time.

In addition to the failure rate, the mean time to failure (MTTF) is a useful parameter to specify the quality of a system. The MTTF is the expected time that a system will operate before the first failure occurs. The MTTF can be calculated by finding the expected value of the time of failure.

From probability theory, we know that the expected value of a random variable, ∞X, is

$$E[X] = \int_{-\infty}^{\infty} xf(x)\,dx$$

where $f(x)$ is the probability density function. In reliability analysis we are interested in the expected value of the time of failure (MTTF), so

$$MTTF = \int_{0}^{\infty} tf(t)\,dt$$

where $f(t)$ is the failure density function, and the integral runs from 0 to ∞ because the failure density function is undefined for times less than 0. We know, however, that the failure density function is

$$f(t) = \frac{dQ(t)}{dt}$$

so, the MTTF can be written as

$$MTTF = \int_{0}^{\infty} t\frac{dQ(t)}{dt}\,dt$$

Using integration by parts and the fact that $dQ(t)/dt = -dR(t)/dt$ we can show that

$$MTTF = \int_{0}^{\infty} t\frac{dQ(t)}{dt}\,dt = -\int_{0}^{\infty} t\frac{dR(t)}{dt}\,dt = \left[-tR(t) + \int R(t)\,dt\right]\Big|_{0}^{\infty} = \int_{0}^{\infty} R(t)\,dt$$

Consequently, the MTTF is defined in terms of the reliability function as

$$MTTF = \int_{0}^{\infty} R(t)\,dt$$

which is valid for any reliability function that satisfies $R(\infty) = 0$.

The mean time to repair (MTTR) is simply the average time required to repair a system. The MTTR is extremely difficult to estimate and is often determined experimentally by injecting a set of faults, one at a time, into a system and measuring the time required to repair the system in each case. The MTTR is normally specified in terms of a repair rate, μ, which is the average number of repairs that occur per time period. The units of the repair rate are normally number of repairs per hour. The MTTR and the rate, μ, are related by

$$MTTR = \frac{1}{\mu}$$

It is very important to understand the difference between the MTTF and the mean time between failure (MTBF). Unfortunately, these two terms are often used interchangeably. While the numerical difference is small in many cases, the conceptual difference is very important. The MTTF is the average time until the first failure of a system, while the MTBF is the average time between failures of a system. If we assume that all repairs to a system make the system perfect once again just as it was when it was new, the relationship between the MTTF and the MTBF can be determined easily. Once successfully placed into operation, a system will operate, on the average, a time corresponding to the MTTF before encountering the first failure. The system will then require some time, MTTR, to repair the system and place it back into operation once again. The system will then be perfect once again and will operate for a time corresponding to the MTTF before encountering its next failure. The time between the two failures is the sum of the MTTF and the MTTR and is the MTBF. Thus, the difference between the MTTF and the MTBF is the MTTR. Specifically, the MTBF is given by

$$\text{MTBF} = \text{MTTF} + \text{MTTR}$$

In most practical applications the MTTR is a small fraction of the MTTF, so the approximation that the MTBF and MTTF are equal is often quite good. Conceptually, however, it is crucial to understand the difference between the MTBF and the MTTF.

An extremely important parameter in the design and analysis of fault-tolerant systems is fault coverage. The fault coverage available in a system can have a tremendous impact on the reliability, safety, and other attributes of the system. Fault coverage is mathematically defined as the conditional probability that, given the existence of a fault, the system recovers [Bouricius et al., 1969]. The fundamental problem with fault coverage is that it is extremely difficult to calculate. Probably the most common approach to estimating fault coverage is to develop a list all of the faults that can occur in a system and to form, from that list, a list of faults from which the system can recover. The fault coverage factor is then calculated appropriately.

Reliability is perhaps one of the most important attributes of systems. The reliability of a system is generally derived in terms of the reliabilities of the individual components of the system. The two models of systems that are most common in practice are the series and the parallel. In a series system, each element of the system is required to operate correctly for the system to operate correctly. In a parallel system, on the other hand, only one of several elements must be operational for the system to perform its functions correctly.

The series system is best thought of as a system that contains no redundancy; that is, each element of the system is needed to make the system function correctly. In general, a system may contain N elements, and in a series system each of the N elements is required for the system to function correctly. The reliability of the series system can be calculated as the probability that none of the elements will fail. Another way to look at this is that the reliability of the series system is the probability that all of the elements are working properly. The reliability of a series system is given by

$$R_{\text{series}}(t) = R_1(t)R_2(t)\ldots R_N(t)$$

or

$$R_{\text{series}}(t) = \prod_{i=1}^{N} R_i(t)$$

An interesting relationship exists in a series system if each individual component satisfies the exponential failure law. Suppose that we have a series system made up of N components, and each component, i, has a constant failure rate of λ_i. Also assume that each component satisfies the exponential failure law. The reliability of the series system is given by

$$R_{\text{series}}(t) = e^{-\lambda_1 t} e^{-\lambda_2 t} \ldots e^{-\lambda_N t}$$

$$R_{\text{series}}(t) = e^{-\sum_{i=1}^{N} \lambda_i t}$$

The distinguishing feature of the basic parallel system is that only one of N identical elements is required for the system to function. The reliability of the parallel system can be written as

$$R_{\text{parallel}}(t) = 1.0 - Q_{\text{parallel}}(t) = 1.0 - \prod_{i=1}^{N} Q_i(t) = 1.0 - \prod_{i=1}^{N} (1.0 - R_i(t))$$

It should be noted that the equations for the parallel system assume that the failures of the individual elements that make up the parallel system are independent.

M-of-N systems are a generalization of the ideal parallel system. In the ideal parallel system, only one of N modules is required to work for the system to work. In the M-of-N system, however, M of the total of N identical modules are required to function for the system to function. A good example is the TMR configuration where two of the three modules must work for the majority voting mechanism to function properly. Therefore, the TMR system is a 2-of-3 system.

In general, if there are N identical modules and M of those are required for the system to function properly, then the system can tolerate $N - M$ module failures. The expression for the reliability of an M-of-N system can be written as

$$R_{\text{M-of-N}}(t) = \sum_{i=0}^{N-M} \binom{N}{i} R^{N-i}(t)(1.0 - R(t))^i$$

where

$$\binom{N}{i} = \frac{N!}{(N-i)!\,i!}$$

The **availability**, $A(t)$, of a system is defined as the probability that a system will be available to perform its tasks at the instant of time t. Intuitively, we can see that the availability can be approximated as the total time that a system has been operational divided by the total time elapsed since the system was initially placed into operation. In other words, the availability is the percentage of time that the system is available to perform its expected tasks. Suppose that we place a system into operation at time $t = 0$. As time moves along, the system will perform its functions, perhaps fail, and hopefully be repaired. At some time $t = t_{\text{current}}$, suppose that the system has operated correctly for a total of t_{op} hours and has been in the process of repair or waiting for repair to begin for a total of t_{repair} hours. The time t_{current} is then the sum of t_{op} and t_{repair}. The availability can be determined as

$$A(t_{\text{current}}) = \frac{t_{\text{op}}}{t_{\text{op}} + t_{\text{repair}}}$$

where $A(t_{\text{current}})$ is the availability at time t_{current}.

If the average system experiences N failures during its lifetime, the total time that the system will be operational is $N(\text{MTTF})$ hours. Likewise, the total time that the system is down for repairs is $N(\text{MTTR})$ hours. In other words, the operational time, t_{op}, is $N(\text{MTTF})$ hours and the downtime, t_{repair}, is $N(\text{MTTR})$ hours. The average, or steady-state, availability is

$$A_{\text{SS}} = \frac{N(\text{MTTF})}{N(\text{MTTF}) + N(\text{MTTR})}$$

We know, however, that the MTTF and the MTTR are related to the failure rate and the repair rate, respectively, for simplex systems, as

$$\text{MTTF} = \frac{1}{\lambda}$$

$$\text{MTTR} = \frac{1}{\mu}$$

Therefore, the steady-state availability is given by

$$A_{ss} - \frac{1/\lambda}{1/\lambda + 1/\mu} = \frac{1}{1 + \lambda/\mu}$$

Defining Terms

Availability, $A(t)$: The probability that a system is operating correctly and is available to perform its functions at the instant of time *t.*

Dependability: The quality of service provided by a particular system.

Error: The occurrence of an incorrect value in some unit of information within a system.

Failure: A deviation in the expected performance of a system.

Fault: A physical defect, imperfection, or flaw that occurs in hardware or software.

Fault avoidance: A technique that attempts to prevent the occurrence of faults.

Fault tolerance: The ability to continue the correct performance of functions in the presence of faults.

Maintainability, $M(t)$: The probability that an inoperable system will be restored to an operational state within the time *t.*

Performability, $P(L,t)$: The probability that a system is performing at or above some level of performance, *L,* at the instant of time *t.*

Reliability, $R(t)$: The conditional probability that a system has functioned correctly throughout an interval of time, $[t_0, t]$, given that the system was performing correctly at time t_0.

Safety, $S(t)$: The probability that a system will either perform its functions correctly or will discontinue its functions in a well-defined, safe manner.

References

A. Avizienis, "Arithmetic error codes: Cost and effectiveness studies for application in digital system design," *IEEE Transactions on Computers,* vol. C-20, no. 11, pp. 1322–1331, November 1971.

W.G. Bouricius, W.C. Carter, and P.R. Schneider, "Reliability modeling techniques for self-repairing computer systems," in *Proceedings of the 24th ACM Annual Conference,* pp. 295–309, 1969.

L. Chen and A. Avizienis, "N-version programming: A fault tolerant approach to reliability of software operation," in *Proceedings of the International Symposium on Fault Tolerant Computing,* pp. 3–9, 1978.

R.W. Hamming, "Error detecting and error correcting codes," *Bell System Technical Journal,* vol. 26, no. 2, pp. 147–160, April 1950.

B.W. Johnson, *Design and Analysis of Fault-Tolerant Digital Systems,* Reading, Mass.: Addison-Wesley, 1989.

J-C. Laprie, "Dependable computing and fault tolerance: Concepts and terminology," in *Proceedings of the 15th Annual International Symposium on Fault-Tolerant Computing,* Ann Arbor, Mich.: pp. 2–11, June 19–21, 1985.

S. Lin and D.J. Costello, Jr., *Error Control Coding: Fundamentals and Applications,* Englewood Cliffs, N.J.: Prentice-Hall, 1983.

V.P. Nelson and B.D. Carroll, *Tutorial: Fault-Tolerant Computing,* Washington, D.C.: IEEE Computer Society Press, 1986.

M.L. Shooman, *Probabilistic Reliability: An Engineering Approach,* New York: McGraw-Hill, 1968.

Further Information

The *IEEE Transactions on Computers, IEEE Computer* magazine, and the *Proceedings of the IEEE* have published numerous special issues dealing exclusively with fault tolerance technology. Also, the IEEE International Symposium on Fault-Tolerant Computing has been held each year since 1971. Finally, the following textbooks are available, in addition to those referenced above:

P.K. Lala, *Fault Tolerant and Fault Testable Hardware,* Englewood Cliffs, N.J.: Prentice-Hall, 1985.

D.K. Pradhan, *Fault-Tolerant Computing: Theory and Techniques,* Englewood Cliffs, N.J.: Prentice-Hall, 1986.

D.P. Siewiorek and R. S. Swarz, *The Theory and Practice of Reliable Systems Design,* 2nd ed., Bedford, Mass.: Digital Press, 1992.

16

Knowledge Engineering

M. Abdelguerfi
University of New Orleans

R. Eskicioglu
University of Alberta

Jay Liebowitz
Johns Hopkins University

16.1 Databases

M. Abdelguerfi and R. Eskicioglu

In the past, file processing techniques were used to design information systems. These systems usually consist of a set of files and a collection of application programs. Permanent records are stored in the files, and application programs are used to update and query the files. The application programs were in general developed individually to meet the needs of different groups of users. In many cases, this approach leads to a duplication of data among the files of different users. Also, the lack of coordination between files belonging to different users often leads to a lack of data consistency. In addition, changes to the underlying data requirements usually necessitate major changes to existing application programs. Among other major problems that arise with the use of file processing techniques are lack of data sharing, reduced programming productivity, and increased program maintenance. Because of their inherent difficulties and lack of flexibility, file processing techniques have lost a great deal of their popularity and are being replaced by **database management systems (DBMS)**.

A DBMS is designed to efficiently manage a shared pool of interrelated data (**database**). This includes the existence of features such as a *data definition language* for the definition of the logical structure of the database (*database schema*), a *data manipulation language* to query and update the database, a *concurrency control* mechanism to keep the database consistent when shared by several users, a *crash recovery* strategy to avoid any loss of information after a system crash, and *safety* mechanisms against any unauthorized access.

Database Abstraction

A DBMS is expected to provide for *data independence*, i.e., user requests are made at a *logical level* without any need for the knowledge of how the data is stored in actual files. This implies that the internal file structure could be modified without any change to the user's perception of the database. To achieve data independence, the Standards Planning and Requirements Committee (SPARC) of the American National Standards Institute (ANSI) in its 1977 report recommended three levels of database abstraction (see Figure 16.1). The lowest level in the abstraction is the internal level. Here, the database is viewed as a collection of files organized according to one of several possible internal data organizations (e.g., B^+-tree data organization). In the conceptual level, the database is viewed at an abstract level. The user at this level is shielded from the internal storage details. At the external level, each group of users has their own perception or *view* of the database. Each view is derived from the conceptual database and is designed to meet the needs of a particular group of users. Such a group can only have access to the data specified by its particular view. This, of course, ensures both privacy and security.

The mapping between the three levels of abstraction is the task of the DBMS. When changes to the internal level (such as a change in file organization) do not affect the conceptual and external levels, the system is said to provide for *physical data independence. Logical data independence* prevents changes to the conceptual level to affect users' views. Both types of data independence are desired features in a database system.

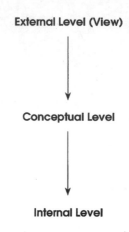

FIGURE 16.1   Data abstraction.

Data Models

A **data model** refers to an integrated set of tools used to describe the data and its structure, data relationships, and data constraints. Some data models provide a set of operators that is used to update and query the database. Data models can be classified in two main categories: *record-based* and *object-based*. Both classes are used to describe the database at the conceptual and external levels. With object-based data models, constraints on the data can be specified more explicitly.

There are three main record-based data models: the *relational, network,* and *hierarchical* models. In the relational model, data at the conceptual level is represented as a collection of interrelated tables. These tables are *normalized* so as to minimize data redundancy and update anomalies. In this model, data relationships are implicit and are derived by matching columns in tables. In the hierarchical and network models, the data is represented as a collection of records and data relationships are explicit and are represented by *links*. The difference between the last two models is that in the hierarchical model, data is represented as a tree structure, while it is represented as a generalized graph in the network model.

In hierarchical and network models, the existence of physical pointers (links) to link related records allows an application program to retrieve a single record at a time by following the pointer's chain. The process of following the pointer's chain and selecting one record at a time is referred to as *navigation.* In nonnavigational models such as the relational model, records are not related through pointer's chains, but relationships are established by matching columns in different tables.

The hierarchical and network models require the application programmer to be aware of the internal structure of the database. The relational model, on the other hand, allows for a high degree of physical and logical data independence. Earlier DBMSs were for the most part navigational systems. Because of its simplicity and strong theoretical foundations, the relational model has since received wide acceptance. Today, most DBMSs are based on the relational model.

Other data models include a popular high level conceptual data model, known as the *Entity-Relationship* (ER) model. The ER model is mainly used for the conceptual design of databases and their applications. The ER model describes data as entities, attributes, and relationships.

An *entity* is an "object" in the real world with an independent existence. Each entity has a set of properties, called *attributes,* that describes it. A *relationship* is an association between entities. For example, a professor entity may be described by its name, age, and salary and can be associated with a department entity by the relationship "works for".

With the advent of advanced database applications, the ER modeling concepts became insufficient. This has led to the enhancement of the ER model with additional concepts, such as generalization, categories, and inheritance, leading to the *Enhanced-ER* or *EER* model.

FAC_INFO (lname,	social_sec#,	street,	city,	dept)
Hosch	383909164	Esplanade	Kenner	CS
Loggins	482233364	Bonnabel	Metairie	EE
Martin	399254402	Williams	Kenner	CH
Krad	100995678	Bourbon	New Orleans	EE
Hanoura	400919945	Bonnabel	Metairie	CH
Prados	388998800	Severn	Matairie	EE
Abdel	389390164	St Charles	New Orleans	CS

DEPT_CHAIR (dept,	chair)
CS	Hosch
EE	Prados
CH	Martin

FIGURE 16.2   An example of two relations: FAC_INFO and DEP_CHAIR.

Relational Databases

The relational model was introduced by E.F. Codd [1970]. Since the theoretical underpinnings of the relational model have been well defined, it has become the focus of most commercial DBMSs.

In the relational model, the data is represented as a collection of relations. To a large extent, each relation can be thought of as a table. The example of Figure 16.2 shows part of a university database composed of two relations. FAC_INFO gives personal information (last name, social security, street and city of residence, and department) of a faculty. DEP_CHAIR gives the last name of the chairman of each department. A faculty is not allowed to belong to two departments. Each row in a relation is referred to as a *tuple*. A column name is called an *attribute name.* The data type of each attribute name is known as its *domain*. A *relation scheme* is a set of attribute names. For instance, the relation scheme (or scheme for short) of the relation FAC_INFO is (lname, social_sec(, street, city, dept). A *key* is a set of attribute names whose composite value is distinct for all tuples. In addition, no proper subset of the key is allowed to have this property. It is not unusual for a scheme to have several possible keys. In FAC_INFO, both lname and social_sec(are possible keys. In this case, each possible key is known as a *candidate key,* and the one selected to act as the relation's key, say, lname, is referred to as the *primary key*. A *superkey* is a key with the exception that there is no requirement for minimality. In a relation, an attribute name (or a set of attribute names) is referred to as a *foreign key,* if it is the primary key of another relation. In FAC_INFO, the attribute name dept is a foreign key, since the same attribute is a key in DEP_CHAIR. Because of updates to the database, the content of a relation is dynamic. For this reason, the data in a relation at a given time instant is called an *instance* of the relation.

There are three integrity constraints that are usually imposed on each instance of a relation: primary key integrity, entity integrity, and referential integrity. The key integrity constraint requires that no two tuples of a relation have the same key value. The entity integrity constraint specifies that the key value of each tuple should have a known value (i.e., no *null* values are allowed for primary keys). The referential integrity constraint specifies that if a relation r_1 contains a foreign key that matches the primary key of a relation r_2, then each value of the foreign key in r_1 must either match a value of the primary key in r_2 or must be null. For the database of Figure 16.2 to be consistent, each value of dept in FAC_INFO must match a value of dept in DEP_CHAIR.

Relational Database Design

The relational database design [Maier, 1983] refers to the process of generating a set of relation schemes that minimizes data redundancy and removes update anomalies. One of the most popular approaches is the use of the *normalization theory*. The normalization theory is based on the notion of *functional dependencies*.

PRODUCT (supplier_name,	product_name,	price,	location)
Martin	sofa	500.99	Kenner
Martin	bed	100.99	Kenner
Martin	desk	150.99	Kenner
Evans	sofa	600.99	Metairie
Evans	desk	250.99	Metairie
Rudd	bed	110.99	Metairie

FIGURE 16.3   Instance of PRODUCT (supplier_name, product_name, price, quantity).

Functional dependencies are constraints imposed on a database. The notion of superkey, introduced in the previous section, can be formulated as follows: A subset of a relation scheme is a superkey if, in any instance of the relation, no two distinct tuples have the same superkey value. If $r(R)$ is used to denote a relation r on a schema R, $K \subseteq R$ a superkey, and $t(k)$ the K-value of tuple t, then no two tuples t_1 and t_2 in $r(R)$ are such that $t_1(K) = t_2(K)$.

The notion of a functional dependency can be seen as a generalization of the notion of superkey. Let X and Y be two subsets of R; the functional dependency $X \rightarrow Y$ exists in $r(R)$ if whenever two tuples in $r(R)$ have the same X-value, their Y-value is also the same. That is, if $t_1(X) = t_2(X)$, then $t_1(Y) = t_2(Y)$. Using functional dependencies, one can define the notion of a key more precisely. A key k of a relation $r(R)$ is such that $k \rightarrow R$ and no proper subset of k has this property. Note that if the schema R is composed of attribute names $\{A_1, A_2, ..., A_n\}$, then each attribute name A_i is functionally determined by the key k, i.e., $k \rightarrow A_i$, $i = 1, ..., n$. An attribute name that is part of a key is referred to as a *prime attribute*. In the example of Figure 16.2, both attribute names street and city are nonprime attributes.

The normalization process can be thought of as the process of decomposing a scheme with update anomalies and data redundancy into smaller schemes in which these undesirable properties are to a large extent eliminated. Depending on the severity of these undesirable properties, schemes are classified into *normal forms*. Originally, Codd defined three normal forms: *first normal form* (1NF), *second normal form* (2NF), and *third normal form* (3NF). Thereafter, a stronger version of the 3NF, known as Boyce-Codd normal form (BCNF), was suggested. These four normal forms are based on the concept of functional dependencies.

The 1NF requires that attribute name values be *atomic*. That is, composite values for attribute names are not allowed. A 2NF scheme is a 1NF scheme in which all nonprime attributes are fully dependent on the key. Consider the relation of Figure 16.3. Each tuple in PRODUCT gives the name of a supplier, a product name, its price, and the supplier's location. The scheme (supplier_name, product_name, price, quantity) is in 1NF since each attribute name is atomic. It is assumed that many products can be supplied by a single supplier, that a given product can be supplied by more than one supplier, and that a supplier has only one location. So, (supplier_name, product_name) is the relation's key and the functional dependency supplier_name → location should hold for any instance of PRODUCT.

The structure of the relation of Figure 16.3 does not allow a supplier to appear in the relation unless it offers at least one product. Even the use of null values is not of much help in this case as product_name is part of a key and therefore cannot be assigned a null value. Another anomaly can be encountered during the deletion process. For instance, deleting the last tuple in the relation results in the loss of the information that Rudd is a supplier located in Metairie. It is seen that the relation PRODUCT suffers from insertion and deletion anomalies.

Modifications can also be a problem in the relation PRODUCT. Suppose that the location of the supplier Martin is moved from Kenner to Slidell. In order not to violate the functional dependency supplier_name location, the location attribute name of all tuples where the supplier is Martin needs to be changed from Kenner to Slidell. This modification anomaly has a negative effect on performance.

In addition, the relation PRODUCT suffers from data redundancy. For example, although Martin has only one location "Kenner", such a location appears in all three tuples where the supplier_name is Martin.

The update anomalies and data redundancy encountered in PRODUCT are all due to the functional dependency supplier_name → location. The right-hand side of this dependency "location" is a nonprime attribute, and the left-hand side represents part of the key. Therefore, we have a nonprime attribute that is

PRO_INFO (*supplier_name,*   *product_name,*   *price*)

supplier_name	product_name	price
Martin	sofa	500.99
Martin	bed	100.99
Martin	desk	150.99
Evans	sofa	600.99
Evans	desk	250.99
Rudd	bed	110.99

SUP_LOC (*supplier_name,*   *location*)

supplier_name	location
Martin	Kenner
Evans	Metairie
Rudd	Metairie

FIGURE 16.4   Decomposition of PRODUCT into PRO_INFO and SUP_LOC.

SUPPLIES (*client_name,*   *supplier_name,*   *location*)

client_name	supplier_name	location
Hosch	Martin	Kenner
Krad	Martin	Kenner
Shengru	Evans	Metairie
Tillis	Rudd	Metairie
Greene	Evans	Metairie

FIGURE 16.5   Instance of SUPPLIES.

only partially dependent on the key (supplier_name, product_name). As a consequence, the schema (supplier_name, product_name, price, location) is not in 2NF. The removal of the partial dependency supplier_name → location will eliminate all the above anomalies. The removal of the partial dependency is achieved by decomposing the scheme (supplier_name, product_name, price, quantity) into two 2NF schemes: (supplier_name, product_name, price), and (supplier_name, location). This decomposition results in relations PRO_INFO and SUP_LOC shown in Figure 16.4. The keys of PRO_INFO and SUP_LOC are (supplier_name, product_name), and supplier_name, respectively.

Normalizing schemes into 2NF removes all update anomalies due to nonprime attributes being partially dependent on keys. Anomalies of a different nature, however, are still possible.

Update anomalies and data redundancy can originate from *transitive dependencies*. A nonprime attribute A_i is said to be transitively dependent on a key k via attribute name A_j, if $k \rightarrow A_j$, $A_j \rightarrow A_i$, and A_j does not functionally determine A_k. A 3NF is a 1NF where no nonprime attribute is transitively dependent on a key.

The relation of Figure 16.5, which is in 2NF, highlights update anomalies and data redundancy due to the transitive dependency of a nonprime attribute on a key. The relation gives the name of a client (client_name), the corresponding supplier (supplier_name), and the supplier's location. Each client is assumed to have one supplier. The relation's key is client_name, and each supplier has only one location. A supplier and his location cannot be inserted in SUPPLIES unless the supplier has at least one client. In addition, the relation has a deletion anomaly since if Tillis is no longer a client of Rudd, the information about Rudd as a supplier and his location is lost. A change to a supplier's location may require updating the location attribute name of several tuples in the relation. Also, although each supplier has only one location, such a location is sometimes repeated several time unnecessarily, leading to data redundancy.

The relation exhibits the following transitive dependency: client_name → supplier_name, supplier_name → location (but not the inverse). The relation CLIENT is clearly in 2NF, but because of the transitive dependency of the nonprime attribute location on the key, it is not in 3NF. This is the cause of the anomalies mentioned above. Eliminating this transitive dependency by splitting the schema into two components will remove these anomalies. Clearly, the resulting two relations SUP_CLI and SUP_LOC are in 3NF (see Figure 16.6).

SUP_CLI (client_name,	supplier_name)	SUP_LOC (supplier_name,	location)
Hosch	Martin	Martin	Kenner
Krad	Martin	Evans	Metairie
Shengru	Evans	Rudd	Metairie
Tillis	Rudd		
Greene	Evans		

FIGURE 16.6 Decomposition of SUPPLIES into SUP_CLI and SUP_LOC.

Each partial dependency of a nonprime attribute on a key can be expressed as a transitive dependency of a nonprime attribute on a key. Therefore, a scheme in 3NF is also in 2NF.

BCNF is a stricter form of 3NF, where a relation r on a schema R is in BCNF if whenever a functional dependency $X \rightarrow Y$ exists in $r(R)$, then X is a superkey of R. The condition of 3NF, which allows Y to be prime if X is not a superkey, does not exist in BCNF. Thus, every scheme in BCNF is also in 3NF, but the opposite is not always true.

A detailed discussion of higher level normalizations, such as 4NF and 5NF, which are based on other forms of dependencies, can be found in Elmasri and Navathe [1994].

Data Definition and Manipulation in Relational Databases

Upon completion of the relational database design, a descriptive language, usually referred to as Data Definition Language (DDL), is used to define the designed schemes and their relationships. The DDL can be used to create new schemes or modify existing ones, but it cannot be used to query the database. Once DDL statements are compiled, they are stored in the *data dictionary*. A data dictionary is a repository where information about database schemas, such as attribute names, indexes, and integrity constraints are stored. Data dictionaries also contain other information about databases, such as design decisions, usage standards, application program descriptions, and user information. During the processing of a query, the DBMS usually checks the data dictionary. The data dictionary can be seen as a relational database of its own. As a result, data manipulation languages that are used to manipulate databases can also be used to query the data dictionary.

An important function of a DBMS is to provide a Data Manipulation Language (DML) with which a user can retrieve, change, insert, and delete data from the database. DMLs are classified into two types: *procedural* and *nonprocedural*. The main difference between the two types is that in procedural DMLs, a user has to specify the desired data and how to obtain it, while in nonprocedural DMLs, a user has only to describe the desired data. Because they impose less burden on the user, nonprocedural DMLs are normally easier to learn and use.

The component of a DML that deals with data retrieval is referred to as *query language*. A query language can be used interactively in a stand-alone manner, or it can be embedded in a general-purpose programming language such as C and Cobol.

One of the most popular query languages is SQL (Structured Query Language). SQL is a query language based to a large extent on Codd's *relational algebra*. SQL has additional features for data definition and update. Therefore, SQL is a comprehensive relational database language that includes both a DDL and DML.

SQL includes the following commands for data definition: CREATE TABLE, DROP TABLE, and ALTER TABLE. The CREATE TABLE is used to create and describe a new relation. The two relations of Figure 16.4 can be created in the following manner:

```
CREATE TABLE PRO_INFO   (supplier_name  VARCHAR(12)   NOT NULL,
                         product_name   VARCHAR(8)    NOT NULL,
                         price          DECIMAL(6,2));
CREATE TABLE SUP_LOC    (supplier_name  VARCHAR(12)   NOT NULL,
                         location       VARCHAR(10));
```

The CREATE TABLE command specifies all the attribute names of a relation and their data types (e.g., INTEGER, DECIMAL, fixed length character "CHAR", variable length character "VARCHAR", DATE).

The constraint NOT NULL is usually specified for those attributes that cannot have null values. The primary key of each relation in the database is usually required to have a nonnull value.

If a relation is created incorrectly, it can be deleted using the DROP TABLE command. The command is DROP TABLE followed by the name of the relation to be deleted. A variation of DROP command, DROP SCHEMA, is used if the whole schema is no longer needed.

The ALTER TABLE is used to add new attribute names to an existing relation, as follows:

```
ALTER TABLE SUP_LOC ADD zip_code CHAR(5);
```

The SUP_LOC relation now contains an extra attribute name, zip_code. In most DBMSs, the zip_code value of existing tuples will automatically be assigned a null value. Other DBMSs allow for the assignment of an initial value to a newly added attribute name. Also, definitions of attributes can be changed and new constraints can be added, or current constraints can be dropped.

The DML component of SQL has one basic query statement, sometimes called a mapping, that has the following structure:

```
SELECT   <attribute_name list>
FROM     <relation_list>
WHERE    <restriction>
```

In the above statement, the SELECT clause specifies the attribute names that are to be retrieved, FROM gives the list of the relations involved, and WHERE is a Boolean predicate that completely specifies the tuples to be retrieved.

Consider the database of Figure 16.4, and suppose that we want the name of all suppliers that supply either beds or desks. In SQL, this query can be expressed as:

```
SELECT   supplier_name
FROM     PRO_INFO
WHERE    product_name = ''bed'' OR product_name = ''sofa''
```

The result of an SQL command "may contain duplicate values and is therefore not always a true relation. In fact, the result of the above query, shown below, has duplicate entries.

```
supplier_name
Martin
Martin
Rudd
```

The entry Martin appears twice in the result, because the supplier Martin supplies both beds and sofas. Removal of duplicates is usually a computationally intensive operation. As a result, duplicate entries are not automatically removed by SQL. To ensure uniqueness, the command DISTINCT should be used. In the above query, if we want the supplier names to be listed only once, the above query should be modified as follows:

```
SELECT DISTINCT  supplier_name
FROM             PRO_INFO
WHERE            product_name = ''bed'' OR product_name = ''sofa''
```

In SQL, a query can involve more than one relation. Suppose that we want the list of all suppliers from Metairie who supply beds. Such a query, shown below, involves both PRO_INFO and SUP_LOC.

```
SELECT   supplier_name
FROM     PRO_INFO, SUP_LOC
WHERE    PRO_INFO.supplier_name = SUP_LOC.supplier_name
              AND product_name = ''bed''
```

PROFESSOR (faculty,	department,	salary)
Smith	Electrical Eng.	$39,000
Joe	Mechanical Eng.	$35,000
Susan	Computer Sci.	$36,000
Erick	Electrical Eng.	$38,000
Paul	Electrical Eng.	$37,000
Johannes	Computer Sci.	$65,000
Rick	Computer Sci.	$32,000
Gerard	Computer Sci.	$43,000
Kenneth	Mechanical Eng.	$40,000

FIGURE 16.7   Instance of the relation PROFESSOR.

When an SQL expression, such as the one above, involves more than one relation, it is sometimes necessary to qualify attribute names, that is, to precede an attribute name by the relation (a period is placed between the two) it belongs to. Such a qualification removes possible ambiguities.

In SQL, it is possible to have several levels of query nesting; this is done by including a SELECT query statement within the WHERE clause.

The output data can be presented in sorted order by using the SQL ORDER BY clause followed by the attribute name(s) according to which the output is to be sorted.

In database management applications it is often desirable to categorize the tuples of a relation by the values of a set of attributes and extract an aggregated characteristic of each category. Such database management tasks are referred to as *aggregation functions*. For instance, SQL includes the following built-in aggregation functions: SUM, COUNT, AVERAGE, MIN, MAX. The attribute names used for the categorization are referred to as GROUP BY columns. Consider the relation PROFESSOR of Figure 16.7. Each tuple of the above relation gives the name of a faculty and his department and academic year salary.

Suppose that we want to know the number of faculty in each department and the result to be ordered by department. This query requests for each department a count of the number of faculty. Faculty are therefore categorized according to the attribute name department. As a result, department is referred to as a GROUP BY attribute. In SQL, the above query is formulated as follows:

```
SELECT      department, COUNT (faculty)
FROM        PROFESSOR
GROUP BY    department
ORDER BY    department
```

The result of applying the COUNT aggregation function is a new relation with two attribute names. They are a GROUP BY attribute (department in this case) and a new attribute called COUNT. The tuples are ordered lexicographically in ascending order according to the ORDER BY attribute, which is department in this case:

```
department              COUNT (faculty)
Computer Sc.                 4
Electrical Eng.              3
Mechanical Eng.              2
```

The relations created through the CREATE TABLE command are known as *base relations*. A base relation exists physically and is stored as a file by the DBMS. SQL can be used to create views using the CREATE VIEW command. In contrast to base relations, the creation of a view results in a *virtual relation*, that is, one that does not necessarily correspond to a physical file. Consider the database of Figure 16.4, and suppose that we want to create a view giving the name of all suppliers located in Metairie, the products each one provides, and the corresponding prices. Such a view, called METAIRIE_SUPPLIER, can be created as follows:

```
CREATE VIEW    METAIRIE_SUPPLIER
AS SELECT      PRO_INFO.supplier_name, product_name, price
```

FROM	PRO_INFO, SUP_LOC
WHERE	PRO_INFO.supplier_name = SUP_LOC.supplier_name
	AND location = ``Metairie''

Because a view is a virtual relation that can be constructed from one or more relations, updating a view may lead to ambiguities. As a result, when a view is generated from more than one relation, there are, in general, restrictions on updating such a view.

Hierarchical Databases

The hierarchical data model [Elmasri and Navathe, 1994] uses a tree data structure to conceptualize associations between different record types. In this model, record types are represented as nodes and associations as links. Each record type, except the root, has only one parent; that is, only parent-child (or one-to-many) relationships are allowed. This restriction gives hierarchical databases their simplicity. Since links are only one way, from a parent to a child, the design of hierarchical database management systems is made simpler, and only a small set of data manipulation commands are needed.

Because only parent-child relationships are allowed, the hierarchical model cannot efficiently represent two main types of relationships: many-to-many relationships and the case where a record type is a child in more than one *hierarchical schema*. These two restrictions can be handled by allowing redundant *record instances*. However, such a duplication requires that all the copies of the same record should be kept consistent at all times.

The example of Figure 16.8 shows a hierarchical schema. The schema gives the relationship between a DEPARTMENT, its employees (D_EMPLOYEE), the projects (D_PROJECT) handled by the different departments, and how employees are assigned to these projects. It is assumed that an employee belongs to only one department, a project is handled by only one department, and an employee can be assigned to several projects. Notice that since a project has several employees assigned to it, and an employee can be assigned to more than one project, the relationship between D_PROJECT and D_EMPLOYEE is many-to-many. To model this relationship multiple instances of the same record type D-EMPLOYEE may appear under different projects.

Such redundancies can be reduced to a large extent through the use of *logical links*. A logical link associates a *virtual record* from a hierarchical schema with an actual record from either the same schema or another schema. The redundant copy of the actual record is therefore replaced by a virtual record, which is nothing more than a pointer to the actual one.

Hierarchical DLLs are used by a designer to declare the different hierarchical schemas, record types, and logical links. Furthermore, a root node must be declared for each hierarchical schema, and each record type declaration must also specify the parent record type.

Unlike relational DMLs, hierarchical DMLs such as DL/1 are record at-a-time languages. DL/1 is used by IBM's IMS hierarchical DBMS. In DL/1 a tree traversal is based on a preorder algorithm, and within each tree, the last record accessed through a DL/1 command can be located through a *currency indicator*.

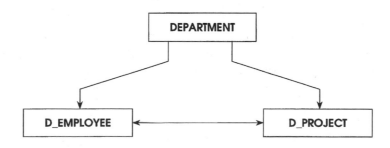

FIGURE 16.8 A hierarchical schema.

Retrieval commands are of three types:

GET UNIQUE <record type> **WHERE** <restrictions>
Such a command retrieves the leftmost record that meets the imposed restrictions. The search always starts at the root of the tree pointed to by the currency indicator.

GET NEXT [<record type> **WHERE** <restrictions>]
Starting from the current position, this command uses the preodrer algorithm to retrieve the next record that satisfies the restrictions. The clause enclosed between brackets is optional. GET NEXT is used to retrieve the next (preorder) record from the current position.

GET NEXT WITHIN PARENT [<record type> **WHERE** <restrictions>]
It retrieves all records that have the same parent and that satisfy the restrictions. The parent is assumed to have been selected through a previous GET command.
Four commands are used for record updates:

INSERT
Stores a new record and links it to a parent. The parent has been already selected through a GET command.

REPLACE
The current record (selected through a previous GET command) is modified.

DELETE
The current record and all its descendants are deleted.

GET HOLD
Locks the current record while it is being modified.

The DL/1 commands are usually embedded in a general-purpose (host) language. In this case, a record accessed through a DL/1 command is assigned to a program variable.

Network Databases

In the network model [Elmasri and Navathe, 1994] associations between record types are less restrictive than with the hierarchy model. Here, associations among record types are represented as graphs.

One-to-one and one-to-many relationships are described using the notion of *set type*. Each set type has an *owner* record type and a *member* record type. In the example of Figure 16.8, the relationship between DEPARTMENT and employee (D_EMPLOYEE) is one-to-many. This relationship defines a set type where the owner record type is DEPARTMENT and the member record type is D_EMPLOYEE. Each instance of an owner record type along with all the corresponding member records represents a set instance of the underlying set type. In practice, a set is commonly implemented using a circular-linked list which allows an owner record to be linked to all its member records. The pointer associated with the owner record is known as the FIRST pointer, and the one associated with a member record is known as a NEXT pointer.

In general, a record type cannot be both the owner and a member of the same set type. Also, a record cannot exist in more than one instance of a specific set type. The latter requirement implies that many-to-many relationships are not directly implemented in the network data model.

The relationship between D_PROJECT and D-EMPLOYEE is many-to-many. In the network model, this relationship is represented by two set types and an intermediate record type. The new record type could be named ASSIGNED_TO (see Figure 16.9). One set has D_EMPLOYEE as owner and ASSIGNED_TO as member record type, and the other has D_PROJECT as owner and ASSIGNED_TO as member record type.

Standards for the network model's DDL and DML were originally proposed by the CODASYL (Conference On Data SYstems Languages) committee in 1971. Several revisions to the original proposal were made later.

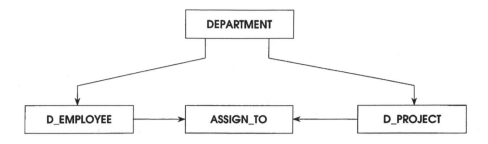

FIGURE 16.9   Representing many-to-many relationships in the network model.

In a network DDL, such as that of the IDMS database management system, a set declaration specifies the name of the set, its owner record type, and its member record type. The insertion mode for the set members needs to be specified using combinations of the following four commands:

AUTOMATIC
An inserted record is automatically connected to the appropriate set instance.

MANUAL
In this case, records are inserted into the appropriate set instance by an application program.

OPTIONAL
A member record does not have to be a member of a set instance. The member record can be connected to or disconnected from a set instance using DML commands.

MANDATORY
A member record needs to be connected to a set instance. A member record can be moved to another set instance using the network's DML.

FIXED
A member record needs to be connected to a set instance. A member record cannot be moved to another set instance.

The network's DDL allows member records to be ordered in several ways. Member records can be sorted in ascending or descending order according to one or more fields. Alternatively, a new member record can be inserted next (prior) to the current record (pointed to by the currency indicator) in the set instance. A newly inserted member record can also be placed first (or last) in the set instance. This will lead to a chronological (or reverse chronological) order among member records.

As with the hierarchy model, network DMLs are record-at-a-time languages, and currency indicators are necessary for navigation through the network database. For example, the IDMS main data manipulation commands can be summarized as follows:

CONNECT
Connects a member record to the specified set instance.

DISCONNECT
A member record is disconnected from a set instance (set membership must be manual in this case).

STORE, MODIFY, and **DELETE**
These commands are used for data storage, modification, and deletion.

FIND
Retrieval command based on set membership.

GET
Retrieval command based on key values.

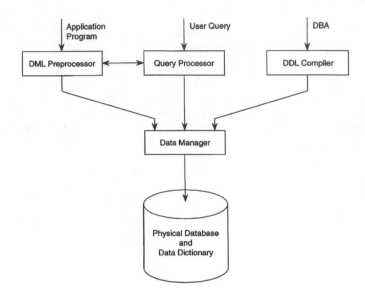

FIGURE 16.10   Simplified architecture of a DBMS.

Architecture of a DBMS

A DBMS is a complicated software structure that includes several components (see Figure 16.10). The DBMS has to interact with the operating system for secondary storage access. The *data manager* is usually the interface between the DBMS and the operating system. The *DDL compiler* converts schema definitions, expressed using DDL statements, into a collection of metadata tables that are stored in the data dictionary. The design of the schemas is the function of the *database administrator* (DBA). The DBA is also responsible for specifying the data storage structure and access methodology and granting and revoking access authorizations. The *query processor* converts high-level DML statements into low-level instructions that the database manager can interpret. The *DML preprocessor* separates embedded DML statements from the rest of an application program. The resulting DML commands are processed by a DML compiler, and the rest of the application program is compiled by a host compiler. The object codes of the two components are then linked.

Data Integrity and Security

Data Integrity

In general, during the design of a database schema several integrity constraints are identified. These constraints may include the uniqueness of a key value, restrictions on the domain of an attribute name, and the ability of an attribute to have a null value. A DBMS includes mechanisms with which integrity constraints can be specified. Constraints such as key uniqueness and the admissibility of null values can be specified during schema definition. Also, more elaborate integrity constraints can be specified. For example, constraints can be imposed on the domain of an attribute name, and any transaction that violates the imposed constraints is aborted. In some cases, it is useful to specify that the system take some actions, rather than just have the transaction responsible for the constraint violation being aborted. A mechanism called *trigger* can be used for that purpose. A trigger specifies a condition and an action to be taken when the condition is met.

Transactions and Data Integrity

In a multiuser DBMS, the database is a shared resource that can be accessed concurrently by many users. A *transaction* usually refers to the execution of a retrieval or an update program. A transaction performs a single logical operation in a database application. Therefore, it is an *atomic* unit of processing. That is, a

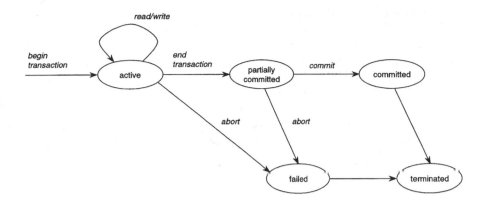

FIGURE 16.11 State transition diagram for transaction execution.

transaction is either performed in its entirety or is not performed at all. Basically, a transaction may be in one of the following states (Figure 16.11):

- active — where read and write operations are performed.
- partially committed — when the transaction ends and various checks are made to ensure that the transaction did not interfere with other transactions.
- failed — when one of the checks failed or the transaction is aborted during the active state.
- committed — when the execution was successfully completed.
- terminated — when the transaction leaves the system.

Transactions originating from different users may be aimed at the same database records. This situation, if not carefully monitored, may cause the database to become *inconsistent*. Starting from a database in a consistent state, it is obvious that if all transactions are executed one after the other, then the database will remain in a consistent state. In a multiuser DBMS, serial execution of transactions is wasteful of system resources. In this case, the solution is to interleave the execution of the transactions. However, the interleaving of transactions has to be performed in a way that prevents the database from becoming inconsistent. Suppose that two transactions T_1 and T_2 proceed in the following way:

Time	T_1	T_2
		read_account(X)
	read_account(X)	X := X − 20
	X := X − 10	write_account(X)
	write_account(X)	
	read_account(Y)	
	Y := Y + 10	
	write_account(Y)	

The first transaction transfers $10 from bank account X to bank account Y. The second transaction withdraws $20 from bank account X. Assume that initially there was $200 in X and $100 in Y. When the two transactions are performed serially, the final amounts in X and Y are $170 and $110, respectively. However, if the two transactions are interleaved as shown, then after the completion of both transactions, there will be $190 in X and $110 in Y. The database is now in an inconsistent state.

It is therefore important to ensure that the interleaving of the execution of transactions leaves the database in a consistent state. One way of preserving data consistency is to ensure that the interleaved execution of transactions is equivalent to their serial execution. This is referred to as *serializable* execution. Therefore, an interleaved execution of transactions is said to be serializable if it is equivalent to a serial execution.

Locking is one of the most popular approachs to achieving serializability. Locking is the process of ensuring that some actions are not performed on a data item. Therefore, a transaction may request a lock on a data item to prevent it from being either accessed or modified by other transactions. There are two basic types of locks. A *shared lock* allows other transactions to read but not write to the data item. An *exclusive lock* allows only a single transaction to read and write a data item. To achieve a high degree of concurrency, the locked data item size must be as small as possible. A data item can range from the whole database to a particular field in a record. Large data items limit concurrency, while small data items result in a large storage overhead and a greater number of lock and unlock operations that the system will have to handle.

Transactions scheduling based on locking achieves serializability in two phases. This is known as *two-phase locking*. During the first phase, the growing phase, a transaction can only lock new data items, but it cannot release any locked ones. During the second phase, the shrinking phase, existing locks can be released, but no new data item can be locked. The two-phase locking scheme guarantees the serializability of a schedule.

Because of its simplicity, the above scheduling method is very practical. However, it may lead to a *deadlock*. A deadlock occurs when two transactions are waiting for each other to release locks and both cannot proceed. A deadlock prevention (or detection) strategy is needed to handle the situation. For example, this can be achieved by requiring that a transactions locks all data items it needs for its execution before it can proceed; when the transaction finds that a needed data item is already locked, then it releases all locks.

If a transaction fails for whatever reason after (partially committed) or (active) while updating the database, it may be necessary to bring the database to its previous (original) state by undoing the transaction. This operation is called *roll-back*. A roll-back operation requires some information about the changes made on the data items during a transaction. Such information is usually kept outside the database in a system log. Generally, roll-back operations are part of the techniques used to recover from transaction failures.

Database Security

A database needs to be protected against unauthorized access. It is the responsibility of the DBA to create account numbers and passwords for legitimate users. The DBA can also specify the type of privileges a particular account has. In relational databases, this includes the privilege to create base relations, create views, alter relations by adding or dropping a column, and delete relations. The DBA can also revoke privileges that were granted previously. In SQL, the command GRANT is used to grant privileges and the REVOKE command to revoke privileges that have been granted.

The concept of views can serve as a convenient security mechanism. Consider a relation EMPLOYEE that gives the name of an employee, date of birth, the department worked for, address, phone number, and salary. A database user who is not allowed to have access to the salary of employees from his own department can have this portion of the database hidden from him. This can be achieved by limiting his access to a view obtained from the relation EMPLOYEE by selecting only those tuples where the department attribute is different from his.

Database security can be enhanced by using *data encryption*. The idea here is to encrypt the data using some coding technique. An unauthorized user will have difficulty deciphering the encrypted data. Only authorized users are provided with keys to decipher the encoded data.

Emerging Trends

Object-Oriented Databases

Object-oriented database systems (OODBMSs) [Brown, 1991] are one of the latest trends in database technology. The emergence of OODBMS is in response to the requirements of advanced applications. In general, traditional commercial and administrative applications can be effectively modeled using one of the three record-based data models. These applications are characterized by simple data types. Furthermore, for such applications, access and relationships are based on data values. Advanced database applications such as those found in engineering CAD/CAM require complex data structures. When these applications are modeled using the relational model, they require an excessive number of relations. In addition, a large number of complex operations are usually needed to produce an answer. This leads, in most cases, to unacceptable performance levels.

The notion of "object" is central to OODBMS. An object can be seen as being an entity consisting of its own *private memory* and *external interface* (or protocol). The private memory is used to store the state of the object, and the external interface consists of a set of operations that can be performed on the object. An object communicates with other objects through messages sent to its external interface. When an object receives a message, it responds by using its own procedures, known as *methods*. The methods are responsible for processing the data in the object's private memory and sending messages to other objects to perform specific tasks and possibly send back appropriate results.

The object-oriented approach provides for a high level of *abstraction*. In addition, this model has constructs that can be used to define new data types and specialized operators that can be applied to them. This feature is known as *encapsulation*.

An object is usually a member of a class. The class specifies the internal structure and the external interface of an object. New object classes can be defined as a *specialization* of existing ones. For example, in a university environment, the object type "faculty" can be seen as a specialization of the object type "employee." Since a faculty is a university employee, it has all the properties of a university employee plus some of its own. For example, some of the general operations that can be performed on an employee could be "raise_salary," "fire_employee," "transfer_employee." For a faculty, specialized operations such as "faculty_tenure" could be defined. Faculty can be viewed as a subclass of employee. As a result, faculty (the subclass) will respond to the same messages as employee (the superclass) in addition to those defined specifically for faculty. This technique is known as *inheritance*. A subclass is said to inherit the behavior of its superclass.

Opponents to the object-oriented paradigm point to the fact that while this model has greater modeling capability, it lacks the simplicity and the strong theoretical foundations of the relational model. Also, the reappearance of the navigational approach is seen by many as a step backward.

Supporters of the object-oriented approach believe that a navigational approach is a necessity in several applications. They point to the rich modeling capability of the model, its high level of abstraction, and its suitability for modular design.

Distributed Databases

A **distributed database** [Ozsu and Valdurez, 1991] is a collection of interrelated databases spread over the nodes of a computer network. The management of the distributed database is the responsibility of a software system usually known as distributed DBMS (DDBMS). One of the tasks of the DDBMS is to make the distributed nature of the database transparent to the user. A distributed database usually reflects the distributed nature of some applications. For example, a bank may have branches in different cities. A database used by such an organization is usually distributed over all these sites. The different sites are connected by a computer network. A user may access data stored locally or access data stored at other sites through the network.

Distributed databases have several advantages. In distributed databases, the effect of a site failure or data loss at a particular node can be minimized through data replication. However, data replication reduces security and makes the process of keeping the database consistent more complicated.

In distributed databases, data is decomposed into fragments that are allocated to the different sites. A fragment is allocated to a site in a way that maximizes local use. This allocation scheme, which is known as *data localization*, reduces the frequency of remote access. In addition, since each site deals with only a portion of the database, local query processing is expected to exhibit increased performance.

A distributed database is inherently well suited for parallel processing at both interquery and intraquery levels. Parallel processing at the interquery level is the ability to have multiple queries executed concurrently. Parallelism at the intraquery level results from the possibility of a single query being simultaneously handled by many sites, each site acting on a different portion of the database.

The data distribution increases the complexity of DDBMS over a centralized DBMS. In fact, in distributed databases, several research issues in distributed query processing, distributed database design, and distributed transaction processing remain to be solved. It is only then that the potential of distributed databases can be fully appreciated.

Parallel Database Systems

There has been a continuing increase in the amount of data handled by database management systems (DBMSs) in recent years. Indeed, it is no longer unusual for a DBMS to manage databases ranging in sizes from hundreds of gigabytes to terabytes. This massive increase in database sizes is coupled with a growing need for DBMSs to exhibit more sophisticated functionality such as the support of object-oriented, deductive, and multimedia applications. In many cases, these new requirements have rendered existing DBMSs unable to provide the necessary system performance, especially given that many mainframe DBMSs already have difficulty meeting the I/O and CPU performance requirements of traditional information systems that service large numbers of concurrent users and/or handle massive amounts of data [DeWitt and Gray, 1992].

To achieve the required performance levels, database systems have been increasingly required to make use of parallelism. Two approaches were suggested to provide parallelism in database systems [Abdelguerfi and Lavington, 1995]. The first approach uses massively parallel general-purpose hardware platforms. Commercial systems, such as Intel's nCube and IBM's SP2 follow this approach and support Oracle's Parallel Server. The second approach makes use of arrays of off-the-shelf components to form custom massively parallel systems. Usually, these hardware systems are based on MIMD parallel architectures. The NCR 3700 and the Super Database Computer II (SDC-II) are two such systems. The NCR 3700 now supports parallel version of Sybase relational DBMS.

The number of general purpose or dedicated parallel database computers is increasing each year. It is not unrealistic to envisage that most high performance database management systems in the year 2000 will support parallel processing. The high potential of parallel databases in the future urges both the database vendors and practitioners to understand the concept of parallel database system in depth.

It is noteworthy that in recent years, popularity of the client/server architecture has increased. This architecture is practically a derivative of shared-nothing case. In this model, clients' nodes access data through one or more servers. This approach derives its strength from an attractive price/performance ratio, a high level of scalability, and the ease with which additional remote hosts can be integrated into the system. Another driving force of the client/server approach is the current trend toward corporate downsizing.

Multimedia

Yet another new generation database application is multimedia, where non-text forms of data, such as voice, video, and image, are accessed via some form of a user interface. Hypermedia interfaces are becoming the primary delivery system for the multimedia applications. These interfaces, such as Mosaic, allow users to browse through an information base consisting of many different types of data. The basis of hypermedia is the hypertext, where some text based information is accessed in a non-sequential manner. Hypermedia is an extension of hypertext paradigm into multimedia.

Defining Terms

Database: A shared pool of interrelated data.

Database computer: A special hardware and software configuration aimed primarily at handling large databases and answering complex queries.

Database management system (DBMS):   A software system that allows for the definition, construction, and manipulation of a database.

Data model: An integrated set of tools to describe the data and its structure, data relationships, and data constraints.

Distributed database: A collection of multiple, logically interrelated databases distributed over a computer network.

References

M. Abdelguerfi and A.K. Sood, Eds., Special Issue on Database Computers, *IEEE Micro,* December 1991.

M. Abdelguerfi and S. Lavingston, Eds., *Emerging Trends in Database and Knowledge Base Machines,* IEEE Computer Science Press, 1995.

A. Brown, *Object-Oriented Databases: Applications in Software Engineering,* New York: McGraw-Hill, 1991.

E.F. Codd, "A relational model of data for large shared data banks," *Communications of the ACM,* pp. 377–387, June 1970.

D. DeWitt and J. Gray, "Parallel database systems: The future of high performance database systems", *Communications of the ACM,* pp. 85–98, June 1992.

R. Elmasri and S.B. Navathe, *Fundamentals of Database Systems,* Redwood City, Calif.: Benjamin/Cummings, 2nd ed., 1994.

D. Maier, *The Theory of Relational Databases,* New York: Computer Science Press, 1983.

M.T. Ozsu and P. Valdurez, *Principles of Distributed Database Systems,* Englewood Cliffs, N.J.: Prentice-Hall, 1991.

16.2   Rule-Based Expert Systems

Jay Liebowitz

Expert systems is probably the most practical application of artificial intelligence (AI). Artificial intelligence, as a field, has two major thrusts: (1) to supplement human brain power with intelligent computer power and (2) to better understand how we think, learn, and reason. Expert systems are one application of AI, and they are being developed and used throughout the world [Feigenbaum et al., 1988; Liebowitz, 1990]. Other major applications of AI are robotics, speech understanding, natural-language understanding, computer vision, and neural networks.

Expert systems are computer programs that emulate the behavior of a human expert in a well-bounded domain of knowledge [Liebowitz, 1988]. They have been used in a number of tasks, ranging from sheep reproduction management in Australia, hurricane damage assessment in the Caribbean, boiler plant operation in Japan, computer configuration in the United States, to strategic management consulting in Europe [Liebowitz, 1991b]. Expert systems technology has been around since the late 1950s, but it has been only since 1980–1981 that the commercialization of expert systems has emerged [Turban, 1992].

An expert system typically has three major components: the dialog structure, inference engine, and knowledge base [Liebowitz and DeSalvo, 1989]. The dialog structure is the user interface that allows the user to interact with the expert system. Most expert systems are able to explain their reasoning, in the same manner that one would want human experts to explain their decisions. The inference engine is the control structure within the expert system that houses the search strategies to allow the expert system to arrive at various conclusions. The third component is the **knowledge base**, which is the set of facts and heuristics (rules of thumb) about the specific domain task. The knowledge principle says that the power of the expert system lies in its knowledge base. Expert system shells have been developed and are widely used on various platforms to help one build an expert system and concentrate on the knowledge base construction. Most operational expert systems are integrated with existing databases, spreadsheets, optimization modules, or information systems [Mockler and Dologite, 1992].

The most successful type of expert system is the rule-based, or production, system. This type of expert system is chiefly composed of IF-THEN (condition-action) rules. For example, the infamous MYCIN expert system, developed at Stanford University for diagnosing bacterial infections in the blood (meningitis), is rule-based, consisting of 450–500 rules. XCON, the expert system at Digital Equipment Corporation used for configuring VAX computer systems, is probably the largest rule-based expert system, consisting of over 11,000 rules. There are other types of expert systems that represent knowledge in ways other than rules or in conjunction with rules. Frames, scripts, and semantic networks are popular knowledge representation methods that could be used in expert systems.

The development of rule-based systems is typically called **knowledge engineering**. The knowledge engineer is the individual involved in the development and deployment of the expert system. Knowledge engineering, in rule-based systems, refers primarily to the construction of the knowledge base. As such, there are six major steps in this process, namely (1) problem selection, (2) knowledge acquisition, (3) knowledge representation, (4) knowledge encoding, (5) knowledge testing and evaluation, and (6) implementation and maintenance. The knowledge engineering process typically uses a rapid prototyping approach (build a little, test a little). Each of the six steps in the knowledge engineering process will be briefly discussed in turn.

Problem Selection

In selecting an appropriate application for expert systems technology, there are a few guidelines to follow:

- Pick a problem that is causing a large number of people a fair amount of grief.
- Select a "doable," well-bounded problem (i.e., task takes a few minutes to a few hours to solve)—this is especially important for the first expert system project for winning management's support of the technology.
- Select a task that is performed frequently.
- Choose an application where there is a consensus on the solution of the problem.
- Pick a task that utilizes primarily symbolic knowledge.
- Choose an application where an expert exists and is willing to cooperate in the expert systems development.
- Make sure the expert is articulate and available and a backup expert exists.
- Have the financial and moral support from management.

The problem selection and scoping are critical to the success of the expert systems project. As with any information systems project, the systems analysis stage is an essential and crucial part of the development process. With expert systems technology, if the problem domain is not carefully selected, then difficulties will ensue later in the development process.

Knowledge Acquisition

After the problem is carefully selected and scoped, the next step is knowledge acquisition. Knowledge acquisition involves eliciting knowledge from an expert or multiple experts and also using available documentation, regulations, manuals, and other written reports to facilitate the knowledge acquisition process. The biggest bottleneck in expert systems development has, thus far, been in the ability to acquire knowledge. Various automated knowledge acquisition tools, such as Boeing Computer Services' AQUINAS, have been developed to assist in this process, but there are very few knowledge acquisition tools on the market. The most commonly used approaches for acquiring/eliciting knowledge include: interviewing (structured and unstructured), protocol analysis, questionnaires (structured and open-ended), observation, learning by example/analogy, and other various techniques (Delphi technique, statistical methods).

To aid the knowledge acquisition process, some helpful guidelines are:

- Before interviewing the expert, make sure that you (as the knowledge engineer) are familiar/comfortable with the domain.
- The first session with the expert should be an introductory lecture on the task at hand.
- The knowledge engineer should have a systematic approach to acquiring knowledge.
- Incorporate the input and feedback from the expert (and users) into the system—get the expert and users enthusiastic about the project.
- Pick up manuals and documentation on the subject material.
- Tape the knowledge acquisition sessions, if allowed.

Knowledge Representation

After acquiring the knowledge, the next step is to represent the knowledge. In a rule-based expert system, the IF-THEN (condition-action) rules are used. Rules are typically used to represent knowledge if the preexisting knowledge can best be naturally represented as rules, if the knowledge is procedural, if the knowledge is mostly context-independent, and if the knowledge is mostly categorical ("yes-no" type of answers). Frames, scripts, and semantic networks are used as knowledge representation schemes for more descriptive, declarative knowledge. In selecting an appropriate knowledge representation scheme, try to use the representation method which most closely resembles the way the expert is thinking and expressing his/her knowledge.

Knowledge Encoding

Once the knowledge is represented, the next step is to encode the knowledge. Many knowledge engineers use expert system shells to help develop the expert system prototypes. Other developers may build the expert system from scratch, using such languages as Lisp, Prolog, C, and others. The following general guidelines may be useful in encoding the knowledge:

- Remember that for every shell there is a perfect task, but for every task there is NOT a perfect shell.
- Consider using an expert system shell for prototyping/proof-of-concept purposes—remember to first determine the requirements of the task, instead of force-fitting a shell to a task.
- Try to develop the knowledge base in a modular format for ease of updating.
- Concentrate on the user interface and human factors features, as well as the knowledge base.
- Use an incremental, iterative approach.
- Consider whether uncertainty should play a part in the expert system.
- Consider if the expert reasons in a data-driven manner (forward chaining) or a goal-directed manner (backward chaining), or both.

Knowledge Testing and Evaluation

Once the knowledge is encoded in the system, testing and evaluation need to be conducted. Verification and validation refers to checking for the consistency of the knowledge/logic and checking the quality/accuracy of advice reached by the expert system. Various approaches to testing can be used, such as: performing "backcasting" by running the expert system (using a representative set of test cases) against documented cases and comparing the expert system-generated results with the historical results, using blind verification tests (modified Turing test), having the expert and other experts test the system, using statistical methods for testing, and others. In evaluating the expert system, the users should evaluate the design of the human factors in the system (i.e., instructions, free-text comments, ease of updating, exiting capabilities, response time, display and presentation of conclusions, ability to restart, ability for user to offer degree of certainty, graphics, utility of the system, etc.).

Implementation and Maintenance

Once the system is ready to be deployed within the organization, the knowledge engineer must be cognizant of various institutionalization factors [Liebowitz, 1991a; Turban and Liebowitz, 1992]. Institutionalization refers to implementing and transitioning the expert system into the organization. Frequently, the technology is not the limiting factor—the *management* of the technology is often the culprit. An expert system may be accurate and a technical success, but without careful attention to management and institutionalization considerations, the expert system may be a technology transfer failure. There are several useful guidelines for proper institutionalization of expert systems:

- Know the corporate culture in which the expert system is deployed.
- Planning for the institutionalization process must be thought out well in advance, as early as the requirements analysis stage.

- Through user training, help desks, good documentation, hotlines, etc., the manager can provide mechanisms to reduce "resistance to change."
- Solicit and incorporate users' comments during the analysis, design, development, and implementation stages of the expert system.
- Make sure there is a team/individual empowered to maintain the expert system.
- Be cognizant of possible legal problems resulting from the use and misuse of the expert system.
- During the planning stages, determine how the expert system will be distributed.
- Keep the company's awareness of expert systems at a high level throughout the system's development and implementation, and even after its institutionalization.

Defining Terms

Expert systems: A computer program that emulates a human expert in a well-bounded domain of knowledge.

Knowledge base: The set of facts and rules of thumb (heuristics) on the domain task.

Knowledge engineering: The process of developing an expert system.

References

E.A. Feigenbaum, P. McCorduck, and P. Nii, *The Rise of the Expert Company,* New York: Times Books, 1988.

J.K. Lee, J. Liebowitz, and Y.M. Chae, Eds., *Proceedings of the Third World Congress on Expert Systems,* New York: Cognizant Communication Corp., 1996.

J. Liebowitz, *Introduction to Expert Systems,* New York: Mitchell/McGraw-Hill Publishing, 1988.

J. Liebowitz, Ed., *Expert Systems for Business and Management,* Englewood Cliffs, N.J.: Prentice-Hall, 1990.

J. Liebowitz, *Institutionalizing Expert Systems: A Handbook for Managers,* Englewood Cliffs, N.J.: Prentice-Hall, 1991a.

J. Liebowitz, Ed., *Operational Expert System Applications in the United States,* New York: Pergamon Press, 1991b.

J. Liebowitz, and D. DeSalvo, Eds., *Structuring Expert Systems: Domain, Design, and Development,* Englewood Cliffs, N.J.: Prentice-Hall, 1989.

R. Mockler and D. Dologite, *An Introduction to Expert Systems,* New York: Macmillan Publishing, 1992.

E. Turban, *Expert Systems and Applied Artificial Intelligence,* New York: Macmillan Publishing, 1992.

E. Turban and J. Liebowitz, Eds., *Managing Expert Systems,* Harrisburg, Pa.: Idea Group Publishing, 1992.

Further Information

There are several journals and magazines specializing in expert systems that should be consulted:

Expert Systems with Applications: An International Journal, New York/Oxford: Pergamon Press, Elsevier.

Expert Systems, Medford, N.J.: Learned Information, Inc.

IEEE Expert, Los Alamitos, Calif.: IEEE Computer Society Press.

AI Expert, San Francisco: Miller Freeman Publications.

Intelligent Systems Report, Atlanta: AI Week, Inc.

17
Parallel Processors

Tse-yun Feng
Pennsylvania State University

Miro Kraetzl
Defence Science and Technology Organisation

Young Choon Lee
University of Sydney

Albert Y. Zomaya
University of Sydney

17.1 Parallel Processors

Tse-yun Feng and Miro Kraetzl

Introduction

A computer usually consists of four major components: the arithmetic-logic unit (ALU), the main memory unit (MU), the input/output unit (I/O), and the control unit (CU). Such a computer is known as a uniprocessor since the processing is achieved by operating on one word or word pair at a time. In order to increase the computer performance, we may improve the device technology to reduce the switching (gate delay) time. Indeed, for the past half century we have seen switching speeds improve from 200 to 300 ms for relays to present-day subnanosecond very large scale integration (VLSI) circuits. As the switching speeds of computer devices approach a limit, however, any further significant improvement in performance is more likely to be in increasing the number of words or word pairs that can be processed simultaneously. For example, we may use one ALU to compute N sets of additions N times in a uniprocessor, or we may design a computer system with N ALUs to add all N sets once. Conceptually, such a computer system may still consist of the four major components mentioned previously except that there are N ALUs. An organization with multiple ALUs under the control of a single CU is called a **parallel processor.** To make a parallel processor more efficient and cost-effective, a fifth major component, called the **interconnection network,** is usually required to facilitate the interprocessor and processor-memory communications. In addition, each ALU requires not only its own registers but also network interfaces; the expanded ALU is then called a **processing element** (PE). Figure 17.1 shows a block diagram of a parallel processor.

Classifications

Flynn has classified computer systems according to the multiplicity of instruction and data streams, where computers are partitioned into four groups [Flynn, 1966]:

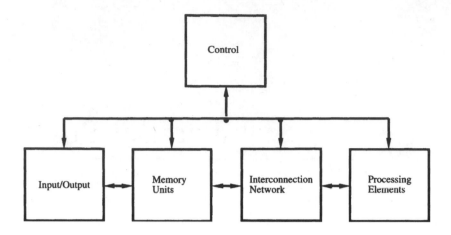

FIGURE 17.1   A basic parallel processor organization.

1. Single instruction stream, single data stream (SISD): The conventional, word-sequential architecture including pipelined computers (usually with parallel ALU).
2. Single instruction stream, multiple data stream (SIMD): The multiple ALU-type architectures (e.g., parallel/array processor). The ALU may be either bit-serial or bit-parallel.
3. Multiple instruction stream, single data stream (MISD): Not as practical as the other classes.
4. Multiple instruction stream, multiple data stream (MIMD): The multiprocessor system.

As a general rule, one could conclude that SISD and SIMD machines are single CU systems, whereas MIMD machines are multiple CU systems. Flynn's classification does not address the interactions among the processing modules and the methods by which processing modules in concurrent system are controlled. As a result, one can classify both uniprocessors and pipelined computers as SISD machines, because both instructions and data are provided sequentially.

We may also classify computer systems according to the number of bits or bit pairs a computer executes at any instant [Feng, 1972]. For example, a computer may perform operations on one bit or bit pair at a time through the use of a simple serial ALU. For an M-bit word or operand, the operation repeats M times (Point A in Figure 17.2). To speed up the processing, a parallel ALU is usually used so that all bits of a word can be operated on simultaneously. This is how a conventional word-sequential computer executes on its operands (point B in Figure 17.2). In a parallel processor, it may execute either (a) all the ith bits of N operands or operand pairs (i.e., bit slice or bis) or (b) all N M-bit operands or operand pairs simultaneously (points C and d in Figure 17.2, respectively). Figure 17.2 also shows some of the systems in this classification. It is seen from this classification that the performance of a computer is proportional to the total number of bits or bit pairs it can execute simultaneously.

Feng's classification [Hwang and Briggs, 1984] was originally intended for parallel processors, and as a result, the number of CUs in a computer system was not specified. Händler extended Feng's classification by adding a third dimension, namely, the number of CUs. Pipelined systems are also included in this classification through additional parameters [Händler, 1977].

Types of Parallel Processors

Ensemble Processors

An ensemble system is an extension of the conventional uniprocessor systems. It is a collection of N PEs (a PE here consists of an ALU, a set of local registers, and limited local control capability) and N MUs, under the control of a single CU. Thus, the organization of an **ensemble processor** is similar to that shown in Figure 17.1 except that there are no direct interprocessor and processor-memory communications, i.e., no interconnection

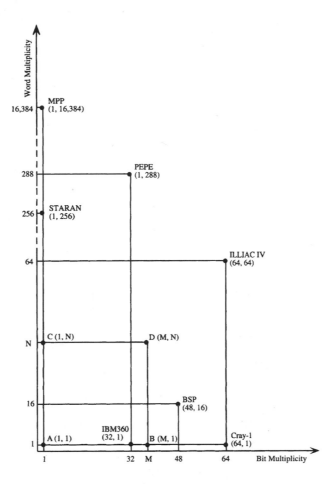

FIGURE 17.2 Feng's classification.

networks. When the need for communication arises, it is done through the CU. This slows down the system for applications requiring extensive interprocessor and processor-memory communications. For example, the sum of two matrices A and B can be executed in one step, if R^2 PEs are available in an ensemble processor, where R is the rank of the matrices. On the other hand, the product of the same two matrices requires extensive data alignment between the elements of A and B. As a result, it is ineffective for performing matrix multiplications with an ensemble processor. Therefore, while the ensemble processors are capable of executing up to N identical jobs simultaneously, they have very limited applications. Parallel element processing ensemble (PEPE) [Evensen and Troy, 1973] is an example of such parallel processors.

Array Processors

Because of the need for interprocessor and processor-memory communication for most applications, a parallel processor usually has one or more circuits (known as interconnection networks) to support various applications for efficient processing. In general, an **array processor** may consist of N identical PEs under the control of a single CU and a number of MUs. Within each PE there are circuits for network interface as well as its own local memories. The PEs and MUs communicate with each other through an interconnection network. A typical array processor organization is shown in Figure 17.3. Depending on the design, each PE may perform serial-by-bit (as in MPP) or parallel-by-bit (as in ILLIAC IV) operations.

As can be seen from Figure 17.3, the interconnection networks play a very important role in parallel processors. The network usually provides a uniform interconnection among PEs on one hand, and PEs and

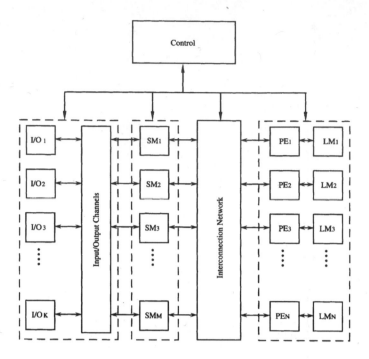

FIGURE 17.3   An array processor organization. I/O, input/output devices; LM, local memory; PE, processing element; SM, shared memory.

MUs on the other. Different array processor organizations might use different interconnection networks [Grammatikakis et al., 2001]. In general, the interconnection networks can be classified into two categories: static and dynamic, as shown in Figure 17.4.

ILLIAC IV [Barnes et al., 1968] and MPP [Batcher, 1979] are examples of parallel processors using static interconnections, while STARAN [Batcher, 1973] and BSP [Kuck and Stokes, 1982] are examples using dynamic interconnections.

The CU usually has its own high-speed registers, local memory, and arithmetic unit. Thus, in many cases, it is a conventional computer and the instructions are stored in a main memory, together with data. However, in some machines such as ILLIAC IV, programs are distributed among the local memories of the PEs. Hence, the instructions are fetched from the processors' local memories into an instruction buffer in the CU. Each instruction is either a local type instruction, where it is executed entirely within the CU, or it is a parallel instruction and is executed in the processing array. The primary function of the CU is to examine each instruction as it is to be executed and to determine where the execution should take place.

Associative Processor

Associative memories, also known as content-addressable memories, retrieve information on the basis of data content rather than addresses. An associative memory performs comparison (i.e., exclusive-OR or equivalence) operations at its bit level. The results of the comparison on a group of bits in a word for all words in the memory are transmitted to a register called a response register or flag. In addition, there are circuits such as multiple match resolver, enable/disable register, and a number of temporary registers, as well as appropriate logic gates for resolving multiple responses and information retrieval. For **associative processors,** arithmetic capabilities are added to this unit. The unit can be viewed as consisting of a number of bit-serial PEs. Furthermore, the bit-level logic is moved out of the memory so that the memory part of the processor consists of a number of random-access memories called word modules. A typical associative processor is shown in Figure 17.5. STARAN and MPP (Figure 17.2) are representative of this bit-serial, word-parallel SIMD organization. In Figure 17.5 the common register is where the common operand is stored and the mask

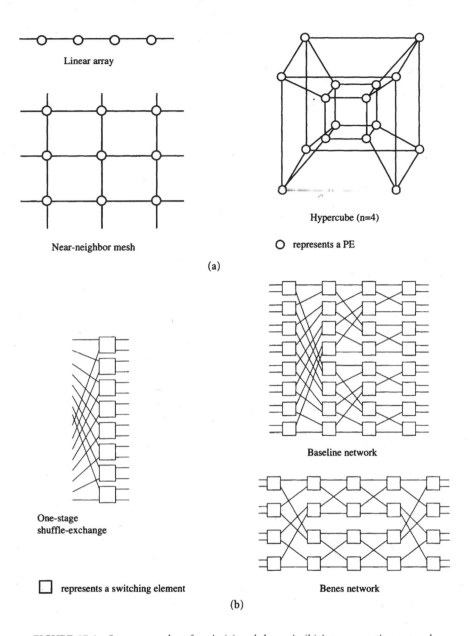

Linear array

Near-neighbor mesh

Hypercube (n=4)

O   represents a PE

(a)

One-stage
shuffle-exchange

Baseline network

Benes network

□   represents a switching element

(b)

FIGURE 17.4   Some examples of static (a) and dynamic (b) interconnection networks.

register defines the bit positions requiring operation. The enable/disable register provides local control of individual PEs. Because of its simplicity in design the per-PE cost of an associative processor is much lower, but the bit-serial operations slow down the system drastically. To compensate for this, these systems are useful only for applications requiring a large number of PEs.

System Utilization

As discussed previously, for any computer there is a maximum number of bits or bit pairs that can be processed concurrently, whether it is under single-instruction or multiple-instruction control [Feng, 1972, 1973]. This maximum degree of concurrency, or maximum concurrency (C_m), is an indication of the

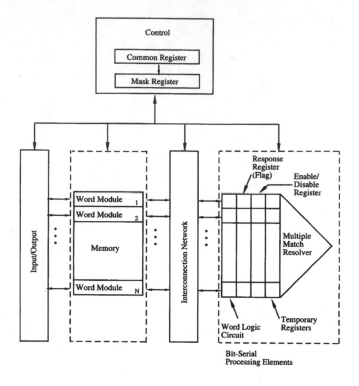

FIGURE 17.5   An associative processor organization.

computer-processing capability. The actual utilization of this capability is indicated by the average concurrency defined to be

$$C_a = \frac{\Sigma c_i \Delta t_i}{\Sigma \Delta t_i}$$

where c_i is the concurrency at Δt_i. If Δt_i is set to one time unit, then the average concurrency over a period of T time units is

$$C_a = \frac{\sum\limits_{i=1}^{T} c_i}{T}$$

The average hardware utilization is then

$$\mu = \frac{C_a}{C_m} = \frac{\sum\limits_{i=1}^{T} c_i}{T C_m} = \frac{1}{T} \sum\limits_{i=1}^{T} \sigma_i$$

where σ_i is the hardware utilization at time i. Whereas C_m is determined by the hardware design, C_a or μ is highly dependent on the software and applications. A general-purpose computer should achieve a high μ for as many applications as possible, whereas a special-purpose computer would yield a high μ for at least the intended applications. In either case, maximizing the value of μ for a computer design is important. This equation can also be used to evaluate the relative effectiveness of machine designs.

For a parallel processor, the degree of concurrency is called the degree of parallelism. A similar discussion can be used to define the average hardware utilization of a parallel processor. The maximum parallelism is then P_m, and the average parallelism is

$$P_a = \frac{\Sigma p_i \Delta t_i}{\Sigma \Delta t_i}$$

or

$$P_a = \frac{\displaystyle\sum_{i=1}^{T} p_i}{T}$$

for T time units. The average hardware utilization becomes

$$\upsilon = \frac{P_d}{P_m} = \frac{\displaystyle\sum_{i=1}^{T} p_i}{TP_m} = \frac{1}{T}\sum_{i=1}^{T} \rho_i$$

where ρ_i is the hardware utilization for parallel processors at time i. With appropriate instrumentation, the average hardware utilization of a system can be determined.

In practice, however, it is not always true that every bit or bit pair that is being processed would be productive. Some of the bits produce only repetitious (superfluous) or even meaningless results. This happens more often and more severely in a parallel processor than in a word-sequential processor. Consider, for example, performing a maximum search operation in a mesh-connected parallel processor (such as ILLIAC IV). For N operands, it takes $(N/2)\log_2 N$ comparisons ($N/2$ comparisons for each of $\log_2 N$ iterations) instead of the usual $N-1$ comparisons in word-sequential machines. Thus, in effect there are

$$\left(\frac{N}{2}\log_2 N\right) - (N-1) = \frac{N}{2}(\log_2 N - 2) + 1$$

comparisons that are nonproductive. If we let $\hat{P}_a$ be the effective parallelism over a period of T time units and $\hat{\upsilon}$, $\hat{p}_i$, and $\hat{\rho}_i$ be the corresponding effective values, the effective hardware utilization is then

$$\hat{\upsilon} = \frac{\hat{P}_a}{P_m} = \frac{\displaystyle\sum_{i=1}^{T} \hat{p}_i}{TP_m} = \frac{1}{T}\sum_{i=1}^{T} \hat{\rho}_i$$

A successful parallel processor design should yield a high $\hat{\upsilon}$, as well as the required throughput for, at least, the intended applications. This not only involves a proper hardware and software design but also the development of efficient parallel algorithms for these applications.

Suppose T_u is the execution time of an application program using a conventional word-sequential machine, and T_c is the execution time of the same program using a concurrent system; the speed-up ratio is then defined as

$$S = \frac{T_u}{T_c}$$

Naturally, for a specific parallel organization, the speed-up ratio determines how well an application program can utilize the hardware resources. Supporting software has a direct effect on the speed-up ratio.

Defining Terms

Array processor: A parallel processor consisting of a number of processing elements, memory modules, and input/output devices as well as interconnection networks under a single control unit.

Associative processor: A parallel processor consisting of a number of processing elements, memory modules, and input/output devices under a single control unit. The capability of the processing elements is usually limited to the bit-serial operations.

Ensemble processor: A parallel processor consisting of a number of processing elements, memory modules, and input/output devices under a single control unit. It has no interconnection network to provide interprocessor or processor-memory communications.

Interconnection network: A network of interconnections providing interprocessor and processor-memory communications. It may be static or dynamic, distributed, or centralized.

Parallel processor: A computing system consisting of a number of processors, memory modules, input/out-put devices, and other components under the control of a single control unit. It is known to be a single-instruction-stream, multiple-data-stream (SIMD) machine.

Processing element: A basic processor consisting of an arithmetic-logic unit, a number of registers, network interfaces, and some local control facilities.

References

G.H. Barnes, R.M. Brown, M. Kato, D.J. Kuck, D.L. Slotnick, and R.A. Stokes, "The ILLIAC IV computer," *IEEE Trans. Comput.,* vol. C-7, pp. 746–757, 1968.

K.E. Batcher, "STARAN/RADCAP hardware architecture," *Proc. Sagamore Computer Conf. on Parallel Processing,* pp. 147–152, 1973.

K.E. Batcher, "MPP — A massively parallel processor," *Proc. Int. Conf. on Parallel Processing,* p. 249, 1979.

A.J. Evensen and J.L. Troy, "Introduction to the architecture of a 288-element PEPE," *Proc. Sagamore Computer Conf. on Parallel Processing,* pp. 162–169, 1973.

T. Feng, "An overview of parallel processing systems," *1972 WESCON Tech. Papers,* Session 1— "Parallel Processing Systems," pp. 1–2, 1972.

T. Feng, Parallel Processing Characteristics and Implementation of Data Manipulating Functions, Dept. of Electrical and Computer Engineering, Syracuse University, RADC-TR-73–189, July 1973.

M.J. Flynn, "Very high speed computing systems," *Proc. IEEE,* vol. 54, (12), pp. 1901–1909, 1966.

M.D. Grammatikakis, D.F. Hsu, and M. Kraetzl, *Parallel System Interconnections and Communications,* Boca Raton, London: CRC Press LLC, 2001.

W. Händler, "The impact of classification schemes on computer architecture," *Proc. Int. Conf. on Parallel Processing,* pp. 7–15, 1977.

K. Hwang and F.A. Briggs, *Computer Architecture and Parallel Processing,* New York, NY: McGraw-Hill, 1984.

D.J. Kuck and R.A. Stokes, "The Borroughs Scientific Processor (BSP)," *IEEE Trans. Comput.,* vol. C-31(5), pp. 363–376, 1982.

Further Information

Proceedings of International Conference on Parallel Processing: An annual conference held since 1972. Recent proceedings published by CRC Press.

IEEE Transactions on Parallel and Distributed Systems: Started in 1990 as a quarterly, now a monthly, published by the IEEE Computer Society.

Journal of Parallel and Distributed Computing: A monthly published by Academic Press.

The Journal of Interconnection Networks (JOIN): Published by World Scientific Publishing Co.

17.2 Parallel Computing

Young Choon Lee and Albert Y. Zomaya

Introduction

In general a task processed by two people is completed faster than when it is processed by one person. The task may get completed even faster if more than two people carry it out. Obviously, this performance gain is due to processing certain parts of the task in parallel. More specifically, the task is partitioned into a series of sub-tasks. These partitioned tasks may include sub-tasks that are not dependent on each other at certain points of processing. Therefore, these independent sub-tasks can be handled by different people at the same time. This is the notion of parallel computing, also called parallel processing.

It is clearly evident that serial computing has certain performance limits, in particular when large problems with tight time constraints and/or a massive amount of computation are to be tackled. One of the most challenging limits to overcome is the speed of a single processor because it is increasingly difficult and expensive to make the processor faster within a limited size. In addition to the speed limit of single processors, an architectural limitation is typically present in serial computers since most modern serial computers are equipped with processors based on von Neumann architecture in which central processing unit, the CPU, accesses memory for both instructions and data. The low data transfer rate between the CPU and memory, known as the von Neumann bottleneck [1], and the speed disparity between them slow down processing speed. Parallel computing is a widely accepted and well-studied solution to overcome many of these limits in serial computing; moreover, it has gained great importance as an increasing number of problems become nearly unsolvable in a reasonable amount of time using serial computing [2]. These challenging problems include both computation and data-intensive problems such as human genome mapping, climate modeling, data mining, and web search engines.

Note, however, that the speed of a parallel computer with *n* processors is rarely, if not never, *n* times faster than that of a serial computer containing a single processor. The fact that the speedup proportional to the number of processors is practically not possible to achieve is due to both software and hardware issues. These include extracting parallelism from a task, scheduling parallelized tasks, and communication overheads between sub tasks running on different processors.

Since the early days of modern computing history, many different architectures and techniques of parallel computing have been studied, proposed, and developed while the parallel random access machine (PRAM) model [3] has remained the most influential theoretical model of parallel computers [2]. These technologies include pipelining, array processors, vector processors, multiprocessors, clustering, shared memory, and distributed memory. Using these architectures and techniques, parallel computers are built in various forms such as computer clusters, parallel supercomputers, massively parallel processor systems, and Grid computing.

Classification of Parallel Computers

The architecture of a computer can be classified in different ways [4–7] according to various characteristics of the computer. The taxonomy of computer architectures proposed by Flynn is the most well-known among the schemes. Flynn's taxonomy classifies computers based on two types of streams, instruction and data, and the singularity or multiplicity of each; hence four classifications as shown in Figure 17.6. Note that computers classified in SISD are serial computers. The majority of parallel computers have used either SIMD or MIMD architectures, whereas no commercial parallel computers exist with MISD architecture.

SISD

Most single-processor computers such as personal computers and workstations fall into SISD architecture. SISD machines do not support any real parallel computing. In other words, a SISD computer executes every algorithm sequentially. One way to imitate parallel computing in these machines is multitasking with the support of operating systems.

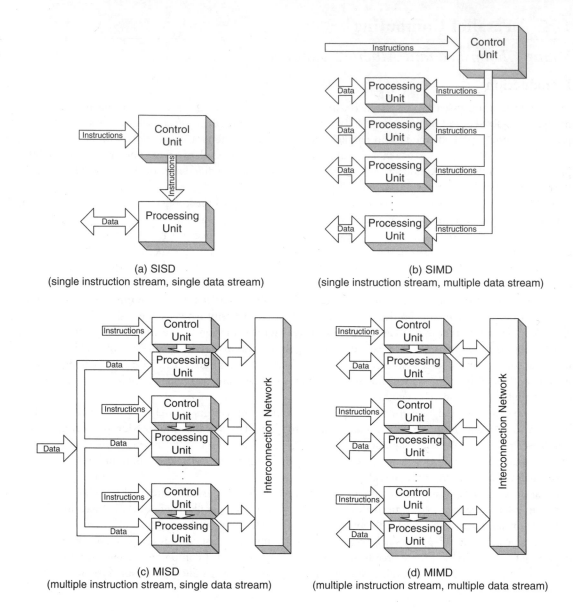

FIGURE 17.6   Flynn's taxonomy.

SIMD

A SIMD computer consists of multiple processors under the control of a single central control unit. The control unit feeds the same instruction to all of these processors at any given clock cycle. Each of these processors executes this instruction on a different datum. That is, SIMD machines are designed and operate based on the paradigm of synchronous data parallelism. The processors may need to communicate with each other in order to exchange data. This inter-processor communication can be performed by either accessing the same memory location in the shared-memory model or passing messages through some form of interconnection network in the distributed-memory model. Array processors and pipelined vector processors are two typical types of SIMD architecture. Problems with a high degree of regularity and data parallelism, such as image processing and data mining, can be most effectively solved on SIMD machines.

SPMD

SPMD (single program, multiple data) is a software paradigm exploiting data parallelism. The processors of a parallel computer execute the same program on different data asynchronously. Since SPMD is a software model it is less architecture dependent [8]. Although SPMD does not belong to Flynn's taxonomy it is often referred to since many scientific and engineering applications can be classified into this category. Parameter sweep applications, such as Monte Carlo simulations, are a typical example class of the SPMD paradigm.

MISD

MISD architecture is the least popular model used to build parallel computers in that the number of problems that are best suited for MISD machines is quite limited. A MISD machine consists of n processor control unit pairs and a common memory shared by the processors. A single datastream can either flow to the processors to deliver the same datum as shown Figure 17.6(c), or pass through the processors such that each processor may manipulate the datum before sending to the next processor.

MIMD

MIMD machines are capable of accommodating a broad range of parallel problems. Most contemporary parallel computers are based on this architecture. A MIMD machine contains n processor control unit pairs and either a single shared memory block, or a memory module for each processor. Each processor operates on its own instruction and data streams. The processors can operate asynchronously or synchronously. MIMD computers with shared memory, i.e., tightly-coupled machines, are further classified as multiprocessors. The other type of MIMD machine is the multicomputer, the loosely coupled model of MIMD in which processors have their own local memories. These two different models are shown in Figure 17.7. It can be easily noticed that multicomputers have better scalability.

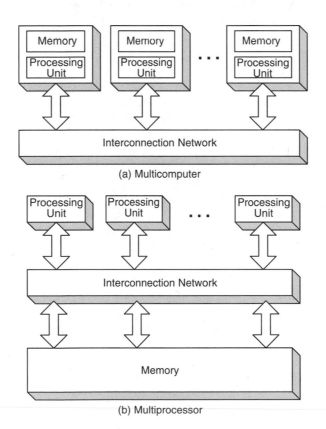

FIGURE 17.7 MIMD models.

Parallel Computer Models

Since the advent of parallel computing, various models of parallel computers have been developed. These include vector processors, symmetric multiprocessing (SMP), massively parallel processing (MPP), cluster computing, distributed computing, and grid computing. While early parallel computers were built using custom designed components, many recent ones have tended to be developed with off-the-shelf processors or commodity PCs. This implies that parallel computers have become affordable for more organizations, e.g., small and medium size companies, university research laboratories, etc.; hence the use of parallel machines is no longer limited to special purposes.

Vector Processors

The vector processor was developed based on the SIMD paradigm, i.e., data parallelism. Unlike a scalar processor that operates on individual elements, a vector processor handles vectors of elements at a time. More specifically, a single instruction is executed on vectors concurrently. For example, an addition of two arrays of numbers can be processed at a single clock cycle.

Typically vector processors use the pipelining technique for both instructions and data, meaning that tasks involved in processing a single instruction and datum are handled by separate parts of the processor. These tasks include decoding the address of the instruction or data, fetching the value and processing it. Pipelining, therefore, improves performance.

Another technique used in vector processors to achieve better performance is the use of vector registers that can be accessed much faster than memory. In vector machines built with this technique vector instructions operate on vectors residing in these registers. Variants of this vector register architecture have been adopted in many vector processors since the advent of the Cray-1 [9]. This architecture is a significant advance over the vector memory–memory architecture in which vectors are accessed to and from main memory. The majority of early vector machines such as the CDC STAR-100 [10] and the TI Advanced Scientific Computer [11] used the vector memory–memory architecture.

Symmetric Multiprocessing (SMP)

An SMP system is a shared-memory multiprocessor in which the same memory is equally accessible by all processors that are under the control of a single operating system. SMP architecture tends to be limited to building parallel computers with a small number of processors due to the difficulty of coordinating memory access between processors [12]. Nonuniform memory access (NUMA) is a memory model that can relieve this memory and network contention by associating each processor with its own local memory; hence the name nonuniform. The access time of nonlocal memory in NUMA architecture is slower than that of local memory.

A global view of memory in SMP systems makes information exchange between processors easy and fast. This makes multithreading a popular choice of programming model in SMP machines. Some of the well-known multithreading packages include the OpenMP application programming interface [13] and POSIX pThreads [14]. By contrast to the efficiency of data exchange, the use of the global address space raises two data integrity issues: synchronization and cache coherency, if local memory caches are used [15,16].

Massively Parallel Processing (MPP)

MPP has been a popular design choice to build high performance parallel computers in the past couple of decades. The primary building block of MPP systems is referred to as a node, which typically has one or more processors and its own local memory. Nodes in MPP machines are often powered by less expensive off-the-shelf processors such as Intel Xeon and Itanium 2, AMD Opteron, and IBM PowerPC processors. As the name implies MPP systems may consist of up to thousands of nodes with some form of interconnection network, such as completely connected, n-dimensional mesh, and hypercube networks. An additional node to a MPP system can be easily connected, in contrast to adding a processor to a SMP machine, hence the better scalability.

The most common programming model in MPP machines is message-passing using either machine-dependent libraries or message-passing interface standards such as the message passing interface (MPI) [17].

Cluster Computing

A cluster is composed of a collection of nodes connected by an interconnection network such as Ethernet [18] or Myrinet [19]. Each of these nodes is typically an inexpensive commodity computer, e.g., a PC or a workstation that contains at least one processor, its own memory, and other local resources. SMP systems may be used as nodes in order to achieve better performance. In most cases, clusters are built with homogenous computers.

Cluster computing can be viewed from four different perspectives:

- *High performance.* Clusters as parallel computers provide suitable platforms for running parallel applications.
   High availability. Clusters tend to be consistently functional tolerating exceptions such as node failure.
- *High throughput.* The effective management and efficient utilization of clusters, i.e., good resource management systems, which enables large amounts of computing power to be delivered over long periods of time. Some well-known examples of such resource management systems include Condor [20] from the computer science department at the University of Wisconsin-Madison, and the Maui scheduler [21].
- *Affordability/Cost-effectiveness.* Clusters built with inexpensive off-the-shelf components can outperform more expensive supercomputers. A cluster is commonly constructed by either a group of individuals, or a cluster vendor [22]. The former normally consists of the users, e.g., academics or researchers who will actually use the cluster.

Distributed Computing/Internet Computing

A distributed computing system is a collection of heterogenous computers, typically PCs and workstations owned by independent individuals. These machines are dispersed in geographically different locations, hence the name distributed. The most typical medium used to connect them is the Internet. The rationale behind distributed computing is to utilize these personally owned PCs and workstations to solve large computational problems while they are not in use.

It is often the case that a distributed computing system with many thousands or even millions of desktop PCs is solely used to tackle a single, large computation-intensive application. In this situation the application is partitioned into a number of sub-tasks so that they can be distributed to the participating computers. Some famous examples of such applications include the SETI@home project [23] and the Folding@home project [24]. The number of applications is somewhat limited mainly due to the difficulties of efficiently orchestrating the wide range of heterogeneous components, e.g., network bandwidths and computing capacities. However, distributed computing is becoming an attractive computing model as the computing power of PCs and workstations is constantly increasing.

Grid Computing

As the demand for more powerful computing resources keeps increasing, especially from scientists and engineers, many organizations have built a myriad of supercomputers with various different architectures. In addition to supercomputers, some other forms of high performance computing systems, such as computer clusters, have been built by a number of institutions, e.g., universities, research laboratories, etc. In general, the use of these specialist computing systems is confined to specific groups of people. Moreover, each of these systems is generally restricted to independent use; that is, it is highly unlikely that a user of one system can access other organizations' systems. A solution to this is grid computing. A grid enables a virtual computing system, interconnecting these geographically distributed, heterogeneous computing systems with a variety of resources, to be constituted. Here, resources refer not only to physical computers, networks, and storage systems, but to much broader entities such as databases, data transfer, and simulation [25]. The grid creates the illusion that its users are accessing a single ultra powerful supercomputer. The user can transparently access resources in the grid by a single login.

A vast number of researchers have been putting in a lot of effort to facilitate building and efficiently utilizing grids. Some significant results for grid computing include the Globus toolkit [26], Legion [27], and

GrADS [28]. These tools, especially the Globus toolkit, have been used to build many grids [29–32]. The Globus toolkit is quite mature and is the *de facto* standard for grid computing.

In a grid it is unlikely that an application with a fine-grained parallel algorithm in which sub-tasks communicate with each other frequently will run across multiple computing systems. Rather, a single computing system with multiple processing units is allocated. Conversely, applications with coarse-grained

```c
#include <stdio.h>
#include <stdlib.h>
#include <string.h>
#include <mpi.h>

long ComputeFactorial(int start, int end);


int main(int argc, char **argv)
{
int myID;
int numProcs;
int num;
int start, end;
long myRes = 1, lRes, total = 1;

MPI_Init(&argc, &argv);
MPI_Comm_size(MPI_COMM_WORLD, &numProcs);
MPI_Comm_rank(MPI_COMM_WORLD, &myID);


if (myID == 0)
{
/* the number should be 3 or greater
since the number of threads is fixed to 3 */
printf("Enter a positive integer (>= 3): ");
scanf("%d", &num);
}

MPI_Bcast(&num, 1, MPI_INT, 0, MPI_COMM_WORLD);
if (myID != 0)
{
start = (myID - 1) * (num / (numProcs - 1));
end = start + (num / (numProcs - 1)) - 1;

/* the last thread takes whatever left */
if (myID == (numProcs - 1))
{
end += num % (numProcs - 1);
}
myRes = ComputeFactorial(start, end );
}

MPI_Reduce(&myRes, &lRes, 1, MPI_INT, MPI_PROD, 0, MPI_COMM_WORLD);

if (myID == 0)
{
total *= lRes;
printf("%d factorial is %d\n", num, total);
}
```

FIGURE 17.8 A C program for computing factorial using MPI.

```
MPI_Finalize();

return 0;
}

long ComputeFactorial(int start, int end)
{
int i;
long result = 1;

for (i = start; i <= end; i++)
result *= (i + 1);

return result;
}
```

FIGURE 17.8   Continued.

parallel algorithms and applications of the SPMD model may be processed by more than one computing system often dispersed in multiple administrative domains.

Parallel Programming Paradigms

There is little benefit from running applications on a parallel computer unless the applications are developed with parallel algorithms. In some parallel computing systems compilers may automatically detect parallelism in applications and parallelize them [16]. However, exploiting parallelism from applications is generally left to programmers. Programmers are often required to explicitly specify which parts of applications are parallelizable and how those parallelized segments interact or communicate with each other. Although there are various approaches to parallelizing programs they can be categorized into three general programming models: (1) message-passing, (2) share-memory, and (3) data parallel. A language-centric classification of parallel programming models can be found in Ref. [33].

Message-Passing

Programming parallel applications with the message-passing paradigm imposes on the programmer the necessity to explicitly specify how data transfer takes place between processes, each of which contains its own memory space. The processes can be created either before or during execution, i.e., static or dynamic process creation. The data transfer is carried out by sending and receiving messages. A process may send a message to a certain number or an arbitrary number of processes. When the message is to pass to one or more particular processes, the sending process has to specify each receiving process identifier, which is unique.

One of the most common ways to implement message-passing programs is using message-passing libraries in combination with standard high-level languages such as C or Fortran. Two well-known message-passing libraries are MPI and the parallel virtual machine (PVM) [34]. Since they are architecture independent, parallel applications written using these libraries are more portable compared to those programmed with vendor-supplied programming facilities [8]. A noticeable difference between these two libraries is that they each are used for the parallel computing model for which they are best suited. While applications in the SPMD model are generally implemented using MPI, PVM is typically the choice for those in the multiple program multiple data (MPMD) model in which different programs run on different data [35].

A typical type of application in the message-passing model is master-slave in which a master process distributes a number of tasks to slave processes. Once the slave processes complete their tasks they send results back to the master process. The master one may further process the received results before producing the final result. An example of MPI code that computes factorial is shown in Figure 17.8.

```
#include <stdio.h>
#include <pthread.h>

#define NB_THREADS 3
#define NB_ARGS 3

void *ComputeFactorial(void * args);

int args[NB_THREADS][NB_ARGS];

int main (int argc, char *argv[])
{
pthread_t threads[NB_THREADS];
pthread_attr_t attr;
int num, i, start, end, rc;
long total = 1;

/* the number should be 3 or greater
since the number of threads is fixed to 3 */
printf("Enter a positive integer (>= 3): ");
scanf("%d", &num);

pthread_attr_init(&attr);

for(i = 0; i < NB_THREADS; i++)
{
start = i * (num / NB_THREADS);
end = start + (num / NB_THREADS) - 1;

/* the last thread takes whatever left */
if (i == (NB_THREADS - 1))
end += num % NB_THREADS;

args[i][0] = start;
args[i][1] = end;
args[i][2] = 1;

rc = pthread_create(&threads[i], &attr,
        ComputeFactorial, args[i]);

if (rc)
{
fprintf(stderr, "ERROR[%d]:while creating thread\n", rc);
exit(-1);
}
}

/* free attribute and wait for the other threads */
pthread_attr_destroy(&attr);

for(i = 0; i < NB_THREADS; i++)
{
rc = pthread_join(threads[i], NULL);
if (rc)
{
fprintf(stderr, "ERROR[%d]: while joining thread\n", rc);
```

Figure 17.9 A C program for computing factorial using pThreads.

```
exit(-1);
}
total *= args[i][2];
}

printf("%d factorial is %ld\n", num, total);

return 0;
}


void *ComputeFactorial(void *args)
{
int i, *_args = (int *) args;

for (i = _args[0]; i <= _args[1]; i++)
_args[2] *= (i + 1);

pthread_exit(NULL);
}
```

FIGURE 17.9 Continued.

Shared Memory/Threads

Unlike the message-passing paradigm, no explicit interaction details between tasks need to be specified in the shared-memory model. Rather, the tasks implicitly communicate with each other via the same shared memory space. Since the tasks share the common memory area, synchronization of the access to this shared memory must be handled. Synchronization is normally achieved by setting a critical section that only one task is allowed to execute, and to access the resources associated with it at a time. Mutual exclusion is the scheme that enables this process. There are different ways to perform mutual exclusion such as using locks or semaphores. A detailed discussion on mutual exclusion can be found in Ref. [36].

The single most common approach in this model is programming using threads. A thread is a spawned instance of part of the program that shares the program execution environment. Threads are typically used to execute a certain number of separate program segments, such as C functions, simultaneously. It is said that multithread programming is a more convenient way to parallelize tasks compared to the message-passing model due to several significant strong points, e.g., portability, latency hiding, scheduling, and load balancing [8].

There are a number of thread APIs that are mostly vendor-specific. It is very obvious that threaded programs written using these APIs are less portable. The straightforward choice for increasing portability is using a standardized thread implementation. Two such implementations that are commonly used are pThreads and OpenMP. Both packages are widely available on various platforms. While pThreads requires the programmer to explicitly handle the details of coordinating threads such as synchronization, OpenMP offers a high-level API that hides most of these low-level works from him or her. A pThreads example of the same factorial computing problem mentioned in the previous section is shown in Figure 17.9.

Data Parallel

The data parallel paradigm is the programming model equivalent to SIMD. It primarily exploits fine-grained parallelism. In this model the same multiple tasks operate on partitioned data of the identical data structure such as an array, a tree, and a set. For example, in Fortran 90 [37] a statement, c = a * b, where a, b and c are arrays of the same data type and the same size of ten, can be broken into ten identical scalar operations. There are no data dependencies between these ten operations, hence they are parallelizable.

Unlike the previous two parallel programming paradigms, the programmer is relatively free from handling interactions between tasks when writing data parallel programs since a single instruction stream exists in each of these programs and data parallel languages take care of architecture specifics [38]. In short, programming in this model can be easier. In addition to the ease of programming, data parallelism inherently contains high scalability.

Data parallel programs are generally written using data parallel languages such as Fortran 90, 95, and 2003, High Performance Fortran (HPF) [39], Split C [40], HyperC [41], and pC++ [42]. Data parallel language compilers are capable of automatically detecting data parallelism, actually parallelizing tasks without much effort by the programmer.

Parallel Programming Considerations

It is meaningless to develop parallel algorithms if they do not give any significant advantages over their sequential counterparts. There are various aspects that should be taken into consideration when designing and deploying parallel algorithms. Some crucial ones are speedup, efficiency, scalability, and portability. Speedup is usually regarded as the most important factor that is used to measure the performance of parallel algorithms. However, the others should also be seriously considered since they are closely related to the cost of running parallel applications. For example, although a parallel algorithm is ten times faster than the fastest sequential counterpart if the number of processors in the parallel computer used is 100, it is said that the parallel algorithm is very poorly designed in terms of its cost-effectiveness. Some other points such as throughput and responsiveness [16] may be taken into account for parallel programming, depending on characteristics of parallel algorithms.

Speedup

The performance of a parallel algorithm is most commonly evaluated by speedup, which is defined as the ratio of the running time of the parallel algorithm on a single processor to the running time of the parallel algorithm on n processors. Theoretically, the most desirable speedup of a parallel algorithm running on n processors is n. However, such speedup is never obtainable in practice for several reasons [16].

Two well-known definitions of the speedup that a parallel algorithm delivers have been proposed by Amdahl [43] and Gustafson and Barsis [44]. *Amdahl's Law* defines speedup based on the time taken on the uniprocessor machine, while the *Gustafson–Barsis Law* claims speedup is based on the time taken on the parallel computer. In other words, the former interprets the running time of the parallel algorithm on a single processor as 1. Conversely, the running time of the algorithm on multiple processors is treated as 1 by the latter. The speedup formulas of these two laws are

$$S = \frac{1}{f + (1-f)/n} \tag{17.1}$$

and

$$S = n - (1-n)f \tag{17.2}$$

respectively, where f is portions of the parallel algorithm that must be run sequentially and n is the number of processors. It may be noted that Amdahl's perspective is relatively more pessimistic than that of Gustafson and Barsis.

Efficiency

When a parallel algorithm is executed on a parallel computer with n processors, the proportion of the time contributed by these processors solely to the computations of the algorithm compared to the running time of the algorithm is an important indication of how efficiently the processors are used. Efficiency is defined as

$$E = \frac{S}{n} \tag{17.3}$$

where n is the number of processors. As mentioned in the previous section, S cannot be the same as n. Therefore, an efficiency of 100% is unattainable.

Scalability

In addition to speedup, scalability is another frequently used measure of how well the performance of a parallel algorithm scales when problem size and parallel computing system size change. Another definition proposed in [45] is: *Scalability is a property which exhibits performance linearly proportional to the number of processors employed.* Note that the growth of a parallel computing system size, i.e., more processors does not always produce a higher speedup of a parallel algorithm due to limits to the scalability of the parallel computing system and/or the parallel algorithm. That is, the more processors, the higher the overheads and the number of sub-tasks partitioned from the algorithm, which may not be as many as the number of processors [46]. Therefore, the scalability analysis helps design parallel machines and parallel algorithms effectively.

One scalability metric often used is *isoefficiency* [47] in which the primary focus is measuring the scalability of parallel algorithms. The rationale behind the isoefficiency concept is that the scalability of a parallel algorithm is measured based on efficiency. More specifically, the scalability of the algorithm is determined by whether its efficiency can be maintained consistently when both the problem size and the number of processors increase.

The reader can find a comprehensive study carried out on various scalability schemes in Kumar and Gupta [48].

Portability

The development of parallel algorithms requires a lot of effort and resources. This factor consequently highlights the importance of the portability of the algorithms. As an increasing number of standardized parallel programming facilities such as MPI and pThreads emerge, the use of these facilities instead of vendor-specific tools does much to alleviate portability problems with parallel programs. However, there is a trade-off between portability and performance. That is, programming parallel programs with an architecture specific tool offers typically better performance compared to parallel programs written with standard parallel APIs, for example.

References

1. J. Backus, "Can programming be liberated from the von Neuman style a functional style and its algebra of programs," *Comm. ACM*, 21, 613, 1978.
2. B. Wilkinson and M. Allen, *Parallel Programming*, 1st ed., Englewood Cliffs, NJ: Prentice-Hall, 1999, p. 5.
3. S. Fortune and J. Wyllie, "Parallelism in random access machines," in *Proc. Tenth Ann. ACM Symp. Theory Comput.*, New York, NY: ACM Press, 1977, p. 114.
4. R.W. Hockney and C.R. Jesshope, *Parallel Computers 2*, 2nd ed., Bristol, England: Adam Hilger/IOP Publishing, 1988, chap. 1.
5. K.M. Kavi and H.G. Gragon, "A conceptual framework for the description and classification of computer architecture," in *Proc. IEEE Int. Workshop Comput. Syst.*, Los Angeles, CA: IEEE Computer Society Press, 1983, p. 10.
6. D.B. Skillicorn, "A taxonomy for computer architectures," *IEEE Comput.*, 21, 46, 1988.
7. M.J. Flynn, "Very high-speed computing systems," in *Proc. IEEE*, vol. 54, no. 12, p. 1966, 1901.
8. A. Grama, A. Gupta, G. Karypis and V. Kumar, *Introduction to Parallel Computing: Design and Analysis of Algorithms*, 2nd ed., Reading, MA: Addison Wesley, Essex, 2003, chaps. 2, 6, and 7.
9. R.M. Russel, "The Cray-1 computer system," *Comm. ACM*, 21, 63, 1978.

10. R.G. Hintz and D.P. Tate, "Control data STAR-100 processor design," in *Proc. COMPCON*, IEEE, 1972, p. 1.
11. W. Watson, "The TI-ASC, a highly modular and flexible super computer architecture," in *Proc. AFIPS*, 1972, p. 221.
12. T.G. Lewis and H. El-Rewini, *Introduction to Parallel Computing*, Englewood Cliffs, NJ: Prentice-Hall, 1992, p. 94.
13. *The OpenMP: A Proposed Industry Standard API for Shared Memory Programming*, OpenMP Architecture Review Board, http://www.openmp.org, 1997.
14. IEEE, Threads Extension for Portable Operating Systems (Draft 10), 1996.
15. A.Y. Zomaya, Ed., *Parallel Computing: Paradigms and Applications*, London, England: International Thomson Computer Press, 1995, p. 14.
16. A.Y. Zomaya, Ed., *Parallel and Distributed Computing Handbook*, New York, NY: McGraw-Hill, 1996, p. 682.
17. Message Passing Interface Forum: MPI: A Message-Passing Interface Standard, *Int. J. Supercomput. Appl.*, 8, 165, 1994.
18. R. Metcalfe and D. Boggs, "Ethernet: distributed packet-switching for local computer networks," *Comm. Assoc. Comput. Mach.*, 19, 395, 1976.
19. N.J. Boden, D. Cohen, R.E. Felderman, A.E. Kulawik, C.L. Seitz, J.N. Seizovic, and W.-K. Su, "Myrinet: a gigabit-per-second local-area network," *IEEE Micro.* 15, 29, 1995.
20. M.J. Litzkow, M. Livny, and M.W. Mutka, "Condor — a hunter of idle workstations," in *Proc. Eighth Int. Conf. Distrib. Comput. Syst.*, San Jose, CA: IEEE Computer Society Press, 1988, p. 104.
21. The Maui scheduler, http://www.mhpcc.edu/maui.
22. W. Gropp, E. Lusk, and T. Sterling, *Beowulf Cluster Computing with Linux*, 2nd ed., Cambridge, MA: The MIT Press, 2003, p. 3.
23. SETI@home, http://setiathome.ssl.berkeley.edu/.
24. Folding@home, http://folding.stanford.edu/.
25. I. Foster and C. Kesselman, Eds., *The Grid: Blueprint for a Future Computing Infrastructure*, Los Altos, CA: Morgan Kaufmann, 1999.
26. I. Foster and C. Kesselman, "Globus: a metacomputing infrastructure toolkit," *Int. J. Supercomput. Appl.*, 11, 115, 1997.
27. A. Grimshaw and W. Wulf, "The legion vision of a worldwide virtual computer," *Comm. ACM*, 40, 39, 1997.
28. F. Berman et al., "The GrADS project: software support for high-level grid application development," *Int. J. High Perform. Comput. Appl.*, 15, 327, 2001
29. R. Stevens et al., "From the I-WAY to the National Technology Grid," *Comm. ACM*, 40, 50, 1997.
30. W.E. Johnston, D. Gannon, and B. Nitzberg, "Grids as production computing environments: the engineering aspects of NASA's Information Power Grid," in *Proc. Eighth Int. Symp. High Perform. Distrib. Comput.*, San Jose, CA: IEEE Computer Society Press, 1999, p. 197.
31. D.A. Reed, "Grids, the TeraGrid and beyond," *Computer*, 36, 62, 2003.
32. P. Eerola et al., "The Nordugrid production grid infrastructure, status and plans," in *Proc. Fourth Int. Workshop Grid Comput.*, San Jose, CA: IEEE Computer Society Press, 2003, p. 158.
33. T.J. Fountain, *Parallel Computing: Principles and Practice*, Cambridge, MA: Cambridge University Press, 1994, p. 84.
34. A. Geist et al., *PVM: Parallel Virtual Machine. A Users' Guide and Tutorial for Networked Parallel Computing*, Cambridge, MA: The MIT Press, 1994.
35. B. Wilkinson and M. Allen, *Parallel Programming: Techniques and Applications Using Networked Workstations and Parallel Computers*, Englewood Cliffs, NJ: Prentice-Hall, 1999, chap. 2.
36. H. Attiya and J. Welch, *Distributed Computing: Fundamentals, Simulations and Advanced Topics*, Hoboken, NJ: Wiley, 2004, chap. 3.
37. M. Metcalf and J. Reid, *Fortran 90 Explained*, Oxford, England: Oxford University Press, 1990.

38. V. Kumar et al., *Introduction to Parallel Computing: Design and Analysis of Algorithms*, Redwood City, CA: Benjamin/Cummings Publishing Company, Inc., 1994, chap. 13.

39. High Performance Fortran Forum, *High Performance Fortran Language Specification*, Technical Report, Rice University, 1992.

40. D.E. Culler et al., "Parallel programming in Split-C," in *Proc. Supercomput. 1993*, Oregon, OR: IEEE Computer Society Press, 1993, p. 262.

41. P. Clermont and N. Paris, "HyperC: portable parallel programming in C," in *Proc. Eighth Int. Parallel Process. Symp.*, H.J. Siegel, Ed., Cancún, NM: IEEE Computer Society Press, 1994, p. 682.

42. F. Bodin et al., "Implementing a parallel C++ runtime system for scalable parallel systems," in *Proc. Supercomput. 1993*, Eugene, OR: IEEE Computer Society Press, 1993, p. 588.

43. G.M. Amdahl, "Validity of the single processor approach to achieving large scale computing capabilities," *Proc. AFIPS Conf.*, 30, 483, 1967.

44. J.L. Gustafson, "Reevaluating Amdahl's Law," *Comm. ACM*, 31, 532, 1988.

45. X. Wu and W. Li, "Scalability of parallel algorithm implementation," in *Proc. Second Int. Symp. Parallel Architect. Algorithms Networks*, Beijing: IEEE Computer Society Press, 1996, p. 559.

46. V. Kumar and A. Gupta, "Analysis of scalability of parallel algorithms and architectures: a survey," in *Proc. Fifth Int. Conf. Supercomput.*, Cologne, West Germany: ACM Press, 1991, p. 396.

47. V. Kumar and V.N. Rao, "Parallel depth-first search, part II: analysis," *Int. J. Parallel Program.*, 16, 501, 1987.

48. V. Kumar and A. Gupta, "Analyzing scalability of parallel algorithms and architectures," *J. Parallel Distrib. Comput.*, 22, 379, 1994.

18
Operating Systems

Raphael Finkel
University of Kentucky

18.1 Introduction

An operating system is the set of programs that control a computer. Some operating systems you may have heard of are Unix (including SCO Unix, Linux, Solaris, Irix, NetBSD, and FreeBSD), the Microsoft family (MS-DOS, MS-Windows, Windows/NT, Windows 2000, Windows 2003, Windows XP), IBM operating systems (MVS, VM, CP, OS/2), Macintosh operating systems (Mac OS), Mach, and VMS. Some of these (Mach and Unix) have been implemented on a wide variety of computers, but most are specific to one or two particular architectures, such as the Digital Equipment Corporation Vax (VMS), the Intel 8086 and successors (the Microsoft family, OS/2), the Motorola 68000 and successors (Mac OS), and the IBM 360 and successors (MVS, VM, CP).

Controlling the computer involves software at several levels. We can distinguish kernel services, library services, and application-level services, all of which are part of the operating system. These services can be pictured as in Figure 18.1. Applications are programs linked to libraries of program units that perform common services like formatting output or presenting information on a display. As these programs run, they are called processes. The kernel supports processes by providing resources (such as computing time, access to physical computer memory, and access to peripheral devices), security (preventing inter-process snooping or interference), and enhancement (such as files and network communication protocols). The kernel becomes active in response to **system calls** (requests for service) from processes and interrupts from devices.

This chapter discusses how operating systems have evolved, often in response to architectural advances. It then examines the goals and organizing principles of current operating systems. Many books describe operating systems concepts [3–5,19–20] and specific operating systems [1,2,9–11].

18.2 Historical Perspective

Operating systems have undergone enormous change over the years. The changes have been driven primarily by hardware facilities and their cost and secondarily by the applications that users have wanted to run on the computers.

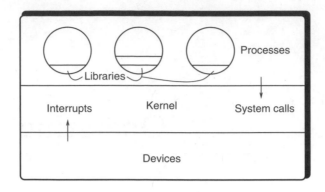

FIGURE 18.1   Operating system services.

Open Shop Organization

The earliest computers were massive, expensive, and difficult to use. Users would sign up for blocks of time during which they were allowed "hands-on" exclusive use of the computer. The user would repeatedly load a program into the computer through a device such as a card reader, watch the results, and then decide what to do next.

A typical session on the IBM 1620, a computer in use around 1960, involved several steps in order to compile and execute a program. First, the user would load the first pass of the Fortran compiler. This operation involved clearing memory by typing a cryptic instruction on the console typewriter; putting the compiler, a 10-inch stack of punched cards, in the card reader; placing the program to be compiled after the compiler in the card reader; and then pressing the "load" button on the reader. The output would be a set of punched cards called "intermediate output." If there were any compilation errors, a light would flash on the console, and error messages would appear on the console typewriter. If everything had gone well so far, the next step would be to load the second pass of the Fortran compiler just like the first pass, putting the intermediate output in the card reader as well. If the second pass succeeded, the output was a second set of punched cards called the "executable deck." The third step was to shuffle the executable deck slightly, load it along with a massive subroutine library (another 10 inches of cards), and observe the program as it ran.

The facilities for observing the results were limited: console lights, output on a typewriter, punched cards, and line-printer output. Frequently, the output was wrong. Debugging often took the form of peeking directly into memory and even patching the executable program by using console switches. If there was not enough time to finish, a frustrated user might get a line-printer dump of memory to puzzle over at leisure. If the user finished before the end of the allotted time, the machine might sit idle until the next reserved block of time.

The IBM 1620 was quite small, slow, and expensive by our standards. It came in three models, ranging from 20K to 60K digits of memory (each digit was represented by 4 bits). Memory was built from magnetic cores, which required approximately 10 microseconds for a read or a write. The machine cost hundreds of thousands of dollars and was physically fairly large, covering about 20 square feet.

Operator-Driven Shop Organization

The economics of massive mainframe computers made idle time very expensive. In an effort to avoid such idleness, installation managers instituted several modifications to the open shop mechanism just outlined. An *operator* was hired to perform the repetitive tasks of loading jobs, starting the computer, and collecting the output. The operator was often much faster than ordinary users at such chores as mounting cards and magnetic tapes, so the setup time between job steps was reduced. If the program failed, the operator could have the computer produce a dump. It was no longer feasible for users to inspect memory or patch programs directly. Instead users would submit their runs, and the operator would run them as soon as possible. Each user was charged only for the amount of time the job required.

The operator often reduced setup time by batching similar job steps. For example, the operator could run the first pass of the Fortran compiler for several jobs, save all the intermediate output, then load the second pass and run it across all the intermediate output that had been collected. In addition, the operator could run jobs out of order, perhaps charging more for giving some jobs priority over others. Jobs that were known to require a long time could be delayed until night. The operator could always stop a job that was taking too long.

The operator-driver shop organization prevented users from fiddling with console switches to debug and patch their programs. This stage of operating system development introduced the long-lived tradition of the users' room, which had long tables often overflowing with oversized fan-fold paper and a quietly desperate group of users debugging their programs until late at night.

Offline Loading

The next stage of development was to automate the mechanical aspects of the operator's job. First, input to jobs was collected *offline* by a separate computer (sometimes called a "satellite") whose only task was the transfer from cards to tape. Once the tape was full, the operator mounted it on the main computer. Reading jobs from tape is much faster than reading cards, so less time was occupied with input/output. When the computer finished the jobs on one tape, the operator would mount the next one. Similarly, output was generated on to tape, an activity that is much faster than punching cards. This output tape was converted to line-printer listings offline.

A small *resident monitor* program, which remained in memory while jobs were executing, reset the machine after each job was completed and loaded the next one. Conventions were established for control cards to separate jobs and specify their requirements. These conventions were the beginnings of command languages. For example, one convention was to place an asterisk in the first column of control cards, to distinguish them from data cards. The compilation job we just described could be specified in cards that looked like this:

```
* JOB SMITH             The user's name is Smith.
*    PASS CHESTNUT      Password so others can't use Smith's account
*    OPTION TIME = 60   Limit of 60 seconds
*    OPTION DUMP = YES  Produce a dump if any step fails.
* STEP FORT1            Run the first pass of the Fortran compiler.
*    OUTPUT TAPE1       Put the intermediate code on tape 1.
*    INPUT FOLLOWS      Input to the compiler comes on the next cards.
    ...                 Fortran program
* STEP FORT2            Run the second pass of the Fortran compiler.
*    OUTPUT TAPE2       Put the executable deck on scratch tape 2.
*    INPUT TAPE1        Input comes from scratch tape 1.
* STEP LINK             Link the executable with the Fortran library.
*    INPUT TAPE2        First input is the executable.
*    INPUT TAPELIB      Second input is a tape with the library.
*    OUTPUT TAPE1       Put load image on scratch tape 1.
* STEP TAPE1            Run whatever is on scratch tape 1.
*    OUTPUT TAPEOUT       Put output on the standard output tape.
*    INPUT FOLLOWS        Input to the program comes on the next cards.
    ...                   Data
```

The resident monitor had several duties:

- Interpret the command language.
- Perform rudimentary accounting.
- Provide device-independent input and output by substituting tapes for cards and line printers.

This last duty is an early example of information hiding and abstraction: programmers would direct output to cards or line printers, but in fact, the output would go elsewhere. Programs called subroutines provided by the resident monitor for input/output to both logical devices (cards, printers) and physical devices (actual tape drives).

The early operating systems for the IBM 360 series of computer used this style of control. Large IBM 360 installations could cost millions of dollars, so it was important not to let the computer sit idle.

Spooling Systems

Computer architecture advanced throughout the 1960s. Input/output units were designed to run at the same time the computer was computing. They generated an interrupt when they finished reading or writing a record instead of requiring the resident monitor to track their progress. An interrupt causes the computer to save some critical information (such as the current program counter) and to jump to a location specific to the kind of interrupt. Device-service routines, known as *device drivers*, were added to the resident monitor to deal with these interrupts.

Drums, and later, disks were introduced as a secondary storage medium. Now the computer could be computing one job while reading another on to the drum and printing the results of a third from the drum. Unlike a tape, a drum allows programs to be stored anywhere, so there was no need for the computer to execute jobs in the same order in which they were entered. A primitive *scheduler* was added to the resident monitor to sort jobs based on priority and amount of time needed, both specified on control cards. The operator was retained to perform several tasks.

- Mount data tapes needed by jobs (specified on control cards, which caused request messages to appear on the console typewriter).
- Decide which priority jobs to run and which to hold.
- Restart the resident monitor when it failed or was inadvertently destroyed by the running job.

This mode of running a computer was known as a *spooling system*, and its resident monitor was the start of modern operating systems. (The word "spool" originally stood for "simultaneous peripheral operations online," but it is easier to picture a spool of thread, where new jobs are wound on the outside, and old ones are extracted from the inside.) One of the first spooling systems was HASP (the Houston Automatic Spooling Program), an add-on to OS/360 for the IBM 360 computer family.

Batch Multiprogramming

Spooling systems did not make efficient use of all of the hardware's resources. The currently running job might not need the entire memory. A job performing input/output causes the computer to wait until the input/output finishes. The next software improvement, which occurred in the early 1960s, was the introduction of *multiprogramming*, a scheme in which more than one job is active simultaneously.

Under multiprogramming, while one job waits for an input/output operation to complete, another can compute. With luck, no time at all is wasted waiting for input/output. The more simultaneous jobs, the better. However, a *compute-bound* job (one that performs little input/output but much computation) can easily prevent *input/output-bound* jobs (those that perform mostly input/output) from making progress.

Multiprogramming also introduces competition for memory. The number of jobs that can be accommodated at one time depends on the size of memory and the hardware available for subdividing that space. In addition, jobs must be secured against inadvertent or malicious interference or inspection by other jobs. It is more critical now that the resident monitor not be destroyed by errant programs, because not one but many jobs suffer if it breaks.

The form of multiprogramming we have been describing is often called *batch multiprogramming* because jobs are grouped into batches: those that need small memory, those that need customized tape mounts, those that need long execution, and so forth. Each batch might have different priorities and fee structures. Some batches (such as large-memory long-execution jobs) can be scheduled for particular times (such as weekends or late at night). Generally, one job from any batch can run at a time.

Each job is divided into discrete steps. Since job steps are independent, the resident monitor can separate them and apply policy decisions to each step independently. Each step might have its own time, memory, and input/output requirements. In fact, two separate steps of the same job can be performed at the same time if they don't depend on each other. The term *process* was introduced in the late 1960s to mean the entity that performs a single job step. The operating system (as the resident monitor may now be called) represents each process by a data structure sometimes called a *process descriptor*, *process control block*, or *context block*. The process control block includes billing information (owner, time used), scheduling information, and the resources the job step needs. While it is running, a process may request assistance from the kernel by submitting a system call across the *process interface*. Executing programs are no longer allowed to control devices directly; otherwise, they could make conflicting use of devices and prevent the kernel from doing its work. Instead, processes must use system calls to access devices, and the kernel has complete control of the *device interface*.

Allocating resources to processes is not a trivial task. A process might require resources (like tape drives) at various stages in its execution. If a resource is not available, the scheduler might block the process from continuing until later. The scheduler must take care not to block any process forever.

Along with batch multiprogramming came new ideas for structuring the operating system. The kernel of the operating system is composed of routines that manage memory, CPU time, devices, and other resources. It responds both to requests from processes and to interrupts from devices. In fact, the kernel runs only when it is invoked either from above, by a process, or below, by a device. If no process is ready to run and no device needs attention, the computer sits idle.

Various activities within the kernel share data, but they must not be interrupted when the data are in an inconsistent state. Mechanisms for *concurrency control* were developed to ensure that these activities do not interfere with each other. The MVS operating system for the IBM 360 family was one of the first to use batch multiprogramming.

Interactive Multiprogramming

The next step in the development of operating systems was the introduction of *interactive multiprogramming*, also called *timesharing*. The principal user-oriented input/output device changed in the late 1960s from cards or tape to an interactive terminal. Instead of packaging all the data that a program might need before it starts running, the interactive user is able to supply input as the program wants it. The data can depend on what the program has produced so far. Among the first terminals were teletypes, which produced output on paper at perhaps 10 characters per second. Later terminals were called "glass teletypes" because they displayed characters on a television screen, substituting electronics for mechanical components. Like a regular teletype, they could not back up to modify data sitting earlier on the screen. Shortly thereafter, terminals gained cursor addressability, which meant that programs could show entire "pages" of information and change any character anywhere on a page.

Interactive computing caused a revolution in the way computers were used. Instead of being treated as number crunchers, computers became information manipulators. Interactive text editors allowed users to construct data files online. These files could represent programs, documents, or data. As terminals improved, so did the text editors, changing from line- or character-oriented interfaces to full-screen interfaces.

Instead of representing a job as a series of steps, interactive multiprogramming identifies a *session* that lasts from initial connection ("login") to the point at which that connection is broken ("logout"). During login, the user typically gives two forms of identification: a name and a password. The password is not echoed at the terminal, or is at least blackened by overstriking garbage, to avoid disclosing it to onlookers. These data are converted into a *user identifier* that is associated with all the processes that run on behalf of this user and all the files they create. This identifier helps the kernel decide whom to bill for services and whether to permit various actions such as modifying files.

During a session, the user imagines that the resources of the entire computer are devoted to this terminal, even though many sessions may be active simultaneously for many users. Typically, one process is created at login time to serve the user. That first process, which is usually a command interpreter, may start others as needed to accomplish individual steps.

Users need to save information from session to session. Magnetic tape is too unwieldy for this purpose. Disk storage became the medium of choice for data storage, both short term (temporary files used to connect steps in a computation), medium term (from session to session), and long-term (from year to year). Issues of disk space allocation and backup strategies needed to be addressed to provide this facility.

Interactive computing was sometimes added into an existing batch multiprogramming environment. For example, TSO ("timesharing option") was an add-on to the OS/360 operating system. The EXEC-8 operating system for Univac computers included an interactive component, too.

Later operating systems were designed from the outset to support interactive use, with batch facilities added when necessary. TOPS-10 and Tenex (for the Digital PDP-10) and almost all operating systems developed since 1975, including Unix (first on the Digital PDP-11), MS-DOS (Intel 8086), OS/2 (Intel 286 family [10]), VMS (Digital VAX [9]), and all their descendents, are designed mainly for interactive use.

Graphical User Interfaces (GUIs)

As computers became less expensive, the time cost of switching from one process to another (which happens frequently in interactive computing) became insignificant. Idle time also became unimportant. Instead, the goal became helping users get their work done efficiently. This goal led to new software developments, enabled by improved hardware.

Graphics terminals, first introduced in the mid-1970s, have led to the video monitors that are now ubiquitous and inexpensive. These monitors allow individual control of multicolored pixels; a high-quality monitor (along with its video controller) can display millions of pixels in an enormous range of colors. Pointing devices, particularly the mouse, were developed in the late 1970s. Software links them to the display so that a visible cursor reacts to physical movements of the pointing device.

The earliest GUIs were just rectangular regions of the display that contained, effectively, a cursor-addressable glass teletype. These regions are called "windows." The best-known windowing packages were those pioneered by Mac OS [15] and the later ones introduced by MS-Windows, OS/2 [10], and X Windows (for Unix, VMS, and other operating systems [12]). Each has developed from simple rectangular models of a terminal to significantly more complex displays.

Programs interact with the hardware by invoking routines in libraries that know how to communicate with the display manager, which itself knows how to place bits on the screen. The early libraries were fairly low-level and hard to use; toolkits (in the X Windows environment), especially ones with a fairly small interpreted language (such as Tcl/Tk [13] or Visual Basic) have eased the task of building good GUIs. Early operating systems that supported graphical interfaces, such as Mac OS and MS-Windows, provided interactive computing but not multiprogramming. Modern operating systems all provide multiprogramming as well as interaction, allowing the user to start several activities and to switch attention to whichever one is currently most interesting.

Distributed Computing

At the same time that displays were improving, networks of computers were being developed. A network requires not only hardware to physically connect machines, but also protocols to use that hardware effectively, operating system support to make those protocols available to processes, and applications that make use of these protocols.

Computers can be connected together by a variety of devices. The spectrum ranges from tight coupling, where several processing units share memory, to very loose coupling, where a number of computers belong to the same international network and can send one another messages.

The ability to send messages between computers opened new opportunities for operating systems. Individual machines become part of a larger whole, and in some ways, the operating system begins to span networks of machines. Cooperation between machines takes many forms.

- Each machine may offer *network services* to others, such as accepting mail, providing information on who is currently logged in, telling what time it is (important in keeping clocks synchronized), allowing users to access machines remotely, and transferring files.
- Machines within the same *site* (typically, those under a single administrative control) may *share file systems* in order to reduce the amount of disk space needed and to allow users to have accounts on multiple machines. Novell nets (MS-DOS), the Sun and Andrew network file systems (Unix), and the Microsoft File-Sharing Protocol (Windows XP) are examples of such arrangements. Shared file systems are an essential component of a *networked operating system*.
- Once users have accounts on several machines, they want to associate graphical windows with sessions on different machines. The machine on which the display is located is called a *thin client* of the machine on which the processes are running. Thin clients have been available from the outset for X Windows; they are also available under Windows 2000 and successors.
- Users want to execute computationally intensive algorithms on many machines in parallel. *Middleware*, usually implemented as a library to be linked into distributed applications, helps programmers build such applications. PVM [6] and MPI [14] are examples of such middleware.

Standardized data formats and conversation rules, together called *protocols*, developed rapidly starting in the 1970s. Low-level protocols such as internet protocol (IP) and transmission-control protocol (TCP) define services for addressing data packets between machines and reliable, in-order routing and delivery of message streams. Higher-level protocols typically use TCP and IP. The file-transfer protocol (FTP) service was developed in the early 1970s as a way of transferring files between machines connected on a network. The simple mail-transfer protocol (SMTP) originated at about the same time. In those days, electronic mail was limited to academic institutions, and the protocol developers were not worried about malicious misuse of the protocol. SMTP is still in use, and many e-mail attacks and spam dissemination methods depend on its lack of source authentication. In the early 1990s, the *gopher* service was developed to create a uniform interface for accessing information across the internet. Information is more general than just files; it can be a request to run a program or to access a database. Each machine that wishes to can provide a server that responds to connections from any site and communicate a menu of available information. This service was superseded in 1995 by the World Wide Web which supports a GUI to gopher, FTP, and HTTP, the hypertext transfer protocol (for retrieving documents with links internally and to other documents, often at other sites, and including text, pictures, video, audio, and remote execution of packaged commands).

All these forms of cooperation introduce security concerns. Each site has a responsibility to maintain security if for no other reason than to prevent malicious users across the network from using the site as a breeding ground for nasty activity, such as attacking other sites.

18.3 Goals of an Operating System

During the evolution of operating systems, their purposes have also evolved. The best-known operating systems are intended for individual users or communities of users performing interactive work. Such operating systems have four major goals.

- Hide details of hardware by creating abstractions that application programmers can use.
- Manage resources that processes need.
- Provide a pleasant and effective user interface.
- Provide a secure and reliable computing environment.

Specialty operating systems have other goals as well:

- Combine the computing power of many computers, possibly geographically remote, into an integrated whole. Such *distributed operating systems* must deal with issues such as synchronization, distributed file systems, distributed shared memory, and process placement and migration. Distributed operating systems, such as Mach [17], Locus [16], and MOSIX [7], are primarily research tools.
- Continue to function even if some components fail. Such *fault-tolerant operating systems* attempt to continue to function even when hardware or software components fail. Fault-tolerant operating systems are especially important in banking and online business applications. Fault tolerance is based on redundancy, both to detect failure and to adjust to it. The RAID organization of disks, although primarily a means to improve performance, is also fault tolerant. Checkpoints of programs and data are a common form of redundancy; they allow computations to be restored if they are interrupted by failures.
- Handle time-critical operations. *Real-time operating systems* schedule processes based on how responsive they must be, typically by associating deadlines with processes, but also by using priority and time-slice reservation schemes. Real-time operating systems are important for process control, such as controlling chemical-manufacturing plants or the control surfaces of airplanes in flight.
- Control a specialized device. *Embedded computers* are used to control automobile engines, industrial robots, network routers, video cassette recorders, and far more. Many peripheral devices attached to ordinary computers, such as disk drives and video cards, have embedded computers. Although the simplest controllers don't need much of an operating system, many embedded computers run operating systems such as Linux or QNX [8]. Operating systems for embedded computers often have real-time aspects.

Abstracting Hardware

We can distinguish between the *physical* world of devices, instructions, memory, and time, and the *virtual* world that is the result of abstractions built by the operating system. An *abstraction* is software (often implemented as a subroutine or as a library of subroutines) that hides lower-level details and provides a set of higher-level functions. Programs that use the abstraction can safely ignore the lower-level (physical) details; they need only deal with the higher-level (virtual) structures.

Why is abstraction important in operating systems? First, the code needed to control peripheral devices is often not standardized; it can vary from brand to brand, and it certainly varies between, say, IDE disks, SCSI tape drives, and USB keyboards. Input/output devices are complex to program efficiently and correctly. Abstracting devices with a uniform interface makes programs easier to write and to modify (for example, to use a different device). Operating systems provide subroutines called *device drivers* that perform input/output operations on behalf of programs. The operations are provided at a much higher level than the device itself provides. For example, a program may wish to write a particular block on a disk. Low-level methods involve sending commands directly to the disk to move the read-write head to the right block and then to undertake memory-to-disk data transfer. When the transfer is complete, the disk interrupts the running program. A low-level program needs to know the format of disk commands, which vary from manufacturer to manufacturer, and must deal with interrupts. In contrast, a program using a high-level routine in the operating system might only need to specify the memory location of the data block and where it belongs on the disk; all the rest of the machinery is hidden.

Second, the operating system introduces new functions as it abstracts the hardware. In particular, operating systems introduce the "file" abstraction. Programs do not need to deal with disks at all; they can use high-level routines to read and write disk files (instead of disk blocks) without needing to design storage layouts, worry about disk geometry, or allocate free disk blocks. The "file" abstraction can then apply to other storage devices, including tapes, compact disks, and network-accessible data. Programs usually don't need to know what the physical device is, only the fact that data can be accessed according to a standard set of routines.

Third, the operating system transforms the computer hardware into multiple virtual computers, each belonging to a different process. Each process views the hardware through the lens of abstraction: memory, time, and other resources are all tailored to the needs of the process. Processes see only as much memory as they need, and that memory does not contain the other processes (or the operating system) at all. They can behave as if they have all the CPU cycles on the machine, although other processes and the operating system itself are competing for those cycles. System calls allow processes to start other processes and to communicate with other processes, either by sending messages or by sharing memory.

Fourth, the operating system can enforce security through abstraction. The operating system must secure both itself and its processes against accidental or malicious interference. Certain instructions of the machine, notably those that halt the machine and those that perform input and output, are moved out of the reach of processes. Memory is partitioned so that processes cannot access each other's memory. Time is partitioned so that even a run-away process will not prevent others from making progress.

Managing Resources

An operating system is not only an abstractor of information, but also an allocator that controls how processes (the active agents) may access resources (passive entities).

A *resource* is a commodity necessary to get work done. The computer's hardware provides a number of low-level resources. Working programs need to reside somewhere in memory, must execute instructions, and need some way to accept data and present results. These needs are related to the fundamental resources of *memory*, *CPU time*, and *input/output*. The operating system abstracts these resources to allow them to be shared.

In addition to these physical resources, the operating system creates virtual, abstract resources. For example, *files* are able to store data. They abstract the details of disk storage. *Pseudo-files* (that is, objects that appear to be data files on disk but are in fact stored elsewhere) can also represent devices, processes, communication ports, and even data on other computers. *Sockets* are process-to-process communication channels that can cross machine boundaries, allowing communication through networks like the Internet. Sockets abstract the details of transmission media and network protocols.

Still higher-level resources can be built on top of abstractions. A *database* is a collection of information, stored in one or more files with structure intended for easy access. A *mailbox* is a file with particular semantics. A *remote file*, located on another machine but accessed as if it were on this machine, is built on both file and network abstractions.

The resource needs of processes often interfere with each other. Resource managers in the operating system include policies that try to be fair in giving resources to the processes and allow as much computation to proceed as possible. These goals often conflict.

Each resource has its own manager, typically in the kernel.

The memory manager allocates regions of main memory for processes. Modern operating systems use address translation hardware that maps between a process's *virtual addresses* and the underlying *physical addresses*. This hardware usually partitions address spaces into equal-sized *pages*, where a page is typically 4KB. When a page in a process's virtual address is in use, all its addresses are mapped to equivalent offsets within a page of the computer's physical memory. Only the currently active part of a process's virtual space needs to be physically resident; the memory manager keeps the rest on backing store (usually a disk) and brings it in as the process needs to access it. Some operating systems also provide *light-weight processes* that share a single virtual space. In addition, parts of virtual space used for programs (as opposed to data) are often shared among processes that are using the same program or library routine. This sharing reduces the contention for main memory and can also make it faster to start up new processes. The memory manager includes policies that determine how much physical memory to grant to each process and which region of physical memory to swap out to make room for other memory that must be swapped in.

The CPU-time manager is called the *scheduler*. Schedulers usually implement a preemptive policy that forces processes to take turns running. Schedulers categorize processes according to whether they are currently runnable (they may not be if they are waiting for other resources) and their priority.

The file manager mediates process requests such as creating, reading, and writing files. It validates access based on the identity of the user running the process and the permissions associated with the file. The file manager also prevents conflicting accesses to the same file by multiple processes. It translates input/output requests into device accesses, usually to a disk, but often to networks (for remote files) or other devices (for pseudo-files).

The device managers convert standard-format requests into the particular commands appropriate for individual devices, which vary widely among device types and manufacturers. Device managers may also maintain caches of data in memory to reduce the frequency of access to physical devices.

Although we usually treat processes as autonomous agents, it is often helpful to remember that they act on behalf of a higher authority: the human *users* who are physically interacting with the computer. Each process is usually "owned" by a particular user. Many users may be competing for resources on the same machine. Even a single user can often make effective use of multiple processes.

Each user application is performed by a process. A word-processing application runs as a process that receives keystrokes (either from the kernel or from a window-server process), converts them into changes in the document, and displays the modified document (either by calls to the kernel or by communicating with a window-server process). A mail application runs as a process that uses SMTP to send documents to mailboxes.

To service requests effectively, the operating system must satisfy two conflicting goals:

- To let each process have whatever resources it wants
- To be fair in distributing resources among the processes

If the active processes cannot all fit in memory, for example, it is impossible to satisfy the first goal without violating the second. If there is more than one process, it is impossible on a single CPU to give all processes as much time as they want; CPU time must be shared.

To satisfy the computer's owner, the operating system must also satisfy a different set of goals:

- To make sure the resources are used as much as possible
- To complete as much work as possible

These latter goals were once more important than they are now, when computers were all expensive mainframes, and it seemed wasteful to let any time pass without some process performing computations, to let any memory sit unoccupied by a process, or to let a tape drive sit idle. The measure of success of an operating system was how much work (measured in "jobs") could be finished and how heavily resources were used. Computers are now far less expensive; responsiveness is a more important criterion than resource usage level.

User Interface

We have seen how operating systems are creators of abstractions and allocators of resources. Both these aspects are centered on the needs of programmers and the processes that execute programs. But many users are not programmers and are uninterested in the process abstraction and in the interplay between processes and the operating system. They don't care about system calls, interrupts, and devices. Instead, they are interested in what might be termed the "look and feel" of the operating system.

The user interacts with the operating system through the *user interface*. Here we will only point out some highlights.

The hardware for user interfaces has seen rapid change over the last 50 years, ranging over plugging wires into a plug board (example: IBM 610, 1957), punching cards and reading printouts (IBM 1620, 1959), remote teletype (DEC PDP-10, 1967), monochrome glass teletypes (around 1973), monochrome graphics terminals with pointing devices (Xerox PARC's Alto computer, around 1974), color video CRTs (around 1980), and LCD monitors (late 1980s).

User-interface software has steadily changed as well. Interactive text editors (WYLBUR and TECO, around 1975) replaced punched paper cards. Interactive command languages replaced job-control languages. Programming environments integrating editing, compiling, and debugging were introduced as early as 1980 (Smalltalk) and are still in heavy use (MetroWerks CodeWarrior; Microsoft Visual Studio). Data entry moved from line-oriented to forms-based (by 1980) to web-based (1995). Many user interfaces are now navigated without needing a keyboard at all; the user clicks a mouse to move to the next step in a process. Voice-activated commands are also gaining in popularity.

The look and feel of an operating system is influenced by many components of the user interface. Some of the most important are the process launcher (a command interpreter, a menu-driven GUI, or clickable icons), the file system (including remote files), online help, and application integration (such as ability to move pictures from a photo editor into a word processor).

Security and Reliability

Security comprises four general aspects.

- Authentication and access control: Parties can be reliably identified (authentication) in order to decide if they are authorized to access data (access control). In addition, it is sometimes important that the party that creates or otherwise accesses data cannot deny that access (nonrepudiation).
- Secrecy (confidentiality): Sensitive information is not divulged to unauthorized parties.
- Protection (integrity): Sensitive information cannot be modified by unauthorized parties. In some situations, such as network communication, it is only possible to guarantee that modifications are detectable.
- Availability: Malicious or malfunctioning parties cannot interfere with proper functioning of the computer.

These concerns pervade operating systems, both at the kernel and application levels. For example, secrecy concerns suggest that we be able to prevent one process from looking at the virtual memory of another process, one user from reading files written by another user, and one computer from interpreting communication between other computers.

Security depends ultimately on hardware assistance. All modern computers provide several *processor states*. Most architectures provide at least two states, called *privileged state* and *nonprivileged state*.

Processes always run in nonprivileged state. Instructions such as those that perform input/output and those that change processor state cause traps when executed in nonprivileged state. Just like interrupts, traps save the current execution context (perhaps on a stack), force the processor to jump to the kernel, and enter privileged state. Once the operating system has finished servicing the trap or interrupt, it returns control to the same process or perhaps to a different one, resetting the computer into nonprivileged state.

The kernel of the operating system runs in privileged state. All instructions have their usual, physical meanings in this state. The kernel only runs when a process has caused a trap or when a peripheral device has generated an interrupt. Traps do not necessarily represent errors; they can also be system calls or attempts to address legitimate virtual memory that is currently swapped to backing store. Interrupts often indicate that a device has finished servicing a request and is ready for more work. The clock interrupts at a regular rate in order to let the kernel make scheduling decisions.

If the operating system makes use of this dichotomy of states, the abstractions that the operating system provides are presented to processes as system calls, which are like new CPU instructions. (Each operating system defines its own set of system calls. Unix variants typically provide about 200 different system calls; Microsoft Windows, which includes both kernel and windowing components, has at least an order of magnitude more.) The physical devices are completely hidden from process view. A program can perform high-level operations (like reading a file) with a single system call. Executing the system call generates a trap, which causes a switch to the privileged state of the kernel. The advantage of the service-call design over a procedure-call design is that it allows access to kernel operations and data only through well-defined entry points.

Not all operating systems make use of non-privileged state. MS-DOS, for example, runs all applications in privileged state. System calls are essentially subroutine calls. Although the operating system provides device and file abstractions, processes may interact directly with disks and other devices. One advantage of this choice is that device drivers can be loaded after the operating system starts; they do not need special privilege. One disadvantage is that viruses can thrive because nothing prevents a program from placing data anywhere it wishes.

Because it runs in privileged state, the kernel must be programmed carefully to avoid taking actions on behalf of a process that violate security. Otherwise, a malicious process might fool the kernel into modifying some other process's memory, scribbling bad data on a disk, or revealing the content of incoming network traffic.

Each process is granted software privileges, usually based on the user on whose behalf it is running. For example, a particular user's processes generally have the privilege to read and write that user's files. These software privileges are distinct from the hardware privileged state, which processes never have. Instead, software privileges are policed by the kernel, which decides whether or not to honor a service request from a process based on the privileges of the requesting process and the nature of the request.

Most operating systems recognize specially privileged pseudo-users, such as "root" (Unix) and "Administrator" (Windows 2000). The kernel allows processes owned by these privileged pseudo-users a wider set of abilities, such as accessing any file, creating raw (protocol-independent) network packets, and shutting down the operating system. Instead of a single all-powerful pseudo-user, operating systems can categorize the sorts of special privileges that a process might need and grant them individually to special processes.

A user who logs into an operating system first interacts with a privileged login process, which authenticates the user and then starts other processes owned by that user to accept further interaction. Authentication typically is based on a user name and a password, although biometric devices and more complex challenge-response protocols can increase the reliability of authentication. Cryptographic techniques allow the login process to authenticate a user name and password without actually storing the passwords in a file. Such a file would be a target of attacks attempting to break its secrecy.

It is quite common for computers to run background processes such as web and mail servers to provide network-accessible functions. Such processes are usually called *daemons*. Daemons often need to run as privileged processes in order to open sensitive network connections and to access arbitrary files. It is therefore important that daemons be written with care to prevent them from being confused into misapplying their privilege. The worst security holes allow a remote intruder access to a privileged process running a general-purpose command interpreter on the attacked machine. Such an intruder can potentially subvert the computer to do any action, possibly hiding this subversion from casual detection. Unfortunately, daemons have historically been quite susceptible to such attacks, often because they rely on library packages that are insufficiently careful with overflows (a programming error) or because the daemons are willing to treat incoming data as a set of commands.

Computer administrators can take several steps to ensure daemon security: (1) stay tuned to news sources announcing discovered daemon flaws, (2) upgrade daemons when a security flaw is announced, (3) only run those daemons that are needed on a particular computer, (4) route all network traffic through *firewalls*, which are configurable rule-based tools that only allow permissible network traffic through.

Even unprivileged, ordinary applications must be written with care. Web browsers and mail readers can be configured to accept programs from the network and run them. The programs might be malicious, in which case we call them *malware*. Even though the malware only runs with the user's privilege, it can still misbehave, modifying or divulging the user's files, tracking the user's actions, sending out unsolicited mass mail, or participating in a distributed denial-of-service (overloading) attack against a target computer somewhere on the network. Malware spreads primarily via e-mail; many mail readers allow users to accidentally run incoming programs. Computer administrators often route incoming e-mail through scanning programs that detect and remove potentially malicious attachments.

In addition to privilege control, applications often use cryptography to achieve security goals with respect to data to be stored on computers and transmitted over the network.

- Authentication and nonrepudiation: Data can be marked in such a way that identifies the party that generated the data.
- Secrecy: Data can be encrypted so only an authorized party can decrypt it; encrypted data carries no information to unauthorized parties.
- Integrity: Data can be marked in such a way that any modification to the data is apparent and can cause the data to be rejected. This technique can be used to assure a computer administrator that important programs have not been replaced with counterfeits.

Cryptographic methods are often complex, involving multi-step protocols that experts must scrutinize for susceptibility to attacks. There are currently no provably unbreakable encryption methods, so experts rely on a fundamental tenet: An encryption method is considered trustworthy in proportion to the amount of effort that researchers with full knowledge of the inner workings of the method have spent trying, unsuccessfully, to break it. The RSA and AES encryption methods, for instance, are considered very secure by this measure; unpublished methods are not.

18.4   Implementing an Operating System

As mentioned earlier, the core of the operating system is the *kernel*, a control program that functions in privileged state, reacting to interrupts from external devices and to service requests and traps from processes. Generally, the kernel is a permanent resident of the computer. It creates and terminates processes and responds to their requests for service.

Processes

Each process is represented in the kernel by a collection of data called the *process descriptor*. A process descriptor includes such information as the following.

- Processor state: stored values of the program counter and registers, needed to resume execution of the process.
- Scheduling statistics, needed to determine when to resume the process and how much time to let it run.
- Memory allocation, both in main memory and backing store (disk), needed to accomplish memory management.
- Other resources held, such as locks or semaphores, needed to manage contention for such resources.
- Open files and pseudo-files (devices, communication ports), needed to interpret service requests for input and output.
- Accounting statistics, needed to bill users and determine hardware usage levels.
- Privileges, needed to determine if activities such as opening files and executing potentially dangerous system calls should be allowed.
- Scheduling state: running, ready, waiting for input/output or some other resource, such as memory.

The process descriptors are indexed by a unique identifier, usually called a *process number*. Some of the information in the process descriptor can be bulky, such as the memory-management information, which includes process-specific tables for converting virtual to physical addresses. Bulky information for idle processes might be stored on disk in order to save space in main memory. Some information may be shared by several processes, such as a case where one process opens a file and then starts a second process, which also may access the open file.

Resuming a process, that is, switching control from the kernel back to the process, is a form of *context switching*. It requires that the processor move from privileged to unprivileged state, that the registers and program counter of the process be restored, and that the address-translation hardware be set up to accomplish the correct mappings for this process. Switching back to the kernel is also a context switch; it can happen when the process tries to execute a privileged instruction (including the system-call instruction) or when a device generates an interrupt.

Hardware is designed to switch context rapidly. For example, the hardware may maintain two sets of registers and address translation data, one for each privilege level. Context switches into the kernel just require moving to the kernel's set of registers. Resuming the most-recently running process is also fast. Resuming a different process requires that the kernel load all the information for the new process into the second set of registers; this activity takes longer. For that reason, a *process switch* is often more expensive than two context switches.

Components of the Kernel

Originally, operating systems were written as a single large program encompassing hundreds of thousands of lines of assembly-language instructions. Two trends have made the job of implementing operating systems less difficult. First, high-level languages have made programming much easier. For instance, over 99% of Linux is written in the C language. Complex algorithms can be expressed in a structured, readable fashion, code can be partitioned into modules that interact with each other in a well-defined manner, and compile-time type checking can catch most programming errors. Only a few parts of the kernel, such as those that switch context or modify execution priority, need to be written in assembly language.

Second, the discipline of structured programming has suggested a layered approach to designing the kernel. Each layer provides abstractions needed by the layers above it. For example, the kernel can be organized as follows:

- Context-switch and process-switch services (lowest layer)
- Device drivers
- Resource managers for memory and time
- File-system support
- Service-call interpreter (highest layer)

For example, the MS-DOS operating system provides three levels: (1) device drivers (the BIOS section of the kernel), (2) a file manager, and (3) an interactive command interpreter. It supports only one process and provides no security, so there is no need for context-switch services. Because system calls do not need to cross protection boundaries, they are implemented as subroutine calls.

The concept of layering allows the kernel to be small, since much of the work of the operating system need not operate in a protected and hardware-privileged environment. When all the layers listed above are privileged, the organization is called a *macrokernel*. Unix is often implemented as a macrokernel.

Even a macrokernel is structured internally into modules. Typical modules include:

- Interrupt and trap handling, including initialization to ensure that each kind of interrupt and trap is directed to the right part of the kernel.
- Synchronization, protecting kernel data structures from access by one part of the kernel when another part is still using them.
- Timing, both to maintain information about the current time and to set alarms for particular intervals in the future.
- Memory, to dynamically allocate physical memory to the kernel, to devices, and to processes, including decisions concerning swapping.
- Device handling, comprising drivers to control peripheral devices.
- Networking, including implementation of essential low-level protocols such as IP and TCP.
- Process handling, including routines for starting and scheduling processes.
- System-call handling, responding to requests from processes.
- File-system manipulation, possibly including multiple kinds of file types (like VFAT and NTFS) on the same computer.

Because any particular computer typically has only a few of a wide variety of devices, it only needs those drivers that deal with the devices that are actually present. Similarly, only it only needs file-system modules that deal with the kinds of file systems for which this particular machine is configured. During initialization, kernel modules register themselves, which means that they insert entries into kernel tables indicating their presence and where their code resides. Some operating systems, such as Linux, allow modules to be added to and subtracted from a running kernel. When a device is plugged into a USB port, a daemon identifies it and then loads the appropriate module into the kernel. When the device is unplugged, the daemon unloads the module. This dynamic choice of modules allows distributors of the operating system to provide a single, fairly small kernel that runs on a wide variety of computers. In particular, a Linux kernel is typically only 2MB before modules.

If the kernel only contains code for process creation, inter-process communication, the mechanisms for memory management and scheduling, and the lowest level of device control, the result is a *microkernel*, also called a "communication kernel." Mechanisms are distinct from policies, which can be outside the kernel. Policies decide which resources should be allocated in cases of conflict, whereas mechanisms carry out those decisions. Mach [17] and QNX [8] follow the microkernel approach. In this organization, services such as the file system and policy modules for scheduling and memory are relegated to processes. These processes are often referred to as **servers**; the ordinary processes that need those services are called their **clients**. The microkernel itself acts as a client of the policy servers. Servers need to be trusted by their clients, and sometimes they need to execute with some degree of hardware privilege (for example, if they access devices).

The microkernel approach has some distinct advantages.

- It imposes uniformity on the requests that a process might make. Processes need not distinguish between kernel-level and process-level services, since all are provided via messages to servers.
- It allows easier addition of new services, even while the operating system is running, as well as multiple services that cover the same set of needs, so that individual users (and their agent processes) can choose whichever seems best. For example, different file organizations for diskettes are possible; instead of having many file-level modules in the kernel, there can be many file-level servers accessible to processes.
- It allows an operating system to span many machines in a natural way. As long as inter-process communication works across machines, it is generally immaterial to a client where its server is located.
- Services can be provided by teams of servers, any one of which can help any client. This organization relieves the load on popular servers, although it often requires a degree of coordination among the servers on the same team.

Microkernels also have some disadvantages. It is generally slower to build and send a message, accept and decode the reply (taking about $10\mu s$), than to make a single system call (taking about $0.1\mu s$). However, other aspects of service tend to dominate the cost, allowing microkernels to be similar in speed to macrokernels. Keeping track of which server resides on which machine can be complex. This complexity may be reflected in the user interface. The perceived complexity of an operating system has a large effect on its acceptance by the user community.

The trend toward microkernels is apparent only in academia and embedded systems. Macrokernels are likely to remain popular for the forseeable future.

18.5   Research Issues and Summary

Operating systems have developed enormously in the last 45 years. Modern operating systems generally have three goals: To hide details of hardware by creating abstractions, to allocate resources to processes, and to provide a effective user interface. Operating systems generally accomplish these goals by running processes in low privilege and providing system calls that invoke the operating system kernel in high privilege state. The recent trend has been toward increasingly integrated graphical user interfaces that encompass the activities of multiple processes on networks of computers.

Current research issues revolve mostly around networked operating systems, including network protocols, distributed shared memory, distributed file systems, mobile computing, and distributed application support. There is also active research in kernel structuring, file systems, and virtual memory.

Defining Terms

The following terms may have more general definitions than shown here, and often have other narrow technical definitions. This list indicates how the terms have been used in this chapter.

Abstraction: An interface that hides lower-level details and provides a set of higher-level functions.

Batch multiprogramming: Grouping jobs into batches based on characteristics such as memory requirements.

Client: A process that requests services by sending messages to server processes.

Command interpreter: A program (usually not in the kernel) that interprets user requests and starts computations to fulfill those requests.

Commands: Instructions in a job-control language.

Compute-bound: A process that performs little input/output but needs significant execution time.

Concurrency control: Means to mediate conflicting needs of simultaneously executing threads.

Context block: Process descriptor.

Context switching: The action of directing the hardware to execute in a different context (kernel or process) from the current context.

Daemon: A process, usually privileged, that provides a network-accessible service.

Database: A collection of files for storing related information.

Device driver: An operating-system module (usually in the kernel) that deals directly with a device.

Device interface: The means by which devices are controlled.

Distributed operating system: An operating system that integrates many computers, perhaps geographically distributed, into a single whole.

Fault-tolerant operating system: An operating system that attempts to continue to function even when hardware or software components fail.

File: A named, long-term repository for data.

Firewall: A configurable rule-based tool that only allows permissible network traffic through.

FTP: The file-transfer protocol service.

Gopher: A network service that connects information providers to their users.

Graphical user interface: Interactive program that makes use of a graphic display and a mouse.

HTTP: Hypertext transfer protocol.

Input/output: A resource: the ability to interact with peripheral devices.

Input/output-bound: A process that spends most of its time waiting for input/output.

Integrated application: An application that agrees on data formats with other applications so they can use each other's outputs.

Interactive multiprogramming: Multiprogramming in which each user deals interactively with the computer.

Job-control language: A way of specifying the resource requirements of various steps in a job.

Job: A set of computational steps packaged to be run as a unit.

Kernel: The privileged core of an operating system, responding to system calls from processes and interrupts from devices.

Light-weight process: A thread.

Macrokernel: A large operating-system core that provides a wide range of services.

Mailbox: A file for saving messages between users.

Malware: Malicious program that attempts to defeat security.

Memory: A resource: the ability to store programs and data.

Microkernel: A small privileged operating-system core that provides process scheduling, memory management, and communication services.

Middleware: Program that provides high-level communication facilities to allow distributed computation.

Multiprogramming: Scheduling several competing processes to run at essentially the same time.

Nanokernel: A very small privileged operating-system core that provides simple process scheduling and communication services.

Network services: Services available through the network, such as mail and file transfer.

Network service: A facility offered by one computer to other computers connected to it by a network.

Networked operating system: An operating system that uses a network for sharing files and other resources.

Nonprivileged state: An execution context that does not allow sensitive hardware instructions to be executed, such as the halt instruction and input/output instructions.

Offline: Handled on a different computer.

Operating system: A set of programs that control a computer.

Operator: An employee who performs the repetitive tasks of loading and unloading jobs.

Physical: The material upon which abstractions are built.

Physical address: A location in physical memory.

Pipeline: A facility that allows one process to send a stream of information to another process.

Privileged state: An execution context that allows all hardware instructions to be executed.

Process: A program in execution.

Process control block: Process descriptor.

Process descriptor: A data structure in the kernel that represents a process.

Process interface: The set of system calls available to processes.

Process number: An identifier that represents a process by acting as an index into the array of process descriptors.

Process switch: The action of directing the hardware to run a different process from the one that was previously running.

Processor state: Privileged or nonprivileged state.

Process: A program being executed; an execution context that is allocated resources such as memory, time, and files.

Protocol: A standardized data format and conversation rules for inter-computer communication on a network.

Pseudo-file: An object that appears to be a file on the disk but is actually some other form of data.

Real-time operating system: An operating system that schedules processes based on how responsive they must be.

Remote file: A file on another computer that appears to be on the user's computer.

Resident monitor: A precursor to kernels; a program that remains in memory during the execution of a job to handle simple requests and to start the next job.

Resource: A commodity necessary to get work done.

Scheduler: An operating system module that manages the time resource.

Server: A process that responds to requests from clients via messages.

System call: The means by which a process requests service from the kernel, usually implemented by a trap instruction.

Session: The period during which a user interacts with a computer.

Shared file system: Files residing on one computer that can be accessed from other computers.

Site: The set of computers, usually networked, under a single administrative control.

SMTP: The simple mail-transfer protocol.

Socket: An abstraction for communication between two processes, not necessarily on the same machine.

Spooling system: Storing newly arrived jobs on disk until they can be run, and storing output of old jobs on disk until it can be printed.

TCP: The transmission-control protocol.

Thin client: A program that runs on one computer that allows the user to interact with a session on a second computer.

Thread: An execution context that is independently scheduled, but shares a single address space with other threads.

Time: A resource: the ability to execute instructions.

Timesharing: Interactive multiprogramming.

User: A human being physically interacting with a computer.

User identifier: A number or string that is associated with a particular user.

User interface: The facilities provided to let the user interact with the computer.

Virtual: The result of abstraction; opposite of physical.

Virtual address: An address in memory as seen by a process, mapped by hardware to some physical address.

Virtual machine: An abstraction produced by a virtualizing kernel, similar in every respect but performance to the underlying hardware.

Virtualizing kernel: A kernel that abstracts the hardware to multiple copies that have the same behavior (except for performance) of the underlying hardware.

World-wide web: A network service that allows users to share multimedia information.

References

1. E. Bott, C. Siechert, and C. Stinson, *Microsoft Windows XP Inside Out, Deluxe Edition*, Redmond, WA: Microsoft Press, 2002.
2. D.P. Bovet and M. Cesati, *Understanding the LINUX Kernel*, 2nd ed., Sebastopol, CA: O'Reilly & Associates, 2002.
3. W.S. Davis and T.M. Rajkumar, *Operating Systems: A Systematic View*, 5th ed., Reading, MA: Addison-Wesley, 2000.
4. R.A. Finkel, *An Operating Systems Vade Mecum*, 2nd ed., Englewood Cliffs, NJ: Prentice-Hall, 1988.
5. I.M. Flynn and A.M. McHoes, *Understanding Operating Systems*, Belmont, CA: Brooks-Cole, 2000.
6. A. Geist, A. Beguelin, and J. Dongarra, Eds., *PVM: Parallel Virtual Machine: A Users' Guide and Tutorial for Network Parallel Computing (Scientific and Engineering Computation)*, Cambridge, MA: MIT Press, 1994.
7. I.F. Haddad and E. Paquin, "MOSIX: a cluster load-balancing solution for Linux," *Linux J.*, 85, 120–122, 124, 2001.
8. D. Hildebrand, "An architectural overview of QNX," *Proc. Usenix Workshop Micro-Kernels Other Kernel Architect.*, pp. 113–126, 1992.
9. L.J. Kenah and S.F. Bate, *VAX/VMS Internals and Data Structures*, Bedford, MA: Digital Equipment Corporation, 1984.
10. M.S. Kogan and F.L. Rawson, "The design of operating system/2" *IBM J. Res. Dev.*, vol. 27, no. 2, pp. 90–104, 1988.
11. S.J. Leffler, M.K. McKusick, M.J. Karels, and J.S. Quarterman, *4-3BSD UNIX Operating System*, Reading, MA: Addison-Wesley, 1989.
12. A. Nye, *Xlib Programming Manual*, 3rd ed., Sebastopol, CA: O'Reilly & Associates, 1992.
13. J.K. Ousterhout, *Tcl and the Tk Toolkit*, Reading, MA: Addison-Wesley, 1994.
14. P. Pacheco, *Parallel Programming With MPI*, Los Altos, CA: Morgan Kaufmann, 1997.
15. D. Pogue and J. Schorr, *Macworld Macintosh Secrets*, IDG Books Worldwide, 1993.
16. G.J. Popek and B.J. Walker, *The Locus Distributed System Architecture*, Cambridge, MA: MIT Press, 1986.
17. R. Rashid, "Threads of a new system," *Unix Rev.*, August, 37–49, 1986.
18. A. Silberschatz, P.B. Galvin, and G. Gagne, *Operating Systems Concepts*, 6th ed., New York, NY: Wiley, 2001.
19. W. Stallings, *Operating Systems: Internals and Design Principles*, 4th ed., Englewood Cliffs, NJ: Prentice-Hall, 2000.
20. A.S. Tanenbaum, *Modern Operating Systems*, 2nd ed., Englewood Cliffs, NJ: Prentice-Hall, 2001.

19

Computer and Communications Security

J. Arlin Cooper
Sandia National Laboratories

Anna M. Johnston
Sandia National Laboratories

19.1 Introduction

Computer security is the protection of computing assets and computer network communication assets against abuse, unauthorized use, and unavailability caused by intentional or unintentional actions, and protection against undesired information disclosure, alteration, or misinformation. In today's environment, the subject encompasses computers ranging from supercomputers to microprocessor-based controllers and micro-computers, personal digital assistants (**PDAs**), software, peripheral equipment (including terminals, printers), communication media (e.g., cables, antennas, satellites), people who use computers or control computer operations, and networks (some of global extent) that interconnect computers, terminals, and other peripherals.

Widespread publicity about computer crimes (losses estimated at more than $500 billion per year), **hacker** (cracker) penetrations, and **viruses** has given computer security a high profile in the public eye [Stallings, 1998]. The same sorts of technologies that have made computers and computer network communications essential tools for information and control in almost all businesses and organizations have provided new opportunities for adversaries as well as for accidents or natural occurrences to interfere with crucial functions. Some of the important aspects are industrial/national espionage, terrorism attacks [Verton, 2003], loss of functional integrity (e.g., in air traffic control, monetary transfer, and national defense systems), and violation of society's desires (e.g., compromise of privacy). The World Wide Web access to the **Internet** has created financial transaction vulnerabilities, crypto system weaknesses, and privacy issues [Ning, 2004]. Infrastructure networks for control of power grids and supervisory control and data acquisition (SCADA) water system control have introduced another class of vulnerabilities.

Fortunately, technological developments also provide controls (proactive and follow-up) for computer security. These include personal transaction devices (e.g., **smart cards**, and **tokens**), **biometric verifiers**, **port protection devices**, encryption, authentication, and digital signature techniques using symmetrical (single-key) or asymmetrical (**public-key**) approaches, automated auditing, formal evaluation of security features and security products, and decision support through comprehensive system analysis techniques. Although the available technology is sophisticated and effective, no computer security protective measures are perfect, so the goal of prevention (security assurance) is almost always accompanied by detection (early discovery of security penetration), by layered approaches, and by penalties (denial of goal, e.g., information destruction; or response, e.g., prosecution and punishment).

The information in this section is intended to survey the major contemporary computer security threats, vulnerabilities, and controls. A general overview of the security environment is shown in Figure 19.1. The oval in the figure contains an indication of some of the crucial concentrations of resources that exist in many facilities, including digital representations of money; representations of information about operations, designs, software, and people; hardware for carrying out (or peripheral to) computing and communications; people involved in operating the facility; utility connections (e.g., power); and interconnection paths to outside terminals and users, including hard-wired connections, modems for computer (and fax) communication over telephone lines, and electromagnetic links (e.g., to satellite links, to ground antenna links, and to aircraft, spacecraft, and missiles). Each of these points of termination is also likely to incorporate computer (or controller) processing.

Other factors implied include profit-motivated or malicious adversaries, line taps or **TEMPEST emanations** interception, probes through known or unknown dial-up connections, unauthorized physical entry, unauthorized actions by authorized personnel, and delivery through ordinary channels (e.g., mail) of information (possibly misinformation) and software (possibly containing embedded threat programs), and the threats of fire, water damage, loss of climate control, electrical disturbances (e.g., due to lightning or power loss). Also indicated is guidance for personnel about acceptable and unacceptable actions through policy and regulations. The subject breadth can be surveyed by categorizing into physical security, cryptology techniques, software security, hardware security, network security, and personnel security (including legal and ethical issues). Because of the wide variety of threats, vulnerabilities, and assets, selections of controls and performance assessment typically are guided by security-specific decision-support analyses, including risk analysis and probabilistic risk assessment (PRA).

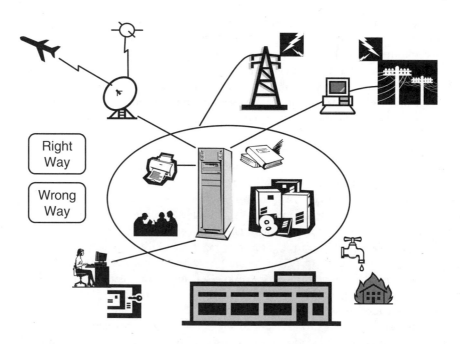

FIGURE 19.1   An overview of the computer and communications security environment.

19.2 Physical Security

Physical access security ranges from facility access control (usually through personal identification or authentication) to access (or antitheft) control for individual items (e.g., diskettes and personal computers). Techniques used generally center around intrusion "prevention" (or invoking a significant time delay for an adversary) and intrusion detection, which allows a response through security guard, legal or administrative action, or automatic devaluation of the penetration goal (e.g., through information destruction) [Cooper, 1989].

Physical environmental security protects against natural threats, such as power anomalies or failures, water damage, fire, earthquake, and lightning damage, among others. Note that some of the "natural" threats (e.g., power failure) can also be caused by adversaries. Since there is potential (in spite of protection) for a loss, contingency planning is essential. This includes provisions for software backup (usually off-site), hardware backup (e.g., using reciprocal agreements, hot sites, or cold sites [Cooper, 1989]), and disaster recovery, guided by a structured team that has prepared through tests (most typically simulated).

An example of power protection technology is the widely used uninterruptible power system (UPS). An online UPS implementation is shown in Figure 19.2. Utility power is shown passed through a switch to a rectifier and gated to an inverter. The inverter is connected to the critical load to be protected. In parallel, continuous charge for a battery bank is provided. Upon loss of utility power, the battery bank continues to run the inverter, thereby furnishing power until graceful shutdown or switching to an auxiliary engine generator can be accomplished. The switch at the lower right protects the UPS by disconnecting it from the load in case of a potentially catastrophic (e.g., short) condition.

19.3 Cryptology

Cryptology, literally translated, is the study of buried things; the "things" refer to writings — cipher systems. Cryptography and cryptanalysis are its two branches: the making and breaking (respectively) of cryptographic systems.

Cryptology has seen major advances in the last quarter century. Before 1976, its sole functions were privacy and simple data authentication; its sole users, government organizations. With the rapid advance of information technologies and the explosion of information in our society, cryptographic systems have become vital in the private sector. The role of cryptology has expanded to include user authentication (signatures, often with non-repudiation); copyright protection; electronic cash, key exchanges and voting — to name but

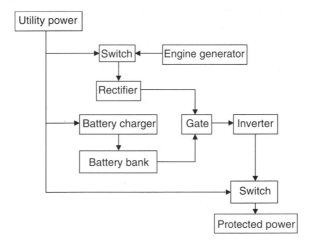

FIGURE 19.2 Uninterruptible power system.

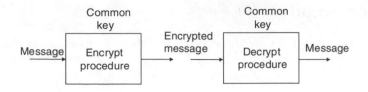

FIGURE 19.3. Symmetric cryptography.

FIGURE 19.4 Public key cryptography.

a few. These new functions were made possible with the advent of public key cryptography [Diffie and Hellman, 1976; Rivest, Shamir, and Adleman, 1978].

There are two major branches in modern cryptography: symmetric or private key cryptography (Figure 19.3) and non-symmetric or public key cryptography (Figure 19.4). Symmetric (private key) cryptography uses a single key and can be used for encryption or secure hash functions. This single key must be kept secret. All cryptography before 1976 was symmetric, and it is still the most heavily relied on type of algorithm for data encryption (privacy and secrecy). Public key cryptography uses different keys: one which must be kept secret and one which may be made public. The two keys are related mathematically by a "one-way" function. A one-way function is easy to compute in one direction but very hard (computationally infeasible) to compute in the other direction. An example of a one-way function is multiplication versus factorization. It is simple to multiply two large prime numbers, but very hard to factor the result.

There are two basic types of symmetric systems: block ciphers and stream ciphers. Stream ciphers encrypt one character (generally one bit) at a time and are rarely used. Block ciphers encrypt a block of characters (usually a fixed length) at a time and are the most commonly used form of symmetric cryptosystem. Many of these block ciphers (such as DES — the Data Encryption Standard) are based on Feistel ciphers (see Figure 19.5). A Feistel cipher splits the data to be encrypted into two blocks. Using a function and secret key (F_k), random looking data would be obtained from one block half and added to the other. The data is then shifted (the two blocks swapped) and the process repeated. Many iterations, or rounds, of this process are done in a typical Feistel cipher: DES uses 16 rounds, enciphering 64-bits.

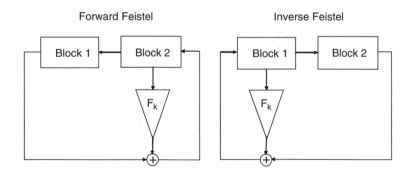

FIGURE 19.5 Feistel cipher.

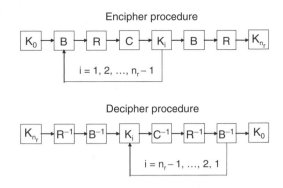

FIGURE 19.6   AES cryptosystem.

DES has one major weakness: it does not use enough secret keys to protect it from the strength of modern computing. There are only 2^{56} possible keys for DES. The National Institute of Standards and Technology (NIST) officially adopted the Rijndael algorithm as the new Advanced Encryption Standard (AES) in May 2002 [Advanced Encryption Standard, 2001]. It is not based on the Feistel structure, but instead uses layers of mixing functions and allows for variable input block and key sizes: 128, 192 or 256 bits each, for 9 possible block/key size combinations. Like Feistel systems, it repeats the mixing layers in rounds, adding "round key" (random data obtained by expanding the original secret key). The number of rounds specified varies from 10 to 14, depending on the block and key lengths. This cryptographic structure has not been studied as much as the Feistel structure, but in the few years it has been studied (the call for a new standard came in 1997), no great weakness has been found.

There are four basic operations done in an AES round: A permutation on individual state bytes (B), shifting of the bytes in an individual row (R), a permutation of the data in each column (C), and finally the addition of round key (K_i). Round key is also added before the start of the rounds, and a final round is done without the column permutation step. Labeling the four individual functions as B, R, C, K_i, AES looks like Figure 19.6.

Besides the basic cryptographic algorithm, cryptosystems use a mode of operation dependant on feedback (see Figure 19.7). In its simplest form, no feedback is used. Block ciphers with no feedback are said to be in codebook mode. Other forms of feedback use past cipher text (cipher block chaining, cipher feedback, etc.), plain text or intermediate states internal to the system for feedback. In Figure 19.7, P_i, C_i represent the plain and cipher text at time i with the box ENC being the encryption box.

Public key cryptography uses asymmetric keys to perform functions not possible with symmetric cryptography. Built on mathematically difficult problems, such as factoring[1] and finding discrete logarithms[2], they are computationally expensive, but essential in information security today. Because of the cost, public key cryptography is generally limited to the functions only it can do. It makes large-scale encryption simpler by enabling electronic key exchanges and user authentication. Once a shared symmetric key has been exchanged and the users' authenticity has been verified, data are protected with a symmetric system.

The two major uses of public key cryptography are electronic key exchanges and signatures; the best-known algorithms for doing this are RSA and the Diffie–Hellman key exchange. RSA is based on the difficulty of factoring while Diffie–Hellman is based on the discrete logarithm problem. RSA has broader functionality — it allows for public encryption, signatures and key exchanges — but requires more keying material (individual

---

[1] Factoring techniques have broken 173 digit (576 bits) RSA moduli (December 2003, see http://www.rsa security.com/rsalabs/challenges/factoring/rsa576.html).

[2] Techniques using special approaches have found 120 digit (399 bit) discrete logarithms for prime moduli (see http://www.medicis.polytechnique.fr/"lecier/english/dlog.html).

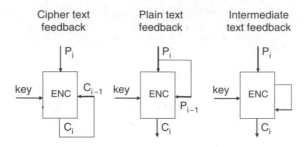

FIGURE 19.7   Feedback modes.

composite numbers, public and secret keys). Diffie–Hellman is strictly for electronic key exchanges, but has much more flexibility in its design. Authenticated versions of Diffie–Hellman have been designed, allowing both the key exchange and signature of the key (authentication) to be performed at the same time, improving its efficiency. It has also been successfully adapted to work over elliptic curves, further reducing the computational cost and transmission requirements.

Figure 19.8 depicts public encryption using RSA. Two large primes, P, Q are chosen and kept secret. Two values, e, d are created such that $ed \equiv 1 \bmod (P-1)(Q-1)$. The values N, e are released to the public. Anyone who wishes to encrypt a message, m, raises m to the e power modulo N. Decryption is only possible if d is known, as raising m^e to the d power modulo N cancels the e value, returning the original message.

Figure 19.9 depicts a simple key exchange with Diffie–Hellman. A group in which discrete logarithms are difficult is chosen along with an element of large order, α. Both sides of the key exchange choose a random secret value r_i and exchange their computed values of α^{r_i}. Computing r_i from α^{r_i} is the discrete logarithm problem. The shared key will be $\alpha^{r_a r_b}$.

19.4   Software Security

A number of techniques that are commonly implemented in software can contribute to protection against adversaries. These include password authentication; memory, file, and database access restrictions; restrictions on processing actions; development and maintenance controls; and auditing.

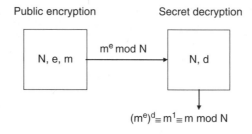

FIGURE 19.8   Rivest–Shamir–Adleman public encryption.

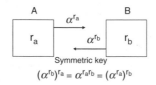

FIGURE 19.9   Diffie–Hellman key exchange.

Passwords, which are intended to authenticate a computer user in a cost-effective way, are sometimes user-selected (a technique resulting in a relatively small potential population), sometimes user-selected from a computer-generated collection, sometimes randomly generated, and sometimes randomly generated from a phonetic construction for pronounceability and memorization ease [Cooper, 1989].

Security control can be physical, temporal, logical, or procedural. Two important logical or procedural control principles are part of fundamental multilevel security (multiple levels of sensitivity and multiple user clearance levels on the same system), as described by part of the Bell–La Padula model. The simple security principle restricts users of a particular clearance level from reading information that is of a more sensitive (more highly classified) level. The star property prohibits information flow from the level at which its sensitivity has been determined to any lower level (write-down). Analogous integrity protection is provided by the Biba integrity model [Gasser, 1988].

Protection rules can be mandatory (used mainly by the government or military) or discretionary (compartmented according to need-to-know regimes of trust typically determined by file owners). The combination of security levels and protection rules at the same level can be associated with a lattice model. In addition to matching the security controls, the lattice model facilitates mathematical verification of security implementations.

A common logical protection rule specification gives the rights of subjects (action initiators) to act on objects (action targets) at any particular time. One way to view these rules (although seldom implemented in this manner) is to consider an access matrix (Table 19.1) containing rows for subject indicators and columns for object indicators. The matrix entries are the rights of subjects to objects. Actual implementation may differ, e.g., by using directories, or capability lists, or capability tokens (row designations for rights of subjects) or access control lists (column designation for rights to objects).

These types of rules can be augmented by software (and/or hardware) memory protection through techniques including fences, base/bounds registers, tagged registers, and paging [Gasser, 1988].

Database management system (DBMS) security and integrity protections include access controls but generally require finer granularity and greater protection (especially for relational databases) against subtle forms of information deduction such as inference and aggregation. Integrity protection mechanisms include field checks, change logs, two-phase updates, error protection codes, range comparisons, and query controllers [Pfleeger, 1989]. Secrecy depends on access control (e.g., file passwords), query controllers, and encryption.

Processing restrictions can, in addition to those implied by memory, file, and database controls, limit the ability of users to, for example, try multiple passwords or multiple user IDs; make financial transactions; change security parameters; move, rename, or output information; and deliver covert channel information (signaling systematically using authorized actions to codify unauthorized data delivery).

Software development and maintenance controls include standards under which programs (including security features) are designed to meet requirements, coded in structured or modular form, reviewed during development, tested, and maintained. Configuration or change control is also important. Computer auditing is intended to provide computer records about user actions for routine review (a productive application for expert systems) and for detailed investigation of any incidents or suspicious circumstances. It is essential that audit records be tamper-proof.

Software security features (including auditing) can be provided as part of the computer operating system or they can be added to an operating system as an add-on product. A U.S. government multilevel *trusted*

TABLE 19.1 An Access Matrix

Subjects/Objects	O_1	O_2	O_3	O_4	O_5
S_1	Own, read, write	Own, read, execute	Own, read, delete	Read, write, execute	Read
S_2		Read	Execute		Read
S_3	Write		Read		Read

TABLE 19.2   NCSC Security Evaluation Ratings

Class Name	Summary of Salient Features
Class A1	Formal top-level specification and verification of security features, trusted software distribution, covert channel formal analysis
Class B3	Tamper-proof kernelized security reference monitor (tamper-proof, analyzable, testable), structured implementation
Class B2	Formal security model design, covert channel identification and tracing, mandatory controls for all resources (including communication lines)
Class B1	Explicit security model, mandatory (Bell–La Padula) access control, abels for internal files and exported files, code analysis and testing
Class C2	Single-level protection for important objects, log-in control, auditing features, memory residue erasure
Class C1	Controlled discretionary isolation of users from data, authentication, testing
Class D	No significant security features identified

computing base development program through NSA's National Computer Security Center (NCSC) resulted in a well known security methodology and assessment scheme for these types of software (and hardware) products [DOD, 1985]. A significant number of operating systems and software security packages have been evaluated and given ratings by NCSC, in addition to hardware–software combinations, encryption devices, and network security systems. The basic evaluation determines the degree of confidence that the system will be resistant to external penetration and internal unauthorized actions. The most secure systems known are classified A1 and utilize a reference monitor (checking every request for access to every resource), a security kernel (concentration of all security-related functions into a module that facilitates protection and validation), and protection against covert channels. Formal analysis is used to assure that the implementation correctly corresponds to the intended security policy. There is an operational efficiency penalty associated with secure multilevel operating systems. Other classes (in order of progressively fewer security features, which results in decreasing security) are B3, B2, B1, C2, C1, and D (see Table 19.2, where security features generally accumulate, reading up from the table bottom).

In addition to computer activity directly controlled by personnel, a family of software threats can execute without direct human control. These techniques include the **Trojan horse**, the virus, the **worm**, the **logic bomb**, and the **time bomb**. The virus and worm (because they copy themselves and spread) are both capable of global-spanning attacks over relatively short time frames. Protection against these threats includes limiting user threats through background screening, using expert system software scanners that search for adversarial program characteristics, comparators, and authenticators or digital signatures that facilitate detection of software tampering. Other software-intensive threats include trapdoors, superzapping, browsing, asynchronous attacks, and the salami attack [Cooper, 1989]. These all usually involve unauthorized actions by authorized people and are most effectively counteracted by insider personnel controls (see Section 19.7, "Personnel Security").

19.5   Hardware Security

In addition to personal authentication through something known (e.g., passwords or PINs), users can be authenticated through something possessed or by something inherent about the user (or by combinations of the three). Hardware devices that contribute to computer security using the approach of something possessed include tokens and smart cards. Biometric verifiers authenticate by measuring human characteristics. Other hardware security devices include encryptor/decryptor units and port protection devices (to make dial-up attacks by hackers more difficult). A generic diagram depicting some of these applied to control of users is shown in Figure 19.10. The controls can be used individually or in various combinations.

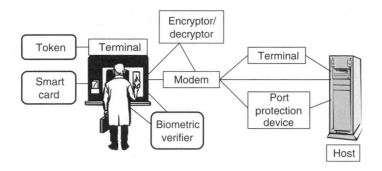

FIGURE 19.10 Depiction of hardware controls.

Tokens are devices that can be hand-carried by authorized computer users and are intended to increase password security by assuring that passwords are used only once, thereby reducing the vulnerability to password compromise. The devices contain an internal algorithm, which either works in synchronization with an identical algorithm in the host computer or transforms an input derived from a computer prompt into a password that matches the computer-transformed result. In order to protect against loss most also require a user password for token access.

Smart cards are credit-card-sized devices intended to facilitate secure transactions, such as credit card purchases, purchases or cash withdrawals that result in bank account debits, or information interchanges. The most common application uses a card reader/network that exchanges data with the smart card over a serial data bus. User information and security information are stored in encrypted form in the card, and physical access to the internal card circuitry is protected by tamper-proof (self-destructive) sealing. Use of the card is controlled by password access.

Because of the vulnerability of passwords to compromise by disclosure or various forms of information tapping, and because of the vulnerability of loss of carried items (e.g., ROM keys, magnetic stripe cards), biometric devices have been developed to measure human characteristics in ways that are resistant to counterfeiting. These devices include signature verifiers (for examining the velocity, acceleration, and pressure characteristics imparted during signing as a function of time), fingerprint and palmprint readers (for examining print pattern characteristics, for example, with the flesh applied to a glass platen), voice verifiers (which evaluate speech characteristics, usually in response to system prompts), hand geometry (including some three-dimensional aspects), eye retina vessel pattern examination (through infrared reflection), and typing rhythm assessment (for user keyboard inputs).

Systematic cracker attacks on dial-up computer ports frequently include searches for modem tones followed by attempts to guess passwords. In response, port protection devices (PPDs) enhance dial-up security. The basic feature of many PPDs is that no modem tone is provided until an additional security barrier (or barriers) is overcome. Most PPDs require a code before computer port connection. Some also identify the user by the code entered, disconnect the call, and dial the number at which the user is expected to be (typically using a separate line to avoid dial-in intercept of the outgoing call).

Personal computer (PC) security is of contemporary interest because these relatively new tools have contributed to a set of security vulnerabilities that differs substantially from conventional computer security concerns. For example, PC users may be more naive about security in general, PC hardware and software and administrative controls are generally more primitive, the PC physical environment is generally less controlled, and PCs are generally more easily misused (e.g., company PCs used for personal benefit).

An additional hardware security topic is associated with TEMPEST (a program to assess the potential for data processing equipment to inadvertently generate "compromising emanations" that convey information to a surreptitious remote sensor). Although originally of concern because of requirements to protect government and military classified data, industrial espionage is now also a concern. Various forms of protection can be used, such as electromagnetic shielding, physical separation of processing equipment from potential adversary locations, fiber-optic communication, and encrypted data transmission. Some commercial equipment has been certified by NSA to have low emanations.

19.6   Network Security

Many business, informational, and scientific interchanges take place nationally and internationally over networks under computer control. Management of network security is exacerbated by physical dispersal and security philosophy disparity. For example, network adversaries may be harder to identify and locate than local computer adversaries. (For an interesting account of overcoming this problem, see Stoll [1989].) As another example, a user at a location that would not allow some form of activity (e.g., copying information from one level of security to a lower level) might find a network connection to a facility for which the activity was accepted. The intended local restriction might thereby be circumvented by conducting the restricted activity at the more permissive location. Opportunities for passive interception (tapping or emanations), or for active spoofing (involving misinformation, replay, etc.), or for disruption (including jamming) are also generally greater owing to network utilization and the potential for distributed attacks [De Decker, 2001]. The emergence of wireless networks has introduced new classes of vulnerabilities [Lioy, 2003].

There are many network topologies, but they can be decomposed into four basic canonical types (Figure 19.11). The star topology has been traditionally used in centrally controlled networks (e.g., the control portion of the telephone system), and security is typically within the central control. Use of star topology in local-area networks (LANs) is increasing. Mesh topology is not readily amenable to central control but is well tailored to protect wide-area network integrity. Mesh topology accommodates variable-path routing schemes, such as packet transmission. The bus topology is commonly used in complex physically constrained systems, such as computing and communication processor interconnection, and missiles and aircraft. The ring and bus topologies are frequently used in LANs. The shared communication media for LANs can jeopardize secrecy unless communications are encrypted.

Network security considerations include secrecy, integrity, authenticity, and covert channels. Potential controls include cryptosystems (for secrecy, integrity, and authentication); error-protection codes, check sums (and other "signatures"), and routing strategies (for integrity); protocols (for key distribution and authentication); access control (for authentication); electronic "agents" to report on network activities [Ning, 2004], and administrative procedures (for integrity and covert channel mitigation). Where encryption is used, network key distribution can be difficult if the physical dimensions of the network are large. Note that several techniques described under hardware security (smart cards, tokens, biometrics, PPDs) are useful for network authentication.

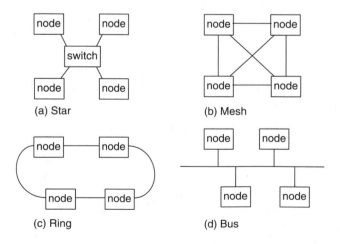

FIGURE 19.11   Basic network topologies.

Various network security approaches have been used; they can be basically classified into centralized security and distributed security (although the use of combinations of the two is common). It is difficult to maintain effective centralized administrative control in a network larger than a LAN, because of the logistics of maintaining current security and authentication data. Network efficiency, reliability, and integrity are also limited by the performance of the central security controller. The major weaknesses of distributed control are associated with inconsistent security enforcement and the security-relevant communication burden.

Networks frequently comprise networks of networks (the Internet is a worldwide interconnection of networks) and firewalls (filter between outside and internal networks), using bridges or routers for protocol pass-through and filtering and gateways for protocol translation and buffering. Bridges, routers, gateways, and firewalls may also have the role of distributed network security controllers.

Various security protocols (orderly coordinated sequences of steps) are used to authenticate network users to each other. The basic purpose of security protocols is to assure all of the parties involved that the parties with whom they are communicating are behaving as expected. Protocols can be arbitrated, adjudicated, or self-enforcing [Pfleeger, 1989]. Analogously, cryptographic sealing and cryptographic time stamps (or other sequence identifiers) can prevent message (or message segment) reuse. These approaches can authenticate the message contents as well as the communicating party. Secure key distribution protocol (e.g., by a network key server or key distribution center) is an important application for protocols.

Network communication involves several levels of nonsecurity protocol for the purpose of allowing users to communicate with integrity. A number of network standards have been developed (e.g., TCP/IP, ISDN, GOSIP, SNMP, SMTP, SSL, VPN, and http and https for the World Wide Web). An illustrative example is the International Standards Organization Open Systems Interconnection model (OSI). The basic OSI structure is shown in Figure 19.12 [Stallings, 1995].

The physical layer mediates access to the transmission medium. Network systems such as token ring, token bus, and carrier sense multiple access with collision detection (CSMA/CD) work at this level. The data link layer can be used to enhance transmission quality through framing, error correction, and check sums. Link (point-to-point) encryption is typically implemented in hardware at this level. The network layer handles network routing functions. This is the highest layer necessary for an intermediate node, as shown in the figure. Correct routing (and protection for pass-through information from users at the intermediate node) is important to security. The transport layer provides end-to-end (initial source to final destination) interprocess communication facilities. End-to-end encryption can be implemented in software at this level. The session

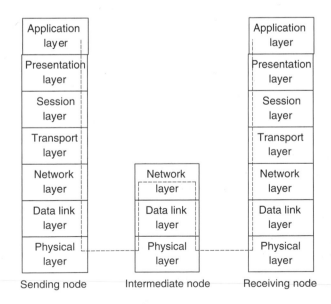

FIGURE 19.12 The ISO network protocol model.

layer manages the overall network connection during a user activity. This connectivity is also important to security. The presentation layer converts between user syntax and network syntax, and the application layer provides user services such as electronic mail and data file transfers. Encryption can be implemented at either of the latter two layers.

Link encryption requires exposure of information at intermediate nodes. While this is essential (at least for routing data) if further routing is required, it may be required to protect information from exposure at intermediate nodes. In addition to routing controls, this may require end-to-end encryption and separation of routing information from protected information. End-to-end encryption is generally performed at the higher levels (e.g., the transport layer). It is not unusual to use both link and end-to-end encryption.

Like isolated computers, networks can have multilevel security. However, implementation and verification of security features are more difficult. For example, covert channel protection is currently quite problematic in networks.

19.7   Personnel Security

Personnel security involves protecting personnel as assets and protecting against personnel because of the threat potential. The basic aspects of the latter topic are motivations that cause various types of threats, the approaches most likely to be used by adversaries, techniques for assessing the threat potential, and techniques for protecting against adversaries.

One motivation for computer/communication attacks is financial gain. This motivation ranges from career criminals who could view the volume of monetary transactions as tempting, and the potential detachment from personnel confrontations as affording low risk of apprehension, to basically honest people who have trouble resisting what they consider to be less serious temptations. Industrial espionage is one of many examples of financial motivation that may result in attempts to evade computer security control.

Another important motivation is information gain or information modification, which could represent no direct financial gain. Some people are curious about information to which they have no right. Some want to modify information (e.g., grades, criminal records, medical records, personnel records) because it reflects an image they want to change.

The motivation of causing personal or public outrage in order to advance a cause is a common motivation. Terrorism is an example, and many acts of terrorism against computers have occurred [Cooper, 1989], especially in Europe. Sometimes the cause is related to revenge, which may be manifested through vandalism.

Espionage activities can be motivated by financial gain, national or company loyalty, blackmail, or even love. Hackers are frequently motivated by the challenge of overcoming security barriers. Usually, self-image and image with peers (e.g., through electronic bulletin board proclamations of breakthroughs) is a strong factor.

Personnel adversaries most commonly choose what they perceive to be the easiest and/or the safest avenue to achieve the desired objective. This is analogous to looking for the weakest link to break a chain. Because these avenues may be either inherent or unknown, security barrier uniformity is frequently sought through the application of basic principles (e.g., separation of duties). Some adversaries are motivated enough and skilled enough to use ingenious approaches, and these provide a warning about what unexpected approaches might succeed. One of the most interesting and informative examples was the break by William Friedman of the Vernam cipher used by the Germans in World War II [Cooper, 1989]. Another was the use of an unintended electronic mail program feature that allowed an adversary to plant a Trojan horse in a privileged area of a computer, which resulted in system privileges when the computer ran routine periodic housekeeping. This was the genesis of the title of the book *The Cuckoo's Egg* [Stoll, 1989]. The same adversary was one of several known to have broken one-way transformed password encryption by downloading the transform and the transformed outputs to his own computer and then exhaustively encrypting a dictionary of words and potential passwords, noting where matches to transformed outputs were obtained.

Approaches used to assess the types of unexpected attacks that might be used are mainly to catalog past approaches that have been used and to foresee new approaches through adversarial simulation. An example of this simulation is the "tiger team" approach, where a collection of personnel of various backgrounds synergistically brainstorm approaches, at least some of which may be tested or actually carried out. Tiger teams have a long history of finding unexpected approaches that can be successful, and thereby identifying protection needs.

Protection against personnel attacks generally falls into two categories: protection against insiders (those having some authorization to use resources) and against outsiders (those who gain access to resources in unauthorized ways). Some protective measures are tailored to one or the other of these two groups; some are applicable in either case.

Typical protections against unauthorized insider activities include preemployment screening and background investigation, polygraph examinations (within legal limits), administrative controls (e.g., security plans and access logs), routine examination and monitoring of activities through audit records, ethics and motivational training, and the threat of legal or job-related punishment for improper activities. The ethics of computer use varies from organization to organization because of society's inexperience in weighing the moral aspects of the topic.

Protection against outsiders includes physical and logical access control using the various forms of hardware and software authentication discussed previously and the threat of legal prosecution for transgressions. This threat depends in large measure on the available laws covering computer security violations. Computer security laws and law enforcement have traditionally been weak relative to laws covering other types of activities (largely because of new legal aspects associated with computing), but a large number of legal approaches are now possible because of laws enacted during the past two decades.

Computer laws, laws that apply to computing, and applicable statutes include the Copyright Act (amended in 1980 to allow software copyrights and help protect against software piracy), the Patent Act (adding firmware and software coverage), "shrinkwrap licenses" (some legal protection, some deterrent), various U.S. Crime Statutes (applicable to U.S. government computers), the Privacy Act (for U.S. government applications and similar to privacy laws in a number of other countries), National Security Decision Directives (NSDD 145 was for NSA-enhanced protection of "sensitive unclassified" information, largely intended to prevent technology drain to unfriendly countries), the Computer Security Act of 1987 (restoring NIST as the primary agency responsible for sensitive unclassified security), the Right to Financial Privacy Act, the Freedom of Information Act, the Electronic Funds Transfer Act, the Fair Credit Reporting Act, the Crime Control Act, the Electronic Communications Privacy Act, the Computer Fraud and Abuse Act, and the Foreign Corrupt Practices Act.

There is now considerable interest in international legal computer communication agreements, especially among countries that interchange significant amounts of computer data. International forums have brought many countries together with the intents of regulatory commonality and transborder data communication control. The U.S. Digital Millennium Copyright Act of 1998 had an international flavor, following along with the international WIPO Copyright Treaty of 1996.

Defining Terms

Biometric verifiers: Devices that help authenticate by measuring human characteristics.

Hacker: Person who explores computer and communication systems, usually for intellectual challenge, commonly applied to those who try to circumvent security barriers (crackers).

Internet: An international connection of networks which can be navigated by the World Wide Web protocols, and over which e-mail, information transfers, and credit card orders can transverse.

Logic bomb: Destructive action triggered by some logical outcome.

PDA: Small (pocket-sized) device for data (e.g., phone numbers, addresses, appointments, memos) and some data processing.

Port protection device: Device in line with modem that intercepts computer communication attempts and requires further authentication.

Public-key cryptosystem: System that uses a pair of keys, one public and one private, to simplify the key distribution problem.

Smart cards: Credit-card-sized devices containing a microcomputer, used for security-intensive functions such as debit transactions.

Software piracy: Unauthorized copying of software for multiple uses, thereby depriving software vendor of sales.

TEMPEST emanations: Electromagnetic, conductive, etc. leakage of information that can be recovered remotely.

Time bomb: Destructive action triggered by computer calendar/clock reading.

Token: Device that generates or assists in generation of one-time security code/passwords.

Trojan horse: Implanted surreptitious code within an authorized program.

Virus: A propagating, self-replicating program that is inserted in and can make changes in application programs or other executable routines.

Worm: Independent self-replicating code that, once initiated, propagates across networks, consuming memory resources.

References

J.A. Cooper, *Computer and Communications Security*, New York, NY: McGraw-Hill, 1989.

B. De Decker, F. Piessens, J. Smits, and E. Van Herreweghen, Eds., *Advances in Distributed Systems Security*, Hingham, MA: Kluwer Academic Publishers, 2001.

Department of Defense Trusted Computer System Evaluation Criteria, DOD 5200.28-STD, Fort Meade, MD: National Computer Security Center, 1985.

W. Diffie and M. Hellman, "New directions in cryptography," Trans. Inf. Theory, IT-22, 6, 644–654, 1976 IEEE, November.

Federal Information Processing Standards Publication, Announcing the Advanced Encryption Standard (AES), November 2001.

M. Gasser, *Building a Secure Computer System*, New York, NY: Van Nostrand Reinhold, 1988.

A. Lioy and D. Mazzocchi, Eds., *Communications and Multimedia Security*, Berlin, Germany: Springer, 2003.

A.J. Menezes, P.C. van Oorschot, and S.A. Vanstone, *Handbook of Applied Cryptography* Boca Raton, FL: CRC Press, 1996.

P. Ning, S. Jadodia, X.S. Wang, *Intrusion Detection in Distributed Systems*, Hingham, MA: Kluwer Academic Publishers, 2004.

C.P. Pfleeger, *Security in Computing*, Englewood Cliffs, NJ: Prentice-Hall, 1989.

R. Rivest, A. Shamir, and L. Adleman, "A method for obtaining digital signatures and public-key cryptosystems," Commun. ACM, 21, 120–126, 1978.

W. Stallings, *Cryptography and Network Security: Principles and Practice*, Englewoood Cliffs, NJ: Prentice-Hall, 1998.

W. Stallings, *Networks and Internet Security*, Englewood Cliffs, NJ: Prentice-Hall, 1995.

C. Stoll, *The Cuckoo's Egg*, New York, NY: Doubleday, 1989.

D. Verton, *Black Ice — The Invisible Threat of Cyber-Terrorism*, New York, NY: McGraw-Hill, 2003.

Further Information

B. Bloombecker, *Spectacular Computer Crimes*, Homewood, IL: Dow-Jones-Irwin, 1990.

D.E.R. Denning, *Cryptography and Data Security*, Reading, MA: Addison-Wesley, 1982.

P.J. Denning, *Computers Under Attack*, New York, NY: ACM Press, 1990.

J. Ellis and T. Speed, *The Internet Security Guidebook—From Planning to Deployment*, San Diego, CA: Academic Press, 2001.

P. Fites, P. Johnston, and M. Kratz, *Computer Virus Crisis*, New York, NY: Van Nostrand Reinhold, 1989.

O. Goldreich, *Modern Cryptography, Probabilistic Proofs and Pseudo-randomness*, Berlin: Springer, 1999.

K. Hafner and J. Markoff, *Cyberpunk*, New York, NY: Simon and Schuster, 1991.

F. Hayes, "Secure your users," *Computerworld*, April 21, 2003.

C. Kaufman, R. Perlman, and M. Speciner, *Network Security*, Englewood Cliffs, NJ: Prentice-Hall, 1995.

J.A. Lewis, Ed., *Cyber Security*, Washington, DC: CSIS Press, 2003.

J.L. Mayo, *Computer Viruses*, Blue Ridge Summit, PA: Windcrest, 1989.

National Research Council, *Computers at Risk*, Washington, DC: National Academy Printers, 1991.

G.J. Simmons, *Contemporary Cryptology*, Piscataway, NJ: IEEE Press, 1992.

E.F. Troy, "Security for dialup lines," Washington, DC: U.S. Department of Commerce, 1986.

"Trusted network interpretation," NCSC-TG-005, Fort Meade, MD: National Computer Security Center, 1987.

20

Computer Reliability

Chris G. Guy
University of Reading

20.1 Introduction

This chapter outlines the knowledge needed to estimate the **reliability** of any electronic system or subsystem within a computer. The word *estimate* was used in the first sentence to emphasize that the following calculations, even if carried out perfectly correctly, can provide no guarantee that a particular example of a piece of electronic equipment will work for any length of time. However, they can provide a reasonable guide to the probability that something will function as expected over a given time period. The first step in estimating the reliability of a computer system is to determine the likelihood of failure of each of the individual components such as resistors, capacitors, integrated circuits, and connectors, which make up the system. This information can then be used in a full system analysis.

20.2 Definitions of Failure, Fault, and Error

A *failure* occurs when a system or component does not perform as expected. Examples of failures at the component level could be a base-emitter short in a transistor somewhere within a large integrated circuit or a solder joint going open-circuit because of vibrations. If a component experiences a failure, it may cause a fault leading to an error, which may lead to a system failure.

A *fault* may be either the outward manifestation of a component failure or a design fault. Component failure may be caused by internal physical phenomena or by external environmental effects such as electromagnetic fields or power supply variations. Design faults may be divided into two classes. The first class of design fault is caused by using components outside their rated specification. It should be possible to eliminate this class of faults by careful design checking. The second class, which is characteristic of large digital circuits such as those found in computer systems, is caused by the designer not taking into account every logical condition that could occur during system operation. All computer systems have a software component as an integral part of their operation, and software is especially prone to this kind of design fault.

A fault may be permanent or transitory. Examples of *permanent faults* are short or open circuits within a component caused by physical failures. *Transitory faults* can be subdivided further into two classes. The first, usually called *transient faults*, are caused by such things as alpha-particle radiation or power supply variations. Large random access memory circuits are particularly prone to this kind of fault. By definition a transient fault is not caused by physical damage to the hardware. The second class are usually called *intermittent faults*. These faults are temporary but reoccur in an unpredictable manner. They are caused by loose physical connections between components or by components used at the limits of their specification. Intermittent faults often become permanent faults after a period of time. A fault may be *active* or *inactive*. For example, if a fault causes the output of a digital component to be stuck at logic 1, and the desired output is logic 1, then this would be classed as an inactive fault. Once the desired output becomes logic 0, then the fault becomes active.

The consequence for the system operation of a fault is an *error*. As the error may be caused by a permanent or by a transitory fault, it may be classed as a *hard error* or a *soft error*. An error in an individual subsystem may be due to a fault in that subsystem or to the propagation of an error from another part of the overall system.

The terms *fault* and *error* are sometimes interchanged. The term *failure* is often used to mean anything covered by these definitions. The definitions given here are those in most common usage.

Physical faults within a component can be characterized by their external electrical effects. These effects are commonly classified into *fault models*. The intention of any fault model is to take into account every possible failure mechanism, so that the effects on the system can be worked out. The manifestation of faults in a system can be classified according to the likely effects, producing an *error model*. The purpose of error models is to try to establish what kinds of corrective action need be taken in order to effect repairs.

20.3   Failure Rate and Reliability

An individual component may fail after a random time so it is impossible to predict any pattern of failure from one example. It is possible, however, to estimate the rate at which members of a group of identical components will fail. This rate can be determined by experimental means using accelerated life tests. In a normal operating environment, the time for a statistically significant number of failures to have occurred in a group of modern digital components could be tens or even hundreds of years. Consequently, the manufacturers must make the environment for the tests extremely unfavorable in order to produce failures in a few hours or days and then extrapolate back to produce the likely number of failures in a normal environment. The *failure rate* is then defined as the number of failures per unit time, in a given environment, compared with the number of surviving components. It is usually expressed as a number of failures per million hours.

If $f(t)$ is the number of components that have failed up to time t, and $s(t)$ is the number of components that have survived, then $z(t)$, the *failure rate* or *hazard rate*, is defined as

$$z(t) = \frac{1}{s(t)} \frac{df(t)}{dt}$$
(20.1)

Most electronic components will exhibit a variation of failure rate with time. Many studies have shown that this variation can often be approximated to the pattern shown in Figure 20.1. For obvious reasons this is known as a *bathtub* curve. The first phase, where the failure rate starts high but is decreasing with time, is where the components are suffering infant mortality; in other words, those that had manufacturing defects are failing. This is often called the *burn-in* phase. The second part, where the failure rate is roughly constant, is the useful life period of operation for the component. The final part, where the failure rate is increasing with time, is where the components are starting to wear out.

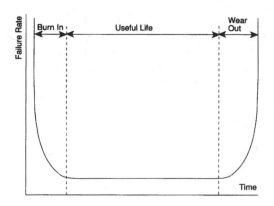

FIGURE 20.1   Variation of failure rate with time.

Using the same nomenclature as before, if

$$s(t) + f(t) = N \tag{20.2}$$

i.e., N is the total number of components in the test, then the *reliability r(t)* is defined as

$$r(t) = \frac{s(t)}{N} \tag{20.3}$$

or in words, and using the definition from the *IEEE Standard Dictionary of Electrical and Electronic Terms*, reliability is the probability that a device will function without failure over a specified time period or amount of usage, under stated conditions.

20.4   Relationship between Reliability and Failure Rate

Using Equation (20.1), Equation (20.2), and Equation (20.3), then

$$z(t) = -\frac{N}{s(t)} \cdot \frac{dr(t)}{dt} \tag{20.4}$$

λ is commonly used as the symbol for the failure rate $z(t)$ in the period where it is a constant, i.e., the useful life of the component. Consequently, we may write Equation (20.4) as

$$\lambda = -\frac{1}{r(t)} \frac{dr(t)}{dt} \tag{20.5}$$

Rewriting, integrating, and using the limits of integration as $r(t) = 1$ at $t = 0$ and $r(t) = 0$ at $t = \infty$ gives the result

$$r(t) = e^{-\lambda t} \tag{20.6}$$

This result is true only for the period of operation where the failure rate is a constant. For most common components, real failure rates can be obtained from such handbooks as the U.S. military's MIL-HDBK-217, as explained in Section "Reliability Calculations for Real Systems".

It must also be borne in mind that the calculated reliability is a probability function based on lifetime tests. There can be no guarantee that any batch of components will exhibit the same failure rate and hence reliability as those predicted because of variations in manufacturing conditions. Even if the components were made at the same factory as those tested, the process used might have been slightly different and the equipment will be older. Quality assurance standards are imposed on companies to try to guarantee that they meet minimum manufacturing standards, but some cases in the United States have shown that even the largest plants can fall short of these standards.

20.5   Mean Time to Failure

A figure that is commonly quoted because it gives a readier feel for the system performance is the **mean time to failure** (MTTF). This is defined as

$$MTTF = \int_0^\infty r(t)dt \tag{20.7}$$

Hence, for the period where the failure rate is constant

$$MTTF = \frac{1}{\lambda} \tag{20.8}$$

20.6   Mean Time to Repair

For many computer systems it is possible to define a **mean time to repair** (MTTR). This will be a function of a number of things, including the time taken to detect the failure, the time taken to isolate and replace the faulty component, and the time taken to verify that the system is operating correctly again. While the MTTF is a function of the system design and the operating environment, the MTTR is often a function of unpredictable human factors and, hence, is difficult to quantify. Figures used for MTTR for a given system in a fixed situation could be predictions based on the experience of the reliability engineers, or could be simply the maximum response time given in the maintenance contract for a computer. In either case, MTTR predictions may be subject to some fluctuations. To take an extreme example, if the service engineer has a flat tire while on the way to effect the repair, then the repair time may be many times the predicted MTTR. For some systems no MTTR can be predicted as they are in situations that make repair impossible or uneconomic. Computers in satellites are a good example. In these cases and all others where no errors in the output can be allowed, fault tolerant approaches must be used in order to extend the MTTF beyond the desired system operational lifetime.

20.7   Mean Time between Failures

For systems where repair is possible, a figure for the expected time between failures can be defined as

$$MTBF = MTTF + MTTR \tag{20.9}$$

The definitions given for MTTF and MTBF are the most commonly accepted ones. In some texts, MTBF is wrongly used as mean time before failure, confusing it with MTTF. In many real systems MTTF is very much greater than MTTR so the values of MTTF and MTBF will be almost identical, in any case.

20.8   Availability

Availability is defined as the probability that the system will be functioning at a given time during its normal working period.

$$Av = \frac{\text{total working time}}{\text{total time}} \tag{20.10}$$

This can also be written as

$$Av = \frac{\text{MTTF}}{\text{MTTF} + \text{MTTR}} \tag{20.11}$$

Some systems are designed for extremely high availability. For example, the computers used by AT&T to control its telephone exchanges are designed for an availability of 0.9999999, which corresponds to an unplanned downtime of two minutes in 40 years. In order to achieve this level of availability, fault tolerant techniques have to be used from the design stage accompanied by a high level of monitoring and maintenance.

20.9   Calculation of Computer System Reliability

For systems that have not been designed to be fault tolerant it is common to assume that the failure of any component implies the failure of the system. Thus, the system failure rate can be determined by the so-called parts count method. If the system contains m types of component, each with a failure rate λ_m, then the system failure rate λ_s can be defined as

$$\lambda_s = \sum_1^m N_m \lambda_m \tag{20.12}$$

where N_m is the number of each type of component.

The system reliability will be

$$r_s(t) = \prod_1^m N_m r_m \tag{20.13}$$

If the system design is such that the failure of an individual component does not necessarily cause system failure, then the calculations of MTTF and $r_s(t)$ become more complicated.

Consider two situations where a computer system is made up of several subsystems. These may be individual components or groups of components, e.g., circuit boards. The first is where failure of an individual subsystem implies system failure. This is known as the series model and is shown in Figure 20.2. This is the same case as considered previously, and the parts count method, Equation (20.12) and Equation (20.13), can be used. The second case is where failure of an individual subsystem does not imply system failure. This is shown in Figure 20.3. Only the failure of every subsystem means that the system has failed, and the system reliability can be evaluated by the following method. If $r(t)$ is the reliability (or probability of not failing) of each subsystem, then $q(t) = 1 - r(t)$ is the probability of an individual subsystem fail-

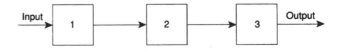

FIGURE 20.2   Series model.

ing. Hence, the probability of them all failing is

$$q_s(t) = [1 - r(t)]^n \qquad (20.14)$$

for n subsystems.

Hence the system reliability will be

$$r_s(t) = 1 - [1 - r(t)]^n \qquad (20.15)$$

In practice, systems will be made up of differing combinations of parallel and series networks; the simplest examples are shown in Figure 20.4 and Figure 20.5.

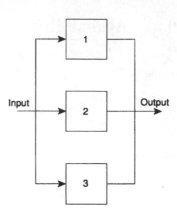

FIGURE 20.3   Parallel model.

Parallel-Series System

Assuming that the reliability of each subsystem is identical, then the overall reliability can be calculated thus. The reliability of one unit is r; hence the reliability of the series path is r^n. The probability of failure of each path is then $q = 1 - r^n$. Hence, the probability of failure of all m paths

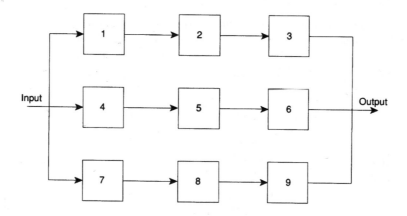

FIGURE 20.4   Parallel series model.

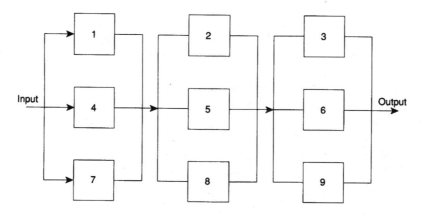

FIGURE 20.5   Series-parallel model.

is $(1 - r^n)^m$, and the reliability of the complete system is

$$r_{ps} = 1 - (1 - r^n)^m \tag{20.16}$$

Series-Parallel System

Making similar assumptions, and using a similar method, the reliability can be written as

$$r_{sp} = [1 - (1 - r)^n]^m \tag{20.17}$$

It is straightforward to extend these results to systems with subsystems having different reliabilities and in different combinations. It can be seen that these simple models could be used as the basis for a fault tolerant system, i.e., one that is able to carry on performing its designated function even while some of its parts have failed.

Practical Systems Using Parallel Subsystems

A computer system that uses parallel subsystems to improve reliability must incorporate some kind of **arbitrator** to determine which output to use at any given time. A common method of arbitration involves adding a **voter** to a system with N parallel modules, where N is an odd number. For example, if $N = 3$, a single incorrect output can be masked by the two correct outputs outvoting it. Hence, the system output will be correct even though an error has occurred in one of the subsystems. This system would be known as **triple-modular-redundant** (TMR) (Figure 20.6).

The reliability of a TMR system is the probability that any two out of the three units will be working. This can be expressed as

$$r_{tmr} = r_1 r_2 r_3 + r_1 r_2 (1 - r_3) + r_1 (1 - r_2) r_3 + (1 - r_1) r_2 r_3 \tag{20.18}$$

where r_n ($n = 1, 2, 3$) is the reliability of each subsystem. If $r_1 = r_2 = r_3 = r$ this reduces to

$$r_{tmr} = 3r^2 - 2r^3 \tag{20.19}$$

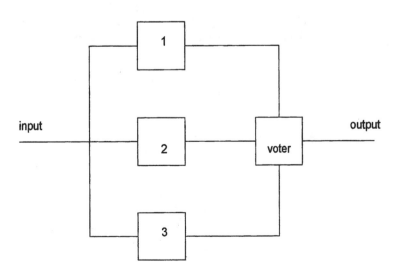

FIGURE 20.6 Triple-modular-redundant system.

The reliability of the voter must be included when calculating the overall reliability of such a system. As the voter appears in every path from input to output, it can be included as a series element in a series-parallel model. This leads to

$$r_{\text{tmr}} = r_{\text{v}}[3r^2 - 2r^3]$$

(20.20)

where r_{v} is the reliability of the voter.

More information on methods of using redundancy to improve system reliability can be found in the appropriate chapter of this handbook.

20.10   Markov Modeling

Another approach to determining the probability of system failure is to use a Markov model of the system rather than the combinatorial methods outlined previously. Markov models involve the defining of *system states* and *state transitions*. The mathematics of Markov modeling are well beyond the scope of this brief introduction, but most engineering mathematics textbooks will cover the technique.

To model the reliability of any system it is necessary to define the various fault-free and faulty states that could exist. For example, a system consisting of two identical units (A and B), either of which has to work for the system to work, would have four possible states. They would be (1) A and B working; (2) A working, B failed; (3) B working, A failed; and (4) A and B failed. The system designer must assign to each state a series of probabilities that determine whether it will remain in the same state or change to another after a given time period. This is usually shown in a state diagram, as in Figure 20.7. This model does not allow for the possibility of repair, but this could easily be added.

20.11   Software Reliability

One of the major components in any computer system is its software. Although software is unlikely to wear out in a physical sense, it is still impossible to prove that anything other than the simplest of programs is totally free from bugs. Hence, any piece of software will follow the first and second parts of the normal bathtub curve (Figure 20.1). The burn-in phase for hardware corresponds to the early release of a complex program, where bugs are commonly found and have to be fixed. The useful life phase for hardware corresponds to the time when the software can be described as stable, even though bugs may still be found. In this phase, where the failure rate can be characterized as constant (even if it is very low), the hardware performance criteria such as MTTF and MTTR can be estimated. They must be included in any estimation of the overall availability for

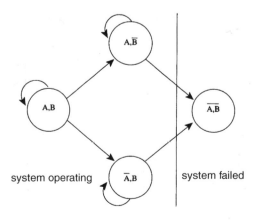

FIGURE 20.7   State diagram for two-unit parallel system.

the computer system as a whole. Just as with hardware techniques redundancy can be used to improve the availability through fault tolerance.

20.12   Reliability Calculations for Real Systems

The most common source of basic reliability data for electronic components and circuits is the military handbook *Reliability Prediction of Electronic Equipment*, published by the U.S. Department of Defense. It has the designation MIL-HDBK-217 and is regularly updated. This handbook provides both the basic reliability data and formulae to modify those data for the application of interest. For example, the formula for predicting the failure rate, λp, of a bipolar or MOS microprocessor is given as

$$\lambda_p = \pi_Q(C_1\pi_T\pi_V + C_2\pi_E)\pi_L \quad \text{failures per } 10^6 \text{ hours}$$

where π_Q is the part quality factor, with several categories, ranging from a full mil-spec part to a commercial part; π_T is the temperature acceleration factor, related to both the technology in use and the actual operating temperature; π_V is the voltage stress derating factor, which is higher for devices operating at higher voltages; π_E is the application environment factor (the handbook gives figures for many categories of environment, ranging from laboratory conditions up to the conditions found in the nose cone of a missile in flight); π_L is the device learning factor, related to how mature the technology is and how long the production of the part has been going on; C_1 is the circuit complexity factor, dependent on the number of transistors on the chip; and C_2 is the package complexity, related to the number of pins and the type of package.

The following figures are given for a 16-bit microprocessor, operating on the ground in a laboratory environment, with a junction temperature of 51°C. The device is assumed to be packaged in a plastic, 64-pin dual in-line package and to have been manufactured using the same technology for several years.

$$\pi_Q = 20 \quad \pi_T = 0.89 \quad \pi_v = 1 \quad \pi_E = 0.38$$
$$\pi_L = 1 \quad C_1 = 0.06 \quad C_2 = 0.033$$

Hence, the failure rate λp for this device operating in the specified environment, is estimated to be 1.32 failures per 10^6 hours. To calculate the predicted failure rate for a system based around this microprocessor would involve similar calculations for all the parts, including the passive components, the PCB, and connectors, and multiplying all the resultant failure rates together. The resulting figure could then be inverted to give a predicted MTTF. This kind of calculation is repetitive, tedious, and therefore prone to errors, so many companies now provide software to perform the calculations.

In any case, this way of calculating systems reliability has been largely discredited as it builds upon data of dubious merit. Several standards bodies (for example the IEEE and the British Standards Institute) now provide comprehensive guidelines for performing system reliability calculations and it is recommended that their methods are followed.

Defining Terms

Availability: This figure gives a prediction for the proportion of time that a given part or system will be in full working order. It can be calculated from

$$A_v = \frac{\text{MTTF}}{\text{MTTF} + \text{MTTR}}$$

Failure rate: The failure rate, λ, is the (predicted or measured) number of failures per unit time for a specified part or system operating in a given environment. It is usually assumed to be constant during the working life of a component or system.

Mean time to failure: This figure is used to give an expected working lifetime for a given part, in a given environment. It is defined by the equation

$$\text{MTTF} = \int_0^\infty r(t)\mathrm{d}t$$

If the failure rate λ is constant, then

$$\text{MTTF} = \frac{1}{\lambda}$$

Mean time to repair: The MTTR figure gives a prediction for the amount of time taken to repair a given part or system.

Reliability: Reliability $r(t)$ is the probability that a component or system will function without failure over a specified time period, under stated conditions.

References

J.-C. Jeffroy and G. Motet, *Design of Dependable Computer Systems*, Hingham, MA: Kluwer Academic Publishers, 2002, ISBN: 1402004370.

B.W. Johnson, *Design and Analysis of Fault Tolerant Digital Systems*, Reading, MA: Addison-Wesley, 1989, ISBN: 0201075709.

P. Lala, *Self-Checking and Fault-Tolerant Digital Design*, Los Altos, CA: Morgan Kaufmann, 2000, ISBN: 0124343708.

D.P. Sieworiek and R.S. Swarz, *Reliable Computer Systems: Design and Evaluation*, Natick, MA: A.K. Peters, 1998, ISBN: 156881092X.

BS 5760, *Reliability of Constructed or Manufactured Products, Systems, Equipments and Components*, London, U.K.: British Standards Institute, 1998.

IEEE Standard 1413, *Standard Methodology for Reliability Prediction and Assessment of Electronic Systems and Equipment*, 1998.

Further Information

The quarterly magazine *IEEE Transactions on Reliability* contains much of the latest research on reliability estimation techniques.

The monthly magazine *Microelectronics and Reliability* covers the field of reliability estimation and also includes papers on actual measured reliabilities.

Sometimes manufacturers make available measured failure rates for their devices.

Mathematics, Symbols, and Physical Constants

Ronald J. Tallarida
Temple University

THE GREAT ACHIEVEMENTS in engineering deeply affect the lives of all of us and also serve to remind us of the importance of mathematics. Interest in mathematics has grown steadily with these engineering achievements and with concomitant advances in pure physical science. Whereas scholars in nonscientific fields, and even in such fields as botany, medicine, geology, etc., can communicate most of the problems and results in nonmathematical language, this is virtually impossible in present-day engineering and physics. Yet it is interesting to note that until the beginning of the twentieth century, engineers regarded calculus as something of a mystery. Modern students of engineering now study calculus, as well as differential equations, complex variables, vector analysis, orthogonal functions, and a variety of other topics in applied analysis. The study of systems has ushered in matrix algebra and, indeed, most engineering students now take linear algebra as a core topic early in their mathematical education.

This section contains concise summaries of relevant topics in applied engineering mathematics and certain key formulas, that is, those formulas that are most often needed in the formulation and solution of engineering problems. Whereas even inexpensive electronic calculators contain tabular material (e.g., tables of trigonometric and logarithmic functions) that used to be needed in this kind of handbook, most calculators do not give symbolic results. Hence, we have included formulas along with brief summaries that guide their use. In many cases we have added numerical examples, as in the discussions of matrices, their inverses, and their use in the solutions of linear systems. A table of derivatives is included, as well as key applications of the derivative in the solution of problems in maxima and minima, related rates, analysis of curvature, and finding approximate roots by numerical methods. A list of infinite series, along with the interval of convergence of each, is also included.

Of the two branches of calculus, integral calculus is richer in its applications, as well as in its theoretical content. Though the theory is not emphasized here, important applications such as finding areas, lengths, volumes, centroids, and the work done by a nonconstant force are included. Both cylindrical and spherical polar coordinates are discussed, and a table of integrals is included. Vector analysis is summarized in a separate section and includes a summary of the algebraic formulas involving dot and cross multiplication, frequently needed in the study of fields, as well as the important theorems of Stokes and Gauss. The part on special functions includes the gamma function, hyperbolic functions, Fourier series, orthogonal functions, and both Laplace and z-transforms. The Laplace transform provides a basis for the solution of differential equations and is fundamental to all concepts and definitions underlying analytical tools for describing feedback control systems. The z-transform, not discussed in most applied mathematics books, is most useful in the analysis of discrete signals as, for example, when a computer receives data sampled at some prespecified time interval. The Bessel functions, also called cylindrical functions, arise in many physical applications, such as the heat transfer in a "long" cylinder, whereas the other orthogonal functions discussed—Legendre, Hermite, and Laguerre polynomials—are needed in quantum mechanics and many other subjects (e.g., solid-state electronics) that use concepts of modern physics.

The world of mathematics, even applied mathematics, is vast. Even the best mathematicians cannot keep up with more than a small piece of this world. The topics included in this section, however, have withstood the test of time and, thus, are truly *core* for the modern engineer.

This section also incorporates tables of physical constants and symbols widely used by engineers. While not exhaustive, the constants, conversion factors, and symbols provided will enable the reader to accommodate a majority of the needs that arise in design, test, and manufacturing functions.

Mathematics, Symbols, and Physical Constants

Greek Alphabet

Greek Letter		Greek Name	English Equivalent	Greek Letter		Greek Name	English Equivalent
A	α	Alpha	a	N	ν	Nu	n
B	β	Beta	b	Ξ	ξ	Xi	x
Γ	γ	Gamma	g	O	o	Omicron	ŏ
Δ	δ	Delta	d	Π	π	Pi	P
E	ε	Epsilon	ĕ	P	ρ	Rho	r
Z	ζ	Zeta	z	Σ	σ	Sigma	s
H	η	Eta	ē	T	τ	Tau	t
Θ	θ ϑ	Theta	th	Y	υ	Upsilon	u
I	ι	Iota	i	Φ	φ φ	Phi	ph
K	κ	Kappa	k	X	χ	Chi	ch
Λ	λ	Lambda	l	Ψ	ψ	Psi	ps
M	μ	Mu	m	Ω	ω	Omega	ō

International System of Units (SI)

The International System of units (SI) was adopted by the 11th General Conference on Weights and Measures (CGPM) in 1960. It is a coherent system of units built form seven *SI base units,* one for each of the seven dimensionally independent base quantities: they are the meter, kilogram, second, ampere, kelvin, mole, and candela, for the dimensions length, mass, time, electric current, thermodynamic temperature, amount of substance, and luminous intensity, respectively. The definitions of the SI base units are given below. The *SI derived units* are expressed as products of powers of the base units, analogous to the corresponding relations between physical quantities but with numerical factors equal to unity.

In the International System there is only one SI unit for each physical quantity. This is either the appropriate SI base unit itself or the appropriate SI derived unit. However, any of the approved decimal prefixes, called *SI prefixes,* may be used to construct decimal multiples or submultiples of SI units.

It is recommended that only SI units be used in science and technology (with SI prefixes where appropriate). Where there are special reasons for making an exception to this rule, it is recommended always to define the units used in terms of SI units. This section is based on information supplied by IUPAC.

Definitions of SI Base Units

Meter: The meter is the length of path traveled by light in vacuum during a time interval of 1/299,792,458 of a second (17th CGPM, 1983).

Kilogram: The kilogram is the unit of mass; it is equal to the mass of the international prototype of the kilogram (3rd CGPM, 1901).

Second: The second is the duration of 9,192,631,770 periods of the radiation corresponding to the transition between the two hyperfine levels of the ground state of the cesium-133 atom (13th CGPM, 1967).

Ampere: The ampere is that constant current which, if maintained in two straight parallel conductors of infinite length, of negligible circular cross-section, and placed 1 m apart in vacuum, would produce between these conductors a force equal to 2×10^{-7} newton per meter of length (9th CGPM, 1948).

Kelvin: The kelvin, unit of thermodynamic temperature, is the fraction 1/273.16 of the thermodynamic temperature of the triple point of water (13th CGPM, 1967).

Mole: The mole is the amount of substance of a system which contains as many elementary entities as there are atoms in 0.012 kg of carbon-12. When the mole is used, the elementary entities must be specified and may be atoms, molecules, ions, electrons, or other particles or specified groups of such particles (14th CGPM, 1971).

Examples of the use of the mole:

    1 mol of H_2 contains about 6.022×10^{23} H_2 molecules, or 12.044×10^{23} H atoms.

    1 mol of HgCl has a mass of 236.04 g.

    1 mol of Hg_2Cl_2 has a mass of 472.08 g.

    1 mol of Hg_2^{2+} has a mass of 401.18 g and a charge of 192.97 kC.

    1 mol of $Fe_{0.91}S$ has a mass of 82.88 g.

    1 mol of e^- has a mass of 548.60 μg and a charge of -96.49 kC.

    1 mol of photons whose frequency is 10^{14} Hz has energy of about 39.90 kJ.

Candela: The candela is the luminous intensity in a given direction of a source that emits monochromatic radiation of frequency 540×10^{12} hertz and that has a radiant intensity in that direction of (1/683) watt per steradian (16th CGPM, 1979).

Names and Symbols for the SI Base Units

Physical Quantity	Name of SI Unit	Symbol for SI Unit
Length	meter	m
Mass	kilogram	kg
Time	second	s
Electric current	ampere	A
Thermodynamic temperature	kelvin	K
Amount of substance	mole	mol
Luminous intensity	candela	cd

SI Derived Units with Special Names and Symbols

Physical Quantity	Name of SI Unit	Symbol for SI Unit	Expression in Terms of SI Base Units	
Frequency[1]	hertz	Hz	s^{-1}	
Force	newton	N	$m\ kg\ s^{-2}$	
Pressure, stress	pascal	Pa	$N\ m^{-2}$	$= m^{-1}\ kg\ s^{-2}$
Energy, work, heat	joule	J	$N\ m$	$= m^2\ kg\ s^{-2}$
Power, radiant flux	watt	W	$J\ s^{-1}$	$= m^2\ kg\ s^{-3}$
Electric charge	coulomb	C	$A\ s$	
Electric potential, electromotive force	volt	V	$J\ C^{-1}$	$= m^2\ kg\ s^{-3}\ A^{-1}$
Electric resistance	ohm	Ω	$V\ A^{-1}$	$= m^2\ kg\ s^{-3}\ A^{-2}$
Electric conductance	siemens	S	Ω^{-1}	$= m^{-2}\ kg^{-1}\ s^3\ A^2$
Electric capacitance	farad	F	$C\ V^{-1}$	$= m^{-2}\ kg^{-1}\ s^4\ A^2$
Magnetic flux density	tesla	T	$V\ s\ m^{-2}$	$= kg\ s^{-2}\ A^{-1}$
Magnetic flux	weber	Wb	$V\ s$	$= m^2\ kg\ s^{-2}\ A^{-1}$
Inductance	henry	H	$V\ A^{-1}\ s$	$= m^2\ kg\ s^{-2}\ A^{-2}$
Celsius temperature[2]	degree Celsius	°C	K	

(continued)

SI Derived Units with Special Names and Symbols (continued)

Physical Quantity	Name of SI Unit	Symbol for SI Unit	Expression in Terms of SI Base Units	
Luminous flux	lumen	lm	cd sr	
Illuminance	lux	lx	cd sr m^{-2}	
Activity (radioactive)	becquerel	Bq	s^{-1}	
Absorbed dose (of radiation)	gray	Gy	J kg^{-1}	$= $ m^2 s^{-2}
Dose equivalent (dose equivalent index)	sievert	Sv	J kg^{-1}	$= $ m^2 s^{-2}
Plane angle	radian	rad	1	$= $ m m^{-1}
Solid angle	steradian	sr	1	$= $ m^2 m^{-2}

[1] For radial (circular) frequency and for angular velocity the unit rad s^{-1}, or simply s^{-1}, should be used, and this may not be simplified to Hz. The unit Hz should be used only for frequency in the sense of cycles per second.

[2] The Celsius temperature θ is defined by the equation:

$$\theta/°C = T/K - 273.15$$

The SI unit of Celsius temperature interval is the degree Celsius, °C, which is equal to the kelvin, K. °C should be treated as a single symbol, with no space between the ° sign and the letter C. (The symbol °K and the symbol ° should no longer be used.)

Units in Use Together with the SI

These units are not part of the SI, but it is recognized that they will continue to be used in appropriate contexts. SI prefixes may be attached to some of these units, such as milliliter, ml; millibar, mbar; megaelectronvolt, MeV; kilotonne, ktonne.

Physical Quantity	Name of Unit	Symbol for Unit	Value in SI Units
Time	minute	min	60 s
Time	hour	h	3600 s
Time	day	d	86,400 s
Plane angle	degree	°	$(\pi/180)$ rad
Plane angle	minute	$'$	$(\pi/10,800)$ rad
Plane angle	second	$''$	$(\pi/648,000)$ rad
Length	ångstrom[1]	Å	10^{-10} m
Area	barn	b	10^{-28} m^2
Volume	liter	l, L	dm^3 $=$ 10^{-3} m^3
Mass	tonne	t	Mg $=$ 10^3 kg
Pressure	bar[1]	bar	10^5 Pa $=$ 10^5 N m^{-2}
Energy	electronvolt[2]	eV $(= e \times V)$	$\approx 1.60218 \times 10^{-19}$ J
Mass	unified atomic mass unit[2,3]	u $(= m_a(^{12}C)/12)$	$\approx 1.66054 \times 10^{-27}$ kg

[1] The ångstrom and the bar are approved by CIPM for "temporary use with SI units," until CIPM makes a further recommendation. However, they should not be introduced where they are not used at present.

[2] The values of these units in terms of the corresponding SI units are not exact, since they depend on the values of the physical constants e (for the electronvolt) and N_a (for the unified atomic mass unit), which are determined by experiment.

[3] The unified atomic mass unit is also sometimes called the dalton, with symbol Da, although the name and symbol have not been approved by CGPM.

Conversion Constants and Multipliers

Recommended Decimal Multiples and Submultiples

Multiples and Submultiples	Prefixes	Symbols	Multiples and Submultiples	Prefixes	Symbols
10^{18}	exa	E	10^{-1}	deci	d
10^{15}	peta	P	10^{-2}	centi	c
10^{12}	tera	T	10^{-3}	milli	m
10^{9}	giga	G	10^{-6}	micro	μ (Greek mu)
10^{6}	mega	M	10^{-9}	nano	n
10^{3}	kilo	k	10^{-12}	pico	p
10^{2}	hecto	h	10^{-15}	femto	f
10	deca	da	10^{-18}	atto	a

Conversion Factors—Metric to English

To Obtain	Multiply	By
Inches	centimeters	0.3937007874
Feet	meters	3.280839895
Yards	meters	1.093613298
Miles	kilometers	0.6213711922
Ounces	grams	$3.527396195 \times 10^{-2}$
Pounds	kilogram	2.204622622
Gallons (U.S. liquid)	liters	0.2641720524
Fluid ounces	milliliters (cc)	$3.381402270 \times 10^{-2}$
Square inches	square centimeters	0.155003100
Square feet	square meters	10.76391042
Square yards	square meters	1.195990046
Cubic inches	milliliters (cc)	$6.102374409 \times 10^{-2}$
Cubic feet	cubic meters	35.31466672
Cubic yards	cubic meters	1.307950619

Conversion Factors—English to Metric*

To Obtain	Multiply	By
Microns	mils	**25.4**
Centimeters	inches	**2.54**
Meters	feet	**0.3048**
Meters	yards	**0.9144**
Kilometers	miles	**1.609344**
Grams	ounces	28.34952313
Kilograms	pounds	**0.45359237**
Liters	gallons (U.S. liquid)	**3.785411784**
Millimeters (cc)	fluid ounces	29.57352956
Square centimeters	square inches	**6.4516**
Square meters	square feet	**0.09290304**
Square meters	square yards	**0.83612736**
Milliliters (cc)	cubic inches	**16.387064**
Cubic meters	cubic feet	$2.831684659 \times 10^{-2}$
Cubic meters	cubic yards	0.764554858

*Boldface numbers are exact; others are given to ten significant figures where so indicated by the multiplier factor.

Conversion Factors—General*

To Obtain	Multiply	By
Atmospheres	feet of water @ 4°C	2.950×10^{-2}
Atmospheres	inches of mercury @ 0°C	3.342×10^{-2}
Atmospheres	pounds per square inch	6.804×10^{-2}
BTU	foot-pounds	1.285×10^{-3}
BTU	joules	9.480×10^{-4}
Cubic feet	cords	**128**
Degree (angle)	radians	57.2958
Ergs	foot-pounds	1.356×10^{7}
Feet	miles	**5280**
Feet of water @ 4°C	atmospheres	33.90
Foot-pounds	horsepower-hours	1.98×10^{6}
Foot-pounds	kilowatt-hours	2.655×10^{6}
Foot-pounds per min	horsepower	3.3×10^{4}
Horsepower	foot-pounds per sec	1.818×10^{-3}
Inches of mercury @ 0°C	pounds per square inch	2.036
Joules	BTU	1054.8
Joules	foot-pounds	1.35582
Kilowatts	BTU per min	1.758×10^{-2}
Kilowatts	foot-pounds per min	2.26×10^{-5}
Kilowatts	horsepower	0.745712
Knots	miles per hour	0.86897624
Miles	feet	1.894×10^{-4}
Nautical miles	miles	0.86897624
Radians	degrees	1.745×10^{-2}
Square feet	acres	**43,560**
Watts	BTU per min	17.5796

*Boldface numbers are exact; others are given to ten significant figures where so indicated by the multiplier factor.

Temperature Factors

$$°F = 9/5 \; (°C) + 32$$

Fahrenheit temperature $= 1.8$ (temperature in kelvins) $- 459.67$

$$°C = 5/9 \; [(°F) - 32)]$$

Celsius temperature $=$ temperature in kelvins $- 273.15$

Fahrenheit temperature $= 1.8$ (Celsius temperature) $+ 32$

Conversion of Temperatures

From	To	
°Celsius	°Fahrenheit	$t_F = (t_C \times 1.8) + 32$
	Kelvin	$T_K = t_C + 273.15$
	°Rankine	$T_R = (t_C + 273.15) \times 18$
°Fahrenheit	°Celsius	$t_C = \dfrac{t_F - 32}{1.8}$
	Kelvin	$T_k = \dfrac{t_F - 32}{1.8} + 273.15$
	°Rankine	$T_R = t_F + 459.67$
Kelvin	°Celsius	$t_C = T_K - 273.15$
	°Rankine	$T_R = T_K \times 1.8$
°Rankine	Kelvin	$T_K = \dfrac{T_R}{1.8}$
	°Fahrenheit	$t_F = T_R - 459.67$

Physical Constants

General

Equatorial radius of the Earth $= 6378.388$ km $= 3963.34$ miles (statute)

Polar radius of the Earth, 6356.912 km $= 3949.99$ miles (statute)

1 degree of latitude at $40°$ $= 69$ miles

1 international nautical mile $= 1.15078$ miles (statute) $= 1852$ m $= 6076.115$ ft

Mean density of the earth $= 5.522$ g/cm^3 $= 344.7$ lb/ft^3

Constant of gravitation $(6.673 \pm 0.003) \times 10^{-8}$ cm^3 gm^{-1} s^{-2}

Acceleration due to gravity at sea level, latitude $45°$ $= 980.6194$ cm/s^2 $= 32.1726$ ft/s^2

Length of seconds pendulum at sea level, latitude $45°$ $= 99.3575$ cm $= 39.1171$ in.

1 knot (international) $= 101.269$ ft/min $= 1.6878$ ft/s $= 1.1508$ miles (statute)/h

1 micron $= 10^{-4}$ cm

1 ångstrom $= 10^{-8}$ cm

Mass of hydrogen atom $= (1.67339 \pm 0.0031) \times 10^{-24}$ g

Density of mercury at $0°$C $= 13.5955$ g/ml

Density of water at $3.98°$C $= 1.000000$ g/ml

Density, maximum, of water, at $3.98°$C $= 0.999973$ g/cm^3

Density of dry air at $0°$C, 760 mm $= 1.2929$ g/l

Velocity of sound in dry air at $0°$C $= 331.36$ m/s $- 1087.1$ ft/s

Velocity of light in vacuum $= (2.997925 \pm 0.000002) \times 10^{10}$ cm/s

Heat of fusion of water $0°$C $= 79.71$ cal/g

Heat of vaporization of water $100°$C $= 539.55$ cal/g

Electrochemical equivalent of silver 0.001118 g/s international amp

Absolute wavelength of red cadmium light in air at $15°$C, 760 mm pressure $= 6438.4696$ Å

Wavelength of orange-red line of krypton 86 $= 6057.802$ Å

π Constants

$\pi = 3.14159\ 26535\ 89793\ 23846\ 26433\ 83279\ 50288\ 41971\ 69399\ 37511$

$1/\pi = 0.31830\ 98861\ 83790\ 67153\ 77675\ 26745\ 02872\ 40689\ 19291\ 48091$

$\pi^2 = 9.8690\ \ 44010\ 89358\ 61883\ 44909\ 99876\ 15113\ 53136\ 99407\ 24079$

$\log_e\pi = 1.14472\ 98858\ 49400\ 17414\ 34273\ 51353\ 05871\ 16472\ 94812\ 91531$

$\log_{10}\pi = 0.49714\ 98726\ 94133\ 85435\ 12682\ 88290\ 89887\ 36516\ 78324\ 38044$

$\log_{10}\sqrt{2\pi} = 0.39908\ 99341\ 79057\ 52478\ 25035\ 91507\ 69595\ 02099\ 34102\ 92128$

Constants Involving e

$e = 2.71828\ 18284\ 59045\ 23536\ 02874\ 71352\ 66249\ 77572\ 47093\ 69996$

$1/e = 0.36787\ 94411\ 71442\ 32159\ 55237\ 70161\ 46086\ 74458\ 11131\ 03177$

$e^2 = 7.38905\ 60989\ 30650\ 22723\ 04274\ 60575\ 00781\ 31803\ 15570\ 55185$

$M = \log_{10}e = 0.43429\ 44819\ 03251\ 82765\ 11289\ 18916\ 60508\ 22943\ 97005\ 80367$

$1/M \cdot = \log_e10 = 2.30258\ 50929\ 94045\ 68401\ 79914\ 54684\ 36420\ 67011\ 01488\ 62877$

$\log_{10}M = 9.63778\ 43113\ 00536\ 78912\ 29674\ 98645\ -10$

Numerical Constants

$\sqrt{2} = 1.41421\ 35623\ 73095\ 04880\ 16887\ 24209\ 69807\ 85696\ 71875\ 37695$

$3\sqrt{2} = 1.25992\ 10498\ 94873\ 16476\ 72106\ 07278\ 22835\ 05702\ 51464\ 70151$

$\log_e2 = 0.69314\ 71805\ 59945\ 30941\ 72321\ 21458\ 17656\ 80755\ 00134\ 36026$

$\log_{10}2 = 0.30102\ 99956\ 63981\ 19521\ 37388\ 94724\ 49302\ 67881\ 89881\ 46211$

$$\sqrt{3} = 1.73205\ 08075\ 68877\ 29352\ 74463\ 41505\ 87236\ 69428\ 05253\ 81039$$
$$\sqrt[3]{3} = 1.44224\ 95703\ 07408\ 38232\ 16383\ 10780\ 10958\ 83918\ 69253\ 49935$$
$$\log_e 3 = 1.09861\ 22886\ 68109\ 69139\ 52452\ 36922\ 52570\ 46474\ 90557\ 82275$$
$$\log_{10} 3 = 0.47712\ 12547\ 19662\ 43729\ 50279\ 03255\ 11530\ 92001\ 28864\ 19070$$

Symbols and Terminology for Physical and Chemical Quantities

Name	Symbol	Definition	SI Unit
Classical Mechanics			
Mass	m		kg
Reduced mass	μ	$\mu = m_1 m_2/(m_1 + m_2)$	kg
Density, mass density	ρ	$\rho = M/V$	kg m^{-3}
Relative density	d	$d = \rho/\rho^{\theta}$	1
Surface density	ρ_A, ρ_S	$\rho_A = m/A$	kg m^{-2}
Momentum	p	$p = mv$	kg m s^{-1}
Angular momentum, action	L	$l = r \yen p$	J s
Moment of inertia	I, J	$I = \Sigma m_i r_i^2$	kg m^2
Force	F	$F = dp/dt = ma$	N
Torque, moment of a force	$T, (M)$	$T = r \times \mathbf{F}$	N m
Energy	E		J
Potential energy	E_p, V, Φ	$E_p = Fds$	J
Kinetic energy	E_k, T, K	$e_k = (1/2)mv^2$	J
Work	W, w	$w = Fds$	J
Hamilton function	H	$H(q, p) = T(q, p) + V(q)$	J
Lagrange function	L	$L(q, \dot{q})T(q, \dot{q}) - V(q)$	J
Pressure	p, P	$p = F/A$	Pa, N m^{-2}
Surface tension	γ, σ	$\gamma = dW/dA$	N m^{-1}, J m^{-2}
Weight	$G, (W, P)$	$G = mg$	N
Gravitational constant	G	$F = Gm_1 m_2/r^2$	N m^2 kg^{-2}
Normal stress	σ	$\sigma = F/A$	Pa
Shear stress	τ	$\tau = F/A$	Pa
Linear strain, relative elongation	ε, e	$\varepsilon = \Delta l/l$	1
Modulus of elasticity, Young's modulus	E	$E = \sigma/\varepsilon$	Pa
Shear strain	γ	$\gamma = \Delta x/d$	1
Shear modulus	G	$G = \tau/\gamma$	Pa
Volume strain, bulk strain	θ	$\theta = \Delta V/V_0$	1
Bulk modulus, compression modulus	K	$K = -V_0(dp/dV)$	Pa
Viscosity, dynamic viscosity	η, μ	$\tau_{x,z} = \eta(dv_x/dz)$	Pa s
Fluidity	ϕ	$\phi = 1/\eta$	m kg^{-1} s
Kinematic viscosity	v	$v = \eta/\rho$	m^2 s^{-1}
Friction coefficient	$\mu, (f)$	$F_{frict} = \mu F_{norm}$	1
Power	P	$P = dW/dt$	W
Sound energy flux	P, P_a	$P = dE/dt$	W
Acoustic factors			
Reflection factor	ρ	$\rho = P_r/P_0$	1
Acoustic absorption factor	$\alpha_a, (\alpha)$	$\alpha_a = 1 - \rho$	1
Transmission factor	τ	$\tau = P_{tr}/P_0$	1
Dissipation factor	δ	$\delta = \alpha_a - \tau$	1

(continued)

Symbols and Terminology for Physical and Chemical Quantities (continued)

Name	Symbol	Definition	SI Unit
Electricity and Magnetism			
Quantity of electricity, electric charge	Q		C
Charge density	ρ	$\rho = Q/V$	C m^{-3}
Surface charge density	σ	$\sigma = Q/A$	C m^{-2}
Electric potential	V, ϕ	$V = dW/dQ$	V, J C^{-1}
Electric potential difference	$U, \Delta V, \Delta\phi$	$U = V_2 - V_1$	V
Electromotive force	E	$E = (F/Q)ds$	V
Electric field strength	$\mathbf{E}$	$\mathbf{E} = \mathbf{F}/Q = -\text{grad } V$	V m^{-1}
Electric flux	Ψ	$\Psi = \mathbf{D}dA$	C
Electric displacement	$\mathbf{D}$	$\mathbf{D} = \varepsilon\mathbf{E}$	C m^{-2}
Capacitance	C	$C = Q/U$	F, C V^{-1}
Permittivity	ε	$D = \varepsilon E$	F m^{-1}
Permittivity of vacuum	ε_0	$\varepsilon_0 = \mu_0^{-1} c_0^{-2}$	F m^{-1}
Relative permittivity	ε_r	$\varepsilon_r = \varepsilon/\varepsilon_0$	1
Dielectric polarization (dipole moment per volume)	$\mathbf{P}$	$\mathbf{P} = \mathbf{D} - \varepsilon_0\mathbf{E}$	C m^{-2}
Electric susceptibility	χ_e	$\chi_e = \varepsilon_r - 1$	1
Electric dipole moment	$\mathbf{p}, \mu$	$\mathbf{p} = Q\mathbf{r}$	C m
Electric current	I	$I = dQ/dt$	A
Electric current density	$\mathbf{j}, \mathbf{J}$	$I = jdxA$	A m^{-2}
Magnetic flux density, magnetic induction	$\mathbf{B}$	$\mathbf{F} = Qv \times \mathbf{B}$	T
Magnetic flux	Φ	$\Phi = \mathbf{B}dA$	Wb
Magnetic field strength	$\mathbf{H}$	$\mathbf{B} = \mu\mathbf{H}$	A M^{-1}
Permeability	μ	$\mathbf{B} = \mu\mathbf{H}$	N A^{-2}, H m^{-1}
Permeability of vacuum	μ_0		H m^{-1}
Relative permeability	μ_r	$\mu_r = \mu/\mu_0$	1
Magnetization (magnetic dipole moment per volume)	$\mathbf{M}$	$\mathbf{M} = \mathbf{B}/\mu_0 - \mathbf{H}$	A m^{-1}
Magnetic susceptibility	$\chi, \kappa, (\chi_m)$	$\chi = \mu_r - 1$	1
Molar magnetic susceptibility	χ_m	$\chi_m = V_m\chi$	m^3 mol^{-1}
Magnetic dipole moment	$\mathbf{m}, \mu$	$E_p = -\mathbf{m} \cdot \mathbf{B}$	A m^2, J T^{-1}
Electrical resistance	R	$\mathbf{P} = \mathbf{Y/I}$	Ω
Conductance	G	$G = 1/R$	S
Loss angle	δ	$\delta = (\pi/2) + \phi_I - \phi_U$	1, rad
Reactance	X	$X = (U/I)\sin\delta$	Ω
Impedance (complex impedance)	Z	$Z = R + iX$	Ω
Admittance (complex admittance)	Y	$Y = 1/Z$	S
Susceptance	B	$Y = G + iB$	S
Resistivity	ρ	$\rho = E/j$	Ω m
Conductivity	κ, γ, σ	$\kappa = 1/\rho$	S m^{-1}
Self-inductance	L	$E = -L(dI/dt)$	H
Mutual inductance	M, L_{12}	$E_1 = L_{12}(Di_2/dt)$	H
Magnetic vector potential	$\mathbf{A}$	$\mathbf{B} = \nabla \times \mathbf{A}$	Wb m^{-1}
Poynting vector	$\mathbf{S}$	$\mathbf{S} = \mathbf{E} \times \mathbf{H}$	W m^{-2}
Electromagnetic Radiation			
Wavelength	λ		m
Speed of light			m s^{-1}
in vacuum	c_0		
in a medium	c	$c = c_0/n$	

(continued)

Symbols and Terminology for Physical and Chemical Quantities (continued)

Name	Symbol	Definition	SI Unit
Electromagnetic Radiation			
Wavenumber in vacuum	V	$V = V/c_0 = 1/n\lambda$	m^{-1}
Wavenumber (in a medium)	σ	$\sigma = 1/\lambda$	m^{-1}
Frequency	v	$v = c/\lambda$	Hz
Circular frequency, pulsatance	ω	$\omega = 2\pi v$	s^{-1}, rad s^{-1}
Refractive index	n	$n = c_0/c$	1
Planck constant	h		J s
Planck constant/2π	$\hbar$	$\hbar = h/2\pi$	J s
Radiant energy	Q, W		J
Radiant energy density	ρ, w	$\rho = Q/V$	$J\ m^{-3}$
Spectral radiant energy density			
in terms of frequency	ρ_v, w_v	$\rho_v = \delta\rho/dv$	$J\ m^{-3}\ Hz^{-1}$
in terms of wavenumber	$\rho_{\bar{v}}, w_{\bar{v}}$	$\rho_{\bar{v}} = d\rho/d\bar{v}$	$J\ m^{-2}$
in terms of wavelength	ρ_λ, w_λ	$\rho_\lambda = \delta\rho/d\lambda$	$J\ m^{-4}$
Einstein transition probabilities			
Spontaneous emission	A_{nm}	$dN_n/dt = -A_{nm}N_n$	s^{-1}
Stimulated emission	B_{nm}	$dn_n/dt = -\rho\bar{v}(\bar{V}_{nm}) \times B_{nm}N_n$	$s\ kg^{-1}$
Radiant power, radiant energy per time	Φ, P	$\Phi = dQ/dt$	W
Radiant intensity	I	$I = d\Phi/d\Omega$	$W\ sr^{-1}$
Radiant exitance (emitted radiant flux)	M	$M = d\Phi/dA_{source}$	$W\ m^{-2}$
Irradiance (radiant flux received)	$E, (I)$	$E = d\Phi/\delta A$	$W\ m^{-2}$
Emittance	ε	$\varepsilon = M/M_{bb}$	1
Stefan–Boltzmann constant	σ	$M_{bb} = \sigma T^4$	$W\ m^{-2}\ K^{-4}$
First radiation constant	c_1	$c_1 = 2\pi hc_0^2$	$W\ m^2$
Second radiation constant	c_2	$c_2 = hc_0/k$	K m
Transmittance, transmission factor	τ, T	$\tau = \Phi_{tr}/\Phi_0$	1
Absorptance, absorption factor	α	$\alpha = \phi_{abs}/\phi_0$	1
Reflectance, reflection factor	ρ	$\rho = \phi_{refl}/\Phi_0$	1
(Decadic) absorbance	A	$A = \lg(1 - \alpha_i)$	1
Napierian absorbance	B	$B = \ln(1 - \alpha_i)$	1
Absorption coefficient			
(Linear) decadic	a, K	$a = A/l$	m^{-1}
(Linear) napierian	α	$\alpha = B/l$	m^{-1}
Molar (decadic)	ε	$\varepsilon = a/c = A/cl$	$m^2\ mol^{-1}$
Molar napierian	κ	$\kappa = \alpha/c = B/cl$	$m^2\ mol^{-1}$
Absorption index	k	$k = \alpha/4\pi\bar{v}$	1
Complex refractive index	$\hat{n}$	$\hat{n} = n + ik$	1
Molar refraction	R, R_m	$R = \frac{(n^2-1)}{(n^2+2)}V_m$	$m^3\ mol^{-1}$
Angle of optical rotation	α		1, rad
Solid State			
Lattice vector	$\mathbf{R}, \mathbf{R}_0$		m
Fundamental translation vectors for the crystal lattice	$\mathbf{a}_1; \mathbf{a}_2; \mathbf{a}_3, \mathbf{a}; \mathbf{b}; \mathbf{c}$	$R = n_1\mathbf{a}_1 + n_2\mathbf{a}_2 + n_3\mathbf{a}_3$	m
(Circular) reciprocal lattice vector	$\mathbf{G}$	$G \cdot R = 2\pi m$	m^{-1}

(continued)

Symbols and Terminology for Physical and Chemical Quantities (continued)

Name	Symbol	Definition	SI Unit
Solid State			
(Circular) fundamental translation vectors for the reciprocal lattice	$\mathbf{b}_1$; $\mathbf{b}_2$; $\mathbf{b}_3$, $a^\star$; $b^\star$; $c^\star$	$\mathbf{a}_i \cdot \mathbf{b}_k = 2\pi\delta_{ik}$	m^{-1}
Lattice plane spacing	d		m
Bragg angle	θ	$n\lambda = 2d\sin\theta$	1, rad
Order of reflection	n		1
Order parameters			
Short range	σ		1
Long range	s		1
Burgers vector	b		m
Particle position vector	r, R_j		m
Equilibrium position vector of an ion	R_o		m
Displacement vector of an ion	$\mathbf{u}$	$\mathbf{u} = \mathbf{R} - \mathbf{R}_0$	m
Debye–Waller factor	B, D		1
Debye circular wavenumber	q_D		m^{-1}
Debye circular frequency	ω_D		s^{-1}
Grüneisen parameter	γ, Γ	$\gamma = \alpha V/\kappa C_V$	1
Madelung constant	α, $\mathscr{M}$	$E_{coul} = \frac{\alpha N_A z_+ z_- e^2}{4\pi\varepsilon_0 R_0}$	1
Density of states	N_E	$N_E = dN(E)/dE$	$J^{-1}\,m^{-3}$
(Spectral) density of vibrational modes	N_ω, g	$N_\omega = dN(\omega)/d\omega$	$s\,m^{-3}$
Resistivity tensor	ρ_{ik}	$E = \rho \cdot j$	$\Omega\,m$
Conductivity tensor	σ_{ik}	$\sigma = \rho^{-1}$	$S\,m^{-1}$
Thermal conductivity tensor	λ_{ik}	$J_q = -\lambda \cdot \operatorname{grad} T$	$W\,m^{-1}\,K^{-1}$
Residual resistivity	ρ_R		$\Omega\,m$
Relaxation time	τ	$\tau = l/v_F$	s
Lorenz coefficient	L	$L = \lambda/\sigma T$	$V^2\,K^{-2}$
Hall coefficient	A_H, R_H	$\mathbf{E} = \rho \cdot \mathbf{j} + R_H(\mathbf{B} \times \mathbf{j})$	$m^3\,C^{-1}$
Thermoelectric force	E		V
Peltier coefficient	Π		V
Thomson coefficient	$\mu,(\tau)$		$V\,K^{-1}$
Work function	Φ	$\Phi = E_\infty - E_F$	J
Number density, number concentration	n, (p)		m^{-3}
Gap energy	E_γ		J
Donor ionization energy	E_δ		J
Acceptor ionization energy	E_α		J
Fermi energy	E_Φ, ε_F		J
Circular wave vector, propagation vector	$\boldsymbol{k}$, $\boldsymbol{q}$	$k = 2\pi/\lambda$	m^{-1}
Bloch function	$u_k(\boldsymbol{r})$	$\psi(\boldsymbol{r}) = u_k(\boldsymbol{r})\exp(i\mathbf{k}\cdot\mathbf{r})$	$m^{-3/2}$
Charge density of electrons	ρ	$\rho(\boldsymbol{r}) = -e\psi^\star(\mathbf{r})\psi(\mathbf{r})$	$C\,m^{-3}$
Effective mass	$m^\star$		kg
Mobility	μ	$\mu = v_{drift}/E$	$m^2\,V^{-1}\,s^{-1}$
Mobility ratio	b	$b = \mu_n/\mu_p$	1
Diffusion coefficient	D	$dN/dt = -DA(dn/dx)$	$m^2\,s^{-1}$
Diffusion length	L	$L = \sqrt{D\tau}$	m
Characteristic (Weiss) temperature	ϕ, ϕ_W		K
Curie temperature	T_C		K
Néel temperature	T_N		K

Credits

Material in Section III was reprinted from the following sources:

D. R. Lide, Ed., *CRC Handbook of Chemistry and Physics,* 76th ed., Boca Raton, FL: CRC Press, 1992: International System of Units (SI), conversion constants and multipliers (conversion of temperatures), symbols and terminology for physical and chemical quantities, fundamental physical constants, classification of electromagnetic radiation.

D. Zwillinger, Ed., *CRC Standard Mathematical Tables and Formulae,* 30th ed., Boca Raton, FL: CRC Press, 1996: Greek alphabet, conversion constants and multipliers (recommended decimal multiples and submultiples, metric to English, English to metric, general, temperature factors), physical constants, series expansion.

Probability for Electrical and Computer Engineers

Charles W. Therrien

The Algebra of Events

The study of probability is based upon experiments that have uncertain outcomes. Collections of these outcomes comprise *events* and the collection of all possible outcomes of the experiment comprise what is called the *sample space*, denoted by S. Outcomes are members of the sample space and events of interest are represented as *sets* of outcomes (see Figure III.1).

The algebra $\mathcal{A}$ that deals with representing events is the usual set algebra. If A is an event, then A^c (the *complement* of A) represents the event that "A did not occur." The complement of the sample space is the *null event*, $\varnothing = S^c$. The event that *both* event A_1 and event A_2 have occurred is the intersection, written as "$A_1 \cdot A_2$" or "$A_1 A_2$" while the event that *either* A_1 or A_2 *or both* have occurred is the union, written as "$A_1 + A_2$."[1]

Table III.1 lists the two postulates that define the algebra $\mathcal{A}$, while Table III.2 lists seven axioms that define properties of its operations. Together these tables can be used to show all of the properties of the algebra of events. Table III.3 lists some additional useful relations that can be derived from the axioms and the postulates.

Since the events "$A_1 + A_2$" and "$A_1 A_2$" are included in the algebra, it follows by induction that for any finite number of events $A_1 + A_2 + \cdots + A_N$ and $A_1 \cdot A_2 \cdots \cdot A_N$ are also included in the algebra. Since problems often involve the union or intersection of an *infinite* number of events, however, the algebra of events must be defined to include these infinite intersections and unions. This extension to infinite unions and intersections is known as a sigma algebra.

A set of events that satisfies the two conditions:

1. $A_i A_j = \varnothing \neq$ for $\neq i \neq j$
2. $A_1 + A_2 + A_3 + \cdots = S$

is known as a *partition* and is important for the solution of problems in probability. The events of a partition are said to be *mutually exclusive* and *collectively exhaustive*. The most fundamental partition is the set outcomes defining the random experiment, which comprise the sample space by definition.

Probability

Probability measures the likelihood of occurrence of events represented on a scale of 0 to 1. We often estimate probability by measuring the *relative frequency* of an event, which is defined as

$$\text{relative frequency} = \frac{\text{number of occurrences of the event}}{\text{number of repetitions of the experiment}}$$

(for a large number of repetitions). Probability can be defined formally by the following axioms:

(I) The probability of any event is nonnegative:

$$\Pr[A] \geqslant 0 \qquad\qquad\qquad\qquad \text{(III.1)}$$

(II) The probability of the universal event (i.e., the entire sample space) is 1:

$$\Pr[S] = 1 \qquad\qquad\qquad\qquad \text{(III.2)}$$

---

[1]Some authors use $\cap$ and $\cup$ rather than $\cdot$ and $+$, respectively.

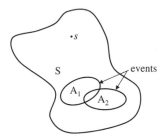

FIGURE III.1 Abstract representation of the sample space S with outcome s and sets A_1 and A_2 representing events.

(III) If A_1 and A_2 are mutually exclusive, i.e., $A_1 A_2 = \varnothing$, then

$$\Pr[A_1 + A_2] = \Pr[A_1] + \Pr[A_2] \tag{III.3}$$

(IV) If $\{A_i\}$ represent a countably infinite set of mutually exclusive events, then

$$\Pr[A_1 + A_2 + A_3 + \cdots] = \sum_{i=1}^{\infty} \Pr[A_i] \quad (\text{if } A_i A_j = \varnothing \quad i \neq j) \tag{III.4}$$

Note that although the additivity of probability for any finite set of disjoint events follows from (III), the property has to be stated explicitly for an infinite set in (IV). These axioms and the algebra of events can be used to show a number of other important properties which are summarized in Table III.4. The last item in the table is an especially important formula since it uses probabilistic information about

TABLE III.1 Postulates for an Algebra of Events

1.	*If $A \in \mathcal{A}$ then $A^c \in \mathcal{A}$*
2.	*If $A_1 \in \mathcal{A}$ and $A_2 \in \mathcal{A}$ then $A_1 + A_2 \in \mathcal{A}$*

TABLE III.2 Axioms of Operations on Events

$A_1 A_1^c = \varnothing$	Mutual exclusion
$A_1 S = A_1$	Inclusion
$(A_1^c)^c = A_1$	Double complement
$A_1 + A_2 = A_2 + A_1$	Commutative law
$A_1 + (A_2 + A_3) = (A_1 + A_2) + A_3$	Associative law
$A_1(A_2 + A_3) = A_1 A_2 + A_1 A_3$	Distributive law
$(A_1 A_2)^c = A_1^c + A_2^c$	DeMorgan's law

TABLE III.3 Additional Identities in the Algebra of Events

$S^c = \varnothing$	
$A_1 + \varnothing = A_1$	Inclusion
$A_1 A_2 = A_2 A_1$	Commutative law
$A_1(A_2 A_3) = (A_1 A_2)A_3$	Associative law
$A_1 + (A_2 A_3) = (A_1 + A_2)(A_1 + A_3)$	Distributive law
$(A_1 + A_2)^c = A_1{}^c A_2{}^c$	DeMorgan's law

TABLE III.4   Some Corollaries Derived from the Axioms
of Probability

$Pr[A^c] = 1 - Pr[A]$
$0 \leq Pr[A] \leq 1$
If $A_1 \subseteq A_2$ then $Pr[A_1] \leq Pr[A_2]$
$Pr[\varnothing] = 0$
If $A_1 A_2 = \varnothing -$ then $= Pr[A_1 A_2] = 0$
$Pr[A_1 + A_2] = Pr[A_1] + Pr[A_2] - Pr[A_1 A_2]$

individual events to compute the probability of the union of two events. The term $Pr[A_1 A_2]$ is referred to as the *joint probability* of the two events. This last equation shows that the probabilities of two events add as in Equation (III.3) only if their joint probability is 0. The joint probability is 0 when the two events have no intersection ($A_1 A_2 = \varnothing$).

Two events are said to be statistically *independent* if and only if

$$Pr[A_1 A_2] = Pr[A_1] \cdot Pr[A_2] \quad \text{(independent events)} \tag{III.5}$$

This definition is not derived from the earlier properties of probability. An argument to give this definition intuitive meaning can be found in Ref. [1]. Independence occurs in problems where two events are not influenced by one another and Equation (III.5) simplifies such problems considerably.

A final important result deals with partitions. *A partition* is a finite or countably infinite set of events $A_1, A_2, A_3, \ldots$ that satisfy the two conditions:

$$A_i A_j = \varnothing \text{ for } i \neq j$$

$$A_1 + A_2 + A_3 + \cdots = S$$

The events in a partition satisfy the relation:

$$\sum_i Pr[A_i] = 1 \tag{III.6}$$

Further, if B is *any* other event, then

$$Pr[B] = \sum_i Pr[A_i B] \tag{III.7}$$

The latter result is referred to as the *principle of total probability* and is frequently used in solving problems. The principle is illustrated by a Venn diagram in Figure III.2. The rectangle represents the sample space and other events are defined therein. The event B is seen to be comprised of all of the pieces

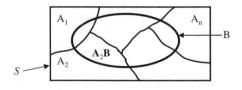

FIGURE III.2   Venn diagram illustrating the principle of total probability.

that represent intersections or overlap of event B with the events A_i. This is the graphical interpretation of Equation (III.7).

An Example

Simon's Surplus Warehouse has large barrels of mixed electronic components (parts) that you can buy by the handful or by the pound. You are not allowed to select parts individually. Based on your previous experience, you have determined that in one barrel, 29% of the parts are bad (faulted), 3% are bad resistors, 12% are good resistors, 5% are bad capacitors, and 32% are diodes. You decide to assign probabilities based on these percentages. Let us define the following events:

Event	Symbol
Bad (faulted) component	B
Good component	G
Resistor	R
Capacitor	C
Diode	D

A Venn diagram representing this situation is shown below along with probabilities of various events as given:

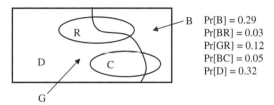

$$\Pr[B] = 0.29$$
$$\Pr[BR] = 0.03$$
$$\Pr[GR] = 0.12$$
$$\Pr[BC] = 0.05$$
$$\Pr[D] = 0.32$$

Note that since any component must be a resistor, capacitor, or diode, the region labeled D in the diagram represents everything in the sample space which is not included in R or C.

We can answer a number of questions.

1. What is the probability that a component is a resistor (either good *or* bad)?
   Since the events B and G form a partition of the sample space, we can use the principle of total probability Equation (III.7) to write:

$$\Pr[R] = \Pr[GR] + \Pr[BR] = 0.12 + 0.03 = 0.15$$

2. Are bad parts and resistors independent?
   We know that $\Pr[BR] = 0.03$ and we can compute:

$$\Pr[B] \cdot \Pr[R] = (0.29)(0.15) = 0.0435$$

   Since $\Pr[BR] \neq \Pr[B] \cdot \Pr[R]$, the events are *not* independent.

3. You have no use for either bad parts or resistors. What is the probability that a part is either bad and/or a resistor?

Using the formula from Table III.4 and the previous result we can write:

$$\Pr[B + R] = \Pr[B] + \Pr[R] - \Pr[BR] = 0.29 + 0.15 - 0.03 = 0.41$$

4. What is the probability that a part is useful to you?

   Let U represent the event that the part is useful. Then (see Table III.4):

$$\Pr[U] = 1 - \Pr[U^c] = 1 - 0.41 = 0.59$$

5. What is the probability of a bad diode?

   Observe that the events R, C, and D form a partition, since a component has to be one and only one type of part. Then using Equation (III.7) we write:

$$\Pr[B] = \Pr[BR] + \Pr[BC] + \Pr[BD]$$

Substituting the known numerical values and solving yields

$$0.29 = 0.03 + 0.05 + \Pr[BD] \quad \text{or} \quad \Pr[BD] = 0.21$$

Conditional Probability and Bayes' Rule

The *conditional* probability of an event A_1 given that an event A_2 has occurred is defined by

$$\Pr[A_1|A_2] = \frac{\Pr[A_1 A_2]}{\Pr[A_2]} \tag{III.8}$$

($\Pr[A_1|A_2]$ is read "probability of A_1 *given* A_2.") As an illustration, let us compute the probability that a component in the previous example is bad given that it is a resistor:

$$\Pr[B|R] = \frac{\Pr[BR]}{\Pr[R]} = \frac{0.03}{0.15} = 0.2$$

(The value for $\Pr[R]$ was computed in question 1 of the example.) Frequently the statement of a problem is in terms of conditional probability rather than joint probability, so Equation (III.8) is used in the form:

$$\Pr[A_1 A_2] = \Pr[A_1|A_2] \cdot \Pr[A_2] = \Pr[A_2|A_1] \cdot \Pr[A_1] \tag{III.9}$$

(The last expression follows because $\Pr[A_1 A_2]$ and $\Pr[A_2 A_1]$ are the same thing.) Using this result, the principle of total probability Equation (III.7) can be rewritten as

$$\Pr[B] = \sum_j \Pr[B|A_j] \Pr[A_j] \tag{III.10}$$

where B is any event and $\{A_j\}$ is a set of events that forms a partition.

Now, consider any one of the events A_i in the partition. It follows from Equation (III.9) that

$$\Pr[A_i|B] = \frac{\Pr[B|A_i] \cdot \Pr[A_i]}{\Pr[B]}$$

Then substituting in Equation (III.10) yields:

$$Pr[A_i|B] = \frac{Pr[B|A_i] \cdot Pr[A_i]}{\sum_j Pr[B|A_j] Pr[A_j]} \tag{III.11}$$

This result is known as *Bayes' theorem* or *Bayes' rule*. It is used in a number of problems that commonly arise in electrical engineering. We illustrate and end this section with an example from the field of communications.

Communication Example

The transmission of bits over a binary communication channel is represented in the drawing below:

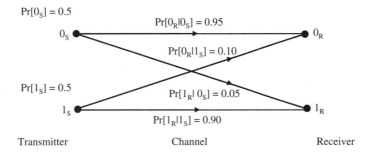

where we use notation like 0_S, 0_R ... to denote events "0 sent," "0 received," etc. When a 0 is transmitted, it is correctly received with probability 0.95 or incorrectly received with probability 0.05. That is, $Pr[0_R|0_S] = 0.95$ and $Pr[1_R|0_S] = 0.05$. When a 1 is transmitted, it is correctly received with probability 0.90 and incorrectly received with probability 0.10. The probabilities of sending a 0 or a 1 are denoted by $Pr[0_S]$ and $Pr[1_S]$. It is desired to compute the *probability of error* for the system.

This is an application of the principle of total probability. The two events 0_S and 1_S are mutually exclusive and collectively exhaustive and thus form a partition. Take the event B to be the event that an error occurs. It follows from Equation (III.10) that

$$Pr[error] = Pr[error|0_S] Pr[0_S] + Pr[error|1_S] Pr[1_S]$$
$$= Pr[1_R|0_S]Pr[0_S] + Pr[0_R|1_S] Pr[1_S]$$
$$= (0.05)(0.5) + (0.10)(0.5) = 0.075$$

Next, given that an error has occurred, let us compute the probability that a 1 was sent or a 0 was sent. This is an application of Bayes' rule. For a 1, Equation (III.11) becomes

$$Pr[1_S|error] = \frac{Pr[error|1_S] Pr[1_S]}{Pr[error|1_S] Pr[1_S] + Pr[error|0_S] Pr[0_S]}$$

Substituting the numerical values then yields:

$$Pr[1_S|error] = \frac{(0.10)(0.5)}{(0.10)(0.5) + (0.05)(0.5)} \approx 0.667$$

For a 0, a similar analysis applies:

$$\begin{aligned}
\Pr[0_S|\text{error}] &= \frac{\Pr[\text{error}|0_S]\,\Pr[0_S]}{\Pr[\text{error}|1_S]\,\Pr[1_S] + \Pr[\text{error}|0_S]\,\Pr[0_S]} \\
&= \frac{(0.05)(0.5)}{(0.10)(0.5) + (0.05)(0.5)} \approx 0.333
\end{aligned}$$

The two resulting probabilities sum to 1 because 0_S and 1_S form a partition for the experiment.

Reference

1.   C. W. Therrien and M. Tummala, *Probability for Electrical and Computer Engineers.* Boca Raton, FL: CRC Press, 2004.

Indexes

Author Index

Subject Index

Page on which term is defined is indicated in bold.